全国高等中医药院  划教材
Bilingual Planned Textbooks for Chinese ~ Colleges and Universities

有机化学
Organic Chemistry

（供中药学类、药学类及相关专业使用）
(For Chinese Materia Medica, Pharmacy and other related majors)

主　编　寇晓娣　陈胡兰

副主编　徐春蕾　胡冬华　刘　华　张园园

编　者（以姓氏笔画为序）

方玉宇（成都中医药大学）　　尹　飞（天津中医药大学）

左振宇（陕西中医药大学）　　朱　静（南京中医药大学）

刘　华（江西中医药大学）　　刘建宇（沈阳药科大学）

刘晓平（沈阳药科大学）　　刘晓芳（山西中医药大学）

杨　静（河南中医药大学）　　杨淑珍（北京中医药大学）

李贺敏（南京中医药大学）　　李嘉鹏（天津中医药大学）

余宇燕（福建中医药大学）　　张园园（北京中医药大学）

陈胡兰（成都中医药大学）　　林玉萍（云南中医药大学）

赵　群（南京中医药大学）　　赵珊珊（长春中医药大学）

胡冬华（长春中医药大学）　　徐春蕾（南京中医药大学）

盛文兵（湖南中医药大学）　　寇晓娣（天津中医药大学）

韩　波（成都中医药大学）　　谢达春（成都中医药大学）

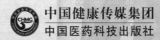
中国健康传媒集团
中国医药科技出版社

内 容 提 要

本书是"全国高等中医药院校中药学类专业双语规划教材"之一。本书共分15章。第一章介绍有机物的特征、共价键的形成，重点讨论共价键的电子效应及其与反应活性间的关系。第二章介绍最简单的有机物烷烃和环烷烃。第三章介绍立体化学的相关内容。从第四章到十四章以官能团为主线，重点介绍各类常见有机物的命名、结构、性质并重点分析结构与性质的关系。第十五章基础有机化学反应机理，作者对反应类型进行总结，同时与各化合物章节进行交叉引用，以机理作为桥梁，将化合物的结构与性质的学习有机地联系起来。本书在语言内容的编排上充分重视了英语学习的特征，希望读者在学习的过程中既能提高语言能力又能掌握专业内容，成为国际化中医药人才。本教材为书网融合教材，即纸质教材有机融合电子教材、教学配套资源（PPT、微课、视频等）、题库系统、数字化教学服务（在线教学、在线作业、在线考试），使教材内容更加立体、生动、形象，便教易学。

本书供全国高等中医药院校中药学、药学及相关专业使用，也可供从事中药研究、生产、销售工作的人员参考。

图书在版编目（CIP）数据

有机化学：汉英对照 / 寇晓娣，陈胡兰主编 .—北京：中国医药科技出版社，2020.9

全国高等中医药院校中药学类专业双语规划教材

ISBN 978-7-5214-1888-0

Ⅰ.①有…　Ⅱ.①寇…②陈…　Ⅲ.①有机化学 - 双语教学 - 中医学院 - 教材 - 汉、英　Ⅳ.①O62

中国版本图书馆 CIP 数据核字（2020）第 100517 号

美术编辑　陈君杞

版式设计　辰轩文化

出版　**中国健康传媒集团** | 中国医药科技出版社

地址　北京市海淀区文慧园北路甲 22 号

邮编　100082

电话　发行：010-62227427　邮购：010-62236938

网址　www.cmstp.com

规格　889 × 1194 mm ¹/₁₆

印张　27½

字数　700 千字

版次　2020 年 8 月第 1 版

印次　2023 年 11 月第 2 次印刷

印刷　三河市万龙印装有限公司

经销　全国各地新华书店

书号　ISBN 978-7-5214-1888-0

定价　86.00 元

获取新书信息、投稿、为图书纠错，请扫码联系我们。

近些年随着世界范围的中医药热潮的涌动，来中国学习中医药学的留学生逐年增多，走出国门的中医药学人才也在增加。为了适应中医药国际交流与合作的需要，加快中医药国际化进程，提高来中国留学生和国际班学生的教学质量，满足双语教学的需要和中医药对外交流需求，培养优秀的国际化中医药人才，进一步推动中医药国际化进程，根据教育部、国家中医药管理局、国家药品监督管理局等部门的有关精神，在本套教材建设指导委员会主任委员成都中医药大学彭成教授等专家的指导和顶层设计下，中国医药科技出版社组织全国 50 余所高等中医药院校及附属医疗机构约 420 名专家、教师精心编撰了全国高等中医药院校中药学类专业双语规划教材，该套教材即将付梓出版。

本套教材共计 23 门，主要供全国高等中医药院校中药学类专业教学使用。本套教材定位清晰、特色鲜明，主要体现在以下方面。

一、立足双语教学实际，培养复合应用型人才

本套教材以高校双语教学课程建设要求为依据，以满足国内医药院校开展留学生教学和双语教学的需求为目标，突出中医药文化特色鲜明、中医药专业术语规范的特点，注重培养中医药技能、反映中医药传承和现代研究成果，旨在优化教育质量，培养优秀的国际化中医药人才，推进中医药对外交流。

本套教材建设围绕目前中医药院校本科教育教学改革方向对教材体系进行科学规划、合理设计，坚持以培养创新型和复合型人才为宗旨，以社会需求为导向，以培养适应中药开发、利用、管理、服务等各个领域需求的高素质应用型人才为目标的教材建设思路与原则。

二、遵循教材编写规律，整体优化，紧跟学科发展步伐

本套教材的编写遵循"三基、五性、三特定"的教材编写规律；以"必需、够用"为度；坚持与时俱进，注意吸收新技术和新方法，适当拓展知识面，为学生后续发展奠定必要的基础。实验教材密切结合主干教材内容，体现理实一体，注重培养学生实践技能训练的同时，按照教育部相关精神，增加设计性实验部分，以现实问题作为驱动力来培养学生自主获取和应用新知识的能力，从而培养学生独立思考能力、实验设计能力、实践操作能力和可持续发展能力，满足培养应用型和复合型人才的要求。强调全套教材内容的整体优化，并注重不同教材内容的联系与衔接，避免遗漏和不必要的交叉重复。

三、对接职业资格考试，"教考""理实"密切融合

本套教材的内容和结构设计紧密对接国家执业中药师职业资格考试大纲要求，实现教学与考试、理论与实践的密切融合，并且在教材编写过程中，吸收具有丰富实践经验的企业人员参与教材的编写，确保教材的内容密切结合应用，更加体现高等教育的实践性和开放性，为学生参加考试和实践工作打下坚实基础。

四、创新教材呈现形式，书网融合，使教与学更便捷更轻松

全套教材为书网融合教材，即纸质教材与数字教材、配套教学资源、题库系统、数字化教学服务有机融合。通过"一书一码"的强关联，为读者提供全免费增值服务。按教材封底的提示激活教材后，读者可通过PC、手机阅读电子教材和配套课程资源（PPT、微课、视频等），并可在线进行同步练习，实时收到答案反馈和解析。同时，读者也可以直接扫描书中二维码，阅读与教材内容关联的课程资源，从而丰富学习体验，使学习更便捷。教师可通过PC在线创建课程，与学生互动，开展在线课程内容定制、布置和批改作业、在线组织考试、讨论与答疑等教学活动，学生通过PC、手机均可实现在线作业、在线考试，提升学习效率，使教与学更轻松。此外，平台尚有数据分析、教学诊断等功能，可为教学研究与管理提供技术和数据支撑。需要特殊说明的是，有些专业基础课程，例如《药理学》等9种教材，起源于西方医学，因篇幅所限，在本次双语教材建设中纸质教材以英语为主，仅将专业词汇对照了中文翻译，同时在中国医药科技出版社数字平台"医药大学堂"上配套了中文电子教材供学生学习参考。

编写出版本套高质量教材，得到了全国知名专家的精心指导和各有关院校领导与编者的大力支持，在此一并表示衷心感谢。希望广大师生在教学中积极使用本套教材和提出宝贵意见，以便修订完善，共同打造精品教材，为促进我国高等中医药院校中药学类专业教育教学改革和人才培养做出积极贡献。

数字化教材编委会

主　编　寇晓娣　陈胡兰

副主编　徐春蕾　胡冬华　刘　华　张园园

编　者（以姓氏笔画为序）

方玉宇（成都中医药大学）　　尹　飞（天津中医药大学）

左振宇（陕西中医药大学）　　朱　静（南京中医药大学）

刘　华（江西中医药大学）　　刘建宇（沈阳药科大学）

刘晓平（沈阳药科大学）　　刘晓芳（山西中医药大学）

杨　静（河南中医药大学）　　杨淑珍（北京中医药大学）

李贺敏（南京中医药大学）　　李嘉鹏（天津中医药大学）

余宇燕（福建中医药大学）　　张园园（北京中医药大学）

陈胡兰（成都中医药大学）　　林玉萍（云南中医药大学）

赵　群（南京中医药大学）　　赵珊珊（长春中医药大学）

胡冬华（长春中医药大学）　　徐春蕾（南京中医药大学）

盛文兵（湖南中医药大学）　　寇晓娣（天津中医药大学）

韩　波（成都中医药大学）　　谢达春（成都中医药大学）

前　言

本教材是"全国高等中医药院校中药学类专业双语规划教材"之一。有机化学为中药学、药学、临床药学及制药相关专业的基础课程，在为后续课程奠定理论基础的同时，更主要的是培养学生对构性关系的理解与认知，从思维方式上让学生为后续生命相关学科的学习做好准备。本教材是由全国十余所高校具有多年药学相关专业双语教学经验的教师编写而成，在编写中充分注意到了有机化学学科学习和语言学习的双重认知规律的要求。我们期望本书的出版不仅能解决原版有机化学教材内容多，难度大，让学生望而却步的问题，更能培养学生的双语科研思维，同时满足双语教学的需要和中医药对外交流的需求，培养优秀的国际化中医药人才，进一步推动中医药国际化进程。为达成以上目标，本教材的编写主要有以下特点。

一、学科内容编排特点

1. 为了突出构性关系的讲解与学习，本教材与以往中文教材相比，虽然仍以化合物官能团类型作为章节的主要划分标准，但更加重视章节之间的反应类型的联系，并在本书的最后加入一章"基础有机化学反应机理"，对反应类型进行总结，同时与各化合物章节进行交叉引用，以机理作为桥梁，将化合物的结构与性质的学习有机地联系起来。

2. 为了更好地适应药学、中药学相关专业的需求，在内容上区别于综合性大学化学类专业的教材，本教材去除了一些过难的机理及过多的反应，在每章的最后讲述本类化合物在医药相关领域的应用。

3. 本书化合物中的命名均以新版命名规则为基础。中文为《有机化学命名原则》（2017年版），英文为"Nomenclature of Organic Chemistry"（IUPAC 2013 edition）。

二、语言内容编排特点

1. 在教材的整体编写中将两种语言内容有机结合，而不是一半中文一半英文的简单加合。在每章前面加入中文学习目标简要介绍本章的主要内容；在章节正文中，采用英文扩展介绍"要点"部分的内容；在每章最后以中文进行总结。

2. 在教材的排版方面，充分注意科技英语词汇的积累，不仅在书后附上词汇表，还在文中随时总结重点英文单词给出中文翻译。并在每页边上留白，学生可以继续对自己不熟悉的单词加注。以此形式促进个性化阅读，提高学生自主学习的能力，让教材成为学生愿意长期保留的参考书。

3. 为提高学生的学习兴趣，不让学生因为英语水平问题过多地影响学科内容的吸收。本教材在编写时句子尽量短小，同时尽量避免与专业无关的生僻字出现。

4. 为在提高英语的阅读能力的同时，辅助提高学生的英语口语交流能力，本教材在每章最后提供讨论题，方便教师组织学生进行讨论式教学。

本书编写中为突出上述特点做出了很多努力，但本书为首次出版，限于学科发展和编者能力，书中不妥之处和疏漏在所难免，敬请同行和读者批评指正，以便再版时进行更正。

编　者
2020 年 5 月

This textbook belongs to the "Bilingual Planned Tetbooks for Chinese Materia Medica Majors in TCM Colleges and Universities". The series are for both Chinese students receiving bilingual education as well as international students. Organic chemistry is the foundation course for majors in pharmacology, Chinese Material Medica, clinical pharmacy and other pharmacy related majors. In addition to building theoretical foundation, more critically, the study of organic chemistry enables students to develop conceptualization and comprehension in structure-activity relationship, thus cultivate scientific thinking which prepares students for further study in other relevant subjects on pharmacy related courses. This particular textbook is composed by experienced bilingual instructors from over ten higher institutions in China teaching various courses in the field of Material Medica. The design of the textbook provides for the needs for learning both organic chemistry and foreign/second language for academic purposes. It is our expectation that this text book, on the one hand, solves the problem that the extensive content of original edition of organic chemistry textbook written in English poses formidable difficulty for students who get discouraged, and on the other hand, get students accustomed to scientific thinking in both languages, and meet the needs for both bilingual education and educating international students. It is our purpose to train and grow global talents and experts in Chinese Material Medica, and to promote the global dissemination and development of Traditional Chinese Medicine. To fulfill these objectives aforementioned, this text book has the following unique features.

1. Special Characteristics of Content Editing

(1) To highlight the explanation of structure-activity relationship, this textbook, while organizes chapters based on types of organic compounds as most Chinese textbooks, attaches special importance to the relationship between the functional groups, and adds the last chapter "Basic Mechanisms of Organic Reactions " to summarize reaction type. The last chapter is cross-referenced with other chapters, using mechanism as the bridge to connect the study of structure and activity.

(2) To better align the needs of students majoring in pharmacology and Chinese Material Medica, and to distinguish from textbooks for chemistry majors in comprehensive universities, this textbook excludes some complicated mechanisms and some reactions not essential for practitioners in the field of traditional Chinese medicine. For each chapter, we use the final section to explain the application of compounds in relevant medical fields.

(3) The naming of the compounds in this book follows the new edition: *Nomenclature of Organic Chemistry* (IUPAC 2013 edition). The Chinese version is (2017 edition).

2. Special Characteristics of Language Choice

(1) Unlike most bilingual text book which is a simple juxtaposition of two languages, this textbook thoughtfully combines and selects Chinese and English in an organic fashion to facilitate learning of both content and languages. Each chapter starts with learning objectives written in Chinese to introduce the major points. The main body of each chapter is written in English. Each chapter is concluded with a summary written in Chinese.

(2) The typesetting of the book facilitates the learning and accumulation of vocabulary in scientific English. In addition to attach a vocabulary list at the end of the book, for key words and phrases, Chinese translation is given in the body of the English text. Extra margin also allows students to add their own notes. This format is chosen to aid personalized reading and promote self-directed learning. We hope the textbook turns into a reference book that students keep for longer terms.

(3) To prevent student enthusiasm being dampened by language obstacles so that students maintain their motivation in learning, authors of chapters do their best to avoid overly complex sentences as well as rarely used words not relevant to the subjects

(4) In addition to improve students' reading comprehension, we also hope this textbook aids student in improving their oral communication on the relevant subjects. Discussion topics are provided at end of each chapter for instructors to engage students in discussion in the target language.

This is the first edition of this book. Despite the intensive efforts we have made to enable and boost the aforementioned features, there are possibly mistakes and errors that we may fail to recognize, due to limited time and knowledge. Comments and suggestions from colleagues and readers are greatly appreciated, and will be used to improve the second edition.

The editor

May 2020

目录 | Contents

第一章 绪论 ⋯⋯⋯⋯⋯⋯⋯⋯⋯⋯⋯⋯⋯⋯⋯⋯⋯⋯⋯⋯⋯⋯⋯⋯⋯⋯⋯⋯⋯ 1

Chapter 1　Introduction ⋯⋯⋯⋯⋯⋯⋯⋯⋯⋯⋯⋯⋯⋯⋯⋯⋯⋯⋯⋯⋯⋯ 1

　1.1　**Organic Compound and Organic Chemistry** ⋯⋯⋯⋯⋯⋯⋯⋯⋯⋯ 1
　　1.1.1　Development of Organic Chemistry ⋯⋯⋯⋯⋯⋯⋯⋯⋯⋯⋯⋯ 1
　　1.1.2　Properties of Organic Compounds ⋯⋯⋯⋯⋯⋯⋯⋯⋯⋯⋯⋯ 3
　1.2　**Chemical Bonds in Organic Compounds** ⋯⋯⋯⋯⋯⋯⋯⋯⋯⋯⋯ 4
　　1.2.1　Valence Bond Theory ⋯⋯⋯⋯⋯⋯⋯⋯⋯⋯⋯⋯⋯⋯⋯⋯⋯ 4
　　1.2.2　Hybrid Orbital Theory ⋯⋯⋯⋯⋯⋯⋯⋯⋯⋯⋯⋯⋯⋯⋯⋯ 5
　　1.2.3　Resonance Theory ⋯⋯⋯⋯⋯⋯⋯⋯⋯⋯⋯⋯⋯⋯⋯⋯⋯⋯ 6
　　1.2.4　Molecular Orbital Theory ⋯⋯⋯⋯⋯⋯⋯⋯⋯⋯⋯⋯⋯⋯⋯ 7
　1.3　**Properties of Covalent Bonds** ⋯⋯⋯⋯⋯⋯⋯⋯⋯⋯⋯⋯⋯⋯⋯ 8
　　1.3.1　Bond Energy, Bond Length, Bond Angle ⋯⋯⋯⋯⋯⋯⋯⋯⋯ 8
　　1.3.2　Bond Polarity and Polarizability ⋯⋯⋯⋯⋯⋯⋯⋯⋯⋯⋯⋯ 9
　　1.3.3　Inductive and Conjugated Effect ⋯⋯⋯⋯⋯⋯⋯⋯⋯⋯⋯⋯ 12
　　1.3.4　Cleavage Modes and Reaction Types of Covalent Bonds ⋯⋯ 14
　1.4　**Intermolecular Forces of Organic Compounds** ⋯⋯⋯⋯⋯⋯⋯⋯ 15
　　1.4.1　Intermolecular Force ⋯⋯⋯⋯⋯⋯⋯⋯⋯⋯⋯⋯⋯⋯⋯⋯⋯ 15
　　1.4.2　Effect of Intermolecular Forces on Physical Properties of Organic Molecules ⋯⋯ 16
　　1.4.3　Effect of Intermolecular Forces on Chemical Properties of Organic Molecules ⋯⋯ 17
　1.5　**Structure and Classification of Organic Compounds** ⋯⋯⋯⋯⋯ 18
　　1.5.1　Structural Representation of Organic Compound ⋯⋯⋯⋯⋯ 18
　　1.5.2　Classification of Organic Compound ⋯⋯⋯⋯⋯⋯⋯⋯⋯⋯ 20
　　1.5.3　Research Methods of Organic Compound ⋯⋯⋯⋯⋯⋯⋯⋯ 22
　1.6　**Brief Introduction of Organic Acid-Base Theory** ⋯⋯⋯⋯⋯⋯ 22
　　1.6.1　Brøsted Acid-Base Theory ⋯⋯⋯⋯⋯⋯⋯⋯⋯⋯⋯⋯⋯⋯ 23
　　1.6.2　Lewis Acid-Base Theory ⋯⋯⋯⋯⋯⋯⋯⋯⋯⋯⋯⋯⋯⋯⋯ 24
　　1.6.3　Electrophile and Nucleophile ⋯⋯⋯⋯⋯⋯⋯⋯⋯⋯⋯⋯⋯ 25
　1.7　**Relation between Organic Chemistry and Pharmacy** ⋯⋯⋯⋯⋯ 26

第二章 烷烃和环烷烃 ⋯⋯⋯⋯⋯⋯⋯⋯⋯⋯⋯⋯⋯⋯⋯⋯⋯⋯⋯⋯⋯⋯ 30

Chapter 2　Alkanes and Cycloalkanes ⋯⋯⋯⋯⋯⋯⋯⋯⋯⋯⋯⋯⋯⋯ 30

　2.1　**Alkanes** ⋯⋯⋯⋯⋯⋯⋯⋯⋯⋯⋯⋯⋯⋯⋯⋯⋯⋯⋯⋯⋯⋯⋯⋯ 30
　　2.1.1　Structure of Alkanes ⋯⋯⋯⋯⋯⋯⋯⋯⋯⋯⋯⋯⋯⋯⋯⋯⋯ 31
　　2.1.2　Constitutional Isomerism in Alkanes ⋯⋯⋯⋯⋯⋯⋯⋯⋯⋯ 32

2.1.3 Nomenclature of Alkanes ⋯⋯⋯⋯⋯⋯⋯⋯⋯⋯⋯⋯⋯⋯⋯⋯⋯⋯⋯ 33
2.1.4 Physical Properties of Alkanes ⋯⋯⋯⋯⋯⋯⋯⋯⋯⋯⋯⋯⋯⋯⋯⋯ 37
2.1.5 Reactions of Alkanes ⋯⋯⋯⋯⋯⋯⋯⋯⋯⋯⋯⋯⋯⋯⋯⋯⋯⋯⋯⋯⋯ 40
2.1.6 Conformations of Alkanes ⋯⋯⋯⋯⋯⋯⋯⋯⋯⋯⋯⋯⋯⋯⋯⋯⋯⋯⋯ 43
2.2 Cycloalkanes ⋯⋯⋯⋯⋯⋯⋯⋯⋯⋯⋯⋯⋯⋯⋯⋯⋯⋯⋯⋯⋯⋯⋯⋯⋯⋯ 47
2.2.1 Structure of Cycloalkanes ⋯⋯⋯⋯⋯⋯⋯⋯⋯⋯⋯⋯⋯⋯⋯⋯⋯⋯⋯ 48
2.2.2 Constitutional Isomerism in Cycloalkanes ⋯⋯⋯⋯⋯⋯⋯⋯⋯⋯⋯ 49
2.2.3 Nomenclature of Cycloalkanes ⋯⋯⋯⋯⋯⋯⋯⋯⋯⋯⋯⋯⋯⋯⋯⋯ 49
2.2.4 Physical Properties of Cycloalkanes ⋯⋯⋯⋯⋯⋯⋯⋯⋯⋯⋯⋯⋯ 50
2.2.5 Reactions of Cycloalkanes ⋯⋯⋯⋯⋯⋯⋯⋯⋯⋯⋯⋯⋯⋯⋯⋯⋯⋯ 51
2.2.6 Conformation of Cyclohexane ⋯⋯⋯⋯⋯⋯⋯⋯⋯⋯⋯⋯⋯⋯⋯⋯ 53
2.3 Application of Alkanes in Pharmacy Area ⋯⋯⋯⋯⋯⋯⋯⋯⋯⋯⋯⋯ 57

第三章 立体化学 ⋯⋯⋯⋯⋯⋯⋯⋯⋯⋯⋯⋯⋯⋯⋯⋯⋯⋯⋯⋯⋯⋯⋯⋯⋯ 63
Chapter 3 Stereoisomerism ⋯⋯⋯⋯⋯⋯⋯⋯⋯⋯⋯⋯⋯⋯⋯⋯⋯⋯⋯⋯⋯ 63
3.1 Introduction to Isomerism ⋯⋯⋯⋯⋯⋯⋯⋯⋯⋯⋯⋯⋯⋯⋯⋯⋯⋯⋯⋯ 63
3.1.1 Constitutional Isomer ⋯⋯⋯⋯⋯⋯⋯⋯⋯⋯⋯⋯⋯⋯⋯⋯⋯⋯⋯⋯ 63
3.1.2 Stereoisomer ⋯⋯⋯⋯⋯⋯⋯⋯⋯⋯⋯⋯⋯⋯⋯⋯⋯⋯⋯⋯⋯⋯⋯⋯ 64
3.2 CIP Sequence ⋯⋯⋯⋯⋯⋯⋯⋯⋯⋯⋯⋯⋯⋯⋯⋯⋯⋯⋯⋯⋯⋯⋯⋯⋯⋯ 65
3.3 Cis-trans Isomer ⋯⋯⋯⋯⋯⋯⋯⋯⋯⋯⋯⋯⋯⋯⋯⋯⋯⋯⋯⋯⋯⋯⋯⋯ 67
3.3.1 Introduction to Cis-trans Isomer ⋯⋯⋯⋯⋯⋯⋯⋯⋯⋯⋯⋯⋯⋯⋯ 67
3.3.2 E-Z Nomenclature ⋯⋯⋯⋯⋯⋯⋯⋯⋯⋯⋯⋯⋯⋯⋯⋯⋯⋯⋯⋯⋯⋯ 68
3.3.3 Significance of cis-trans isomers in the Pharmacy Study ⋯⋯⋯⋯ 69
3.4 Enantiomer ⋯⋯⋯⋯⋯⋯⋯⋯⋯⋯⋯⋯⋯⋯⋯⋯⋯⋯⋯⋯⋯⋯⋯⋯⋯⋯ 69
3.4.1 Introduction to Enantiomer ⋯⋯⋯⋯⋯⋯⋯⋯⋯⋯⋯⋯⋯⋯⋯⋯⋯ 69
3.4.2 R, S Nomenclature of Chiral Carbon Atoms ⋯⋯⋯⋯⋯⋯⋯⋯⋯ 70
3.4.3 Fischer Projection ⋯⋯⋯⋯⋯⋯⋯⋯⋯⋯⋯⋯⋯⋯⋯⋯⋯⋯⋯⋯⋯⋯ 72
3.4.4 D, L Nomenclature of Chiral Carbon Atoms ⋯⋯⋯⋯⋯⋯⋯⋯⋯ 74
3.4.5 Acyclic Molecules with Two or More Stereocenters ⋯⋯⋯⋯⋯⋯ 75
3.4.6 Cyclic Molecules with Two or More Chiral Centers ⋯⋯⋯⋯⋯⋯ 77
3.5 Optical Activity—How Chirality Is Detected in the Laboratory ⋯⋯ 79
3.6 Significance of Chirality in the Pharmacy Study ⋯⋯⋯⋯⋯⋯⋯⋯⋯ 80
3.7 Separation of Enantiomers—Resolution ⋯⋯⋯⋯⋯⋯⋯⋯⋯⋯⋯⋯⋯ 81

第四章 烯烃 ⋯⋯⋯⋯⋯⋯⋯⋯⋯⋯⋯⋯⋯⋯⋯⋯⋯⋯⋯⋯⋯⋯⋯⋯⋯⋯⋯ 87
Chapter 4 Alkenes ⋯⋯⋯⋯⋯⋯⋯⋯⋯⋯⋯⋯⋯⋯⋯⋯⋯⋯⋯⋯⋯⋯⋯⋯ 87
4.1 Structure of Alkenes ⋯⋯⋯⋯⋯⋯⋯⋯⋯⋯⋯⋯⋯⋯⋯⋯⋯⋯⋯⋯⋯⋯ 87
4.2 How to Calculate the Index of Hydrogen Deficiency ⋯⋯⋯⋯⋯⋯⋯ 89
4.3 Nomenclature of Alkenes ⋯⋯⋯⋯⋯⋯⋯⋯⋯⋯⋯⋯⋯⋯⋯⋯⋯⋯⋯⋯ 91
4.4 Physical Properties of Alkenes ⋯⋯⋯⋯⋯⋯⋯⋯⋯⋯⋯⋯⋯⋯⋯⋯⋯ 93
4.5 Reactions of Alkenes ⋯⋯⋯⋯⋯⋯⋯⋯⋯⋯⋯⋯⋯⋯⋯⋯⋯⋯⋯⋯⋯ 94

    4.5.1　Reactions of Alkenes-An Overview ································ 94

    4.5.2　Electrophilic Additions ··· 95

    4.5.3　Hydroboration-Oxidation ·· 101

    4.5.4　Free-Radical Addition of HBr-Peroxide Effect ············· 103

    4.5.5　Oxidation and Reduction ··· 104

  4.6　Application of Alkenes in the Pharmacy Study ···················· 107

第五章　炔烃、二烯烃和周环反应 ····································· 112

Chapter 5　Alkynes, Dienes and Pericyclic Reaction ··············· 112

  5.1　Alkynes ··· 113

    5.1.1　Structure of Alkynes ··· 113

    5.1.2　Isomerism and Nomenclature of Alkynes ·················· 113

    5.1.3　Physical Properties of Alkynes ································· 114

    5.1.4　Chemical Properties of Alkynes ······························ 115

    5.1.5　Synthesis of Alkynes ··· 121

  5.2　Dienes ··· 121

    5.2.1　Structure of Dienes ··· 122

    5.2.2　Nomenclature of Dienes ··· 125

    5.2.3　Reactions of Dienes ·· 125

  5.3　Pericyclic Reaction ··· 128

  5.4　Application of Alkynes in the Pharmacy Study ··················· 130

第六章　芳香化合物 ··· 134

Chapter 6　Aromatic Compound ····································· 134

  6.1　Introduction to Aromatic Compounds ······························ 134

  6.2　Nomenclature of Aromatic Compounds ··························· 139

  6.3　Structure of Aromatic Compound of Hückel's Rule ·············· 142

    6.3.1　Hückel's Rule ··· 142

    6.3.2　Molecular Orbital Derivation of Hückel's Rule ············· 143

    6.3.3　Aromatic Ions ··· 144

    6.3.4　Polynuclear Aromatic Hydrocarbons ······················· 145

  6.4　Properties of Aromatic Compounds ································· 146

    6.4.1　Electrophilic Aromatic Substitution ························· 146

    6.4.2　Reactions at Benzylic Position ································· 154

    6.4.3　Effect of Substitution ·· 155

  6.5　Application of Aromatic Compounds in the Pharmacy Study ··· 158

第七章　卤代烃 ··· 162

Chapter 7　Alkyl Halides ··· 162

  7.1　Structure of Alkyl Halides ··· 162

  7.2　Nomenclature of Alkyl Halides ····································· 163

7.2.1　Common Nomenclature of Alkyl Halides ··· 163

7.2.2　Systematic Nomenclature of Alkyl Halides ··· 163

7.3　Physical Properties of Alkyl Halides ·· 164

7.4　Reactions of Alkyl Halides ··· 165

7.4.1　Nucleophilic Substitution in Alkyl Halides ··· 166

7.4.2　β-Elimination Reaction ·· 175

7.4.3　Competitions between Substitutions and Eliminations ···························· 178

7.4.4　Reaction to Form Organometallic Compounds ·· 180

7.4.5　Reduction Reaction of Alkyl Halides ·· 182

7.5　Application of Alkyl Halides in the Pharmacy Study ·································· 183

第八章　醇、酚和醚 ··· 189

Chapter 8　Alcohols, Phenols and Ethers ··· 189

8.1　Alcohols ·· 189

8.1.1　Structure of Alcohols ·· 189

8.1.2　Nomenclature of Alcohols ··· 190

8.1.3　Physical Properties of Alcohols ·· 191

8.1.4　Reactions of Alcohols ·· 192

8.1.5　Application of Alcohols in the Pharmacy Study ······································· 199

8.2　Phenols ·· 200

8.2.1　Structure of Phenols ··· 200

8.2.2　Nomenclature of Phenols ·· 201

8.2.3　Physical Properties of Phenols ·· 201

8.2.4　Reactions of Phenols ··· 202

8.2.5　Application of Phenols in the Pharmacy Study ·· 206

8.3　Ethers ·· 207

8.3.1　Structure of Ethers ··· 208

8.3.2　Nomenclature of Ethers ·· 208

8.3.3　Physical Properties of Ethers ·· 209

8.3.4　Reactions of Ethers ·· 210

8.3.5　Application of Ethers in the Pharmacy Study ··· 211

第九章　醛和酮 ·· 215

Chapter 9　Aldehydes and Ketones ··· 215

9.1　Structure of Aldehydes and Ketones ·· 215

9.2　Nomenclature of Aldehydes and Ketones ·· 216

9.2.1　IUPAC Names of Aldehydes ·· 216

9.2.2　IUPAC Names of Ketones ·· 217

9.2.3　Common Names ··· 217

9.3　Physical Properties of Aldehydes and Ketones ·· 218

9.4　Reactions of Aldehydes and Ketones ··· 218

9.4.1 Nucleophilic Addition ·· 218

9.4.2 Oxidation ··· 224

9.4.3 Reduction ··· 225

9.4.4 Keto-Enol Tautomerism ·· 226

9.4.5 Reaction at the α-Carbon ······································· 227

9.4.6 Reaction of α, β-unsaturated aldehydes and ketones ············ 229

9.5 Application of Aldehydes in the Pharmacy Study ················ 230

第十章 羧酸及衍生物 ·· 235

Chapter 10 Carboxylic Acids and Derivatives ······················ 235

10.1 Structure of Acids ··· 235

10.2 Nomenclature of Carboxylic Acids ································ 236

10.2.1 Common Name ·· 236

10.2.2 System Nomenclature ·· 236

10.3 Physical Properties of Carboxylic Acids ························ 238

10.3.1 State ··· 238

10.3.2 Boiling Point ··· 238

10.3.3 Melting Point ··· 239

10.3.4 Solubility ·· 240

10.3.5 Relative Density ··· 240

10.4 Chemical Properties of Carboxylic Acid ························· 240

10.4.1 Acidity of Carboxylic Acids ······································ 240

10.4.2 Reaction of The Carbonyl Group in the Carboxyl Group ·········· 245

10.5 Functional Derivatives of Carboxylic Acids ······················ 248

10.5.1 Structure of Carboxylic Acid Derivatives ························· 248

10.5.2 Nomenclature of Acid Derivatives ································ 249

10.6 Reactions of Acid Derivative ······································ 251

10.6.1 Acidity of Amides, Imides, and Sulfonamides ···················· 251

10.6.2 Substitution of Carboxylic Acid Derivatives ······················ 252

10.6.3 Interconversion of Functional Derivatives ························ 256

10.6.4 Reactivity of Acid Derivatives ···································· 256

10.6.5 Reduction of Carboxylic Acid Derivatives ························ 258

10.7 Application of Carboxylic Acids and Derivatives in the Pharmacy Study ·· 259

10.7.1 Application of Carboxylic Acids in the Pharmacy Study ············ 259

10.7.2 Application of Carboxylic Acid Derivatives in the Pharmacy Study ··· 260

第十一章 取代羧酸 ·· 265

Chapter 11 Substituted Carboxylic Acids ··························· 265

11.1 Halogenated acid ·· 266

11.1.1 Structure and Nomenclature ······································ 266

11.1.2 Physical Properties of Halogenated Acids ························· 266

11.1.3　Reactions of Halogenated Acids ················· 266

11.1.4　Preparation of Halogenated Acids ················· 269

11.2　Hydroxy Acid ················· 270

11.2.1　Structure and Nomenclature ················· 270

11.2.2　Physical Properties of Hydroxy Acids ················· 271

11.2.3　Reactions of Hydroxy Acids ················· 271

11.2.4　Preparation of Hydroxy Acid ················· 274

11.3　Carbonyl Acids ················· 274

11.3.1　Structure and Nomenclature ················· 275

11.3.2　Physical Properties of Carbonyl Acids ················· 275

11.3.3　Reactions of Carbonyl Acids ················· 276

11.3.4　Synthesis Using β-dicarbonyl Compounds ················· 278

11.3.5　Preparation of Carbonyl Acids ················· 281

11.4　Amino Acids ················· 282

11.4.1　Structure and Nomenclature ················· 282

11.4.2　Physical Properties of Amino Acid ················· 284

11.4.3　Reactions of Amino Acids ················· 284

11.5　Application of Substituted Carboxylic Acid in the Pharmacy Study ················· 287

第十二章　胺 ················· 293

Chapter 12　Amine ················· 293

12.1　Structure of Amine ················· 294

12.2　Nomenclature of Amine ················· 294

12.3　Chirality of Amines and Quaternary ················· 296

12.4　Physical Properties ················· 297

12.5　Reactions of Amines ················· 298

12.5.1　Basicity ················· 298

12.5.2　Alkylation of Ammonia ················· 299

12.5.3　Acylation of Amines ················· 300

12.5.4　Reaction with Nitrous Acid ················· 301

12.5.5　Reaction of Arenediazonium Salts ················· 302

12.5.6　Azo Coupling ················· 304

12.5.7　Hofmann Elimination ················· 305

12.5.8　Cope Elimination ················· 307

12.6　Applications of Amines in the Pharmacy Study ················· 308

第十三章　糖类化合物 ················· 313

Chapter 13　Carbohydrates ················· 313

13.1　Monosaccharides ················· 314

13.1.1　Classification of Monosaccharides ················· 314

13.1.2　Physical Properties of Monosaccharides ················· 315

13.1.3　Open Chain Structure and Relative Configuration of Hexose ·············· 315

13.2　Cyclic Structure of Monosaccharides ·············318

13.2.1　Oxygen Ring Structure of Monosaccharides ·············· 318

13.2.2　Haworth Projections of Carbohydrates ·············· 320

13.2.3　Conformation Representations and Anomeric Effect ·············· 322

13.3　Reactions of Monosaccharides ·············324

13.3.1　Epimerization ·············· 324

13.3.2　Oxidation Reaction ·············· 325

13.3.3　Formation of Glycosides (Acetals) ·············· 328

13.3.4　Osazone Formation ·············· 329

13.3.5　Reduction to Alditols ·············· 330

13.3.6　Esterification ·············· 331

13.3.7　Formation of Cyclic Acetals or Ketals ·············· 331

13.3.8　Dehydration and Chromogenic Reaction ·············· 332

13.3.9　Important Monosaccharides ·············· 332

13.4　Disaccharides and Oligosaccharide ·············334

13.4.1　Nonreducing disaccharide ·············· 334

13.4.2　Reducing disaccharide ·············· 335

13.4.3　Cyclodextrin ·············· 337

13.5　Polysaccharides ·············338

13.5.1　Starch ·············· 338

13.5.2　Glycogen ·············· 339

13.5.3　Cellulose ·············· 339

13.5.4　Chitin and Chitosan ·············· 340

13.6　Applications of Carbohydrates in the Pharmacy Study ·············341

第十四章　杂环化合物 ·············347

Chapter 14　Heterocyclics ·············347

14.1　Structures and Classifications of Heterocyclics ·············347

14.2　Nomenclature of Heterocyclics ·············348

14.3　Properties of Five-membered Heterocycles ·············350

14.3.1　Furan, Pyrrole and Thiophene ·············· 350

14.3.2　Imidazole, Pyrazole and Thiazole ·············· 356

14.4　Properties of Six-membered Heterocycles ·············358

14.4.1　Pyridine ·············· 359

14.4.2　Pyrimidine ·············· 362

14.5　Properties of Fused Heterocyclics ·············363

14.5.1　Indole ·············· 363

14.5.2　Quinoline and Isoquinoline ·············· 364

14.5.3　Purine ·············· 366

14.6　Applications of Heterocyclics in the Pharmacy Study ·············367

第十五章　基础有机化学反应机理 ··· 373

Chapter 15　Basic Mechanisms of Organic Reactions ································· 373

　15.1　Types of Mechanism and How to Present it ·································373

　　15.1.1　Types of Mechanism of Organic Reactions ··························· 373

　　15.1.2　How to Present a Mechanism ·· 374

　15.2　Types of Intermediates of Organic Reactions ··························· 376

　15.3　Types of Organic Reactions ··379

　　15.3.1　Classification by Structural Difference ·································· 379

　　15.3.2　Classification by Bond-broken Type ····································· 380

　15.4　Addition Reaction ·· 381

　　15.4.1　Electrophilic Addition ·· 381

　　15.4.2　Nucleophilic Addition ·· 383

　　15.4.3　Free Radical Addition ·· 385

　　15.4.4　Addition with Pericyclic Process ·· 386

　15.5　Substitution ··388

　　15.5.1　Electrophilic Substitution ··· 388

　　15.5.2　Nucleophilic Substitution ··· 390

　　15.5.3　Free Radical Substitution ··· 392

　15.6　Elimination ···392

　　15.6.1　Elimination Reactions ·· 393

　　15.6.2　Elimination versus Substitution ·· 394

　15.7　Rearrangement Reactions ··395

　　15.7.1　Nucleophilic Rearrangement ·· 395

　　15.7.2　Electrophilic Rearrangement ·· 397

　　15.7.3　Neutral Rearrangement ··· 397

　15.8　Reaction Types Include Two Kinds of Mechanisms ·····················399

词汇表 ·· 403

参考答案 ·· 415

参考文献 ·· 421

第一章 绪 论
Chapter 1　Introduction

 学习目标

　　1．**掌握**　有机化学结构式的表示方法,共价键的断裂方式和有机反应的类型。
　　2．**熟悉**　共价键的形成、性质以及诱导效应、共轭效应,有机化学常用的酸碱理论、两类重要试剂和有机化合物分子间力。
　　3．**了解**　有机化合物特点、分类、研究方法以及有机化学在医药领域的重要性。

Organic chemistry (有机化学) is a unique subject involving a large number of natural and synthetic substances, which are directly related to human's clothing, food, housing and transportation. **Organic compounds** (有机化合物) give us light, when you look at a book, your eyes are using an organic compound to turn light into nerve stimulation so that you know what you see. Organic compounds protect our health, for example, the discovery of penicillin has opened up new avenues of treatment and saved thousands of lives. Organic compounds such as starch and cellulose can provide energy for life. These organic compounds contain carbon and hydrogen elements, some also contain oxygen, nitrogen, sulfur, phosphorus and halogen. Organic compounds are now defined as compounds that contain carbon. Furthermore, the definition of organic chemistry also emerged: organic chemistry is the study of carbon compounds. This is the definition we still use today. Organic chemistry is a subject studying the origin, structure, property, synthesis, application, theory and method of organic compounds.

1.1　Organic Compound and Organic Chemistry

PPT

1.1.1　Development of Organic Chemistry

The term "organic chemistry" was first coined in 1806 by Swedish scientist (J. J. Berzelius), which is considered as the counterpart of inorganic chemistry. At that time, many chemists believed that organic compounds could only exist in living organisms and inorganic compounds could only exist in lifeless mineral resources, which were called the "vital force" theory.

In 1828, the German chemist (F. Wohler) converted **ammonium cyanate** (氰酸铵) into **urea** (尿素)

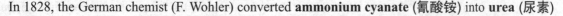

by heating. Urea is organic, while ammonium cyanate is inorganic. Wohler's experiment gave a shock to the "vital force" theory. After that, Wohler's student - German chemist (H. Kolbe) synthesized acetic acid in 1845 by using carbon disulfide, chlorine and water; French chemist (M. E. P. Berthelot) successfully synthesized oil in 1854. With the successive synthesis of sugar, **amine** (胺) and other organic matters, the idea of the mysterious "vital force" creating organic matters was shaken and organic compounds were no longer merely the products of living plants and animals.

By the end of the 18th century, many organic matters had been isolated and described qualitatively. However, how to express the relationship among atoms in organic molecules confused the chemists. By 1830, the German chemist (J. von Liebig) had developed carbon-hydrogen analysis. In 1883, the French chemist (J. B. A. Dumas) established the method of nitrogen analysis. The establishment of these quantitative analytical methods for organic compounds has enabled chemists to obtain empirical formulas of organic compounds.

The **optical isomerism** (旋光异构) of **tartaric acid** (酒石酸) was discovered by the French scientist (L. Pasteur) in 1848. In 1858, the German chemist (F. A. Kekule) proposed the concept that carbon was tetravalence and used the short line "—" to indicate "bond" for the first time. He proposed that carbon atoms could bond to each other in a molecule and they could bond to each other not only in a single bond, but also in a double or triple bond. He also proposed the structure of benzene. In 1874, the Dutch chemist (J. H. van't Hoff) and the French chemist (J. A. Le Bel) proposed the theory of carbon **valence** (化合价) tetrahedron independently, in which the carbon atom occupied the center of the tetrahedron and its four valence bonds pointed to the four vertexes of the tetrahedron. This theory has revealed the cause of organic optical isomerism, laid the foundation of organic **stereochemistry** (立体化学) and promoted the development of organic chemistry.

In this period, great progress had been made in the determination, classification and chemical synthesis of organic compounds, but the nature of the valence bond had not been solved. In 1916, G. N. Lewis *et al.* proposed the electron theory of valence bonds based on the theory of atomic structure. By the 1960s, on the basis of extensive experiences in organic synthesis reactions, R. B. Woodward and R. Hoffman recognized the relationship between the chemical reaction and **molecular orbital** (分子轨道). They studied a series of reactions, such as **electrocyclic** (电环化) reaction, σ bond **rearrangement** (重排) and cyclic **addition reaction** (加成反应) and proposed the conversation of orbital symmetry.

The establishment of organic chemistry theory and the improvement of experimental conditions promoted the development of organic synthesis. In 1958, the first **sequence of amino acids** (氨基酸序列) in the protein was deciphered. In 1965, Chinese chemists (Xing Qiyi and Wang You) completed the synthesis of the first **bovine insulin** (牛胰岛素). In 1972, the complete synthesis of vitamin B12 with 2^9 (512) isomers was achieved by R. B. Woodward and Swiss chemist A. Eschenmoser. In 1982, the complete synthesis of the so-called "despair compound" - erythromycin with 2^{18} isomers was achieved. In 1989, Professor Yoshito Kishi from Harvard University synthesized sea anemone toxin. The total synthesis of sea anemone toxin was once considered to be the Everest in the field of total synthesis due to its large molecular weight, 64 **asymmetric centers** (不对称中心), 7 double bonds (Z, E) and 2^{71} **isomers** (异构体) which was close to the Avogadro constant (6.02×10^{23}). It is the most complex organic compound that has ever been found. These achievements indicate that the organic synthesis has developed from the realm of necessity to the realm of freedom. So far the number of organic compounds is more than 80 million, which is growing rapidly every year.

Nowadays, organic chemistry has developed into a discipline that attaches equal importance to theory and experiment. It has interacted with other subjects, which leads to the emergence of new branches in chemistry. Based on organic chemistry, pharmaceuticals, pesticides and **petrochemical** (石油化学) industries have become the pillar industries of the national economy. In the future, organic chemistry will play an extremely important role in life science, environmental protection and energy development.

1.1.2   Properties of Organic Compounds

The main reason for wide varieties of organic compounds is the structural characteristics of the carbon atom. Carbon is located in the second period and IVA group in the periodic table. The outermost layer of carbon has four electrons, so it is not easy for it to gain or lose electrons and can only form four covalent bonds with other atoms. Carbon atoms can form single, double, or triple bonds with each other atoms. They can be connected in the way of linear chains, branched chains or in rings. Thus, although only few elements (C, H, O, N, P, S, X, etc.) are involved in organic compounds, they can form thousands kinds of organic compounds.

The properties of organic compounds are determined mainly by the types of chemical bonds in molecules. The atoms of organic compounds are mainly connected by covalent bonds, so organic and inorganic compounds have different properties. Organic compounds have the following properties.

1. Combustibility (可燃性)   most organic materials are unstable to heat and easy to burn, but there are a few exceptions that are not easy to burn, such as CCl_4 extinguishing agent.

2. Water Solubility (水溶性)   most organic compounds are difficult to dissolve in water and easy to dissolve in organic solvents. Organic compounds have a lot of covalent bonds and the molecules have no polarity or weak polarity. According to the principle of "like dissolves like", organic compounds are usually difficult to dissolve in water with strong polarity. But there are exceptions, such as low-molecular-weight alcohol, aldehyde, ketone, carboxylic acid, amino acid and carbohydrate, which are all soluble in water.

3. Melting (熔点) and Boiling Point (沸点)   Most organic compounds have low melting and boiling points and are volatile. Organic compounds are usually gas, liquid or solid with low melting points, which are mostly below 300°C. This is because the intermolecular interactions of organic compounds are mainly van der Waals forces.

4. Reaction Rate (反应速率)   The reaction rates of most organic compounds are slow. The essence of chemical reactions is the breaking of old bonds and the formation of new bonds. Compared with organic materials, ionic bonds of inorganic materials are easy to break and form in water, while covalent bonds are difficult to break and form. So the inorganic reaction is fast and organic reaction is slow. Therefore, when organic reactions are carried out, heating, catalyst or light are often applied to speed up the reaction. There are exceptions, for example, the reaction of trinitrotoluene (TNT) is an explosive reaction.

5. Complex Products (复杂产物) and Side Reactions (副反应)   An organic compound is a complex molecule made by many atoms, so when it reacts with a reagent, all parts of the molecule may be affected and each part of the molecule can participate in the reaction. The main reaction is often accompanied by the production of a variety of by-products, so difficult separation and purification should be done to remove these by-products. Inorganic reactions tend to react in the quantitative way.

6. Isomerism (同分异构) Isomerism is common in organic compounds, which is caused by the bonding characteristics of the carbon atom. Organic compounds may contain the same kind of atoms and the same number of each atom. And these atoms may combine in different ways to form molecules with different structures, which is called isomerism.

1.2 Chemical Bonds in Organic Compounds

PPT

The chemical bond is a term used to describe the strong interaction between two or more adjacent atoms (or ions). Chemical bonds can be divided into three main types: **ionic bond (离子键)**, **covalent bond (共价键)** and **metallic bond (金属键)**. Covalent bonds are the most common chemical bonds in organic compounds.

In the early 20th century, chemists combined quantum mechanics and chemistry in a way that led to two theories to explain the nature and formation of covalent bonds. Pauling put forward the **Valence Bond Theory (VB)(价键理论)** and R. S. Mulliken put forward the **Molecular Orbital Theory (MO) (分子轨道理论)**. These two theories are briefly described below.

1.2.1　Valence Bond Theory

Valence Bond Theory, which is also known as **electron pairing** (电子配对). The core content is the energy of the system is reduced by electron pairing so that stable covalent bonds are formed. There are three requirements for electron pairing.

(1) Paired electrons must spin in the opposite direction.

(2) An unpaired electron can only match once, indicating the character of saturation in covalent bonds.

(3) Electron pairing should only be carried out along the direction of large electron cloud density. Effective overlap can be obtained. Thus stable covalent bond can be formed, indicating the character of orientation in covalent bonds.

According to the **greatest orbital overlap principle** (轨道最大重叠原则), orbitals can overlap in two different ways for bonding so two types of covalent bonds can be formed——σ and π bond. The covalent bond that is formed when the two s orbitals overlap is called a σ bond. A σ bond is cylindrically symmetrical—the electrons in the bond are symmetrically distributed about an imaginary line connecting the **nuclei** (原子核) of the two atoms joined by the bond, as shown in Figure 1.1. When two p atomic orbitals overlap, the side of one orbital overlaps the side of the other. The side-to-side overlap of two parallel p orbitals forms a bond that is called a π bond, as shown in Figure 1.2.

Figure 1.1　Two s orbitals form σ bond
图 1-1　两个 s 轨道形成 σ 键

Figure 1.2　Two p orbitals form π bond
图 1-2　两个 p 轨道形成 π 键

1.2.2　Hybrid Orbital Theory

Valence Bond Theory strongly demonstrates the nature and characteristics of covalent bonds, but it has limitations in explaining the number of valence bonds of many molecules and atoms and the molecular spatial structures. For example, the valence electron configuration of the carbon atom is $2s^2 2p^2$. According to the electron configuration rule, only two p electrons should be unpaired, while carbon atoms in many compounds have a valence of four instead of two. To explain these contradictions, Pauling *et al.* proposed the concept of hybrid orbitals to enrich and develop valence bond theory. The key points of the **hybrid orbital theory** (杂化轨道理论) are as follows.

(1) Different types of atomic orbitals with similar energy can be combined linearly to redistribute energy and determine the direction of space, forming equal number of new atomic orbitals. This kind of orbital recombination is called **hybridization** (杂化). The new orbital after hybridization is called hybrid orbital. The hybrid electron cloud is concentrated in one direction like a gourd.

(2) Hybrid orbitals are more helpful to maximum overlap between atomic orbitals, so hybrid orbitals have better bonding ability than the original ones.

(3) In order to minimize the repulsion energy, the hybrid orbitals are distributed at the maximum angle in space. Different types of hybrid orbitals have different angles so the molecules formed after bonding have different spatial configurations.

The formation of hybrid orbitals can be illustrated by the example of the carbon atom. First, one 2s electron in the carbon atom jumps to the 2p orbital, as shown in Figure 1.3, and then the 2s orbital and the 2p orbital are hybridized. There are three hybrid forms: one s orbital and three p orbitals are hybridized to form four sp^3 hybrid orbitals which adopt the **tetrahedral configuration** (四面体构型), as shown in Figure 1.4. The hybridization of one s orbital and two p orbitals results in three sp^2 hybrid orbitals which adopt the **plane triangle configuration** (平面三角形构型). The left p orbital without hybridization is perpendicular to the plane of sp^2 hybrid orbitals, as shown in Figure 1.5. The hybridization of one s orbital and one p orbital results in two sp hybrid orbitals which adopt the linear configuration. The left two p orbitals without hybridization are perpendicular to the straight line on sp hybrid orbitals respectively, as shown in Figure 1.6.

Besides carbon atom, other atomic orbitals, such as O, N, B, S can also hybridize as long as they have the similar atomic orbital energy and sp^3 hybridization adopts the tetrahedral configuration, sp^2 hybridization adopts the plane triangle configuration and sp hybridization adopts the linear configuration.

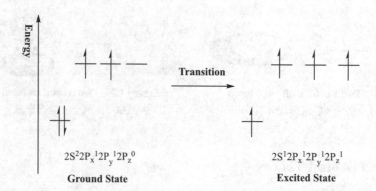

Figure 1.3   One electron in the 2s orbital of carbon jumps to the 2p orbital
图1-3   碳原子2s轨道的一个电子跃迁到2p轨道

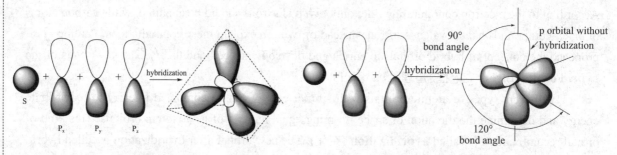

Figure 1.4 sp^3 hybridization of the carbon atom
图1-4   碳原子的sp^3杂化

Figure 1.5 sp^2 hybridization of the carbon atom
图1-5   碳原子的sp^2杂化

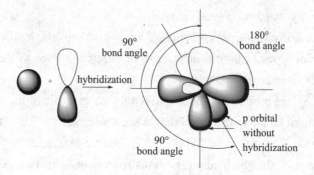

Figure 1.6 sp hybridization of the carbon atom
图1-6   碳原子的sp杂化

1.2.3 Resonance Theory

Resonance Theory (共振理论) is a kind of molecular structure theory proposed by Pauling in 1931, which is the extension and development of valence bond theory. When the structure of a molecule, ion, or radical can not be correctly described by the classical Lewis structural formula, multiple Lewis structural formulas can be used to express the molecular structure. These Lewis structural formulas are called resonance formulas. Double-headed

Figure 1.7   Resonance structures of benzene
图1-7   苯分子的共振结构

arrows are used to connect each resonance formula, as shown in Figure 1.7. The real structure of the molecule is the **resonance hybrid (共振杂化体)** of these resonant structures. The contribution of each resonant structure to the resonance hybrid is different, that is, the degree of participation in resonance hybrid is different. The more stable the resonance structure, the greater the contribution to the resonance hybrid is. The energy of the resonance hybrid is lower than any of the resonance structures. In addition to conforming to the rule of valence bonds, resonance structures must also comply with the principle that the nuclear position of each resonance structure remains unchanged and the number of paired electrons or unshared electrons of each resonance structure remains unchanged.

The advantage of resonance theory is to use electron formula to qualitatively describe the position of charge distribution in the conjugate system, which is convenient and practical to use.

1.2.4 Molecular Orbital Theory

Valence Bond Theory is simple, intuitive and closely connected with classical chemical theory so it developed rapidly in the early days. However, Valence Bond Theory overemphasizes the correlation of electrons but ignores the integrity of the molecule, which makes it difficult to explain some phenomena such as **electron delocalization** (电子离域) and **paramagnetism** (顺磁性) of the oxygen molecule. Molecular Orbital Theory which can explain above phenomena well, has developed greatly in recent years. The key points of the molecular orbital theory are as follows.

(1) When atoms form a molecule, all the electrons contribute and the electrons in the molecule no longer belong to one atom, but move in the range of the whole molecule.

(2) Molecular orbitals can be obtained by the **linear combination** (线性组合) of atomic orbitals (LCAO) in the molecule. The combination of different atomic orbitals should obey the following three principles: ① **Symmetrical matching principle** (对称性匹配原则); ② **Energy similarity** (能量近似); ③ **Maximum overlapping principle** (最大重叠原则). The energy of the molecular orbital is lower than the original atomic orbital energy. When it is filled with electrons, the total molecular energy is reduced, which is good for bonding. Such orbital is called the **bonding orbital (ψMO)** (成键轨道). Some is higher than the original atomic orbital energy, which is not good for bonding when it is filled with electrons. Such orbital is called **anti-bonding orbital (ψ*-MO)** (反键轨道). Some has the same energy as the original atomic orbital, which has no effect on bonding. Such orbital is called **non-bonding orbital** (非键轨道).

(3) After molecular orbitals are formed, they are filled with electrons. The arrangement of electrons in molecular orbitals follows the same principles as in atomic orbitals, which are **Pauli exclusion principle** (泡利不相容原理), **minimum energy principle** (能量最低原理) and **Hund rule** (洪特规则). The molecular orbital energy level diagram is formed in the order of molecular orbital energies and electrons are filled in molecular orbitals according to the rule. The molecular orbital energy level diagram of the oxygen molecule is shown in Figure 1.8. It can be seen from the figure that two s atomic orbitals form two molecular orbitals: σ_s bonding orbital (addition of wave functions) and σ_s^* anti-bonding orbital (subtraction of wave functions); two p atomic orbitals are combined in the "head to head" way to form two molecular orbitals: σ_p and σ_p^*; two p atomic orbitals are combined in the "shoulder to shoulder" way to form two molecular orbitals: π_p bonding orbital and π_p^* anti-bonding orbital. Each of two atoms has three p atomic orbitals, which can form

six molecular orbitals: σ_{p_x}, $\sigma_{p_x}^*$, π_{p_y}, $\pi_{p_y}^*$, π_{p_z}, $\pi_{p_z}^*$, three of them are bonding orbitals and three are anti-bonding orbitals. There are two single electrons after electron filling, so the oxygen molecule is paramagnetic.

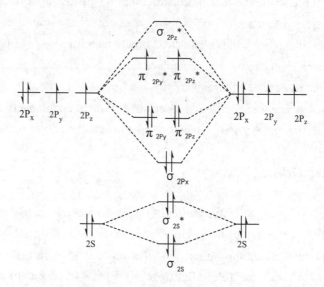

Figure 1.8   Molecular orbital energy level diagram of the oxygen molecule
图1-8　氧气分子的分子轨道能级图

(4) In the molecular orbital theory, **bond order** (键级) is used to indicate the stability of a bond. Bond order is defined as: bond order = (number of electrons in the bonding orbital–number of electrons in the anti-bonding orbital) / 2

PPT

1.3　Properties of Covalent Bonds

The physical quantities that characterize chemical bonds are called bond parameters, which mainly include bond energy, bond length, bond angle, polarity and polarizability.

1.3.1　Bond Energy, Bond Length, Bond Angle

(1) Bond energy　Covalent bond is the strong interaction between two neighboring atoms. Energy is needed for breaking the chemical bond. This energy is expressed as dissociation energy. Energy is released when a new chemical bond is formed. This energy is expressed as **bond energy** (键能). For the diatomic molecule, the bond energy is equal to the dissociation energy. In the condition of 100 kPa and 298.15 kPa, the energy required to dissociate one mole of ideal gaseous molecule AB into atom A and B in the ideal gaseous state is called the dissociation energy of AB, the unit is $kJ \cdot mol^{-1}$. Bond energies of the same covalent bond in different polyatomic molecules vary, but not too much. The average bond energy of the same bond in different molecules can be regarded as the bond energy of one bond. Generally, the

larger the bond energy is, the stronger the bond is. The data of bond energies and lengths of common covalent bonds are shown in Table 1.1.

Table 1.1　Bond energies and lengths of common covalent bonds

表1-1　常见共价键的键能与键长数据

Bond	Bond energy/ $(kJ \cdot mol^{-1})$	Bond length/pm	Bond	Bond energy/ $(kJ \cdot mol^{-1})$	Bond length/pm
C — C	347.27	154	C — F	485.34	142
C = C	605.61	134	C — Cl	338.90	177
C ≡ C	836.80	120	C — Br	300.51	191
C — H	414.22	109	C — I	217.57	22
C — O	359.82	144	C = O	736.38 (aldehyde)	120
C — N	305.43	147	C ≡ N	891.19	115
C — S	27.96	182	O — H	464.40	97

(2) Bond length　The average distance between nuclei of two bonding atoms in a molecule is called the **bond length** (键长). The results of spectra and diffraction experiments show that the bond length of the same bond in different molecules is almost equal. So the average bond length can be regarded as the bond length of one bond. When two atoms form the covalent bond, the shorter the bond length is, the stronger the bond is.

(3) Bond angle　The angle between two bonds formed by the same atom in a molecule is called the **bond angle** (键角). Bond angle is an important parameter to reflect the **steric configuration** (空间构型) of the molecule. In general, the steric configuration of the molecule can be determined by bond angles and bond lengths.

1.3.2　Bond Polarity and Polarizability

(1) Bond polarity　The bond **polarity** (极性) is caused by the different **electronegativity** (电负性) of the bonding atom. If bonding atoms have the same electronegativity, the dense region of electron clouds is in the middle of two nuclei and the center of positive charges from nuclei coincides with the center of negative charges from bonding electron pairs. Such covalent bond is called **nonpolar covalent bond** (非极性共价键). When the electronegativity of the bonding atom is different, the bonding electrons are more attracted to the atom with the greater electronegativity and the center of positive charges of the bond does not coincide with the center of negative charges. Such covalent bond is called **polar covalent bond** (极性共价键). The polarity of the bond depends on the difference in the electronegativity of the bonding atom. The greater the difference in electronegativity is, the stronger the polarity of the bond is. Common atomic electronegativity values are shown in Table 1.2.

Table 1.2   Electronegativities of common elements
表1-2   常见元素的电负性

H 2.1						
Li 1.0	Be 1.5	B 2.0	C 2.5	N 3.0	O 3.5	F 4.0
Na 0.9	Mg 1.2	Al 1.5	Si 1.8	P 2.1	S 2.5	Cl 3.0
K 0.6	Ca 1.0					Br 2.8
						I 2.5

The polarity of a molecule can be measured by the **dipole moment** (偶极矩). The dipole moment of a bond is equal to the magnitude of the charge on either atom (either the partial positive charge or the partial negative charge, because they have the same magnitude) times the distance between the two charges, which is symbolized by the Greek letter "μ"

$$\mu = q \times d$$

A dipole moment is reported in a unit called a debye (D). The dipole moment is directional, indicated by $\longrightarrow$, the direction of the arrow is from positive to negative charge.

For diatomic molecule, the polarity of the bond is the polarity of the molecule. For polyatomic molecule, the polarity is determined by the vector sum of dipole moments of all covalent bonds. For example:

μ=1.86 D chloromethane   μ=1.46 D ammonia   μ=0 tetrachloromethane

A molecule whose vector sum of dipole moments is zero is a **nonpolar molecule** (非极性分子) and a molecule whose vector sum of dipole moments is not equal to zero is a **polar molecule** (极性分子). The larger the dipole moment is, the stronger the polarity of the molecule is. Table 1.3 lists the dipole moments of certain compounds and drugs.

Table 1.3   Dipole moments of certain compounds and drugs
表1-3   一些物质与药物的偶极矩数据

Compound	u /D	Compound	u /D
H_2	0	CH_3CH_2OH	1.70
I_2	0	CH_3Br	1.78
CO_2	0	Benzene	0
CH_4	0	Phenol	1.70

(continued)

Compound	u /D	Compound	u /D
O_2	0	Ethyl ether	1.14
CO	0.12	Aniline	1.51
HI	0.38	Dehydrocholesterol	1.81
HBr	0.78	H_2O	1.84
PCl_3	0.90	CH_3Cl	1.86
HCl	1.03	Cholesterol	1.99
Barbital	1.10	Aspirin	2.80
Phenobarbital	1.16	HCN	2.93
CH_3COOH	1.40	Acetone	2.80
NH_3	1.46	Acetanilide	3.55
CH_3OH	1.68	Androsterone	3.70
Methyltestosterone	4.17	o-dinitrobenzene	0.30
Sulfanilamide	5.37	m-dinitrobenzene	3.80
Phenacetin	5.67	p-dinitrobenzene	6.00
Nitrobenzene	4.19		

(2) Bond polarizability   Under the influence of the external electric field (reagent, solvent, polar container), the polarity of the bond will change, which is called the **polarizability** (极化性) of the bond. For example, under normal circumstances, the Br—Br bond is non-polar ($\mu=0$). However, when the external negative electric field is close to the Br—Br bond, the centers of positive and negative charges are separated from each other due to the induction of E^- and so the dipole moment μ appears:

$$\underset{\mu = 0}{\text{Br—Br}} \longrightarrow \underset{\mu > 0}{\overset{\delta^- \quad \delta^+}{\text{Br—Br}}} \quad E^-$$

Dipole moment caused by the polarization of the molecule (or covalent bonds) due to the influence of the external electric field is called **induced dipole moment** (诱导偶极矩), which is different from the dipole moment of a polar covalent bond. The dipole moment of a polar covalent bond is caused by different electronegativities of bonding atoms and is a permanent property of the covalent bond. The induced dipole moment is generated under the influence of the external electric field, which is a temporary phenomenon. It disappears with the disappearance of the external electric field, so it is also called **instantaneous dipole** (瞬时偶极).

Different covalent bonds have different sensitivities to the effects of the external electric field, which is usually called polarizability. The larger the polarizability of the covalent bond is, the easier it is to be influenced by the electric field. The polarizability of the bond is related to the fluidity of bonding electrons which is determined by the electronegativity of the bonding atom and the atomic radius. The greater electronegativity of the bonding atom is, the smaller the radius of the atom is, the larger the binding force to outer shell electrons is, the worse the electron mobility is, and the worse the polarizability of the covalent bond is. Conversely, the polarizability is greater. The polarizability of the bond plays an important role in the reactivity of the molecule. The polarizability of the covalent bond is one of the main causes of chemical reactions. For example:

C—X　Bond Polarity: C—F > C—Cl > C—Br > C—I

C—X　Bond Polarizability: C—I > C—Br > C—Cl > C—F

C—X　Chemical Activity: C—I > C—Br > C—Cl > C—F

1.3.3　Inductive and Conjugated Effect

The interaction of atoms in organic compounds is an important and ubiquitous phenomenon in organic chemistry. The essence of the interatomic interaction in molecules is generally described by electron effect and steric effect. Electron effect describes the effect of the density distribution of electron cloud on the property of the molecule, which includes inductive effect and conjugated effect. Both different electro negativities of bonding atoms and external electric field can make the covalent bond to produce polarity. This polarity can be transferred in the molecule in two ways: inductive effect and conjugated effect.

（1）Inductive effect　The **inductive effect** (诱导效应) is caused by different electronegativities of bonding atoms, which causes the electron cloud of the whole molecule to shift in one direction along the carbon chain (non-polar bond becomes polar bond), which is the interaction among atoms or atomic groups that are not directly connected to each other in the molecule. The polarity of the bond is transmitted by electrostatic induction along the σ bond in the molecular chain and weakening from the near to the distant gradually, which disappears after three carbon atoms. Inductive effect without external electric field is called **static induction effect** (静态诱导效应), while inductive effect with external electric field is called **dynamic inductive effect** (动态诱导效应). The inductive effect is represented by the symbol I and the direction of electron movement is represented by " → ". For example, in the molecule of $C_3 \rightarrow C_2 \rightarrow C_1 \rightarrow Cl$, the electron cloud between C_1 and Cl is inclined to chlorine atom due to the higher electronegativity of chlorine atom. Although C_2 and C_3 are not directly connected to the chlorine atom, the electron cloud still shifts by the electrostatic inductive effect, which will be weakened with farther distance from the chlorine atom.

The magnitude and direction of the inductive effect are related to the electronegativities of atoms or atomic groups, which is usually defined by the hydrogen in the C-H bond as I=0. If an atom or group which has lower electronegativity than hydrogen replaces the hydrogen in the C-H bond, the electron cloud will shift towards the carbon atom. This phenomenon is called **electron-donating inductive effect (+I)** (给电子诱导效应). If an atom or group which has higher electronegativity than hydrogen replaces the hydrogen in the C-H bond, the electron cloud will shift far away from the carbon atom. This phenomenon is called **electron-withdrawing inductive effect (-I)** (吸电子诱导效应). The following is the order of inductive effects of some atoms or groups.

Electron-Withdrawing Group:　$—NO_2 > —CN > F > Cl > Br > I > —C \equiv CH > —OCH_3 > —C_6H_5 > —CH= CH_2 > H$

Electron-Donating Group: $(CH_3)_3C— > (CH_3)_2CH— > CH_3CH_2— > CH_3— > H$

Inductive effect has a great effect on the **acidity** (酸性) of organic **carboxylic acid** (羧酸). Inductive effect (–I) which can cause the decrease of electron cloud density on the carboxylic carbon atom can enhance acidity. On the contrary, inductive effect (+I) which can cause the increase of electron cloud density on the carboxylic carbon atom can weaken the acidity. And the greater the effect, the greater the change in acidity. The greater the effect is, the greater the change in acidity is. For example, the acidities of following substituted acetic acids increase with the enhancement of electron-withdrawing ability of substituents.

	$H\text{-}CH_2CO_2H$	ICH_2CO_2H	$BrCH_2CO_2H$	$ClCH_2CO_2H$	Cl_2CHCO_2H	Cl_3CCO_2H
pK_a	4.76	3.18	2.90	2.86	1.30	0.64

(2) Conjugated effect   Molecules with conjugated systems exhibit a series of particular characteristics in physical and chemical properties due to the interaction among unsaturated atoms or interaction between unsaturated atoms and other atoms. This phenomenon is called **conjugated effect** (共轭效应), which refers to the transmission of the bond polarity through the conjugated system in the molecule. The conjugated effect without the external electric field is called static conjugated effect and the conjugated effect with the external electric field is called dynamic conjugated effect. The conjugated effect is represented by the symbol C. Conjugated effect is not found in all organic compounds, but only exists in the molecule with the conjugated system. Common conjugated systems include π-π and p-π conjugate. When the C-H σ bond and the p orbital overlap laterally, **hyperconjugated** (超共轭的) σ-π or σ-p system can be formed.

π-π conjugate: The system of alternating single and double bonds is called π-π conjugated system, which can be used to explain some particular properties of conjugated alkenes. The simplest π-π conjugated system is 1, 3-butadiene $CH_2\text{=}CH\text{-}CH\text{=}CH_2$.

Example:

p-π conjugate: The system in which an atom with the lone pair is connected to a double bond by a single bond is called the p-π conjugated system, which can be used to explain some particular characteristics of O, N, X and other atoms with lone pairs attached to the double bond, such as the special stability of chlorine in the molecule of vinyl chloride $CH_2\text{=}CHCl$.

Example:

σ-π hyperconjugation: The system in which the C-H σ bond is connected to a double bond by a single bond is called the σ-π hyperconjugated system, which can be used to explain the stabilities of some alkenes.

Example:

σ-p hyperconjugation: The system in which the C-H σ bond is connected to the p orbital by a single bond is called the σ-p hyperconjugated system, which can be used to explain the stabilities of carbocation and free radical.

Example:

The conjugated effect has directivity as well as the inductive effect. The **electron-donating conjugated effect** (给电子共轭效应) is called positive conjugated effect (+C) and the **electron-withdrawing conjugated effect** (吸电子共轭效应) is called negative conjugated effect (–C). Since the conjugated effect can make electrons transfer on the conjugated chain through overlapping orbitals, the conjugated effect shows the alternations of positive and negative charges, which is not weakened by the growth of the conjugated chain.

The inductive effect and conjugated effect are usually discussed separately, however, sometimes both of them exist in a molecule, which restrict and interact with each other.

1.3.4 Cleavage Modes and Reaction Types of Covalent Bonds

The essence of organic chemical reactions is the breaking of old covalent bonds and the formation of new covalent bonds. In organic reactions, due to different molecular structures and reaction conditions, the cleavage modes of covalent bonds can be divided into the following types.

(1) Homolytic cleavage and free radical reaction   covalent bond cleavage occurs when a pair of bonding electrons are split evenly between two atoms or groups A:B→A·+B·, which is called **homolytic cleavage** (均裂). The atom or atomic group with the single electron generated by homolytic cleavage is called free radical or radical. Reaction initiated by free radical is called radical reaction. Free radical is generally generated under the condition of light, heat or free radical initiator. Most free radicals are unstable and very active.

(2) Heterolytic cleavage and ionic reaction   covalent bond cleavage occurs when a pair of bonding electrons are taken by one atom or atomic group A:B→A$^+$+:B$^-$, which is called **heterolytic cleavage** (异裂). Heterolytic cleavage produces positive and negative ions, which are usually unstable. Reaction initiated by ions generated via heterolytic cleavage is called ionic reaction. Ionic reactions are usually catalyzed by acids, bases or polar solvents, which can be divided into electrophilic reaction and nucleophilic reaction according to the type of reaction reagent.

(3) Pericyclic reaction   The old bond breaking and new bond formation take place simultaneously

without the formation of active intermediates such as radicals or ions, which is called **pericyclic reaction** (周环反应). Generally, it is a reaction between conjugated diene and unsaturated hydrocarbon under the condition of heating or light, which is characterized by a one-step reaction with a **cyclic transition state** (环状过渡态).

PPT

1.4 Intermolecular Forces of Organic Compounds

The atoms in organic molecules are joined together by covalent bonds, which are assembled into polar or nonpolar molecules. There are intermolecular and intramolecular interactions among molecules, which affect the physical and chemical properties of organic compounds, such as melting point, boiling point, solubility, acidity, alkalinity, reaction rate and direction.

1.4.1 Intermolecular Force

(1) Dispersion force   The fact that a substance can change from gas to liquid or from liquid to solid indicates that there is intermolecular force among molecules, which is called intermolecular **van der Waals force** (范德华力). It is a universal force existed as the main form of intermolecular forces, including the **dispersion force** (色散力), **induction force** (诱导力) and **dipole-dipole force** (取向力). The dispersion force is caused by the instantaneous **relative displacement** (相对位移) of moving electrons in the molecule, which causes the temporary **non-coincidence** (不重合) of positive and negative charge centers. The instantaneous dipole-dipole interaction is related to the **deformation** (变形) of the molecule. Only dispersion forces exist among nonpolar molecules. For most molecules, the dispersion force is the main interaction. Generally, the larger the molecular weight is, the larger the deformability of the molecule is, the greater the dispersion force is. Dispersion force exists among all molecules.

Van der Waals forces include attractive and repulsive forces, which are only apparent when the distance of two molecules is 2.1 nm.

(2) Induction force   The induction forces appear from the induction (also known as polarization), which is the attractive interaction between a permanent dipole on one molecule (polar molecule) with an induced (by the former dipole) dipole on another molecule. This interaction is called induction force.

(3) Dipole-Dipole force   It is the interaction among polar molecules. The electronegativities of two atoms in the polar molecule are different, which generates the polarization of valence bond electrons and then the permanent dipole and forces between two dipoles are formed. The magnitude of the dipole-dipole force depends on the magnitude, mutual position and distance of dipoles and this force is ubiquitous in aqueous solutions. Its strength is stronger than that of dipole-inductive dipole interaction, which is very important for the **specificity** (专一性) and **stereoselectivity** (立体选择性) of the drug-receptor interaction.

(4) Hydrogen bond   A **hydrogen bond** (氢键) is an interaction between a hydrogen bonded to an O, N, or F and a lone pair of an O, N, or F in another molecule (X-H……Y, X and Y can be F, O, N), which is regarded as the dipole-dipole interaction.

Example:

Intramolecular Hydrogen Bond

The biggest difference between hydrogen bond and van der Waals force is that the hydrogen bond has **saturability** (饱和性) and **directivity** (方向性的) and the force of hydrogen bond is stronger than van der Waals force. There are **intramolecular** (分子内的) and **intermolecular** (分子间的) hydrogen bonds, which affect not only the physical and chemical properties of organic molecules, but also spatial structures of many molecules.

1.4.2   The Effect of Intermolecular Forces on Physical Properties of Organic Molecules

(1) Boiling point   It is related to the molecular polarity, molecular weight, contact area and hydrogen bond. The higher the polarity, the higher the boiling point is. The mutual attraction between the positive and negative poles of polar molecules enhances the intermolecular interaction, which reduces the molecular vapor pressure and results in the higher boiling point. For example, **propane** (丙烷) and **acetaldehyde** (乙醛) have the same molecular weight, but the difference in boiling point is large, because acetaldehyde is a polar molecule.

$$CH_3CH_2CH_3 \qquad\qquad CH_3CHO$$

propane   acetaldehyde

b.p.–42℃   b.p.20℃

With the same polarity, the higher the molecular weight, the more the number of carbon and hydrogen atoms interacting with each other, the greater the dispersion force, and the higher the boiling point. With the same polarity and molecular weight, the larger the molecular contact area, the higher the boiling point. For example, *n*-pentane (正戊烷) has a higher boiling point than **neopentane** (新戊烷), because neopentane molecules are hindered by branched chains and cannot be closely together, so the molecular contact area is small and the boiling point is lower than that of the straight chain molecule.

$$\begin{array}{c} CH_3 \\ | \\ CH_3CCH_3 \\ | \\ CH_3 \end{array}$$

$$CH_3CH_2CH_2CH_2CH_3$$

n-pentane   neopentane

b.p.36℃   b.p.10℃

The existence of intermolecular hydrogen bond makes the boiling point increase and the stronger the hydrogen bond, the higher the boiling point. The intramolecular hydrogen bond makes the boiling point reduce. For example, **butanol** (丁醇) and **ethyl ether** (乙醚) have the same relative molecular weight, while ethanol molecules have polar hydroxyl groups and intermolecular hydrogen bonds, which make the boiling point much higher than Cheat of ethyl ether.

CH₃CH₂CH₂CH₂OH　　　　　　CH₃CH₂OCH₂CH₃

butanol　　　　　　　　　　ethyl ether

b.p.117.7℃　　　　　　　　b.p.36℃

(2) Melting point　　It is not only related to the polarity, molecular weight, molecular contact area and hydrogen bond, but also related to the symmetry of the molecule.

The higher the molecular polarity, the greater the dipole-dipole force, and the higher the melting point. The higher the polarizability, the larger the contact area and the higher the melting point. The melting point is not only related to the intermolecular force, but also related to the symmetry of the molecule. For example, melting points of *n*-pentane and neopentane are -130℃ and -17℃, respectively. Due to the symmetrical structure of the neopentane molecule and the close arrangement of molecules in the lattice, so the melting point is greatly increased.

Intramolecular hydrogen bond reduces the melting point of organic matter, while intermolecular hydrogen bond increases the melting point. For example:

p-nitrophenol

m.p.114℃

b.p.279℃

o-nitrophenol

m.p.45℃

b.p.216℃

(3) Solubility　　The principle of **"like dissolves like"** (相似相溶).

Solvent
- Polar solvent
  - Protic solvent —— H₂O、ROH、NH₃、RCOOH
  - Aprotic solvent —— acetone、ethyl ether、DMSO、DMF
- Nonpolar solvent —— Hydrocarbon

Since water molecules can form hydrogen bonds with some organic compounds, the presence of hydrogen bond can affect the solubility of organic compound. Intermolecular hydrogen bond can affect not only the boiling point, melting point, density and water solubility of organic matter, but also the viscosity of organic matter. Generally, organic compounds with similar molecular weights and forming intermolecular hydrogen bonds have high viscosities.

1.4.3　The Effect of Intermolecular Forces on Chemical Properties of Organic Molecules

Intermolecular forces can affect the chemical properties of organic compounds, especially for the hydrogen bond which has the greatest effect.

(1) The effect of hydrogen bond on the acidity of organic compound　　Hydrogen bond has an obvious effect on the acidity of all kinds of organic acids and can also stabilize the anion of acid. In the **protic solvent** (质子溶剂), carboxylate anion can be stabilized by solvent through hydrogen bond. In

dicarboxylic acids, when two carboxyl groups are close to each other, the first and second ionization constants can differ by several orders of magnitude. For example, the first and second ionization constants of **oxalic acid** (草酸) are $K_{a1} = 3.5 \times 10^{-2}$ and $K_{a2} = 4 \times 10^{-5}$, respectively. This is because after ionizing one hydrogen ion, the carboxylate anion can form an intramolecular hydrogen bond with another carboxyl group. Due to the stability of the five-membered ring, the second hydrogen ion is difficult to ionize, which leads to the value of K_{a1} is higher than that of K_a of monocarboxylic acid (generally 1.75×10^{-5}) and the value of K_{a2} is lower than that of K_a.

$$\begin{array}{c}COOH\\|\\COOH\end{array} \xrightarrow{-H^+} \quad + \quad H^+$$

(2) The effect of hydrogen bond on the organic reaction   For **nucleophilic substitution** (亲核取代) and molecular elimination reactions on saturated carbon atoms in solvent, different solvents have different effects on the reaction. As for the effect on addition reaction, since the intermediate is the carbon ion, the solvent can strongly affect the rate of addition reaction.

1.5 Structure and Classification of Organic Compounds

The structure of organic compound includes plane and stereochemical structure, so there are two ways to express the structure of organic compound.

1.5.1 Structural Representation of Organic Compound

1. Constitutional Formula   The **constitutional formula** (构造式) represents the connection and arrangement of atoms or atomic groups in a molecule on a two-dimensional plane. The expression of constitutional formula includes **electronic formula** (电子式), **structural formula** (结构式), **condensed structural formula** (结构简式) and **bond line formula** (键线式).

(1) Electronic formula (Lewis structural formula)   The two electrons forming the covalent bond are represented by two dots between the two atoms. The following examples are 1-butene, 1-propanol and cyclopentane.

but-1-ene   propan-1-ol   cyclopentane

(2) Structural formula   Short lines are used to indicate the connection and arrangement of atoms or atomic groups in a molecule.

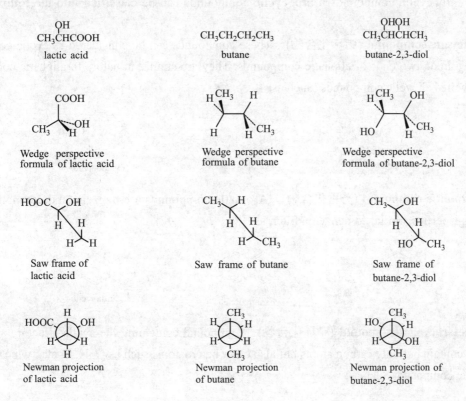

H–C=C–C–C–H

but-1-ene

H–C–C–C–OH

propan-1-ol

cyclopentane

(3) **Condensed structural formula**   To make it easier to write, omitting both the C-H and C-C single bonds to form the condensed structural formula.

$CH_2=CH-CH_2-CH_3$
$CH_2=CHCH_2CH_3$

but-1-ene

$CH_3-CH_2-CH_2-OH$
$CH_3CH_2CH_2OH$

propan-1-ol

cyclopentane

(4) **Bond line structure**   A chain of carbon atoms can be represented by drawing all of the C-C bonds while omitting carbons and hydrogens. In these simplified representations, called bond-line structure or carbon skeleton diagrams, the only atoms specifically written in are those that are neither carbon nor hydrogen bound to carbon. Hydrogens bound to these **heteroatoms** (杂原子) are shown, however.

but-1-ene

OH

propan-1-ol

cyclopentane

2. The Representation of Stereochemical Structure   Many molecules have three-dimensional space structures and need to be represented by the stereochemical structure of the molecule. The commonly used expression methods are **wedge perspective formula** (楔形式), **saw frame projection** (锯架式), **Newman projection** (纽曼投影式) and **Fisher projection** (费歇尔投影式). Detailed introduction of these kinds of structures can be found in chapter three.

$CH_3CHCOOH$ (OH)
lactic acid

$CH_3CH_2CH_2CH_3$
butane

$CH_3CHCHCH_3$ (OH OH)
butane-2,3-diol

Wedge perspective formula of lactic acid

Wedge perspective formula of butane

Wedge perspective formula of butane-2,3-diol

Saw frame of lactic acid

Saw frame of butane

Saw frame of butane-2,3-diol

Newman projection of lactic acid

Newman projection of butane

Newman projection of butane-2,3-diol

$$
\begin{array}{c}
\text{COOH} \\
\text{HO}\!-\!\!\!-\!\!\!-\text{H} \\
\text{CH}_3
\end{array}
\qquad
\begin{array}{c}
\text{CH}_3 \\
\text{H}\!-\!\!\!-\!\!\!-\text{H} \\
\text{H}\!-\!\!\!-\!\!\!-\text{H} \\
\text{CH}_3
\end{array}
\qquad
\begin{array}{c}
\text{CH}_3 \\
\text{H}\!-\!\!\!-\!\!\!-\text{OH} \\
\text{H}\!-\!\!\!-\!\!\!-\text{OH} \\
\text{CH}_3
\end{array}
$$

Fisher projection of lactic acid   Fisher projection of butane   Fisher projection of butane-2,3-diol

1.5.2 Classification of Organic Compound

There are tens of millions of organic compounds which have been characterized. For the convenience of learning and using, chemists divide organic compounds into several or a dozen categories, so as to have a better and deeper understanding of the generality and individuality of each category of organic compounds and corresponding changing rules among them. Organic compounds are usually classified according to the carbon skeleton or the functional group.

1. Classification by Carbon Skeleton According to the classification of carbon skeleton, organic compounds can be divided into three types: **open chain compound** (开链化合物), **carbocyclic compound** (碳环化合物) and **heterocyclic compound** (杂环化合物).

(1) Open chain compound Compounds in which carbon atoms are joined by an open chain skeleton are called open chain compounds. These compounds were first discovered in fats and oils, which are also called **aliphatic compounds** (脂肪族化合物). Such as:

$$\text{CH}_3\text{CH}_2\text{CH}_2\text{CH}_3 \qquad \text{CH}_2\!\!=\!\!\text{CHCH}_2\text{CH}_3 \qquad \text{CH}_3\text{CH}_2\text{CH}_2\text{CH}_2\text{OH}$$

butane   but-1-ene   butan-1-ol

(2) Carbocyclic compound Compounds in which carbon atoms are joined to form a ring skeleton are called carbocyclic compounds. Carbocyclic compounds can be classified into the following two categories:

1) **Alicyclic compound** (脂环化合物) These compounds can be considered as cyclic compounds formed by joining two ends of aliphatic compounds. They are similar in nature to aliphatic compounds, which are called alicyclic compounds. Such as:

cyclopentane   cyclohexene   cyclooctyne

2) **Aromatic compound** (芳香化合物) Aromatic compounds are benzene and compounds with chemical properties similar to benzene. Such as:

benzene   naphthalene   anthracene

(3) **Heterocyclic compound** (杂环化合物) Compound containing the cyclic structure and atoms in the ring contain not only carbon atoms but also other heteroatoms such as O, N, S, etc., which is called heterocyclic compound. Such as:

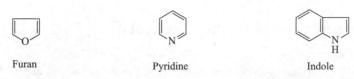

Furan　　　　　　　　　Pyridine　　　　　　　　　Indole

2. Classification by Functional Group　**Functional group** (官能团) is the group of atoms that determines the chemical behavior of the compound. Compounds containing the same functional groups generally have similar properties, which can be studied as a group. Common important functional groups are shown in Table 1.4.

Table 1.4　Some important functional groups
表1-4　一些重要的官能团

Structure	Name	Structure	Name
C=C	Double bond	C(OR)(OR)	Ketal
C≡C	Triple bond	C=N-R	Imino
—OH	Hydroxyl	C=N-NH	Hydrazone
—X	Halogen	C=N-OH	Oximido
-C-O-C-	Ether	-C(=O)-OH	Carboxyl
-C-O-O-C-	Peroxy	R-C(=O)-	Acyl
—OX	Hypohalous	R-C(=O)-X	Acyl halide
—NH$_2$	Amino	R-C(=O)-OR	Ester
—NHR	Secondary amino	-C(=O)-O-C(=O)-	Anhydride
—NHX	Halide amino	-C(=O)-NH$_2$	Amide
—NHOH	Hydroxyamino	-C(=O)-NHR	Secondary amide
—NH—NH$_2$	Hydrazino	-C(=O)-NR$_2$	Tertiary amide
-C(=O)-H	Formyl	—NO$_2$	Nitro
-C(=O)-	Carbonyl	—NO	Nitroso
CH(OR)(OR)	Acetal	—SO$_3$H	Sulfo

1.5.3　Research Methods of Organic Compound

The study of unknown organic compounds generally requires the following process.

1. Separation and Purification　Organic compounds that need to be characterized must be pure. Organic compounds, whether obtained from nature or artificial synthesis, always contain a certain amount of impurities which need to be removed by separation and purification. There are many ways to separate and purify organic compounds. Different methods can be used as needed, such as **recrystallization** (重结晶), distillation and so on.

After purification, to check whether the organic compounds meet the requirements of purity, the purity test needs to be done. Because pure organic compounds have fixed physical constants, such as the melting point, boiling point and so on. The determination of these physical constants is an effective method to test the purity of organic compounds. In addition, gas or liquid **chromatography** (色谱) and physical technology are commonly used in recent years, which have the special feature of high accuracy, rapid in operation, and low consumption of chemical reagents.

2. Determination of Empirical Formula and Molecular Formula

(1) After obtaining the pure substance, the **qualitative analysis** (定性分析) of the elements is carried out to find out which kinds of atoms exist in the molecule.

(2) Then the **quantitative analysis** (定量分析) of elements is carried out to find out the relative number of various atoms and thus the empirical formula can be determined. The empirical formula can only indicate the ratio of atoms of each element in the molecule, but the molecular formula can not be determined.

(3) Molecular weight need to be determined to get the exact number of atoms and the molecular formula. There are many methods to determine molecular weight, such as vapor density method, freezing point depression method, etc. But in recent years, mass spectrometry, gas chromatography-mass spectrometry, liquid chromatography-mass spectrometry and other methods are mostly used.

3. Determination of Structural Formula　Isomerism is common in organic compounds. Organic compounds with the same molecular formula often have more than one structure, including the different arrangement and spatial distribution of atoms or atomic groups. The structure of an organic compound is generally determined by **infrared spectroscopy** (红外光谱), **ultraviolet spectroscopy** (紫外光谱), **nuclear magnetic resonance** (核磁共振), mass spectrometry and single crystal X-ray diffraction.

If a known compound is separated and purified, the structure of the compound can be determined by measuring its physical constants and typical spectral data after purification and then comparing with relevant manuals or literature data.

1.6　Brief Introduction of Organic Acid-Base Theory

The chemical properties of organic compounds are related to the acidity, alkalinity and the transfer of electrons. A brief introduction to the acid-base theory in organic chemistry is necessary to understand organic chemical reactions. The acid-base theories used in organic chemistry include Brøsted acid-base theory (Johann

Nicolaus Brøsted 1879~1947) and Lewis acid-base theory (Gilbert Newton Lewis 1875~1946).

1.6.1　Brøsted Acid–Base Theory

In 1923, the Finnish chemist Brøsted and the American chemist Lowry proposed the acid-base proton theory simultaneously. The theory is that any molecule or ion that can give a proton is an acid and any molecule or ion that can combine with a proton is a base, that is, the acid is the **donor** (供体) of the proton and the base is the **acceptor** (受体) of the proton. When hydrochloric acid dissolves in water, an acid-base reaction occurs.

$$HCl + H_2O \rightleftharpoons H_3O^+ + Cl^-$$

acid　　base　　conjugate　conjugate
　　　　　　　　acid　　　base

Organic acids can be the following organic compounds

acetic acid　　　　　methanol　　　　　acetone

An organic base can be a negative ion (B^-) or a molecule containing lone pair electrons, such as:

methylamine　　　methanol　　　dimethylamine

The acid-base reaction is the proton transfer reaction between the conjugate acid-base pair. When an acid releases a proton, the acid radical is generated, which is the **conjugate base** (共轭碱) of the acid; the protonated compound formed by the combination of base and proton is the **conjugate acid** (共轭酸) of the base, such as:

acid　　　base　　　　　conjugate　conjugate
　　　　　　　　　　　　acid　　　base

$$CH_3\overset{O}{\overset{\|}{C}}OH + H_2O \rightleftharpoons H_3O^+ + CH_3COO^-$$

$$H_2O + CH_3NH_2 \rightleftharpoons CH_3NH_3^+ + OH^-$$

$$H_2SO_4 + CH_3OH \rightleftharpoons CH_3OH_2^+ + HSO_4^-$$

It should be noted that water can be either an acid or a base.

The acid-base strength can be determined in many solvents, but mostly in aqueous solutions, by the dissociation constant Ka of the acid.

$$HA + H_2O \rightleftharpoons A^- + H_3O^+$$

$$Ka = \frac{[H_3O^+][A^-]}{[HA]}$$

Acid strength can be represented as pK_a, $pK_a = -\lg K_a$. The smaller the pK_a, the stronger the acidity is; the larger the pK_a, the weaker the acidity is. The pK_a values of common inorganic acids, such as H_2SO_4, HNO_3 and HCl, are between 2 and 9; the organic acids are normally weaker and pK_a values are between 5 and 15. The acidity of the solution is represented as pH ($pH=-\lg[H^+]$) and the intensity of the acid or base is determined by its structure.

The acidity of a compound depends mainly on the stability of the anion (conjugate base) generated after dissociating the proton. The more stable the anion (conjugate base), the less likely the A^- combines with the H^+, the stronger the acidity of the conjugate acid.

$$HA \rightleftharpoons H^+ + A^-$$

1.6.2 Lewis Acid-Base Theory

Compound that can provide the proton can be considered as acid, which is the Brøsted proton acid-base theory. However, the acid-base theory mostly used in organic chemistry is the electron acid-base theory which is proposed by the American physical chemist Lewis. The theory is that any molecule, ion or atomic group that can provide a pair of electrons, namely the donor of the electron pair, is called **Lewis base** (路易斯碱). Any molecule, ion or atomic group that can accept electrons, namely the acceptor of the electron pair, is called **Lewis acid** (路易斯酸).

occupied   orbital   unoccupied   orbital

$$B \underbrace{}_{\text{Lewis base}} : + \underbrace{}_{\text{Lewis acid}} A \longrightarrow B-A$$

Lewis acids can be neutral molecules, positive ions and **metallic compounds** (金属化合物), such as:

Neutral molecule: H_2O HCl HBr HNO_3 H_2SO_4

$$H_3C-\overset{\overset{O}{\|}}{C}-OH \qquad C_6H_5-OH \qquad CH_3CH_2OH$$

Positive ion: Li^+ Mg^{2+} Br^+ R^+

Metallic compound: $AlCl_3$ BF_3 $TiCl_4$ $FeCl_3$ $ZnCl_2$

Lewis base: the compound having lone pair electrons, anion, alkene or aromatic compound

Neutral electron donor: $CH_3CH_2\ddot{O}H$ $CH_3\ddot{O}CH_3$ $CH_3\overset{:\ddot{O}:}{C}CH_3$ $CH_3\overset{:\ddot{O}:}{C}\ddot{O}H$ $CH_3\overset{:\ddot{O}:}{C}NH_2$

ethanol   methyl ether   acetone   acetic acid   acetamide

$CH_3\ddot{N}CH_3$ $CH_3\ddot{S}CH_3$ (cyclopentadiene)   (benzene)

dimethylamine   dimethyl sulfide   cyclopentadiene   benzene

anion: R^- OH^- RO^-

Lewis acid-base reaction:

$$\overset{\delta^-}{Cl} \overset{\delta^+}{-H} \ + \ :\overset{..}{O}-H \ \Longleftrightarrow \ H-\overset{..}{O}^+-H \ + \ Cl^-$$
$$\qquad\qquad\qquad\ \ H \qquad\qquad\qquad\ H$$

Lewis acid   Lewis base

$$Cl-\overset{\overset{\textstyle Cl}{|}}{\underset{\underset{\textstyle Cl}{|}}{Al}} \ + \ :\overset{\overset{\textstyle CH_3}{|}}{\underset{\underset{\textstyle CH_3}{|}}{N}}-CH_3 \ \Longleftrightarrow \ Cl-\overset{\overset{\textstyle Cl}{|}}{\underset{\underset{\textstyle Cl}{|}}{Al}}-\overset{\overset{\textstyle CH_3}{|}}{\underset{\underset{\textstyle CH_3}{|}}{N}}-CH_3$$

Lewis acid   Lewis base

Lewis acid-base theory expands the variety and range of acid and base, which is useful for organic chemistry. According to Lewis acid-base theory, any ion or molecule that accepts the electron pair is an acid and any that donates the electron pair is a base. Sometimes the product of acid-base neutralization is not water but often a complex, in which the "donor" and "acceptor" of the electron pair are the key factors.

1.6.3 Electrophile and Nucleophile

Lewis acid can accept the electron pair, which is called **electrophile** (亲电试剂) in the organic reaction; Lewis base can donate the electron pair, which is called **nucleophile** (亲核试剂) in the organic reaction.

$$\overset{\delta^+}{CH_3}\overset{\delta^-}{-I} \ + \ OH^- \longrightarrow CH_3OH + I^-$$

In this reaction, OH^- donates the electron pair, which is the nucleophile; CH_3^+ generated by the heterolytic cleavage of CH_3I accepts the electron pair, which is the electrophile.

For a molecule, it usually has both an electrophilic reaction center and a nucleophilic reaction center. In most cases, one of them has the relatively high reaction activity, which determines whether the molecule belongs to the electrophile or the nucleophile. The following reagents Br_2, HBr, H_2O and HCN in the heterolytic reaction are discussed as example.

$$Br:Br \longrightarrow :\overset{..}{\underset{..}{Br}}{}^+ + :\overset{..}{\underset{..}{Br}}{}^-$$

In the heterolytic reaction, the bromine molecule is split into the positive ion and negative ion. The positive ion Br^+ has the higher energy and higher reaction activity, so the bromine molecule is an electrophile.

For the HBr molecule, the electrophilic center is much more active than the nucleophilic center, so the HBr has the electrophilic property and is an electrophile. For the water $\overset{\delta^+}{H}-\overset{\delta^-}{OH}$ molecule, there is no obvious difference between two centers and the electrophilic center is not very active, which explains why the addition reaction of olefin and water must need the catalyst.

The reactivity of the molecule $\overset{\delta^+}{H}-\overset{\delta^-}{CN}$ is determined by the nucleophilic center, because the free electron pair on the carbon atom in the CN^- ion determines that the energy and reactivity of CN^- are greater than that of the H^+ ion. The CN^- ion can attack the positive part of the reactive molecule (electrophilic center) and form the covalent bond by sharing the electron pair with another atom, so the HCN is a nucleophile.

Common electrophile: X_2, HOX, HX, $AlCl_3$, BF_3, ArN_2^+, R_4N^+, NO_2^+.

Common nucleophile: $NaNH_2$, C_2H_5ONa, CH_3COONa, NaOH, NaCN, RMgX, NH_3, RNH_2.

1.7   Relation between Organic Chemistry and Pharmacy

Organic chemistry is an important basic discipline of medicine. Our life is almost inseparable from organic matter. More than 100 years ago, dyes and drugs were extracted from animals and plants. Nowadays western medicines are obtained through organic synthesis. All of these are related to organic chemistry. In ancient times, people knew that certain natural substances could cure diseases and pain, such as drinking alcohol for pain relief, **rhubarb (大黄)** for **catharsis (导泻)**, neem used as insect repellent, willow for abatement of fever, which were all the medical experiences summarized by working people through the long-term practice in our country. Modern research has confirmed that there are some organic active ingredients in rhubarb, neem, willow bark and other natural plants, which have specific actions on certain part of the body and cause a typical reaction.

With the deepening of Chinese medicine research, more emphasis is placed on clarifying the **pharmacodynamic (药效的)** material basis of Chinese medicine, so as to explore the principle of Chinese medicine for prevention and treatment of diseases. The basis of pharmacodynamic substances of traditional Chinese medicine and its compound prescription is the chemical components that exert pharmacodynamic effects, among which organic chemical components account for the majority. Organic chemistry studies the basis of pharmacodynamic substances of traditional Chinese medicine from the perspective of molecules and thus promoting the development of research on traditional Chinese medicine.

For specialties about traditional Chinese medicine, such as pharmacology of traditional Chinese medicine, processing of Chinese herbal medicine, Chinese drugs pharmaceutics, identification of Chinese medicine, quality inspection, storage, which all need the basic theory of organic chemistry and experimental skills, especially for the study of Chinese medicine effective components, through extraction, separation, determination of the structure, synthetic experiments, all these studies are inseparable from the basic theory of organic chemistry and experimental skills.

The basic theory and properties of organic chemistry can explain some special phenomena of drugs. For example, electron effect is an important basic theory in organic chemistry, which is ubiquitous in organic drug molecules and directly affects the properties of drugs. For example, **ethyl *p*-aminobenzoate** (对氨基苯甲酸乙酯) can be used as the local anesthetic, but **ethyl *p*-nitrobenzoate** (对硝基苯甲酸乙酯) cannot, because the electron-donating p-π conjugation effect from the amino group is much greater than its electron-withdrawing inductive effect. Therefore, the total electronic effect is the electron-donating effect. This effect reduces the electropositivity of the carbonyl carbon atom and slows down the hydrolysis rate of the ester, which has the clinical value. The electron-withdrawing inductive effect of the nitro group from ethyl p-nitrobenzoate molecule is the same as its p-π conjugation effect. This effect increases the electropositivity of the carbonyl carbon atom and speeds up the hydrolysis rate of the ester, which is not suitable for the clinical application.

Chemical modification of organic chemistry has important practical significance in improving drug stability, increasing drug solubility, prolonging action time, reducing side effects and eliminating bad taste. The modification of chemical structures of organic drugs includes the formation of salt, **ester** (酯), **amide** (酰胺) and so on, which are the important types of chemical reactions in organic chemistry. For example, **salicylic acid** (水杨酸) has the antipyretic analgesic effect, but the stimulation of the gastrointestinal tract of it is very serious. Therefore, it has been modified by acylation reaction into acetyl salicylic acid, namely aspirin, which has been widely used as the antipyretic analgesic drug in clinic. In addition, if salicylic acid is modified into the methyl salicylate by esterification, namely wintergreen oil, which is the external medicine for sprain.

Organic chemistry is a subject closely related to **pharmacy** (药学). Many important discoveries and breakthroughs in pharmacy include a lot of research work related to organic chemistry. With the development of organic chemistry, **traditional Chinese medicine** (中药) and innovative drugs can be explored on a deeper level and human beings will gradually conquer various diseases.

重 点 小 结

1. **有机化合物的特点**　大多数有机物对热不稳定,易燃烧;大多有机化合物一般难溶于水,易溶解于有机溶剂中;多数有机化合物熔、沸点低,易挥发;大多数有机化合物的反应速度慢且产物复杂,副反应多;有机化合物中普遍存着异构现象。

2. **有机化合物的化学键**　化学键基本上可分为三类:离子键、共价键和金属键。其中共价键是有机化合物中最常见的化学键,对共价键本质与形成进行解释的理论包括价键理论和分子轨道理论。

3. **共价键的性质**　共价键的键参数主要有键能、键长、键角、键的极性及键的极化性;键的极性是由于成键原子的电负性不同而引起的;不同的共价键对外界电场的影响有不同的感受能力,这种感受能力通常叫做可极化性;分子中原子间相互影响问题的实质,一般用电子效应和立体效应来描述,电子效应说明了分子中电子云密度的分布对性质的产生的影响,包括诱导效应与共轭效应,共轭效应包括π-π共轭、p-π共轭、σ-π超共轭和σ-p超共轭。共价键的断裂方式包括均裂、异裂和协同反应。

4. **有机化合物分子间力**　分子间的相互作用力称为分子间的范德华力,包括色散力、诱导力和取向力,色散力存在于一切分子间。

5. **有机化合物结构的表示方法及分类**　构造式是在二维平面上表示分子中原子或原子团的连接方式和排列次序,表示构造式常用到电子式、结构式、结构简式和键线式。分子立体结构的常用表示方法有楔形透视式、锯架投影式、纽曼投影式、费歇尔投影式。有机化合物通常采用按碳骨架分类和按官能团分类两种方法,按碳骨架分类可分为开链化合物、碳环化合物、杂环化合物。

6. **有机酸碱理论简介**　有机化合有机化学中的酸碱理论有勃朗斯特酸碱理论和路易斯酸碱理论。 勃朗斯特酸碱理论认为凡能给出质子的分子或离子都是酸,凡能与质子结合的分子或离子都是碱,即酸是质子的给予体,碱是质子的接受体。路易斯酸碱理论认为凡是能提供一对电子的分子、离子或原子团,即电子对的给予体,称之为路易斯碱;凡是能接受电子的分子、离子或原子团,即电子对的接受体,称之为路易斯酸。

Problems
目 标 检 测

1. Organic compounds have far fewer elements than inorganic compounds, but the kinds of organic compounds are much more than inorganic compounds. Why?

有机物比无机物所含元素种类少，但为什么有机物的种类远多于无机物？

2. What orbitals are used to form the 4 σ bonds in methane?

甲烷用什么轨道形成四个 σ 键？

3. Explain why a s bond formed by overlap of an s orbital with an sp^3 orbital of carbon is stronger than a s bond formed by overlap of an s orbital with a p orbital of carbon.

请解释为什么s轨道与sp^3杂化轨道形成的σ键比s轨道与p轨道形成的σ键更强。

4. Draw a Lewis structure for each of the following species.

请画出下列分子及离子的路易斯结构式。

    a. H_2CO_3　　　　b. CO_3^{2-}　　　　c. CH_2O　　　　d. CO_2

5. Does the molecule with polar covalent bonds have to be a polar molecule? Please give some examples.

含有极性共价键的分子一定是极性分子吗？请举例说明。

6. Decide whether the following compounds are polar molecules.

判断下列化合物是否为极性分子。

Br_2,　HCl,　CH_3OH,　CH_4,　CH_3NH_2,　H_2O,　CH_3OCH_3,　CO_2

7. Which compound has a larger dipole moment, $CHCl_3$ or CH_2Cl_2?

氯仿和二氯甲烷哪个化合物的偶极矩更大？

8. Which of the bonds in a carbon-oxygen double bond has more effective orbital-orbital overlap, the s bond or the p bond?

碳氧双键中哪个键有更好的轨道重叠，s轨道成键还是p轨道成键？

9. NH_2^- is a stronger base than OH^-, and their conjugate acids are NH_3 and H_2O, which one has the stronger acidity? Why?

NH_2^-的碱性比OH^-的碱性强，其共轭酸分别是NH_3和H_2O，哪个酸性强？为什么？

10. Which of the following compounds can be used as an electrophile or a nucleophile?

下列物质中哪个可以作为亲电试剂，哪个可以作为亲核试剂？

    a. Br_2　　　　b. $NaCN$　　　　c. $(CH_3)_2CHOH$　　　　d. $AlCl_3$

Discussion Topic

Discuss the main contribution of R. B. Woodward to organic chemistry.

Famous American organic chemist, professor R. B. Woodward of Harvard University, rose to fame in 1944 by synthesizing quinine and then he synthesized a number of complex natural product molecules, including cholesterol, cortisone, strychnine, lysergic acid, reserpine, chlorophyll, cephalosporin and colchicine. Woodward won the 1965 Nobel Prize in chemistry for his work in organic synthesis. In collaboration with Swiss organic chemist Albert Eschenmoser, he led more than 100 scientists in a 12-year effort to complete the synthesis of vitamin B_{12} in 1973. In the process of synthesizing vitamin B_{12}, Woodward and his student Hoffmann proposed conservation of orbital symmetry to explain the pericyclic reaction in 1965. In 1981, Hoffmann won the Nobel Prize in chemistry for this work.

（李嘉鹏　刘建宇）

第二章 烷烃和环烷烃
Chapter 2 Alkanes and Cycloalkanes

 学习目标

1. **掌握** 烷烃和环烷烃的分类和命名及相关的化学反应。
2. **熟悉** 烷烃和环烷烃的构象。
3. **了解** 烷烃在制药领域的应用。

Organic compounds all contain carbon, but they can also contain a wide variety of other elements. In this chapter, we focus on the simplest organic compounds—those that contain only the elements of carbon and hydrogen, called **hydrocarbons** (烃). Hydrocarbons fall into two broad classes based on the types of bonds between the carbon atoms. A hydrocarbon that has only carbon-carbon single bonds is saturated. Hydrocarbons that contain carbon-carbon multiple bonds are **unsaturated** (不饱和的). Alkanes and cycloalkanes are two types of saturated hydrocarbons. **Alkanes** (烷烃) have carbon atoms bonded in chains; **cycloalkanes** (环烷烃) have carbon atoms bonded to form a ring. Alkanes and cycloalkanes are so similar that many of their properties can be considered side by side.

2.1 Alkanes

PPT

Alkanes are often described as saturated hydrocarbons because they contain the maximum possible number of hydrogens per carbon. They have the general empirical formula C_nH_{2n+2}. Methane, CH_4, is the simplest alkane, followed by ethane, then propane.

$$CH_4 \qquad CH_3CH_3 \qquad CH_3CH_2CH_3$$
Methane   Ethane   Propane

Natural gas is colorless and nearly odorless, such as methane, ethane and propane. The characteristic odor of natural gas which is used for heating our homes and cooking is from sulfur-containing compounds that are deliberately added to warn us of potentially dangerous leaks.

2.1.1   The Structure of Alkanes

We will use simple alkanes as examples to study some of the properties of organic compounds, including the structure of sp^3-hybridized carbon atoms and properties of C-C and C-H single bonds.

Alkanes have only sp^3-hybridized carbons. Methane (CH_4) is a nonpolar molecule, and has four covalent carbon–hydrogen bonds. In methane, all four C-H bonds have the same length (1.10 Å), and all the bond angles (109.5°) are the same. Therefore, all four covalent bonds in methane are identical. Methane has **regular tetrahedral geometry** (正四面体结构). Three different ways to represent a methane molecule are shown in Figure 2.1.

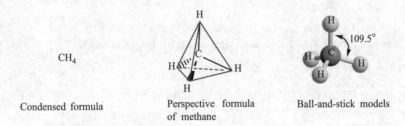

Condensed formula   Perspective formula   Ball-and-stick models
                            of methane

Figure 2.1   Three different ways to represent a methane molecule
图2-1   甲烷分子的三种不同表示方式

In a **perspective formula** (透视式), bonds in the plane of the paper are drawn as solid lines, bonds sticking out of the plane of the paper towards you are drawn as **solid wedges** (实楔线), and those pointing back from the plane of the paper away from you are drawn as **broken wedges** (虚楔线).

Ethane (CH_3CH_3), the two-carbon alkane, is composed of two methyl groups with overlapping sp^3 hybrid orbitals forming a sigma bond (σ) between them. In ethane, the bond between the two carbon atoms is longer than a C-H bond, But, like the C-H bonds, it is a covalent bond in the Lewis sense. In terms of **hybrid orbitals** (杂化轨道), the carbon-carbon bond in ethane consists of two electrons in a bond formed by the overlap of two sp^3 hybrid orbitals, one from each carbon. Thus, the carbon-carbon bond in ethane is an sp^3-sp^3 σ bond. The C-H bonds in ethane are like those of methane. They consist of covalent bonds, each of which is formed by the overlap of a carbon sp^3 orbital with a hydrogen $1s$ orbital; that is, they are sp^3-1s σ bonds. Both the H-C-C and H-C-H bond angles in ethane are approximately tetrahedral because each carbon bears four groups. Three different ways to represent an ethane molecule are shown in Figure 2.2.

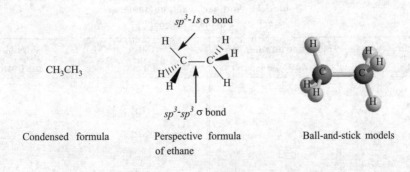

Condensed formula   Perspective formula   Ball-and-stick models
                            of ethane

Figure 2.2   Three different ways to represent an ethane molecule
图2-2   乙烷分子的三种不同表示方式

2.1.2 Constitutional Isomerism in Alkanes

Think about the ways that carbon and hydrogen can combine to make alkanes. With one carbon and four hydrogens, only one structure is possible: methane, CH_4. Similarly, there is only one possible combination of two carbons with six hydrogens (ethane, CH_3CH_3) and only one possible combination of three carbons with eight hydrogens (propane, $CH_3CH_2CH_3$). If larger numbers of carbons and hydrogens combine, however, more than one kind of molecule can result. For example, there are two substances with the formula C_4H_{10}. The four carbons can be in a row (1), or they can branch (2).

$$C_4H_{10} \qquad CH_3-CH_2-CH_2-CH_3 \qquad \begin{array}{c} CH_3-CH-CH_3 \\ | \\ CH_3 \end{array}$$

$$(1) \qquad\qquad\qquad (2)$$

Similarly, there are three C_5H_{12} molecules, and so on for larger alkanes.

$$C_5H_{12} \qquad CH_3CH_2CH_2CH_2CH_3 \qquad \begin{array}{c} CH_3CHCH_2CH_3 \\ | \\ CH_3 \end{array} \qquad \begin{array}{c} CH_3 \\ | \\ CH_3CCH_3 \\ | \\ CH_3 \end{array}$$

$$(3) \qquad\qquad (4) \qquad\qquad (5)$$

Compounds like (1) and (3), whose carbons are connected in a row, are called **straight-chain alkanes (直链烷烃)** or **normal alkanes (*n*-alkane,正烷烃)**. Compounds like (2)、(4) and (5), whose carbon chains branch, are called **branched- chain alkanes (支链烷烃)**.

Compounds like the two C_4H_{10} molecules and the three C_5H_{12} molecules, which have the same formula but different structures, are different compounds with different properties. For example, the boiling point of (1) butane is –0.5℃, whereas that of (2) isobutane is –11.7℃. Different compounds that have the same molecular formula are said to be **isomers (同分异构体)** or isomeric compounds.

There are different types of isomers. Isomers that differ in the connectivity of their atoms, such as (1) butane and (2) isobutane, are called **constitutional isomers (构造异构体)** or **structural isomers (结构异构体)**.

For the higher alkanes (*n*>4), more than two isomers are possible. There are three pentanes, C_5H_{12}, as shown below. There are five hexanes, C_6H_{14}; nine heptanes, C_7H_{16}; and eighteen octanes, C_8H_{18}. You cannot easily calculate the number of isomers. As Table 2.1 shows, the number of possible alkane isomers increases dramatically as the number of carbon atoms increases.

Table 2.1 Number of alkane isomers
表2-1 烷烃同分异构体的数目

Molecular formula	Possible number of constitutional isomers	Molecular formula	Possible number of constitutional isomers
CH_4	1	C_7H_{16}	9
C_2H_6	1	C_8H_{18}	18
C_3H_8	1	C_9H_{20}	35
C_4H_{10}	2	$C_{10}H_{22}$	75
C_5H_{12}	3	$C_{20}H_{42}$	366,319
C_6H_{14}	5	$C_{30}H_{62}$	4,111,846,763

Isomers are different compounds with the same molecular formula. For now, we recognize constitutional isomers. There are several types of isomerism in organic compounds, and we will cover them in detail in Chapter 3 (Stereoisomerism).

2.1.3　Nomenclature of Alkanes

Nomenclature (命名法) in organic chemistry is of two types: common (or "trivial") and systematic. This problem of naming organic molecules has been with organic chemistry from its very beginning, but the initial method was far from systematic. Compounds have been named after their discoverers ("Nenitzescu's hydrocarbon"), after localities ("sydnones"), after their shapes ("cubane," "basketane"), and after their natural sources ("vanillin"). Many of these common or trivial names are still widely used. However, now there is a precise system for naming the alkanes. Nomenclature for other classes of compounds will be presented in individual chapters as the chemistry of the functional groups is covered. The common, nonsystematic names of some compounds have persisted. A few of the most important of these will be covered in individual chapters.

1. Common (trivial) Names　The older names for organic compounds. The names methane, ethane, propane, and butane have historical roots. From pentane on, alkanes are named using the Greek word for the number of carbon atoms, plus the suffix *-ane* to identify the molecule as an alkane.

If all alkanes only had unbranched (straight-chain) structures, their nomenclature would be simple. Most alkanes have structural isomers, however, and we need a way of naming all the different isomers. For example, there are two isomers of formula C_4H_{10}. The unbranched isomer is simply called butane (or **n-butane** (正丁烷) meaning "normal" butane), and the branched isomer is called **isobutane** (异丁烷), meaning an "isomer of butane."

$$C_4H_{10} \qquad CH_3-CH_2-CH_2-CH_3 \qquad\qquad CH_3-\underset{\underset{CH_3}{|}}{CH}-CH_3$$

$$(1)\ \textit{n}\text{-butane} \qquad\qquad (2)\ \text{isobutane}$$

The three isomers of C_5H_{12} are called pentane (or *n*-pentane), isopentane, and **neopentane** (新戊烷).

$$C_5H_{12} \qquad CH_3CH_2CH_2CH_2CH_3 \qquad CH_3\underset{\underset{CH_3}{|}}{CH}CH_2CH_3 \qquad CH_3\overset{\overset{CH_3}{|}}{\underset{\underset{CH_3}{|}}{C}}CH_3$$

$$(3)\ \textit{n}\text{-pentane} \qquad\qquad (4)\ \text{isopentane} \qquad\quad (5)\ \text{neopentane}$$

Isobutane, isopentane, and neopentane are **common names** (普通命名法) or **trivial names** (俗名), meaning historical names arising from common usage. Common names cannot easily describe the larger, more complicated molecules having many isomers, however. The number of isomers for any molecular formula grows rapidly as the number of carbon atoms increases (Table 2.1). This system of common names has no way of handling other branching patterns; therefore, for more complex alkanes, we must use the more flexible IUPAC system.

2. IUPAC Names　The formal system of nomenclature for organic compounds. **Systematic nomenclature (系统命名法)**, in which the name of a compound describes its structure, was first introduced by a chemical congress in Geneva, Switzerland, in 1892. It has continually been revised since then, mostly by the International Union of Pure and Applied Chemistry (**IUPAC**, usually spoken as eye-you-pac). This international group has developed a detailed system of nomenclature that we call the IUPAC rules. The IUPAC rules are accepted throughout the world as the standard method for naming organic compounds. The names that are generated using this system are called IUPAC names or systematic names.

The IUPAC rules for the nomenclature of alkanes form the basis for the substitutive nomenclature of most other compound classes. Hence, it is important to learn these rules and be able to apply them.

(1) Continuous-Chain (Unbranched) alkanes   These are compounds containing only carbon and hydrogen in a continuous, unbranched carbon chain. All carbon-carbon bonds are single bonds. The names of these compounds constitute the basis of the nomenclature of hydrocarbon derivatives and organic compounds in general.

微课

The IUPAC rules assign names to straight-chain alkanes as shown in Table 2.2. The names of the continuous-chain (unbranched) saturated hydrocarbons (alkanes) are derived from the Greek names for the numbers of carbon atoms present. Notice that the prefix *n-* is not part of the IUPAC system. The IUPAC name for $CH_3CH_2CH_2CH_3$ is butane, not *n*-butane.

Table 2.2　IUPAC names of straight-chain alkanes up to 10 carbon atoms
表2-2　10以内碳的直链烷烃的IUPAC命名

No. of carbons	C_nH_{2n+2}	Formula name
1	CH_4	**Meth**ane (甲烷)
2	CH_3CH_3	**Eth**ane (乙烷)
3	$CH_3CH_2CH_3$	**Prop**ane (丙烷)
4	$CH_3CH_2CH_2CH_3$	**But**ane (丁烷)
5	$CH_3CH_2CH_2CH_2CH_3$	**Pent**ane (戊烷)
6	$CH_3CH_2CH_2CH_2CH_2CH_3$	**Hex**ane (己烷)
7	$CH_3CH_2CH_2CH_2CH_2CH_2CH_3$	**Hept**ane (庚烷)
8	$CH_3CH_2CH_2CH_2CH_2CH_2CH_2CH_3$	**Oct**ane (辛烷)
9	$CH_3 CH_2CH_2CH_2CH_2CH_2CH_2CH_2CH_3$	**Non**ane (壬烷)
10	$CH_3 CH_2CH_2CH_2CH_2CH_2CH_2CH_2CH_2CH_3$	**Dec**ane (癸烷)

These alkanes in Table 2.2 differ only by the number of –CH$_2$- groups (**methyl-ene groups,** 亚甲基) in the chain. If the molecule contains *n* carbon atoms, it must contain (*2n+2*) hydrogen atoms. The series of compounds, like the unbranched alkanes that differ only by the number of -CH$_2$- groups is called a **homologous series** (同系列), and the individual members of the series are called **homologs** (同系物). For example, butane is a homolog of propane, and both of these are homologs of hexane and decane.

Although we have derived the C_nH_{2n+2} formula using the unbranched *n*-alkanes, it applies to branched alkanes as well. Any isomer of one of these n-alkanes has the same molecular formula. Just as butane and pentane follow the rule, their branched isomers isobutane, isopentane, and neopentane also follow the rule.

医药大学堂
WWW.YIYAODXT.COM

(2) Branched-Chain alkanes Branching groups are in general called **substituents** (取代基), and substituents derived from alkanes are called **alkyl groups** (烷基). Note that alkyl groups are simply parts of larger compounds. All of these alkyl groups can be designated by the symbol R. Alkyl groups are named by replacing the –*ane* ending of the parent alkane with an –*yl* ending. For example, removal of a hydrogen atom from methane, CH_4, generates a methyl group, methyl. Similarly, removal of a hydrogen atom from the end carbon of any *n*-alkane gives the series of straight-chain alkyl groups shown in Table 2.3.

Table 2.3　Some straight-chain alkyl groups
表2-3　一些直链烷基

Alkane	Alkyl Group	Abbreviation
$CH_3-\vdots-H$ Methane	CH_3- Methyl	Me—
$CH_3CH_2-\vdots-H$ Ethane	CH_3CH_2- Ethyl	Et—
$CH_3CH_2CH_2-\vdots-H$ Propane	$CH_3CH_2CH_2-$ Propyl	Pr—
$CH_3CH_2CH_2CH_2-\vdots-H$ Butane	$CH_3CH_2CH_2CH_2-$ Butyl	Bu—

Just as straight-chain alkyl groups are generated by removing a hydrogen atom from an end carbon, branched alkyl groups are generated by removing a hydrogen atom from an internal carbon.

The alkyl groups with three- and four-carbon chains deserve special mention. A three-carbon alkyl group could be attached to a longer chain at either of the outside carbons or at the middle carbon; the groups are called propyl and **isopropyl** (异丙基).There are two structural isomers of a four-carbon alkyl group, each of which has two different points of connection. The groups are called butyl, *sec*-**butyl** (仲丁基), isobutyl and *tert*-**butyl** (叔丁基).

$CH_3CH_2CH_3$ Propane	$CH_3CH_2CH_2-$ Propyl	and	$\overset{\vert}{CH_3CHCH_3}$ Isopropyl
$CH_3CH_2CH_2CH_3$ Butane	$CH_3CH_2CH_2CH_2-$ Butyl	and	$\overset{\vert}{CH_3CH_2CHCH_3}$ *sec*-Butyl
$CH_3\underset{\underset{CH_3}{\vert}}{CHCH_3}$ Isobutane	$CH_3\underset{\underset{CH_3}{\vert}}{CHCH_2}-$ Isobutyl	and	$CH_3\underset{\underset{CH_3}{\vert}}{CCH_3}$ *tert*-Butyl

(3) Classification of carbon substitution A carbon atom is classified as **primary** (1°, 伯), **secondary** (2°, 仲), **tertiary** (3°, 叔) and **quaternary** (4°, 季) depending on the number of carbon atoms bonded to it. Primary carbon (1°) is a carbon at the end of a chain, which is bonded to one other carbon. Secondary carbon (2°) is a carbon in the middle of a chain, which is bonded to two other carbons. Tertiary carbon (3°) is a carbon with three carbons attached to it and quaternary carbon (4°) is a carbon with four carbons attached to it.

$$R-\overset{\overset{\displaystyle H}{|}}{\underset{\underset{\displaystyle H}{|}}{C^{1^\circ}}}-H \qquad R-\overset{\overset{\displaystyle H}{|}}{\underset{\underset{\displaystyle R}{|}}{C^{2^\circ}}}-H \qquad R-\overset{\overset{\displaystyle R}{|}}{\underset{\underset{\displaystyle R}{|}}{C^{3^\circ}}}-H \qquad R-\overset{\overset{\displaystyle R}{|}}{\underset{\underset{\displaystyle R}{|}}{C^{4^\circ}}}-R$$

Primary carbon(1°)   Secondary carbon(2°)   Tertiary carbon(3°)   Quaternary carbon(4°)

Different types of carbon atom are shown in the following compound.

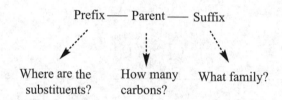

(4) IUPAC rules   A chemical name has three parts in the IUPAC system: prefix, parent, and suffix. The parent selects a main part of the molecule and tells how many carbon atoms are in that part; the suffix identifies the functional-group family the molecule belongs to; and the prefix gives the locations of the functional groups and other substituents on the parent.

Prefix —— Parent —— Suffix

Where are the   How many   What family?
substituents?   carbons?

All of these principles are illustrated in the following examples.

Step1   Find the parent hydrocarbon

$$\overset{6}{C}H_3\overset{5}{C}H_2\overset{4}{C}H_2\overset{3}{C}H_2\overset{2}{C}H\overset{1}{C}H_3$$

substituent -----> CH₃   Named as a substitued hexane

Find the longest continuous chain of carbon atoms present in the molecule, and use the name of that chain as the parent name. The longest chain may not always be apparent from the manner of writing; you may have to "turn corners".

substituent - - - - - -

$$\overset{7}{C}H_3\overset{6}{C}H_2\overset{5}{C}H_2\overset{4}{C}H_2\overset{3}{C}HCH_3$$

           CH₂CH₃   Named as a substitued heptane
           2   1

If two different chains of equal length are present, choose the one with the larger number of branch points as the parent.

substituent ----► CH$_3$
　　　　　　　 $\overset{2}{|}$ $\overset{3}{|}$
CH$_3$CHCHCH$_2$CH$_2$CH$_3$

substituent ----► CH$_2$CH$_3$

NOT

substituent
┊
┌─────────────┐
│ CH$_3$　　　　　│
│ $\overset{4}{|}$　　　│
CH$_3$CHCHCH$_2$CH$_2$CH$_3$
└─────────────┘
　　　　 $|$
　　　 CH$_2$CH$_3$
　　　 $\overset{5}{}$ $\overset{6}{}$

Named as a hexane
with two substituents

Named as a hexane
with one substituent

Step2: Number the atoms in the main chain. Number the longest chain beginning with the end of the chain nearer to the substituent. Use the numbers obtained by application of step 2 to designate the location of the substituent group.

$\overset{6}{C}H_3\overset{5}{C}H_2\overset{4}{C}H_2\overset{3}{C}H_2\overset{2}{C}HCH_3$
　　　　　　　　　 $|$
　　　　　　　 CH$_3$

2-Methyl hexane

$\overset{7}{C}H_3\overset{6}{C}H_2\overset{5}{C}H_2\overset{4}{C}H_2\overset{3}{C}HCH_3$
　　　　　　　　　 $|$
　　　　　　　 CH$_2$CH$_3$
　　　　　　　　 $\overset{}{2}$ $\overset{}{1}$

3-Methyl heptane

When two or more substituents are present, give each substituent a number corresponding to its location on the longest chain. The substituent groups are listed **alphabetically** (按字母顺序排列).

　　　　 CH$_3$
　　　　 $|$
CH$_3$$\overset{2}{C}H\overset{3}{C}H\overset{4}{C}H_2$$\overset{5}{C}H_2$$\overset{6}{C}H_3$
　　　　 $|$
　　　 CH$_2$CH$_3$

3-Ethyl-2-methylhexane

If there are two or more substituents are identical, indicate this by the use of the prefixes **di- (二)**, **tri- (三)**, **tetra- (四)**, and so on. Ignore di-, tri-, etc. for alphabetizing.

CH$_3$　CH$_3$
$|$　　 $|$
CH$_3$$\overset{2}{C}H\overset{3}{C}H\overset{4}{C}H\overset{5}{C}H_2$$\overset{6}{C}H_3$
　　　　 $|$
　　　 CH$_2$CH$_3$

3-Ethyl-2,4-**di**methylhexane

CH$_3$　CH$_3$
$|$　　 $|$
CH$_3$$\overset{2}{C}H\overset{3}{C}H\overset{4}{C}H\overset{5}{C}H_2$$\overset{6}{C}H_3$
　　　　 $|$
　　　 CH$_3$

2,3,4-**Tri**methylhexane

CH$_3$ CH$_3$
$|$　　 $|$
CH$_3$$\overset{2}{C}$$\overset{3}{C}H\overset{4}{C}H\overset{5}{C}H_2$$\overset{6}{C}H_3$
$|$　　 $|$
CH$_3$ CH$_3$

2,2,4,4-**Tetra**methylhexane

If two substituents are at same positions, look for the next closet group. If there are different substituents in equivalent positions on opposite ends of the parent chain, give the substituent of lower alphabetical order the lower number.

CH$_3$　　　 CH$_3$
$|$　　　　 $|$
CH$_3$$\overset{2}{C}H\overset{3}{C}H\overset{4}{C}H_2$$\overset{5}{C}HCH_3$
　　　 $|$
　　 CH$_2$CH$_3$

3-Ethyl-2,5-dimethylhexane

　　　　　　 CH$_3$
　　　　　　 $|$
CH$_3$$\overset{2}{C}H_2$$\overset{3}{C}H\overset{4}{C}HCH_2$$\overset{6}{C}H_3$
　　　　 $|$
　　 CH$_2$CH$_3$

3-Ethyl-4-methylhexane

2.1.4 Physical Properties of Alkanes

Each time we come to a new family of organic compounds, we'll consider the trends in their boiling points, melting points, densities, and solubilities. These physical properties of an organic compound are important because they determine the conditions under which the compound is handled and used.

Alkanes contain only carbon-carbon and carbon-hydrogen bonds. Because carbon and hydrogen have similar electronegativity values, the C-H bonds are essentially nonpolar. Thus, alkanes are nonpolar, and they interact only by dispersion force which are easily overcome. These forces govern the physical properties of alkanes such as low melting and boiling points, and low solubility in polar solvents, e.g. water, but high solubility in nonpolar solvents, e.g. hexane and **dichloromethane** (二氯甲烷).

Your goal should not be to memorize physical properties of individual compounds, but rather to learn to predict trends in how physical properties vary with structure.

1. Boiling Points of Alkanes The boiling point is the temperature at which the vapor pressure of a substance equals atmospheric pressure (which, at sea level, is 760 mmHg).

Alkanes containing 1 to 4 carbons are gases at room temperature. Alkanes containing 5 to 17 carbons are colorless liquids. High-molecular-weight alkanes (those containing 18 or more carbons) are white, waxy solids.

A graph of n-alkane boiling points versus the number of carbon atoms (Figure 2.3) shows that boiling points increase with increasing molecular weight. Each additional group increases the boiling point by about 30℃ up to about ten carbons, and by about 20℃ in higher alkanes.

In general, a branched alkane boils at a lower temperature than the n-alkane with the same number of carbon atoms; the more branching, the lower the boiling point. This difference in boiling points arises because normal alkanes have efficient contact between chains, and the molecules can move close together. Branching in alkanes increases the distance between molecules, and the chains of carbon atoms are less able to come close to one another. A branched alkane is more compact and has a smaller surface area than a normal alkane. The order of boiling point of the isomeric C_5H_{12} compounds illustrates this phenomenon. For any group of isomeric alkanes, the most branched isomer (such as neopentane) has the lowest boiling point. The normal alkane (such as pentane) has the highest boiling point.

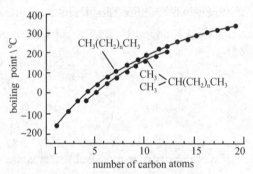

Figure 2.3 Boiling points of some normal alkanes
图2-3 一些直链烷烃的沸点

	$CH_3CH_2CH_2CH_2CH_3$	$CH_3CH_2CHCH_3$ over CH_3	$CH_3-C(CH_3)_2-CH_3$
	n-pentane	iso-pentane	neopentane
b.p./℃	36.1	27.8	9.5

2. Melting Points of Alkanes The melting point of a substance is the temperature above which it is transformed spontaneously and completely from the solid to the liquid state. The melting point is an especially important physical property in organic chemistry because it is used both to **identify** (鉴定) organic compounds and to assess their **purity** (纯度). Melting points are usually depressed, or lowered, by **impurities** (杂质).

The melting points of the normal alkanes increase steadily with increasing molecular weights, as shown in the Table 2.4. Although most alkanes are liquids or gases at room temperature and have relatively low melting points, their melting points illustrate trends that are observed in the melting points of other types of organic compounds (Figure 2.4). Like their boiling points, the melting points increase with increasing molecular weight. The melting point graph is not smooth, however. The **sawtooth-shaped(锯齿状)** graph of melting points is smoothed by drawing separate lines for the alkanes with even and odd numbers of carbon atoms. The melting points of *n*-alkanes with an **even number** (偶数) of carbon atoms lie on a separate, higher curve from those of the alkanes with an **odd number** (奇数) of carbons. The odd-carbon alkane molecules do not "fit together" as well in the crystal as the even-carbon alkanes. In other words, alkanes with even numbers of carbon atoms pack better into a solid structure, so that higher temperatures are needed to melt them. Alkanes with odd numbers of carbon atoms do not pack as well, and they melt at lower temperatures. Similar alternation of melting points is observed in other series of compounds.

Table 2.4   Physical properties of some normal alkanes
表2-4　一些直链烷烃的物理性质

Compound Name	Condensed Structural Formula	Melting point/°C	Boiling point/°C	Density/(g.ml^{-1})
methane	CH_4	−182.5	−161.7	−
ethane	CH_3CH_3	−183.3	−88.6	−
propane	$CH_3CH_2CH_3$	−187.7	−42.1	−
butane	$CH_3(CH_2)_2CH_3$	−138.3	−0.5	−
pentane	$CH_3(CH_2)_3CH_3$	−129.8	36.1	0.6262
hexane	$CH_3(CH_2)_4CH_3$	−95.3	68.7	0.6603
heptane	$CH_3(CH_2)_5CH_3$	−90.6	98.4	0.6837
octane	$CH_3(CH_2)_6CH_3$	−56.8	125.7	0.7026
nonane	$CH_3(CH_2)_7CH_3$	−53.5	150.8	0.7177
decane	$CH_3(CH_2)_8CH_3$	−29.7	174.0	0.7299

* The densities tabulated in this text are of the liquids at 20°C unless otherwise noted.

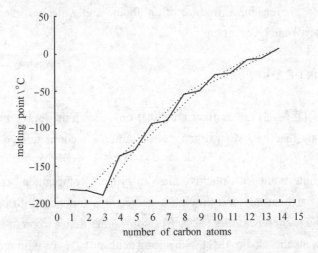

Figure 2.4   Melting points of some normal alkanes
图2-4　一些直链烷烃的熔点

Branched-chain hydrocarbons tend to have lower melting points than linear ones because the branching interferes with regular packing in the crystal. When a branched molecule has a substantial symmetry, however, its melting point is typically relatively high because of the ease with which symmetrical molecules fit together within the crystal. For example, the melting point of the very symmetrical molecule neopentane, −16.8℃, is considerably higher than that of the less symmetrical pentane, −129.8℃.

	$CH_3CH_2CH_2CH_2CH_3$	$CH_3CH_2CHCH_3$ $\overset{|}{CH_3}$	$\overset{CH_3}{\underset{CH_3}{CH_3-\overset{|}{\underset{|}{C}}-CH_3}}$
	n-pentane	*iso*-pentane	neopentane
m.p. /℃	−129.8	−159.9	−16.8

3. Solubilities and Densities of Alkanes   Solubilities are important in determining which solvents can be used to form solutions; most reactions are carried out in solution. Water solubility is particularly important for several reasons. For one thing, water is the solvent in biological systems. For this reason, water solubility is a crucial factor in the activity of drugs and other biologically important compounds.

Alkanes are not soluble in water, a polar substance. The two substances do not meet the usual criterion of solubility "Like dissolves like". Water molecules are too strongly attracted to each other by hydrogen bonds to allow nonpolar alkanes to slip in between them and dissolve. Alkanes dissolve in nonpolar or weakly polar organic solvents, such as **toluene (甲苯) , diethyl ether** (乙醚)**,** hexane and dichloromethane.

Alkanes are said to be **hydrophobic (疏水的) ("water hating"憎水)**. Alkanes are good lubricants and preservatives for metals because they keep water from reaching the metal surface and causing **corrosion** (腐蚀).

The **density** (密度) of a compound is another property. The average density of the liquid alkanes listed in Table 2.4 is about 0.7 g · ml^{-1}; that of higher-molecular-weight alkanes is about 0.8 g · ml^{-1}. All liquid and solid alkanes are less dense than water (1.0 g · ml^{-1}). And because they are insoluble in water, they float on water.

Like boiling point or melting point, the density of a compound determines how the compound is handled. For example, whether a water-insoluble compound is more or less dense than water determines whether it will appear as a lower or upper layer when added to water. Alkanes have considerably lower densities than water. For this reason, a mixture of an alkane and water will separate into two distinct layers with the less dense alkane layer on top.

2.1.5   Reactions of Alkanes

Functional groups (官能团) are distinct chemical units, such as double bonds, hydroxyl groups, or halogen atoms, that are reactive. Most organic compounds are characterized and classified by their functional groups.

Alkanes are the simplest and least reactive class of organic compounds because they contain only hydrogen and sp^3 hybridized carbon, and they have no reactive functional groups. Alkanes contain no double or triple bonds and no **heteroatoms** (杂原子) (atoms other than carbon or hydrogen).

Chemists found that alkanes do not react with strong acids or bases or with most other reagents. They attributed this low reactivity to a lack of affinity for other reagents, so they coined the name "**paraffins**

(石蜡)".Paraffins are used for wax candles, in lubricants, and to seal home-canned jams, jellies, and other preserves. **Petrolatum** (石油冻), so named because it is derived from petroleum refining, is a liquid mixture of high molecular weight alkanes. It is sold as mineral oil and **vaseline** (凡士林) and is used as an **ointment** (油膏) base in **pharmaceuticals** (药物) and **cosmetics** (化妆品) and as a **lubricant** (润滑油) and rust **preventive** (防锈剂).

Most useful reactions of alkanes take place under energetic or high-temperature conditions. Alkanes react with O_2 under certain conditions. They also react with halogens under UV light or at high temperatures, and the reaction is called a **free radical chain reaction** (自由基链锁反应).

1. Reaction with Oxygen: Combustion (燃烧)　Alkanes undergo combustion reaction with oxygen at high temperatures to produce carbon dioxide and water. This is why alkanes are good fuels. Oxidation of saturated hydrocarbons is the basis for their use as energy sources for heat, e.g. natural gas, liquefied petroleum gas (LPG) and fuel oil, and for power, e.g. gasoline, diesel fuel and aviation fuel.

$$C_nH_{2n+2}+O_2 \longrightarrow CO_2+H_2O+Q$$

Following are balanced equations for the complete combustion of methane, the major component of natural gas, and propane, the major component of LPG or bottled gas. The heat liberated when an alkane is oxidized to carbon dioxide and water is called its **heat of combustion** (燃烧热).

$$CH_4+2O_2 \longrightarrow CO_2+2H_2O+212 \text{ kcal} \cdot \text{mol}^{-1}$$

$$CH_3CH_2CH_3+5O_2 \longrightarrow 3CO_2+4H_2O+530 \text{ kcal} \cdot \text{mol}^{-1}$$

2. Reaction with Halogens: Halogenation (卤化反应)　Alkanes can react with halogens（F_2、Cl_2、Br_2、I_2）to form **alkyl halides** (卤代烃). For example, methane reacts with chlorine（Cl_2）to form chloromethane (methyl chloride), dichloromethane (methylene chloride), trichloromethane (**chloroform，氯仿**), and tetrachloromethane (**carbon tetrachloride, 四氯化碳**).

Heat or light is usually needed to initiate this halogenation. If we mix methane with chlorine or bromine in the dark at room temperature, nothing happens. If, however, we heat the mixture to 100℃ or higher or expose it to light, a reaction begins at once. The products of the reaction between methane and chlorine are chloromethane and hydrogen chloride.

What occurs is a substitution reaction—in this case, the substitution of chlorine for hydrogen in methane. If chloromethane is allowed to react with more chlorine, further chlorination produces a mixture of dichloromethane, trichloromethane, and tetrachloromethane. This reaction may continue; heat （△）or light (*hv*) is needed for each step.

Reactions of alkanes with chlorine and bromine proceed at moderate rates and are easily controlled. Reactions with fluorine are often too fast to control, however. Iodine reacts very slowly or not at all.

3. Mechanism of Alkane Chlorination　The **activation energy** (活化能),*Ea*,is the minimum kinetic energy the molecules must have to overcome the repulsions between their electron clouds when

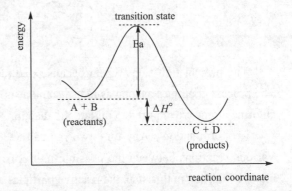

Figure 2.5　Reaction-energy diagram for a one-step exothermic reaction

图2-5　一步放热反应的反应能量图

they collide.The activation energy represents the energy difference between the reactants and **the transition state** (过渡态),the highest-energy state in a molecular collision that leads to reaction.The concepts of transition state and activation energy are easier to understand graphically.Figure 2.5 shows a reaction-energy diagram for a one-step exothermic reaction.The heat of reaction ($\triangle H°$) is the difference in energy between the reactants and the products.

A radical, often called a **free radical** (自由基), is a highly reactive and short lived species with an unpaired electron. Free radicals are electron-deficient species, but usually uncharged. So their chemistry is very different from the chemistry of electron-deficient and even-electron species, e.g. **carbocations** (碳正离子) and **carbanions** (碳负离子). A radical behaves like an electrophile, as it requires only a single electron to complete its **octet** (八隅体).

Radical reactions are often called chain reactions. All chain reactions have three steps: chain **initiation** (引发), chain **propagation** (增长) and chain **termination** (终止).For example, the halogenation of alkane is a free radical chain reaction. The chlorination of alkanes occurs by a series homolytic bond cleavage and homogenic bond formation steps.

(1) Initiation   Initiation generates a reactive intermediate. The chlorine molecule absorbs either heat energy or light energy and the bond breaks to give two chlorine atoms that are radicals.

(2) Propagation   In this step, the intermediate reacts with a stable molecule to produce another reactive intermediate and a product molecule. A chlorine atom abstracts a hydrogen atom from methane, breaking a C-H bond, making an H-Cl bond and yields a new electrophilic species, the methyl radical, which has an unpaired electron. The methyl radical abstracts a chlorine atom, a Cl-Cl bond breaks, a C-Cl bond forms and generates a chlorine radical. A radical reacts and another radical forms. This step, which continues the reaction by generating new radicals, is called a propagation step.

The propagation steps repeat because one radical generates another in this sequence of reactions. The process continues as long as radicals and a supply of both reactants are present. Therefore, only a few chlorine atoms are required to initiate the reaction.

(3) Termination   Any time two radicals recombine, the chain stops. These are termination steps. Various reactions between the possible pairs of radicals allow for the formation of ethane, Cl_2 or the methyl chloride. In this step, the reactive particles are consumed, but not generated.

$$H_3C \cdot \quad + \quad \cdot Cl \longrightarrow CH_3Cl$$

Methyl radical + Chlorine radical → Methyl chloride

$$H_3C \cdot \quad + \quad \cdot CH_3 \longrightarrow CH_3CH_3$$

Methyl radical + Methyl radical → Ethane

$$Cl \cdot \quad + \quad \cdot Cl \longrightarrow Cl_2$$

Chlorine radical + Chlorine radical → Chlorine

Bromination of alkanes follows the same mechanism as chlorination. The only difference is the reactivity of the radical; i.e., the chlorine radical is much more reactive than the bromine radical. Thus, the chlorine radical is much less selective than the bromine radical. In other word, bromination of alkane is a very much more selective than the chlorination reaction. For example, chlorination of 2-methylpropane yields 63% 1-chloro-2-methylpropane and 37% 2-chloro-2-methyl-propane. Bromination of 2-methyl-propane yields 99% 2-bromo-2-methylpropane.

$$H_3C-\overset{CH_3}{\underset{H}{C}}-CH_3 + Cl_2 \xrightarrow[25\,℃]{hv} CH_3\overset{CH_3}{C}HCH_2-Cl + CH_3\overset{CH_3}{\underset{Cl}{C}}CH_3$$

2-methylpropane

1-chloro-2-methylpropane   2-chloro-2-methylpropane
63%   37%

$$H_3C-\overset{CH_3}{\underset{H}{C}}-CH_3 + Br_2 \xrightarrow[127\,℃]{hv} CH_3\overset{CH_3}{C}HCH_2-Br + CH_3\overset{CH_3}{\underset{Br}{C}}CH_3$$

1-bromo-2-methylpropane   2-bromo-2-methylpropane
1%   99%

2.1.6   Conformations of Alkanes

Ethane can exist in various spatial arrangements which result from rotation of the CH_3 groups around the carbon-carbon σ bond. The different arrangements formed by rotations about a single bond are called **conformations (构象异构),** and a specific conformation is called a conformer ("conformational isomer"). When the groups rotate around the C-C bond, the positions of the hydrogen atoms change with respect to one another, but the connectivities of all the bonds remain the same. Thus, various conformations have different shapes, but are not structural isomers. Pure conformers cannot be isolated in most cases, because the molecules are constantly rotating through all the possible conformations. Once we understand the conformations of small molecules, we will be able to apply conforma-tional concepts to much larger molecules, such as carbohydrates (碳水化合物) or complex drugs. The conformation of a molecule accounts for its highly specific biological function.

1. Conformation of Ethane   Ethane, the two-carbon alkane, is composed of two methyl groups with overlap ping sp^3 hybrid orbitals forming a sigma bond between them. The two methyl groups are not fixed in a single position but are relatively free to rotate about the sigma bond connecting the two carbon atoms. Rotation around the C-1 to C-2 bond interconverts conformational isomers. Figure 2.6 shows two such arrangements.

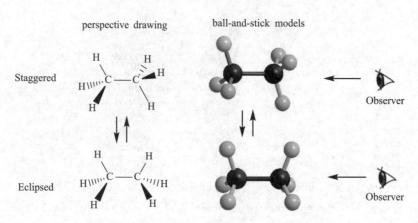

Figure 2.6 Conformations of ethane
图2-6 乙烷的构象

Ethane can exist in an infinite number of conformations. The conformation in which the hydrogen atoms and the bonding electrons are the farthest away from one another has the lowest energy. This conformation is **staggered** (交叉式). The conformation in which the hydrogen atoms are closest to one another has the highest energy. This conformation is **eclipsed** (重叠式).

Figure 2.7 uses Sawhorse structures (锯架结构式) and Newman projections (纽曼投影式) to illustrate some of these ethane conformations. Sawhorse structures picture the molecule looking up at an angle toward the carbon-carbon bond. "Sawho rse" representations of the conformations of ethane are three-dimensional, and show the carbon– carbon bond as well as all of the C-H bonds. Sawhorse structures can be misleading, depending on how the eye sees them. We will generally use perspective or Newman projections to draw molecular conformations.

In drawing conformations, we often use Newman projections, a way of drawing a molecule looking straight down the bond connecting two carbon atoms. Newman projection is devised by Melvin S. Newman (1908~1993). A Newman projection is a type of planar projection along one bond, which we'll call the **projected bond** (投影键). For example, we view the ethane molecule in a Newman projection along the carbon– carbon bond, as shown in Figure 2.7.

The front carbon atom is represented by three lines (three bonds) coming together in a Y shape. The back carbon is represented by a circle with three bonds pointing out from it. Any conformation can be specified by its **dihedral angle(θ)** (二面角)**,** the angle between the C-H bonds on the front carbon atom and the C-H bonds on the back carbon in the Newman projection. Two of the conformations have special names. The conformation with $\theta = 0°$ is called **the eclipsed conformation** (重叠构象) because the Newman projection shows the hydrogen atoms on the back carbon to be hidden (eclipsed) by those on the front carbon. **The staggered conformation** (交叉构象)**,** with $\theta = 60°$, has the hydrogen atoms on the back carbon staggered halfway between the hydrogens on the front carbon. Rotating the methyl group on the right by 60° converts a staggered conformation into an eclipsed conformation.

In a sample of ethane gas at room temperature, the ethane molecules rotate millions of times per second, and their conformations are constantly changing. These conformations are not all equally favored. However. the lowest-energy conformation is the staggered conformation, with the electron clouds in the C-H bonds separated as much as possible. The interactions of the electrons in the bonds make the eclipsed conformation about 12.6 kJ · mol^{-1} higher in energy than the staggered conformation. Three kilocalories are not a large amount of energy, and at room temperature, most molecules have enough **kinetic energy** (动

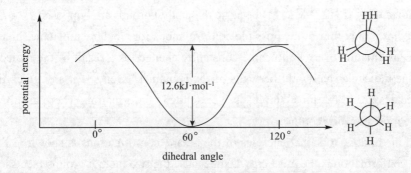

Figure 2.7   Different expresses of ethane conformations
图2-7   乙烷构象的不同表达

能) to overcome this small rotational barrier.

Figure 2.8 shows how **the potential energy** (势能) of ethane changes as the carbon–carbon bond rotates. The y axis shows the potential energy relative to the most stable (staggered) conformation. The x axis shows the dihedral angle as it increases from 0° (eclipsed) through 60° (staggered) and on through additional eclipsed and staggered conformations as continues to increase.

Figure 2.8   Variation of potential energy about the carbon-carbon bond of ethane
图2-8   乙烷碳-碳键的势能变化

As ethane rotates toward an eclipsed conformation, its potential energy increases, and there is resistance to the rotation. This resistance to twisting (torsion) is called **torsional strain** (扭转张力), and the 12.6 kJ · /mol^{-1} of energy required is called **torsional energy** (扭转能).

Conformational analysis (构象分析) is the study of the energetics of different conformations. Many reactions depend on a molecule's ability to twist into a particular conformation; conformational analysis can help to predict which conformations are favored and which reactions are more likely to take place. We will apply conformational analysis to butane.

2. Conformation of Butane   Butane is the four-carbon alkane, with molecular formula C_4H_{10}. We refer to *n*-butane as a straight-chain alkane, but the chain of carbon atoms is not really straight. The angles between the carbon atoms are close to the tetrahedral angle, about 109.5°. Rotations about any of the carbon–carbon bonds are possible. Rotations about either of the end bonds (C2-C3) just rotate a methyl group like in ethane.

Figure 2.9 shows four conformations for a butane molecule. Notice that we have defined the dihedral angle θ as the angle between the two end methyl groups.

45

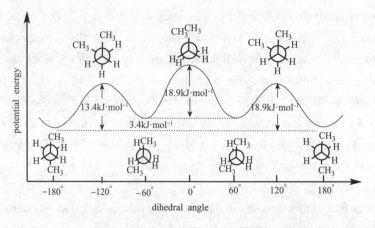

Figure 2.9 Four conformations of a butane molecule
图2-9　丁烷的四种构象

Three of the conformations are given special names. When the methyl groups are pointed in the same direction ($\theta = 0°$), they eclipse each other. This conformation is called **totally eclipsed conformation (全重叠构象)**, to distinguish it from the other eclipsed conformations like the one at $\theta = 120°$. At $\theta = 60°$, the butane molecule is staggered and the methyl groups are toward the left and right of each other. This 60° conformation is called **gauche conformation (邻位交叉构象)**. Another staggered conformation occurs at $\theta = 180°$ with the methyl groups pointing in opposite directions. This conformation is called **anti-staggered conformation (对位交叉构象)** because the methyl groups are "opposed."

Figure 2.9 shows only four of the possible conformations for a butane molecule. In an actual sample of butane, the conformation of each molecule constantly changes as a result of the molecule's collisions with other butane molecules and with the walls of the container. Even so, at any given time, a majority of butane molecules are in the most stable, fully extended conformation. There are the fewest butane molecules in the most crowded conformation.

A graph of the relative torsional energies of the butane conformations is shown in Figure 2.10. The totally eclipsed conformation is the most crowded because the two methyl groups face each other. All the staggered conformations (anti and gauche) are lower in energy than any of the eclipsed conformations. The anti conformation is lowest in energy and is the most stable because the methyl groups at the ends of the four carbon chain are farthest apart. The gauche conformations, with the methyl groups separated by just 60°, are 3.4kJ higher in energy than the anti conformation because the methyl groups are close enough that their electron clouds begin to repel each other.

Figure 2.10 Variation of potential energy about the carbon-carbon bond of butane
图2-10　丁烷碳碳键的势能变化

To summarize, for any alkane (except, of course, for methane), there are an infinite number of conformations. The majority of molecules in any sample will be in the least crowded conformation; the fewest will be in the most crowded conformation.

2.2　Cycloalkanes

Carbon atoms of an alkane joined to form a ring is called a **cycloalkane** (环烷烃). One hydrogen atom has to be removed from each end of the alkane carbon chain to form the cycloalkanes therefore have two less hydrogen atoms than the parent alkane and a generic formula of C_nH_{2n}.

Cycloalkane can be divided into monocyclic, bicyclic and polycyclic according to the number of rings. Monocycloalkane can be divided into small ring (three-or four-membered ring), common ring (five- or six-membered ring), middle ring (seven- to twelve-membered ring) and large ring (ring formed by more than twelve carbon atoms) according to the number of carbon atoms.

When writing structural formulas for cycloalkanes, chemists generally use line-angle formulas to represent cycloalkane rings. Each ring is represented by a **regular polygon** (正多边形) that has the same number of sides as there are carbon atoms in the ring. For example, chemists represent cyclobutane by a square, cyclopentane by a pentagon, and cyclohexane by a hexagon.

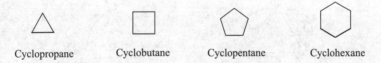

Cyclopropane　　　Cyclobutane　　　Cyclopentane　　　Cyclohexane

There exist **spiro alkane** (螺环烃) **and bridged alkane** (桥环烃) in bicyclic and polycyclic alkanes. An cycloalkane that contains two rings that share one carbon atom is classified as a spiro alkane, and the carbon atom shared by two rings is called **spiro atom** (螺原子). While, an cycloalkane that contains two rings that share two carbon atoms is classified as a bridged alkane, and the shared carbon atoms are called **bridgehead carbons** (桥碳原子), and the carbon chain connecting them is called a **bridge** (桥).

Five-membered (cyclopentane) and six-membered (cyclohexane) rings are especially abundant in nature. Petroleum contains cyclohexane, methylcyclopentane, and other essential oils, such as **camphor** (樟脑) and **menthol** (薄荷醇), also contain cycloalkanes.

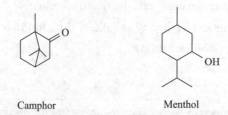

Camphor   Menthol

2.2.1 Structure of Cycloalkanes

The smallest cycloalkane is cyclopropane, in which the three carbon atoms lie in one plane shown in Figure 2.11. The angle between adjacent C–C bonds is only 60° , which is very much smaller than that 109.5° of sp^3 hybrid orbital, so the C–C bond in cyclopropane can not overlap along the bond axis like the alkane, but only partially overlaps to form a bending bond, also known as "**banana bond**" (香蕉键), which is weaker than the normal σ bonds in alkanes. The "banana bond" makes the electron clouds distributing outside of the straight line connecting two carbon atoms, which makes the electrophilic reagent easy to attack. In addition, the C–H bonds are totally eclipsed conformation, and there is a large **torsional strain** (扭转张力) between them. These two factors lead to high **ring strain** (环张力) and instability of cyclopropane, cyclopropane is therefore susceptible to chemical reactions that can open up the three-membered ring.

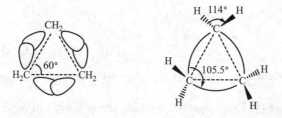

Figure 2.11   The "banana bond" of a cyclopropane molecule
图2-11 环丙烷的"香蕉键"

The small ring (three- or four- membered ring) has lower stability and is easy to react with some reagents. In order to explain this fact, the German chemist **J. Baeyer** (拜尔) (1835~1917) put forward the "strain theory" in 1885. He assumes that all the cycloalkanes are plane structures of regular polygons. If the bond angle of the ring deviates 109.5°, it will produce an inward or outward twisting strain, which is called **angle strain** (角张力) or **Bayer strain** (拜尔张力). Bayer believes that it is the angle strain leads to the instability of the ring. By calculating the deviation between the C–C–C bond angle of various cycloalkanes and the normal bond angle 109.5° of sp^3 hybrid orbital, the stability of cycloalkanes can be judged. The larger the deflection angle of 109.5°, the greater the angle strain and the lower the stability of the ring.

regular polygon inner angle	60°	90°	108°	120°	128° 34'	135°
deflection angle	+24° 44'	+9° 44'	+0° 44'	−5° 16'	−10° 3'	−12° 46'

According to "strain theory", cyclopropane is the most unstable, cyclopentane is the most stable, and the stability of cyclohexane and its subsequent cycloalkanes should be gradually reduced. In fact, cyclohexane is more stable than cyclopentane, and macrocycloalkane are more stable. Therefore, "strain theory" has its limitations. It can well explain the instability of small ring. The main reason for this contradiction is that Bayer's assumption that the carbon atoms in ring are in one plane is incorrect. In fact, the carbon atoms of all the rings are not in one plane except that the three-membered ring shown in Figure 2.12.

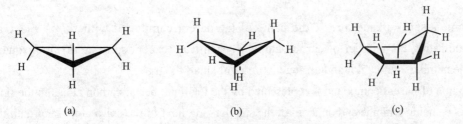

(a)   (b)   (c)

Figure 2.12   The nonplanar structure of cycloalkane molecule
图2-12   环烷烃的非平面结构

2.2.2   Constitutional Isomerism in Cycloalkanes

Cycloalkanes and normal chain **alkenes (烯烃)** with the same number of carbon atoms are functional group isomers. There are also structural isomers due to different number of carbon atoms in the ring, different structures of substituents and different positions of substituents on the ring. The cycloalkane isomers with the molecular formula C_5H_{10} are as follows.

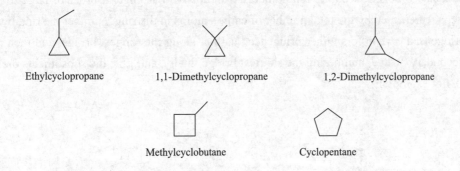

Ethylcyclopropane   1,1-Dimethylcyclopropane   1,2-Dimethylcyclopropane

Methylcyclobutane   Cyclopentane

2.2.3   Nomenclature of Cycloalkanes

To name a cycloalkane, prefix the name of the corresponding open-chain alkane with *cyclo-*. When only one substituent is present on the ring, there is no need to give it a number. For example:

sec -Butylcyclopentane   Isopropylcyclohexane

If there are two substituents on the ring, number the ring by beginning with the substituent of lower alphabetical order and list them in alphabetical order. If there are three or more substituents, number the ring to give all the substituents the lowest set of numbers. For example:

1-Isobutyl-2-methylcyclopentane

4-Isopropyl-1,2-dimethylcyclohexane

To name a spiro alkane, prefix the name of the parent compound with *spiro-*, place the Arabic numbers indicating the number of carbon atoms in each ring (except spiro atom) in a square bracket, separate the numbers with a dot and arrange in order of small to large.

The name of parent compound is represented by the total number of carbon atoms in the ring. Number the ring by beginning with the carbon atom of the smaller ring next to the spiro atom, and then through the spiro atom to the second ring, and give the substituents the lowest set of numbers. For example:

1-Methylspiro[2.4]heptane

6-Ethyl-10-methylspiro[4.5]decane

To name a bridge alkane, prefix the name of the parent compound with *bicyclo-* or *tricyclo-* (the number of carbon-carbon single bonds that need to be broken to convert a bridge alkane into a chain alkane), place the Arabic numbers indicating the number of carbon atoms on each bridge (except bridge atoms) in a square brackets, separate the numbers with dots and arrange in order of large to small. The parent name is represented by the total number of carbon atoms in the ring. Number the ring by beginning with one bridgehead carbon to another bridgehead carbon along the longest bridge, and then numbering along the secondary bridge, numbering the shortest bridge finally, and give the substituents the lowest set of numbers. For example:

3-Ethyl-2-methylbicyclo[3.3.0]octane

1,6-Dimethylbicyclo[3.2.1]octane

2.2.4　Physical Properties of Cycloalkanes

Under normal temperature and pressure, cyclopropane and cyclobutane are gases, cyclopentane to cycloundecane are liquids, and cycloalkanes with more than 12 carbons are solids. The melting point, boiling point and relative density of cycloalkanes are higher than those of corresponding alkanes, but the relative densities are still less than 1, which are difficult to dissolve in water and easy to dissolve in organic solvent. The physical constants of some cycloalkanes are listed in Table 2.5.

Table 2.5   Physical constants of some monocycloalkanes

表2-5   几种单环环烷烃的物理性质

Compound name	Melting point($^\circ$C)	Boiling point($^\circ$C)	Density ($g \cdot ml^{-1}$)
cyclopropane	−127.6	−32.7	0.7200（−79°C）
cyclobutane	−80	12.5	0.7030（0°C）
cyclopentane	−93.9	49.3	0.7457
cyclohexane	6.6	80.7	0.7786
cycloheptane	−12	118.5	0.8098

2.2.5   Reactions of Cycloalkanes

The chemical properties of cycloalkanes are basically the same as those of alkanes. For example, cycloalkanes can react with halogens by radical substitution in light or at higher temperature, but not with oxidants (such as potassium permanganate) at room temperature. For example:

$$\triangleright \xrightarrow[hv]{Cl_2} \triangleright\!\!-Cl \ + \ HCl$$

$$\text{cyclohexyl-ethyl} \xrightarrow[hv]{Br_2} \text{cyclohexyl-CH}_2Br \ + \ HBr$$

Cyclopropane and cyclobutane, the two small cycloalkanes, show certain chemical properties that are entirely different from those of the other cycloalkanes due to ring strain and are prone to ring opening that is called addition reaction. In an **addition reaction** (加成反应), a carbon-carbon bond is broken, and the two atoms of the reagent appear at the ends of the propane chain.

For example, in the presence of a nickel catalyst, the rings open up, and form corresponding open chain alkanes.

$$\triangle \xrightarrow[80^\circ C]{H_2/Ni} CH_3CH_2CH_3$$

$$\square \xrightarrow[120^\circ C]{H_2/Ni} CH_3CH_2CH_2CH_3$$

While, cyclopentane, cyclohexane and macrocycloalkanes require higher temperature than cyclopropane for ring opening.

Cyclopropane can also react with halogen at room temperature; cyclobutane does not react at room temperature, and can only be added by heating.

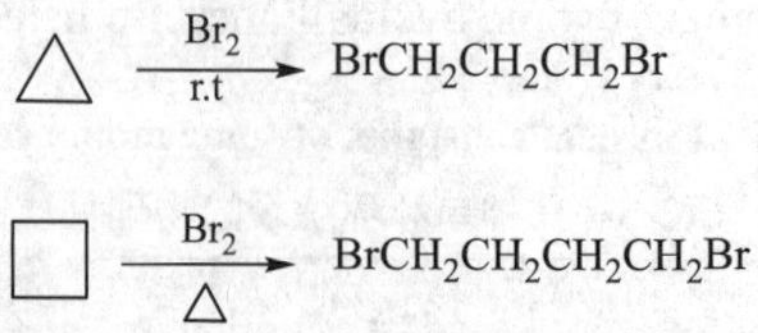

$$\triangle \xrightarrow[\text{r.t}]{\text{Br}_2} BrCH_2CH_2CH_2Br$$

$$\square \xrightarrow[\triangle]{\text{Br}_2} BrCH_2CH_2CH_2CH_2Br$$

It is very difficult for cycloalkanes with five-membered ring or more to react with halogens. With the increase of temperature, they are prone to free radical substitution.

At room temperature, cyclopropane can react with hydrogen halide to form 1-halopropane. For example:

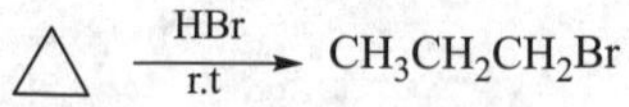

$$\triangle \xrightarrow[\text{r.t}]{\text{HBr}} CH_3CH_2CH_2Br$$

When an asymmetrical cyclopropane reacts with hydrogen bromide, there is regional selectivity. The ring opening position is between the two ring carbon atoms with the most hydrogen atoms and the least hydrogen atoms. For example:

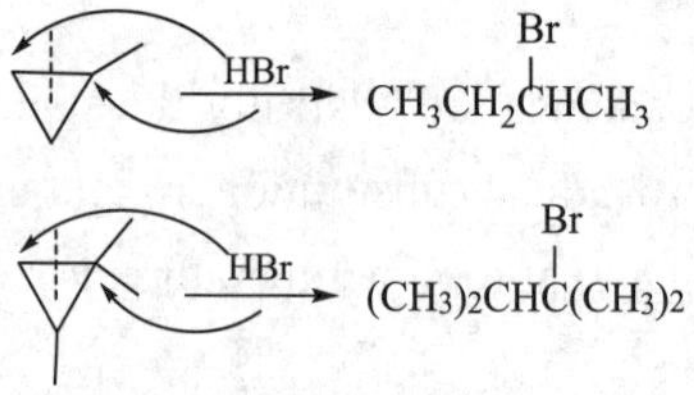

$$\xrightarrow{\text{HBr}} CH_3CH_2\overset{\overset{\displaystyle Br}{|}}{C}HCH_3$$

$$\xrightarrow{\text{HBr}} (CH_3)_2CH\overset{\overset{\displaystyle Br}{|}}{C}(CH_3)_2$$

The above reactions show that the activity of cycloalkanes is related to the ring size. The stability order of cycloalkanes is as follows.

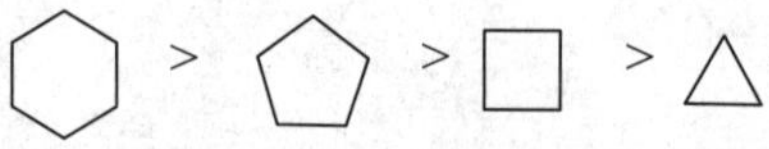

$$\hexagon > \pentagon > \square > \triangle$$

The stability of the ring can also be explained by measuring the combustion heat of the cycloalkanes. The combustion heat reflects the molecular thermodynamic energy and stability. Combustion heat refers to the heat released by the complete combustion of 1 mol organic compounds at standard pressure. But the molecular composition of different cycloalkanes is different, so the relative stability can not be directly compared according to the combustion heat.

However, the relative stability of the rings can be judged by comparing the average combustion heat of each CH_2 with combustion heat data of some monocycloalkanes.

Table 2.6 shows that the average combustion heat of each CH_2 of cyclopropane and cyclobutene is higher than that of alkane, indicating that their internal energy is high and unstable; the average combustion heat of each CH_2 of cyclohexane is same as that of alkane which is the most stable; the average combustion heat of each CH_2 of other rings is close to that of alkane and stable. This is consistent with the fact that people have observed in practice.

Table 2.6   The combustion heat of some monocycloalkanes
表2-6   一些单环环烷烃的燃烧热

Compound	Heat of combustion of each CH_2 / ($kJ \cdot mol^{-1}$)	Difference of combustion heat with corresponding alkanes / ($kJ \cdot mol^{-1}$)
cyclopropane	697.1	38.5
cyclobutane	686.2	27.4
cyclopentane	664.0	5.4
cyclohexane	658.6	0
cycloheptane	662.4	3.8
cyclooctane	663.6	5.0
cyclononane	664.1	5.5
cyclodecane	663.6	5.0
cycloundecane	664.5	5.0
cyclododecane	659.9	1.3
cyclotridecane	660.2	1.7
cyclotetradecane	658.6	0
cyclopentadecane	659.0	0.4
cyclohexadecane	658.7	0.1

2.2.6   Conformation of Cyclohexane

1. Boat Conformation and Chair Conformation   The data of combustion heat shows that cyclohexane is the most stable and important cycloalkane. Many compounds in nature contain cyclohexane like carbon ring compounds.

In the cyclohexane molecule, although the rotation of C–C bond in the ring is limited to some extent, it can still rotate in the range where the ring is not broken, thus producing different conformations. In a series of dynamic equilibrium systems, there are two typical conformations, **boat conformation** (船式构象) and **chair conformation** (椅式构象) shown in Figure 2.13.

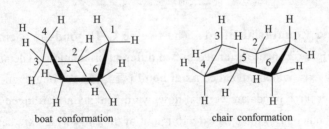

boat conformation   chair conformation

Figure 2.13   Two typical conformations of a cyclohexane molecule
图2-13   环己烷的两种典型构象

In the boat conformation shown in Figure 2.14, the carbon atoms of C_2、C_3、C_5 and C_6 are in one plane, and the carbon atoms C_1 and C_4 are both above this plane. One carbon is "bow" and the other is "stern", the whole molecule is like a boat, so it is called boat conformation. In this conformation, the

微课

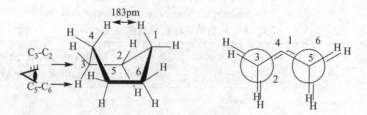

Figure 2.14 Newman projection of boat conformation of cyclohexane
图2-14 环己烷船式构象的纽曼投影式

bond angle is close to 109.5°, and there is no angle strain. However, the C–H bonds between C_2 and C_3, C_5 and C_6 are totally eclipsed conformation, and there is a large distortion strain between them. In addition, the distance between the "bow hydrogen" and the "stern hydrogen" on C_1 and C_4 is relatively close (only 183 pm), which is less than the sum of the van der Waals radius (240 pm) of the two hydrogen atoms, there is **van der Waals strain** (范德华张力) between them. Due to the existence of distortion strain and van der Waals strain, the boat conformation is relatively unstable.

In the chair conformation shown in Figure 2.15, the carbon atoms of C_2、C_3、C_5 and C_6 are also in one plane, and the carbon atoms of C_1 and C_4 are above and below this plane respectively. The whole molecule is like a chair, so it is called chair conformation. In this conformation, the bond angles are all close to 109.5 °, and there is no angle strain. At the same time, it can be seen from the Newman projection that there are no eclipsing C–H bonds in the chair conformation of cyclohexane, all the C–H bonds are fully staggered, cyclohexane is strain-free. The energy of chair conformation is about 30 kJ · mol^{-1} lower than that of boat conformation.

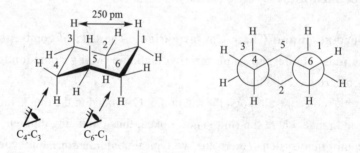

Figure 2.15 Newman projection of chair conformation of cyclohexane
图2-15 环己烷椅式构象的纽曼投影式

2. Axial Bond and Equatorial Bond The twelve C–H bonds in a chair conformation of cyclohexane,shown in Figure 2.16, are arranged in two different orientations. The six C–H bonds that are parallel to the symmetry axis, which are called **axial bond** (直立键). Among them, three C–H bonds are vertical upward, the other three bonds are vertical downward, and six are arranged alternately. The other six C–H bonds are perpendicular to the axis extending out of the ring, which are called **equatorial bond** (平伏键). Three C–H bonds extend upward, the other three bonds extend downward, and six horizontal bonds are arranged alternately. Every carbon atom has an axial bond and an equatorial bond. If the axial bond on the same carbon atom of cyclohexane is upward, then the equatorial bond must be downward; if the axial bond is downward, then the equatorial bond must be upward.

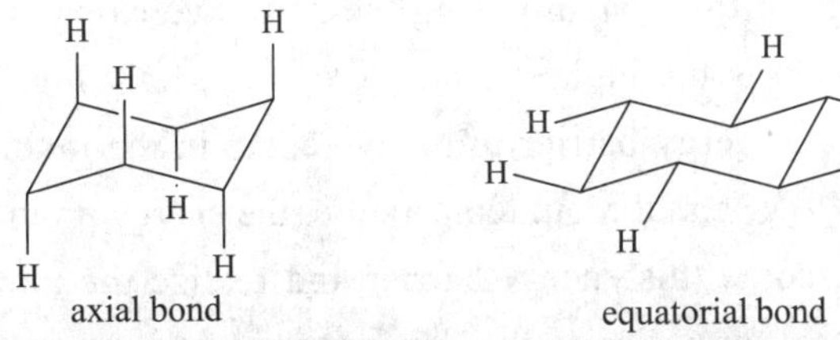

axial bond   equatorial bond

Figure 2.16   Axial bond and equatorial bond of cyclohexane
图2-16   环己烷的直立键和平伏键

3. Conversion of Various Conformations of Cyclohexane   One chair conformation of cyclohexane can be converted to another chair conformation through the twist of C–C bond rapidly at room temperature, shown in Figure 2.17. An important thing to notice is that when the ring flips, all of the bonds that are axial become equatorial and vice versa. Two kinds of chair conformations can form a dynamic equilibrium system.

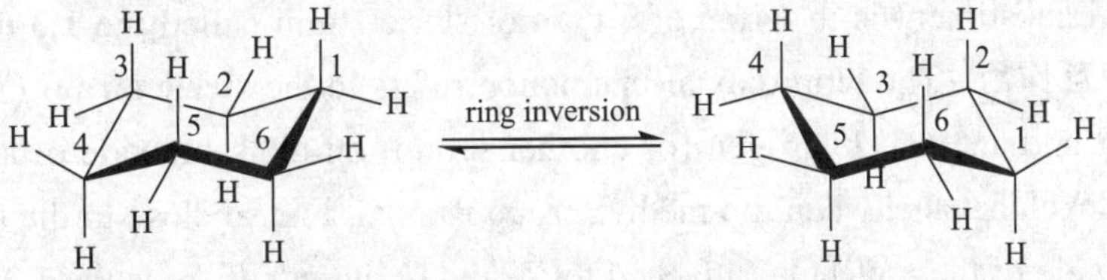

ring inversion

Figure 2.17   Ring inversion of cyclohexane
图2-17   环己烷的翻环作用

In the process of the ring inversion, the chair conformation passes through the **half chair conformation** (半椅式构象) and the **twist boat conformation** (扭船式构象), and then to the boat conformation; the boat conformation passes through the twist boat conformation and the half chair conformation and then becomes another chair conformation. Figure 2.18 shows the conformation change

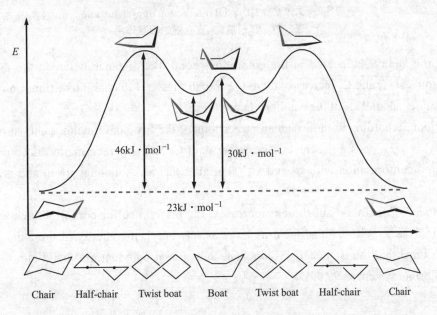

E

46kJ · mol^{-1} 30kJ · mol^{-1}

23kJ · mol^{-1}

Chair   Half-chair   Twist boat   Boat   Twist boat   Half-chair   Chair

Figure 2.18   Potential energy diagrams of various conformations of cyclohexane
图 2-18   环己烷的各种构象势能关系图

and the potential energy change of the system during the ring inversion. Among them, the potential energy of half chair conformation is the highest, and there is not only high torsional strain but also angle strain, which is the highest energy barrier to be overcome in the mutual conversion of various conformations of cyclohexane. However, at room temperature, the energy provided by molecular thermal movement can completely overcome this energy barrier and realize the mutual conversion between different conformations. Chair conformation is the dominant conformation, accounting for more than 99.99% at room temperature. Therefore, cyclohexane and most of its derivatives mainly exist in stable chair conformation.

4. Conformation of Substituted Cyclohexane   If we replace a hydrogen atom of cyclohexane with a substituent, there are two kinds of chair conformations: one with the substituent equatorial, the other with it axial. The two conformations can be in rapid dynamic equilibrium. It has been proved that the dominant conformation of the monosubstituted cyclohexane is that the substituent occupies an equatorial bond.

A simple and convenient way to analyze the relative stabilities of chair conformations with equatorial and axial substituent is in terms of a type of steric strain called the **1,3-diaxial interaction (1,3-二直立键的相互作用)**. The term diaxial interaction refers to the **steric strain** (空间张力) between an axial substituent and an axial hydrogen (or another group) on the same side of a cyclohexane ring. Considering methylcyclohexane, when the methyl group is axial, it is so close to the two axial hydrogen atoms on the same side of the molecule (attached to C_3 and C_5 atoms), there is the 1,3-diaxial interaction shown as in Figure 2.19.

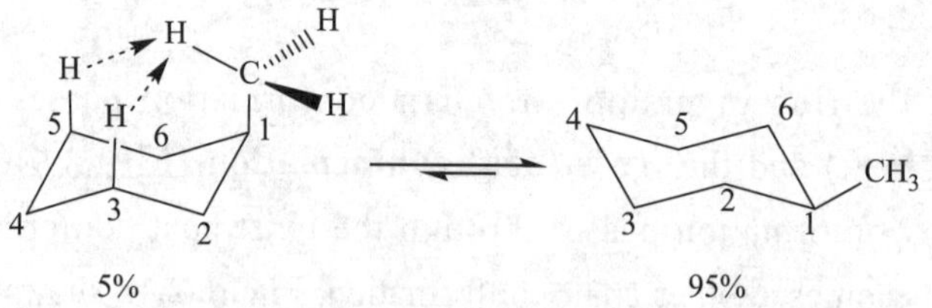

Figure 2.19   1, 3-Diaxial interaction of cyclohexane
图 2-19   环己烷1,3-直立键的相互作用

When the methyl is located in the equational bond, the distance between the methyl and the hydrogen atoms at C_3 and C_5 increases, there is generally less 1,3-diaxial interaction. Calculations show that it constitutes about 95% of the equilibrium mixture.

In addition, according to the Newman projections of the two conformations shown in Figure 2.20, when methyl is axial, it is a gauche conformation with C_3 and C_5; when methyl is equatorial, it is an anti-staggered conformation with C_3 and C_5. The latter has less torsional strain and is the dominant conformation.

As the size of the alkyl substituent increases, the preference for conformations with the group equatorial increases. With a substituent as large as *tert*-butyl, the energy of the axial conformation becomes so large that the isomer of cyclohexane substituted by equatorial bond almost occupies the whole equilibrium system (> 99.99%).

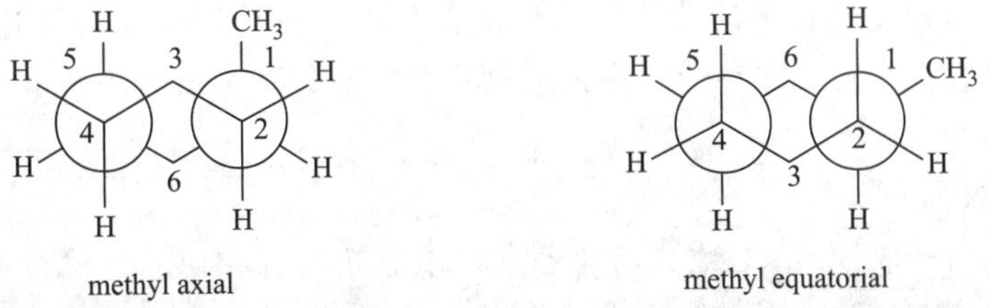

methyl axial

methyl equatorial

Figure 2.20　Newman projection of chair conformation of methylcyclohexane

图2-20　甲基环己烷椅式构象的纽曼投影式

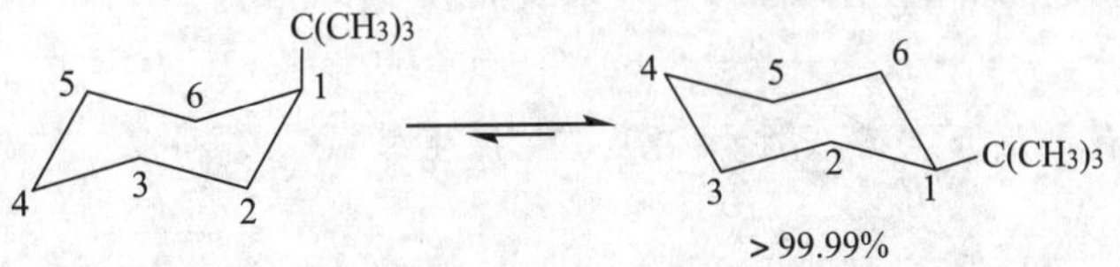

> 99.99%

2.3　Application of Alkanes in Pharmacy Area

Vaseline is a purified mixture of semisolid hydrocarbons obtained from petroleum, which has water-resistant and is a excellent **moisturizing** (保湿) supplies. Vaseline is widely used in the **matrix** (基质) and **lubricant** (润滑剂),**adhesive** (粘合剂),**thickener** (增稠剂) and skin protection cream for raw materials, almost all applications in all varieties of **ointment** (软膏).

重 点 小 结

一、烷烃

只含碳和氢两种元素的化合物称为烃。烃分为饱和烃和不饱和烃。

1. 烷烃的结构　烷烃碳原子发生sp^3杂化,形成碳碳 σ 键和碳氢 σ 键,键角为109.5°。

2. 烷烃的构造异构体　具有相同分子式的不同化合物称为同分异构体。其中连接方式不同的同分异构体称为构造异构体。从四个碳以上的烷烃开始有构造异构体存在。随着碳原子数目的增多,同分异构体的数目显著增加。

3. 烷烃的命名　有机化合物的命名分为普通命名法(俗名)和系统命名法。烷烃命名以"–ane"为后缀。

(1) 普通命名法　直链烷烃多用"正"烷烃表示。注意"正"、"异"、"新"的结构特点。

(2) 系统命名法(IUPAC命名法)　直链烷烃;不用"正 "表示;注意同系列、同系物的概念。支链烷烃:支链烷烃中的支链称为烷基,用"–anyl"为后缀。

碳原子的分级:根据所连碳原子的不同,可以分为伯(1°)、仲(2°)、叔(3°)、季(4°)四种碳。

IUPAC 命名原则:①确定母体和支链。②找到最长的连续的碳链作主链并命名。如果有两条不同的等长主链,优先选择支链最多的为主链。③从最接近支链的端碳开始编号。如果有两个以上不同的取代基时,按取代基英文首字母顺序排列这些取代基。如果有两个或多个相同的取代基,则要用di-(二),tri-(三),tetra-(四)等前缀表示。

4. 烷烃的物理性质

(1) 烷烃的沸点 一般来说,支链烷烃的沸点比碳原子数目相同的正烷烃的沸点低;而且分支越多,沸点越低。

(2) 烷烃的熔点 直链烷烃的熔点随分子量增加而增加;偶数碳的直链烷烃熔点升高的幅度高于奇数碳的直链烷烃,支链烃类的熔点往往比直链烃类低。

(3) 溶解度和密度:烷烃难溶于水,易溶于甲苯、乙醚等有机溶剂。所有烷烃的密度都比水小。

5. 烷烃的反应

(1) 燃烧 烷烃在高温下与氧发生燃烧反应,产生二氧化碳和水,并放出热量,称为燃烧热。烷烃都是很好的燃料。

(2) 卤代反应 烷烃和卤素反应可以发生取代反应生成卤代烃。

(3) 氯代反应的机制 ——自由基链锁反应 反应分为三步,即链引发、链增长和链终止。在烷烃的卤代反应中,氯的反应活性高于溴,但溴的反应选择性高于氯。

6. 烷烃的构象

(1) 乙烷的构象 旋转乙烷的碳碳单键形成了交叉式和重叠式两种构象。重叠构象不如交叉构象稳定。

(2) 丁烷的构象 旋转C_2-C_3单键产生四种典型的构象,即全重叠构象、邻位交叉构象、部分重叠构象和对位交叉构象。其中对位交叉构象能量最低、最稳定,而全重叠构象的能量最高,最不稳定。

二、环烷烃

1. 环烷烃的结构

(1) 除环丙烷外,其余环烷烃均是非平面结构。

(2) 拜尔张力学说 假定所有成环碳原子在同一个平面上,并排成正多边形,然后计算平面正多边形的内角,再与109.5°相比,偏离程度越大,角张力越大,环越不稳定。根据拜耳学说推断,环丙烷、环丁烷角张力大,分子内能高,稳定性差;但是对说明六元环以及大环化合物的稳定性问题有局限。

2. 环烷烃的构造异构体 单环烷烃通式为C_nH_{2n},与碳原子数相同的单烯烃互为构造异构体,还有因成环碳原子数、取代基的构造以及取代基在环上位置等不同而产生的构造异构。

3. 环烷烃的命名

(1) 单环环烷烃命名 与链烃相似,只需在名称前加一个"环"字;环上只有一个取代基时,无需标记取代基的位置;当环上有多个取代基时,要保证所有取代基位次都最小;编号有选择时,按英文名称字母顺序给较低编号。

(2) 螺环烃命名 用"螺"做词头,将每个环中所含的碳原子数目(螺原子除外)由小到大写在方括号中,数字间用下角圆点隔开,母体名称由成环碳原子总数表示。若环上有取代基,编号时从与螺原子紧邻的小环碳原子开始,然后经过螺原子到另一个环,同时保证所有取代基的位次最小。

(3) 桥环烃命名 用"二环""三环"等作词头,将每个桥上的碳原子数目(桥原子除外)由大到小写在方括号中,数字间用下角圆点隔开,母体名称由成环碳原子总数表示。若环上有取代基,编号时从一个桥头碳原子开始,沿最长的桥到达另一个桥头碳原子,再沿次长桥编号,最短桥上的碳原子最后编号,同时保证所有取代基的位次最小。

4. 环烷烃的物理性质　环烷烃中,小环为气态,常见环为液态,中环及大环为固态。环烷烃环中C–C单键旋转受到一定的限制,因此环烷烃分子具有一定的对称性和刚性,沸点、熔点和相对密度都比相应的开链烷烃高。

5. 环烷烃的化学性质

(1)环烷烃在光照或在较高的温度下可与卤素发生自由基取代反应,在常温下与氧化剂(如高锰酸钾)不发生反应。

(2)环丙烷和环丁烷这两个小环烃,由于分子张力较大,表现出不稳定性,容易开环发生加成反应。非对称的烷基取代环丙烷与氢卤酸发生开环反应时,存在区域选择性,通常氢卤酸中的氢原子加在连有氢原子较多的碳原子上,而卤原子则加在连有氢原子较少的碳原子上。

6. 环己烷的构象

(1)旋转环己烷结构中的C–C单键可以产生无数种构象,各构象异构体在室温下无法分离。典型构象是船式和椅式。这两种构象中,环C–C键的键角均为109.5°。

(2)椅式构象中的直立键和平伏键　六个C–H键与C₃轴近似于平行关系的直立键或α键(axial bond),三个指向环的上方,三个指向环的下方,交替排列。六个C–H键与分子的C₃轴近似于垂直关系,都伸出环外的平伏键或e键,三个向上斜伸,三个向下斜伸,交替排列。室温下,环己烷可以通过椅式构象的翻环作用从一种椅式构象转变成另一种椅式构象。环经翻转后,与碳相连的化学键的空间指向不变,但翻环前的α键变成e键,e键变成α键。

(3)椅式构象是优势构象　船式构象中,"船底"的四个碳原子上的C–H键都处于重叠式,具有较大的扭转张力。船头、船尾两个碳原子上的旗杆氢存在范德华排斥力,表现出空间张力。而椅式构象是一个既无角张力,又几乎无扭转张力和空间张力的环,是环己烷所有构象中能量最低的优势构象。

(4)取代环己烷的构象　环己烷的一元取代物是e-取代和α-取代的平衡混合物,e-取代是优势构象。平衡态中的α-取代环己烷,取代基与两个α-氢之间存在1,3-直立键的相互作用,使位能升高而不稳定。而e-取代环己烷,因取代基在水平方向伸出环外,避开了1,3-直立键的相斥作用,成为平衡体系中相对稳定的优势构象。

Problems

目 标 检 测

题库

1. Which of the following compounds contains primary hydrogen, secondary hydrogen and tertiary hydrogen?

下列(　　)化合物含有伯、仲、叔氢。

A. 2,2,4,4-Tetramethylpentane　　　　　　B. 2,3,4-Trimethylpentane

C. Heptane　　　　　　　　　　　　　　D. 2,2,4-Trimethylpentane

2. How many structural isomers are there in the molecular formula of C_7H_{16} containing three methyl?

分子式为C_7H_{16}且含有三个甲基的化合物,有(　　)种构造异构体。

A. 3　　　　　　B. 4　　　　　　C. 5　　　　　　D. 6

3. Which of the following alkanes has the highest boiling point?

下列烷烃，（　　　）沸点最高?

    A. Neopentane   B. Isopentane

    C. Hexane   D. Octane

4. The number of structural isomers of alkanes with molecular formula C_6H_{14} is (　　　).

分子式为C_6H_{14}的烷烃构造异构体的数目是（　　　）。

    A. 6   B. 5   C. 4   D. 3

5. Which of the following alkanes has the highest melting point? (　　　).

下列烷烃，熔点最高的是（　　　）。

    A. 2-Methylheptane   B. 2,2,3,3-Tetrabutane

    C. 3,3-Dimethylhexane   D. Octane

6. The IUPAC name for the compound ⟨structure⟩ is (　　　).

化合物 ⟨structure⟩ 的IUPAC名称是（　　　）。

    A. 3-Ethyl-2,4,6-trimethyloctane   B. 3-Isobutyl-4,6-dimethyloctane

    C.6-Isobutyl-3,5-dimethyloctane   D. 6-Ethyl-3,5,7-trimethyloctane

7. The IUPAC name for the compound ⟨structure⟩ is (　　　).

化合物 ⟨structure⟩ 的IUPAC名称是（　　　）。

    A. 3-Ethyl-6,7-dimethylnonane   B. 6,7-Dimethyl-3-Ethylnonane

    C. 3,4-Dimethyl-7-Ethylnonane   D. 7-Ethyl-3,4-dimethylnonane

8. The IUPAC name for the compound ⟨structure⟩ is (　　　).

化合物 ⟨structure⟩ 的IUPAC名称是（　　　）。

    A. 1-Isobutyl-3-isopropylcyclopentane   B. 3-Isobutyl-1-isopropylcyclopentane

    C. 1-Isobutyl-4-isopropylcyclopentane   D. 1-(*sec*-Butyl)-3-isopropylcyclopentane

9. The IUPAC name for the compound ⟨structure⟩ is (　　　).

化合物 ⟨structure⟩ 的IUPAC名称是（　　　）。

    A. 2-Ethyl-1-methylcyclohexane   B. 1-Ethyl-2-methylcyclohexane

    C. 2-Methyl-1-ethylcyclohexane   D. 1-Ethyl-6-methylcyclohexane

10. The IUPAC name for the compound is (　　　).

化合物 的IUPAC名称是（　　　）。

A. 3-(sec-Butyl)-1-isopropyl-4-methylcyclohexane

B. 2-(sec-Butyl)-4-isopropyl-1-methylcyclohexane

C. 1-(sec-Butyl)-5-isopropyl-2-methylcyclohexane

D. 5-(sec-Butyl)-1-isopropyl-4-methylcyclohexane

11. The IUPAC name for the compound 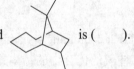 is (　　　).

化合物 的IUPAC名称是（　　　）。

A. 7-(tert-Butyl)-6-methylspiro[3.5]nonane

B. 3-(tert-Butyl)-2-methylspiro[3.5]nonane

C. 7-(tert-Butyl)-8-methylspiro[5.3]nonane

D. 7-(tert-Butyl)-6-methylspiro[4.6]nonane

12. The IUPAC name for the compound is (　　　).

化合物 的IUPAC名称是（　　　）。

A. 7,9,9-trimethylbicyclo[1.2.4]nonane　B. 7,9,9-trimethylbicyclo[6.4.3]nonane

C. 2,9,9-trimethylbicyclo[4.2.1]nonane　D. 7,9,9-trimethylbicyclo[4.2.1]nonane

13. The main product of the following reaction is (　　　).

下列反应的主要产物是（　　　）。

$$\bigcirc\!\!-CH_3 + Cl_2 \xrightarrow{hv} ?$$

A. 　—CH₂Cl

B. Cl / CH₃

C. CH₃ / Cl

D. Cl / —CH₃

14. The main product of the following reaction is (　　　).

下列反应的主要产物是（　　　）。

$$\bigcirc\!\!\triangleright + HBr \longrightarrow ?$$

A.

B. (cyclohexane-CH₂Br)

C. (cyclohexane with Br and CH₃)

D. (cycloheptane-Br)

15. Which of the following conformations for $ClCH_2CH_2Br$ is the most stable ?
下列有关$ClCH_2CH_2Br$的构象中最稳定的是（   ）。

A. (Newman projection)   B. (Newman projection)

C. (Newman projection)   D. (Newman projection)

16. Which of the following free radicals is the most stable?
下列自由基相对最稳定的是（   ）。

A. $\overset{\cdot}{C}H_3$

B. $H_2C—\overset{\cdot}{C}H_2$

C. $H_3C—\overset{\cdot}{C}H—CH_3$

D. $\overset{CH_3\overset{\cdot}{C}CH_3}{\underset{CH_3}{|}}$

17. The formula for a compound that containing six carbons but no secondary hydrogen is (   ).
一个含有六个碳原子但不含有仲氢的化合物的结构式是（   ）。

A. (structure)   B. (structure)

C. (structure)   D. (structure)

18. Isopentane reacts with bromine under light, and the main product is (   ).
异戊烷在光照下和溴反应中，主要产物是（   ）。

A. (structure)   B. (structure)

C. (structure)   D. (structure)

19. The orbit of single electron in carbon radical is (   ).
含碳自由基结构中单电子所处的轨道为（   ）。

A. s   B. p   C. sp^3 D. sp^2

Discussion topic

Please write the dominant conformation of glycol and explain why it is the dominant conformation.

（刘　华　刘晓平）

第三章　立体化学
Chapter 3　Stereoisomerism

学习目标

1. **掌握**　顺反异构、对映异构的构型判定及命名规则。

2. **熟悉**　同分异构体的分类及判断标准、费歇尔投影式的转换原则、顺序规则的应用、分子对映异构与分子对称性的关系、对映体与非对映体、内消旋体与外消旋体的区别。

3. **了解**　顺反异构体和对映异构体的性质及与生理活性的关系、光学纯度和外消旋体的拆分等基本知识。

3.1　Introduction to Isomerism

PPT

Stereoisomerism (立体异构) is the study of the three-dimensional structure of molecules. **Isomers** (同分异构体) are different compounds with the same molecular formula. There are several types of **isomerism** (异构现象) in organic compounds, and isomers are grouped into two broad classes: **constitutional isomers** (构造异构体) and **stereoisomers** (立体异构体).

3.1.1　Constitutional Isomer

Constitutional isomers have the same molecular formula but differ in their bonding sequence; that is, their atoms are connected differently. For example, **butane** (正丁烷) and **isobutene** (异丁烷) have the same molecular formula C_4H_{10}, but they differ in the sequence of bonding. Similarly, **ethanol** (乙醇) and **dimethyl ether** (甲醚) are two constitutional isomers.

$$CH_3CH_2CH_2CH_3 \qquad\qquad \begin{array}{c} CH_3 \\ | \\ CH_3CHCH_3 \end{array}$$

butane　　　　　　　　　isobutane

$$CH_3CH_2OH \qquad\qquad CH_3OCH_3$$

ethanol　　　　　　　　　dimethyl ether

Constitutional isomers may differ in ways other than the **branching** (支链) of their **carbon chain** (碳链). They may also differ in the position of the functional group. For example, **but-1-ene** (丁-1-烯) and **but-2-ene** (丁-2-烯) are structural isomers that differ in the position of a **double bond** (双键).

$$CH_2 = CHCH_2CH_3 \qquad CH_3CH = CHCH_3$$
but-1-ene　　　　　　　but-2-ene

3.1.2　Stereoisomer

Stereoisomers are isomers that have the same molecular formula and the same connectivity of their atoms but differ in how their atoms are oriented in three-dimensional space. For example, *cis*-**butenedioic acid** (顺丁烯二酸) and *trans*-**butenedioic acid** (反丁烯二酸) have the same connections of carbon chain, so they are not constitutional isomers. They are stereoisomers because they are different only in the spatial orientation of the groups attached to the double bond. The *cis* isomer has the two hydrogen atoms on the same sides of the double bond, and the *trans* isomer has them on opposite sides.

cis-butenedioic acid　　　*trans*-butenedioic acid
maleic acid　　　　　　　fumaric acid

Stereoisomer mainly includes **configurational** (构型) and **conformational** (构象) isomers. **Configurational isomers** (构型异构体) are isomers that differ only in the orientation of the atoms in space, that is, the arrangement of their atoms or groups are "fixed" in space within a molecule (e.g. *cis*-butenedioic acid and *trans*-butenedioic acid).

Configurational isomers are further grouped into **enantiomers** (对映异构体) and **diastereomers** (非对映异构体).Enantiomers are mirror-image isomers but **not superimposable** (不能重叠). For example, the two enantiomers of **lactic acid** (乳酸) are mirror-images of each other but not superimposable. Unlike enantiomers, diastereomers are stereoisomers that are not mirror-images of each other.

lactic acid
a pair of enantiomers

The different arrangements of atoms or groups formed by rotations or twists about a single bond are called **conformations** (构象异构), and each specific conformation is called a **conformational isomer** (构象异构体). The molecules are constantly rotating through all the possible conformations. For example, a cyclohexane molecule interconverts from the **chair conformation** (椅式构象) to the **boat conformation** (船式构象).

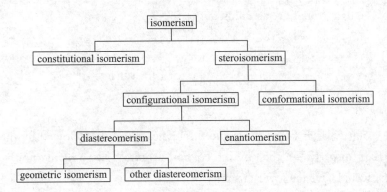

chair   half-chair   twist boat   boat

conformational isomers of cyclohexane

The relationship between isomerism can be expressed as follows.

```
                        isomerism
               ┌────────────┴────────────┐
      constitutional isomerism      steroisomerism
                          ┌────────────────┴────────────────┐
              configurational isomerism        conformational isomerism
              ┌────────────┴────────────┐
        diastereomerism          enantiomerism
        ┌────────┴────────┐
geometric isomerism   other diastereomerism
```

The discovery of stereoisomerism is one of the most important breakthroughs in the **structure theory** (结构理论) of organic chemistry. Stereoisomerism explains why there are several types of isomers, forcing scientists to come up with tetrahedralcarbon atom. In this chapter, we study the three-dimensional structure of molecules to understand their stereoisomerism relationships, which lays a foundation for the future studies on the chemical properties and reactions of organic compounds.

3.2   CIP Sequence

The difference between two enantiomers (e.g. lactic acid) lies in the three-dimensional arrangement of the four groups or atoms around the **asymmetric carbon** (不对称碳原子). However, both of them are named lactic acid, or **2-hydroxy propionic acid**（**2-羟基丙酸**) in the IUPAC system. We need a simple way to distinguish the enantiomers and to give them unique names.

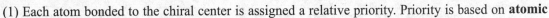

(−)-lacticacid   (+)-lactic acid

The **Cahn-Ingold-Prelog sequence**（**CIP顺序规则**) is widely used in naming the **absolute configurations** (绝对构型) of **chiral centers** (手性中心). To determine the name of a configurational isomer, we need to assign "**priorities** (优先级)" to the four groups bonded to a **chiral carbon** (手性碳), and then we name the configurational isomer based on the relative positions of these groups. The rules are listed as following.

(1) Each atom bonded to the chiral center is assigned a relative priority. Priority is based on **atomic**

number (原子序数). Atoms with higher atomic numbers have higher priorities.

Examples of priority for atoms:

$$I > Br > Cl > S > F > O > N > {}^{13}C > {}^{12}C > Li > {}^{3}H > {}^{2}H > {}^{1}H$$

With different **isotopes** (同位素) of the same element, the heavier isotopes, the higher priorities. For example, tritium (${}^{3}H$) has a higher priority than deuterium (${}^{2}H$), followed by hydrogen (${}^{1}H$).

For example, if the four groups bonded to achiral carbon atom were H, CH_3, NH_2, and Br, the relative priorities of the four groups are as following.

$$-Br > -NH_2 (N) > -CH_3 (C) > -H$$

The **bromine atom** (溴原子) would receive the highest priority, followed by the **nitrogen atom** (氮原子) of the **NH_2 group** (氨基), then by the **carbon atom** (碳原子) of the **methyl group** (甲基). **Hydrogen** (氢原子) would have the lowest priority.

(2) In case that priorities of atoms bonded directly to the chiral center are indistinguishable, use the next atoms along the chain of each group as deciders, and continue until a priority can be assigned.

For example, **isopropyl** (异丙基)-$CH(CH_3)_2$ has a higher priority than **n-propyl** (正丙基)-$CH_2CH_2CH_3$ or **bromoethyl** (溴乙基)-CH_2CH_2Br. The first carbon in the isopropyl group is bonded to two carbons (C, C, H), while the first carbon in the n-propyl group (or the bromoethyl group) is bonded to only one carbon (C, H, H). n-Propyl group and bromoethyl group have identical first atoms and second atoms, but the bromine atom in the third position gives-CH_2CH_2Br a higher priority than-$CH_2CH_2CH_3$.

(3) If a group contains a double or triple bond, both the atoms (participating in the double or triple bond) are doubled or tripled. That is, imagining that each π bond is broken and the atoms at both ends are duplicated. One π bond broken, two imaginary atoms added.

Examples of priority for **unsaturated groups** (不饱和基团):

Thus we assign a higher priority to **ethynyl** (乙炔基)-C≡CH than to **ethenyl** (乙烯基)-CH=CH₂.

Thus we assign a higher priority to carboxyl (羧基)–COOH than to aldehyde (醛基)–CHO.

(4) The *cis*-group has higher priority than the *trans*-group.

(5) The *R*-group has higher priority than the *S*-group.

According to the CIP sequence, the common atoms or groups are arranged in the following order:-I, -Br, -Cl, -SO₃H, -F, -OCOR, -OR, -OH, -NO₂, -NR₂, -NHR, -NH₂, -CCl₃, -CHCl₂, -COCl, -CH₂Cl, -COOR, -COOH, -CONH₂, -COR, -CHO, -CR₂OH, -CHROH, -CH₂OH, -CR₃, -CHR₂, -CH₂R, -CH₃, -H, lone pair electrons.

3.3　Cis-trans Isomer

PPT

3.3.1　Introduction to Cis-trans Isomer

Cis (顺式) and *trans* (反式) isomers are configurational isomers of stereoisomerism. *Cis-trans* isomers are also called **geometric isomers** (几何异构体) because they differ in the geometry of the groups on a double bond. **Alkenes** (烯烃) contain carbon-carbon double bonds, which cannot rotate. Many alkenes show geometric (*cis-trans*) isomerism. If two similar groups bonded to the **double-bond carbons** (双键碳原子) are on the same side of the double bond, the alkene is the *cis* isomer. If the similar groups are on opposite sides of the bond, the alkene is *trans* isomer. For example, two methyl groups are on the same side of the double bond in ***cis*-but-2-ene** (顺-丁-2-烯), but on the opposite side of the double bond in ***trans*-but-2-ene** (反-丁-2-烯). Not all alkenes have *cis-trans* isomers. For example, **but-1-ene** (丁-1-烯) has two identical hydrogens on one end of the double bond, and can not show *cis-trans* isomerism.

微课

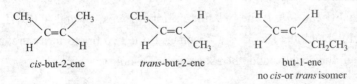

cis-but-2-ene　　　　*trans*-but-2-ene　　　　but-1-ene
no *cis*-or *trans* isomer

To have *cis-trans* isomerism, there must be two different groups on each end of the double bond. If either carbon of the double bond holds two identical groups, the molecule can not have *cis-trans* isomerism.

Alkenes have rigid double bonds that prevent rotation, giving rise to *cis* and *trans* isomers. **Cycloalkanes** (环烷烃) are similar to alkenes, the carbon-carbon bonds can not rotate freely for rigid structure. A cycloalkane has two distinct faces. If two substituents point toward the same face, the cycloalkane is *cis* isomer. If they point toward opposite faces, the cycloalkane is *trans* isomer. For example, two methyl groups point towards the same face in **cis-1,2-dimethylcyclohexane** (顺-1,2-二甲基环己烷), but to the opposite faces in **trans-1,2-dimethylcyclohexane** (反-1,2-二甲基环己烷). These geometric isomers can not interconvert without breaking and reforming of bonds.

cis-1,2-dimethylcyclohexane   *tans*-1,2-dimethylcyclohexane

3.3.2 E-Z Nomenclature

The *cis-trans* nomenclature for geometric isomers sometimes can not give an exact name. For example, the isomers of **2-bromopent-2-ene**(2-溴戊-2-烯) are not clearly *cis* or *trans* because there are no similar groups bonded to the double-bond carbons.

geometric isomers of 2-bromopent-2-ene

To fix this problem, we use the **E-Z system of nomenclature**（*E, Z*命名法) for *cis-trans* isomers, which is based on the CIP sequence for asymmetric carbon atoms. It assigns a unique configuration of either *E* or *Z* to any double-bond geometric isomer.

To name an alkene by the *E-Z* system, we need to separate the double bond into its two ends. Consider each end of the double bond separately, and use the CIP rules to assign the relative priorities to the two substituent groups on that end. Do the same for the other end of the double bond. If the two first-priority groups are together on the same side of the double bond, the alkene is a *Z* isomer, from the German word zusammen, meaning together. If the two first-priority atoms are on opposite sides of the double bond, the alkene is an *E* isomer, from the German word entgegen, meaning opposite.

For example, there are four different groups on the double-bond carbons in **2-bromopent-2-ene**（2-溴戊-2-烯), bromine atom and methyl group-CH₃ on one end, ethyl group-CH₂CH₃ and hydrogen on the other end. Bromine atom has a higher priority（①) than methyl group(②), and ethyl group has a higher priority（①) than hydrogen(②). If the two first-priority groups -Br and -CH₂CH₃ are together on the same side of the double bond, the alkene is a *Z* isomer. If the two first-priority groups -Br and -CH₂CH₃ are on opposite sides of the double bond, the alkene is an *E* isomer.

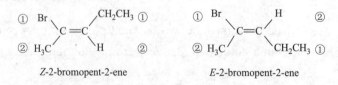

Z-2-bromopent-2-ene   *E*-2-bromopent-2-ene

If an alkene has more than one double bond, the stereoisomerism of each double bond should be specified. The following compound is properly named **(2E, 4Z)-5-chlorohexane-2, 4-dienoicacid（(2E, 4Z)-5-氯己-2, 4-二烯酸)**:

$$H_3\overset{6}{C}$$

(2E,4Z)-5-chlorohexane-2,4-dienoicacid

The *E-Z* system of nomenclature is more widely used than *cis–trans* system. It is required especially when a double bond is not clearly *cis* or *trans*.

3.3.3　Significance of cis-trans isomers in the Pharmacy Study

Base on the structural differences, *cis-trans* isomers often have different physical and chemical properties. Furthermore they exhibit different physiological activities. For example, the two configurations of the **non-steroidal antiestrogen** (非甾体抗雌激素药) **tamoxifen** (他莫西芬) have completely opposite pharmacological effects. The *Z* isomer has **antiestrogen effects** (抗雌激素作用), while the *E* isomer is an **estrogen stimulant** (雌激素兴奋剂). The *trans* isomer of **glimepiride** (格列美脲) is used as a **hypoglycemic agent** (降血糖药), while the *cis* isomer has no physiological activity. The configuration of the carbon-carbon double bonds of **linoleic acid** (亚油酸) and **arachidonic acid** (花生四烯酸) with **hypolipidemic effect** (降血脂作用) are constant. The two double bonds at the 9 and 12 positions in linoleic acid are *cis*, and the double bonds are all *cis* in arachidonic acid. If the configurations change, their physiological activities will be affected. Therefore, whether a mixture of *cis-trans* isomers can be used as a medicine, the pharmacology and toxicology of single *cis* isomer and single *trans* isomer should be considered.

linoleic acid

arachidonic acid

3.4　Enantiomer

PPT

3.4.1　Introduction to Enantiomer

What is the difference between your left hand and your right hand? They look similar, but cannot overlap (重叠). The relationship between your two hands is that they are not superimposable mirror images of each other. Objects that have left-handed and right-handed forms are called **chiral** (手性的), which means "handed" in Greek. **Enantiomers** (对映异构体) are **mirror-image** (镜像) isomers but **not superimposable** (不可重叠的). Any chiral compound must have an enantiomer.

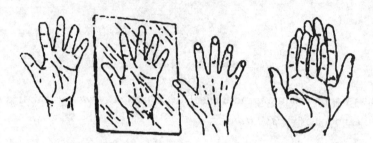

We can tell whether a compound is chiral by looking at its mirror image. Every compound has a mirror image, but a **chiral molecular** (手性分子) has a mirror image that is different from the original one. A chiral compound, for example, **lactic acid** (乳酸)(A) has a mirror image (B), which is different from the original molecule.

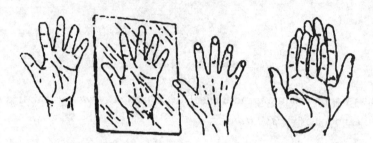

Unlike lactic acid, the mirror image (B) of **propionic acid** (丙酸)(A) is the same as the first structure. Such compound is called **achiral** (非手性的), meaning "not chiral". An achiral compound has a mirror image that is the same as the original one. Any compound that is achiral cannot have an enantiomer.

3.4.2 R, S Nomenclature of Chiral Carbon Atoms

As shown below, these mirror images are different, however, both are named lactic acid or **2-hydroxypropionic acid**（2-羟基丙酸) in the IUPAC system. The difference between the two enantiomers of lactic acid lies in the three-dimensional arrangement of four distinct groups around a carbon atom, which is called an **asymmetric carbon atom** (不对称碳原子) or a **chiral carbon atom** (手性碳原子), and is often designated by an asterisk (*). The enantiomers of lactic acid represent the two possible arrangements of its four groups around the chiral carbon atom.

mirror

enantiomers of lactic acid

Each chiral carbon atom is assigned an *R* or *S* based on its three-dimensional configuration. To determine the name, we need to follow a two-step procedure. First we assign priorities to the four groups bonded to the chiral carbon atom using the CIP sequence. And then we assign the name *R* or *S* based on the relative positions of these groups.

That is, using a three-dimensional drawing, identify the four substituents on the chiral carbon atom, and assign a priority from first-priority to fourth-priority. Put the fourth-priority group (*d*) away from you and view the molecule with the first (*a*), second (*b*), and third priority groups (*c*) radiating toward you like the spokes of a steering wheel. Draw an arrow from the first-priority group (*a*), through the second (*b*), to the third (*c*). If the arrow points **clockwise** (顺时针), the chiral carbon atom is named *R* (rectus in Latin, meaning upright). If the arrow points **counterclockwise** (逆时针), the chiral carbon atom is named *S* (sinister in Latin, meaning left).

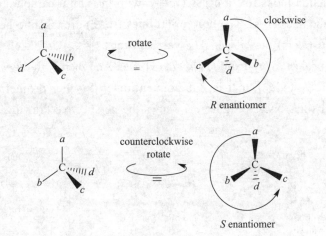

Priority: *a* (1st) > *b* (2nd) > *c* (3rd) > *d* (4th)

Let's use the enantiomers of lactic acid as an example. Of the four atoms attached to the chiral carbon in lactic acid, oxygen atom in -OH has the largest atomic number, giving it the highest priority (①). Next (②) is the carbon atom in -COOH, since it is bonded to oxygen atoms. Third (③) is the methyl group -CH₃, followed by the lowest-priority (④) hydrogen atom.

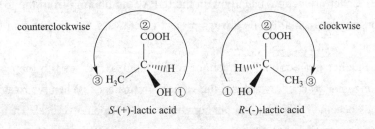

S-(+)-lactic acid　　　　　*R*-(-)-lactic acid

Priority: -OH (1st) > -COOH (2nd) > -CH₃ (3rd) > -H (4th)

The (+)-enantiomer is the one on the left, assigned to have the *S* configuration. When we position the (+)-enantiomer with its fourth-priority hydrogen atom pointing away from us, the arrow from -OH to -COOH to -CH$_3$ points counterclockwise. Thus, the (+)-enantiomer of lactic acid has the *S* configuration. In contrast, the (−)-enantiomer on the right is assigned to have the *R* configuration. When we position the (−)-enantiomer with its fourth-priority hydrogen atom pointing away from us, the arrow from -OH to -COOH to -CH$_3$ points clockwise. Thus, the (−)-enantiomer of lactic acid has the *R* configuration.

3.4.3 Fischer Projection

We generally use **perspective drawing** (透视图) with **dashed lines** (虚线) and **wedges** (楔形线) to indicate the stereochemistry of chiral carbon atoms as above. However, when there are several chiral carbons in a molecule, perspective drawing would be tiring and time-consuming. In addition, the complicated drawings make it difficult to see the similarities and differences in groups of stereoisomers.

To solve the problem, Emil Fischer developed a symbolic way to draw the structure of chiral carbon atoms, namely **Fischer projection** (费歇尔投影式). The Fischer projection makes them to be drawn rapidly and facilitates comparison of stereoisomers.

The Fischer projection looks like a **cross** (十字), with the chiral carbon atom (usually not drawn in) at the point where the lines cross. The **horizontal lines** (横线) are taken to be wedges, that is, bonds that project out **toward** (朝向) the viewer. The **vertical lines** (竖线) are taken to be **dashed lines** (虚线), that is, bonds that project **away from** (远离) the viewer. And then both the wedges and dashed lines are simplified to **full lines** (实线), that is, the convention in Fischer projection is: "The horizontal lines forward, vertical lines back". Figure 3.1 shows the perspective drawing implied by the Fischer projection.

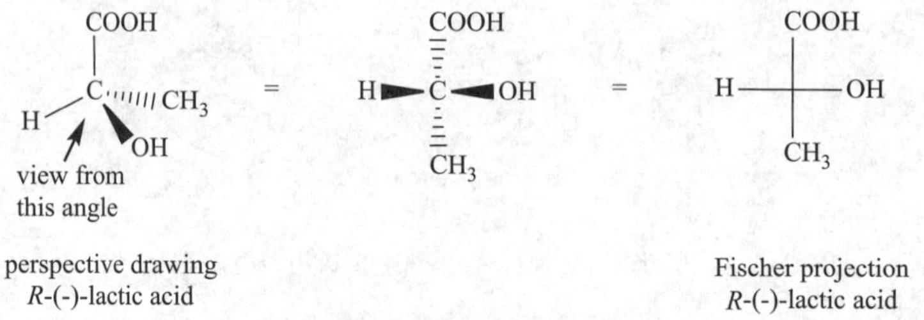

Figure 3.1　Fischer projection of lactic acid
图3-1　乳酸的费歇尔投影式

The final rule for drawing Fischer projections is that the **carbon chain** (碳链) is drawn along the vertical lines of the Fischer projection, usually with the IUPAC numbering from top to bottom. Note that in Fischer projection, the horizontal bonds always project out toward the viewer, and the vertical bonds project away from the viewer.

In comparing Fischer projections, we cannot rotate them by 90° or flip them over. The original projection has the horizontal groups forward and the vertical groups back. When we rotate the projection by 90°, the vertical bonds become horizontal and the horizontal bonds become vertical. Then the viewer would see a different molecule (actually, the enantiomer of the original one). When we flip the projection over, as shown below, the positions of -COOH and -CH$_3$ in *S*-lactic acid are interchanged. *S*-lactic acid converts to its enantiomer *R*-lactic. Either of these operations gives an incorrect representation of the molecule.

$$
\begin{array}{ccc}
& \text{COOH} & \\
\text{HO} & \!\!-\!\!\!\vert\!\!\!-\!\! & \text{H} \\
& \text{CH}_3 &
\end{array}
\xrightarrow{\text{rotate } 90°}
\begin{array}{ccc}
& \text{OH} & \\
\text{H}_3\text{C} & \!\!-\!\!\!\vert\!\!\!-\!\! & \text{COOH} \\
& \text{H} &
\end{array}
\qquad
\begin{array}{ccc}
& \text{COOH} & \\
\text{HO} & \!\!-\!\!\!\vert\!\!\!-\!\! & \text{H} \\
& \text{CH}_3 &
\end{array}
\xrightarrow{\text{flip over}}
\begin{array}{ccc}
& \text{CH}_3 & \\
\text{HO} & \!\!-\!\!\!\vert\!\!\!-\!\! & \text{H} \\
& \text{COOH} &
\end{array}
$$

S-(+)-lactic acid   *R*-(-)-lactic acid   *S*-(+)-lactic acid   *R*-(-)-lactic acid

configuration changed   configuration changed

A 90° rotation (旋转90°) or a **flip over (翻转),** equal to **interchanging (交换)** the groups once, is NOT allowed. The Fischer projection must be kept in the plane of the paper, and it may be rotated only by 180°. Fischer projections that differ by a 180° rotation are the same. When we rotate a Fischer projection by 180°, the vertical bonds still end up vertical, and the horizontal lines still end up horizontal. "The horizontal lines forward, vertical lines back" convention is maintained. A 180° rotation equates with interchanging the groups twice, the latter is also allowed.

$$
\begin{array}{ccc}
& \text{COOH} & \\
\text{HO} & \!\!-\!\!\!\vert\!\!\!-\!\! & \text{H} \\
& \text{CH}_3 &
\end{array}
\xrightarrow{\text{rotate } 180°}
\begin{array}{ccc}
& \text{CH}_3 & \\
\text{H} & \!\!-\!\!\!\vert\!\!\!-\!\! & \text{OH} \\
& \text{COOH} &
\end{array}
\qquad
\begin{array}{ccc}
& \text{COOH} & \\
\text{HO} & \!\!-\!\!\!\vert\!\!\!-\!\! & \text{H} \\
& \text{CH}_3 &
\end{array}
\xrightarrow[\text{groups twice}]{\text{interchange}}
\begin{array}{ccc}
& \text{CH}_3 & \\
\text{H} & \!\!-\!\!\!\vert\!\!\!-\!\! & \text{OH} \\
& \text{COOH} &
\end{array}
$$

S-(+)-lactic acid   *S*-(+)-lactic acid   *S*-(+)-lactic acid   *S*-(+)-lactic acid

configuration kept   configuration kept

The *R* or *S* configuration can also be determined directly from the Fischer projection using the CIP sequence, without having to convert it to a perspective drawing. First we assign the priorities of the four groups, then we can draw an arrow from group 1 to group 2 to group 3 and see which way it goes. The lowest priority atom is usually hydrogen. In the Fischer projection, the carbon chain is along the vertical line, so the hydrogen atom is usually on the horizontal line and projects out in front. In the view of opposite side, the hydrogen would be in back, that is, away from the viewer. As in the definition of *R* and *S*, the arrow would rotate in another direction. By mentally turning the arrow around, we can assign the configuration.

For example, consider the Fischer projection formula of (-)-lactic acid. First priority goes to the -OH group, followed by the -COOH group and the -CH$_3$ group. The hydrogen atom has the lowest priority. The arrow from group 1 to group 2 to group 3 appears counter clockwise in the Fischer projection. If the molecule is turned over (or viewed in the opposite side), the hydrogen is in back. The arrow is clockwise, so this is *R*-lactic acid.

counterclockwise   clockwise

$$
\underset{\substack{\text{Fischer projection}\\ R\text{-(-)-lactic acid}}}{
\begin{array}{ccc}
& \text{②COOH} & \\
\text{H} & \!\!-\!\!\!\vert\!\!\!-\!\! & \text{OH①} \\
& \text{③CH}_3 &
\end{array}}
\;=\;
\underset{\substack{\text{Hydrogen}\\ \text{in front}}}{
\begin{array}{c}
\text{COOH} \\
\text{H}\blacktriangleright\text{C}\!-\!\text{OH} \\
\text{CH}_3
\end{array}}
\;=\;
\underset{\substack{\text{Hydrogen}\\ \text{in back}}}{
\begin{array}{c}
\text{②COOH} \\
\text{H}\backslash\!\backslash\text{C} \\
\text{①HO} \quad \text{CH}_3\text{③}
\end{array}}
$$

perspective drawing

R-(-)-lactic acid

As a result, when naming *R* and *S* from Fischer projections with the hydrogen on a horizontal bond, just apply the normal rules backward. That is, when the arrow from group 1 to group 2 to group 3 appears clockwise in the Fischer projection, it is *S* configuration. On the contrary, when the arrow appears counter clockwise, it is *R* configuration.

In case that the lowest priority group is on the vertical bond and projects out in back, there is no need to turn over the molecule. Draw an arrow from the first-priority group, through the second, to the third. The *R* configuration can be determined when the arrow appears clockwise in the Fischer projection. And the *S* configuration can be determined when the arrow appears counterclockwise.

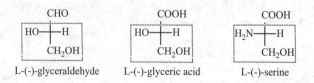

Fischer projection   methyl group   perspective drawing
S configuration   in back   *S* configuration

3.4.4 D, L Nomenclature of Chiral Carbon Atoms

There are two systems to designate configuration of enantiomers: the **absolute configuration** (绝对构型) of *R* and *S* system as mentioned above, and the **relative configuration** (相对构型) of D and L system. Emil Fischer used **glyceraldehyde** (甘油醛) as a standard for the D and L system of designating configuration. He arbitrarily assigned one enantiomer as the **D-glyceraldehyde**（**D-甘油醛**)(a), in which the –OH group on the chiral carbon is on the right in the Fischer projection. The other enantiomer, in which the –OH group on the chiral carbon is on the left in the Fischer projection, was assigned as **L-glyceraldehyde**（**L-甘油醛**)(b).

D-(+)-glyceraldehyde   L-(-)-glyceraldehyde
(a)   (b)

D and L system is common in biology/biochemistry, especially with sugars and amino acids. In the D and L system, structures that are similar to glyceraldehyde (at chiral carbon) could be determined by compared with it. Compounds with the same relative configuration as D-(+)-glyceraldehyde are assigned the D configurations, and those with the relative configuration of L-(-)-glyceraldehyde are given the L ones.

L-(-)-glyceraldehyde   L-(-)-glyceric acid   L-(-)-serine

As we convert one compound into another by a reaction that does not break bonds at the asymmetric carbon atom, the product must have the same relative configuration as the reactant. Most naturally occurring sugars degrade so that they are given the D configuration. This means that the bottom asymmetric carbon of the sugar has its –OH group on the right in the Fischer projection.

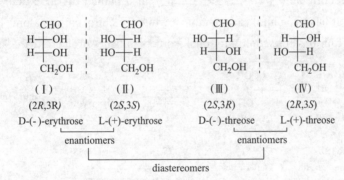

CHO
H——OH
CH₂OH
D-(+)-glyceraldehyde

COOH
H——OH
CH₂OH
D-(+)-glyceric acid

CHO
HO——H
H——OH
CH₂OH
D-(–)-threose

CHO
H——OH
HO——H
H——OH
CH₂OH
D-(+)-glucose

One must remember that there is no correlation between D and L configurations and (+) and (–) rotations. The D-isomer does not have to have a (+) rotation, and similarly the L-isomer does not have to have a (–) rotation. For example, the D-threose has a (–) rotation, instead of a (+) rotation.

3.4.5　Acyclic Molecules（非环状分子）with Two or More Stereocenters

There is one chiral carbon atom in lactic acid. The enantiomers of lactic acid represent two different arrangements of its four groups around the asymmetric carbon atom. In general, the number of stereoisomers depending on the number of chiral carbons is 2^n, in which "n" is the number of chiral carbon atoms.

A compound with two different chiral carbon atoms might represent four possible arrangements of the groups around the asymmetric carbon atoms, that is, there are four configurational isomers. For example, **aldotetrose（丁醛糖）,** namely**2,3,4-trihydroxy butanal（2,3,4-三羟基丁醛）** in the IUPAC system has four configurational isomers, in fisher projection as follows:

CHO
H——OH
H——OH
CH₂OH
（Ⅰ）
(2R,3R)
D-(–)-erythrose

CHO
HO——H
HO——H
CH₂OH
（Ⅱ）
(2S,3S)
L-(+)-erythrose

CHO
HO——H
H——OH
CH₂OH
（Ⅲ）
(2S,3R)
D-(–)-threose

CHO
H——OH
HO——H
CH₂OH
（Ⅳ）
(2R,3S)
L-(+)-threose

enantiomers   enantiomers

diastereomers

There are two pairs of enantiomers in these four configurational isomers. **D-erythrose（D-赤藓糖)** （Ⅰ）and **L-erythrose（L-赤藓糖）**（Ⅱ）are one pair of enantiomers. **D-threose（D-苏阿糖)**（Ⅲ）and **L-threose（L-苏阿糖）**（Ⅳ）are another pair of enantiomers. If we have a mixture of equal amounts of a pair of enantiomers, **racemates（外消旋体）** are formed, also named a（±）pair, or a (d,l) pair. D-erythrose （Ⅰ）and D-threose （Ⅲ）are not enantiomers because they are not mirror images of each other, they are **diastereomer（非对映体）.** Either member of one pair of enantiomers is a diastereomer of either member of the other pair.

In the configurational isomers containing two or more chiral carbon atoms, if only different configuration of one chiral carbon atoms, other chiral carbon atoms are the same, the isomers are called **epimers（差向异构体）.** For example, D-erythrose （Ⅰ）and D-threose （Ⅲ）are C-2 epimers. D-erythrose （Ⅰ）and L-threose （Ⅳ）are C-3 epimers. Aldotetrose is one of the simplest compounds containing two different chiral carbon atoms adjacent to each other. It is customary to compare the configuration of compounds with their isomers. If -a and -a are identical atoms or groups on the same side of the main carbon chain in Fischer projection, those with erythrose configuration are called **erythro-**（赤型). Otherwise, if -a and -a are identical atoms or groups on the opposite side, like configuration of threose, they are called **threo-**(苏型).

For example, the **ephedrine** (麻黄碱) and **pseudoephedrine** (伪麻黄碱) derived from the traditional Chinese medicine **ephedra** (麻黄) have two adjacent but different chiral carbon atoms in their structures. Their configurations can be represented by the erythro- and threo- types.

$$
\begin{array}{cc}
C_6H_5 & C_6H_5 \\
H\!-\!OH & HO\!-\!H \\
H\!-\!NHCH_3 & CH_3NH\!-\!H \\
CH_3 & CH_3
\end{array}
$$

erythro-(+)-ephedrine   erythro-(-)-ephedrine

$$
\begin{array}{cc}
C_6H_5 & C_6H_5 \\
H\!-\!OH & HO\!-\!H \\
CH_3NH\!-\!H & H\!-\!NHCH_3 \\
CH_3 & CH_3
\end{array}
$$

threo-(+)-pseudoephedrine   threo-(-)-pseudoephedrine

Compounds containing one chiral carbon atom have a pair of enantiomers. Compounds containing two different chiral carbon atoms have four enantiomers. And so on, compounds with n different chiral carbon atoms have 2^n stereoisomers. Note that we may not always have 2^n isomers, especially when two chiral carbon atoms have identical substituents.

2,3-dihydroxysuccinic acid（2,3-二羟基丁二酸), also named **tartaric acid** (酒石酸), has fewer than 2^n stereoisomers. It has two chiral carbon atoms, while the chiral carbon atoms have identical groups attached to them. The four possible configurational isomers in fisher projection are shown as follows.

$$
\begin{array}{cccc}
COOH & COOH & COOH & COOH \\
H\!-\!OH & HO\!-\!H & HO\!-\!H & H\!-\!OH \\
HO\!-\!H & H\!-\!OH & HO\!-\!H & H\!-\!OH \\
COOH & COOH & COOH & COOH
\end{array}
$$

σ mirror plane of symmetry

（Ⅰ） （Ⅱ） （Ⅲ） （Ⅳ）

(2R,3R)   (2S,3S)   (2S,3R)   (2R,3S)

(+)-tartaric acid   (-)-tartaric acid

enantiomers   the same compound meso-tartaric acid

diastereomers

There are only three stereoisomers of 2,3-dihydroxysuccinic acid because two of the four structures are identical.**(+)-tartaric acid** (右旋酒石酸)（Ⅰ) and **(–)-tartaric acid** (左旋酒石酸)（Ⅱ) are a pair of enantiomers. The diastereomer（Ⅲ) and（Ⅳ) on the right are the same compound, which is achiral, having a mirror **plane of symmetry** (对称面). Compounds that are achiral even though they have chiral carbon atoms are called **meso compounds** (内消旋化合物) or **mesomers** (内消旋体). Like the **meso-tartaric acid** (内消旋酒石酸), most meso compounds have this kind of symmetric structure, containing two similar halves of the molecule with opposite configurations, so that the optical rotation cancels each other out inside the molecule.

Both mesomers and racemates do not exhibit optical activity, however, they are essentially different. Mesomers are **pure compounds** (纯粹化合物) that cannot be separated into two optically active ones, while racemates are mixture of equal amounts of a pair of enantiomers. Therefore, racemates can be separated into left-handed and right-handed isomers.

Asymmetric carbon atom is the most common example of **stereocenters** (立构中心), most chiral compounds have at least one asymmetric carbon atom. Some compounds are chiral even though they have no asymmetric atoms at all, such as **biphenyl** (联苯) derivatives and **allene** (丙二烯) derivatives.

The two **benzene rings** (苯环) in biphenyl can rotate freely around the single bond, but in **6,6'-dinitro-2,2'-biphenyldicarboxylic acid** (6,6'-二硝基-2,2'-联苯二甲酸), the **carboxyl group** (羧基) and **nitro group** (硝基) are too large to be forced so close together. The free rotation of the benzene ring is hindered, so that the two benzene rings cannot lie in the same plane. These two compounds are nonsuperimposable mirror images, and they cannot interconvert. They are enantiomers, each of which is chiral and optically active.

Allene (丙二烯) derivatives are compounds that contain the C=C=C unit. The carbon atoms on either end of the allene molecule are sp^2 **hybridized**, and the carbon atoms in the middle is sp hybridized. The two p orbitals on the sp hybridized carbon atom are perpendicular, so the two π-bonds in the molecule are perpendicular to each other. Therefore, the four atoms or groups on the two ends of the allene molecule are in two planes perpendicular to each other, and the three-dimensional image is as follows.

When a ≠ b and d ≠ e, the whole molecule has no symmetry plane and is chiral. For example, **penta-2,3-diene** (2,3-戊二烯) is a chiral molecule. Carbon atom 3 is sp hybridized, carbons 2 and 4 are both sp^2 hybridized and planar, but their planes are perpendicular to each other. There is no asymmetric carbon atom. Nevertheless, penta-2,3-diene is chiral, as you should see from the following drawings of the enantiomers.

3.4.6   Cyclic Molecules (环状分子) with Two or More Chiral Centers

Stereoisomerism (立体异构) is also possible in the cyclic molecules. We can analyze stereoisomerism in cyclic compounds in the same way as in acyclic compounds. For example, **2-methylcyclopentanol** (2-甲基环戊醇), a compound with two chiral centers, has four stereoisomers. Both the *cis* and *trans* isomer are chiral: The *cis*-2-methylcyclopentanol has one pair of **enantiomers** (对映异构体), and the *trans*-2-methylcyclopentanol has a second pair of enantiomers. The *cis* and *trans* isomers are stereoisomers that are not mirror images of each other; that is, they are **diastereomers** (非对映体).

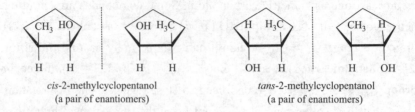

cis-2-methylcyclopentanol   *tans*-2-methylcyclopentanol
(a pair of enantiomers)   (a pair of enantiomers)

1,2-dimerthylcyclopentane（1,2-二甲基环戊烷）also has two **chiral carbons** (手性碳). However, only three stereoisomers exist for this compound. The *cis* and *trans* isomer are geometric isomers, and they are also diastereomers. The *trans*-1,2-dimerthylcyclopentane is chiral and has a pair of enantiomers, while the *cis*-1,2-dimerthylcyclopentane has an internal mirror plane of symmetry, so it is a **meso compound** (内消旋化合物), **achiral**.

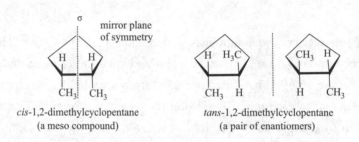

cis-1,2-dimethylcyclopentane   *tans*-1,2-dimethylcyclopentane
(a meso compound)   (a pair of enantiomers)

There are four stereoisomers of **3-methylcyclopentanol**（3-甲基环戊醇）. Both the *cis* and *trans* isomer are chiral. The *cis* isomer has one pair of enantiomers, and the *trans* isomer has a second pair of enantiomers.

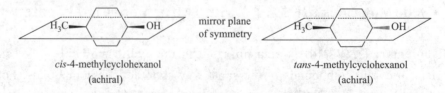

cis-2-methylcyclopentanol   *tans*-2-methylcyclopentanol
(a pair of enantiomers)   (a pair of enantiomers)

There are two **stereocenters** (立体中心) in **4-methylcyclohexanol**（4-甲基环己醇）, but they are not chiral centers because the carbon atoms of the stereocenters do not have four different groups attached. 4-methylcyclohexanol therefore has two diastereomers. The *cis* and *trans* isomer are also geometric isomers. Both of these isomers are achiral. In each, a mirror plane of symmetry runs through the -CH$_3$, and -OH groups and the carbon atoms bonded to them.

cis-4-methylcyclohexanol   *tans*-4-methylcyclohexanol
(achiral)   (achiral)

PPT

3.5　Optical Activity—How Chirality Is Detected in the Laboratory

Enantiomers are different compounds, we need a simple method to distinguish between enantiomers and measure their purity in the laboratory. One property that differs between enantiomers is their effect on the rotation of the **plane of polarized light** (平面偏振光). Each member of a pair of enantiomers rotates the plane of polarized light in opposite directions, and for this reason, enantiomers are said to be **optically active** (光学活性的).

Plane polarized light is composed of waves that vibrate in only one plane. When unpolarized light passes through a **polarizing filter** (偏振光滤光器), the randomly vibrating light waves are filtered so that most of the light passing through is vibrating in one direction. The direction of vibration is called the axis of the filter.

When light passes first through one polarizing filter and then through another, the amount of light emerging depends on the relationship between the axes of the two filters. If the axes of the two filters are lined up (parallel), then nearly all the light that passes through the first filter also passes through the second. If the axes of the two filters are perpendicular (crossed poles), however, all the polarized light that emerges from the first filter is stopped by the second. At intermediate angles of rotation, intermediate amounts of light pass through.

When polarized light passes through a solution containing a chiral compound, the chiral compound causes the plane of vibration to rotate. Rotation of the plane of polarized light is called **optical activity** (光学活性), and substances that rotate the plane of polarized light are said to be optically active.

Two enantiomers have identical physical properties, except for the direction they rotate the plane of polarized light. Enantiomeric compounds rotate the plane of polarized light by exactly the same amount but in opposite directions.

A **polarimeter** (旋光仪) measures the rotation of polarized light. It has a tubular cell filled with a solution of the optically active material and a system for passing polarized light through the solution and measuring the rotation as the light emerges. The light from a **sodium lamp** (钠光灯) is filtered so that it consists of just one wavelength, because most compounds rotate different wavelengths of light by different amounts. The wavelength of light most commonly used for polarimetry is a yellow emission line in the spectrum of sodium, called the **sodium D line** (钠D线).

Monochromatic (单色光) light from the source passes through a polarizing filter, then through the sample cell containing a solution of the optically active compound. On leaving the sample cell, the polarized light encounters another polarizing filter. This filter is movable, with a scale allowing the operator to read the angle between the axis of the second (analyzing) filter and the axis of the first (polarizing) filter. The operator rotates the analyzing filter until the maximum amount of light is transmitted, and then reads the observed rotation from the protractor. The observed rotation is symbolized by the Greek letter alpha. The working principle of polarimeter is shown in Figure 3.2.

source   polarizing filter   sample cell   polarizing filter

Figure 3.2   Working principle of polarimeter
图3-2   旋光仪工作原理

Compounds that rotate the plane of polarized light toward the right (clockwise) are called **dextrorotatory** (右旋), meaning "toward the right". Compounds that rotate the plane toward the left (counterclockwise) are called **levorotatory** (左旋), meaning "toward the left". These terms are sometimes abbreviated by a lowercase d or l. Using IUPAC notation, the direction of rotation is specified by the (+) or (−) sign of the rotation: Dextrorotatory (clockwise) rotations are (+) or *d*. Levorotatory (counterclockwise) rotations are (−) or *l*.

The angular rotation of polarized light is a characteristic physical property of chiral compound. The rotation observed in a polarimeter depends on the concentration of the sample solution and the length of the cell, as well as the optical activity of the compound.

To use the optical activity as a characteristic property of a compound, we must standardize the conditions for measurement. We define a compound's **specific rotation** (比旋度) as the rotation found using a dm sample cell and a concentration of 1 g · ml^{-1}. Other cell lengths and concentrations may be used, as long as the observed rotation is divided by the path length of the cell (*l*) and the concentration (*c*).

$$[\alpha] = \frac{\alpha}{c \cdot l}$$

α is the rotation observed in the polarimeter, *c* is the concentration (g · ml^{-1}); *l* is the length of the sample cell(dm).

3.6 Significance of Chirality in the Pharmacy Study

PPT

Biological systems commonly distinguish between enantiomers. Enantiomers are also closely related to physiological activity. Many optically active compounds have physiological activities, but two enantiomers may have different biological properties. For example, the **pupil dilation** (扩瞳作用) of (−)-**scopolamine** (东莨菪碱) is 20 times greater than that of (+)-isomer. The **vasoconstriction** (血管收缩作用) of (−)-adrenalin is 12~15 times greater than that of (+)-isomer. The **broncho dilation effect** (支气管舒张作用) of (−)-**isoproterenol** (异丙肾上腺素) is 800 times greater than that of (+)-isomer.

The physiological activity of a drug is due to the interaction with the receptors in the organism and the receptors have certain three-dimensional images. When drug interacts with the receptor, its three-dimensional structure should be adapted to that of the receptor. Only in this way can it exert its physiological effects and produce specific pharmacological effects. For example, it is generally believed that **adrenalin** (肾上腺素) forms three-point binding with the receptor: amino, benzene ring and two phenolic hydroxyl groups and alcohol hydroxyl groups in the side chain as shown in Figure 3.3, only two groups of (+)-adrenalin bind to the receptor, while all the three groups of (−)-adrenalin bind to the receptor. So the physiological effect of (+)-adrenalin is weaker than that of (−)-adrenalin. According to this theory, the hydroxyl group of (+)-adrenalin does not work. If the hydroxyl group is removed, it will generate **deoxyadrenalin** (去氧肾上腺素). Its effect is similar to that of (+)-adrenalin.

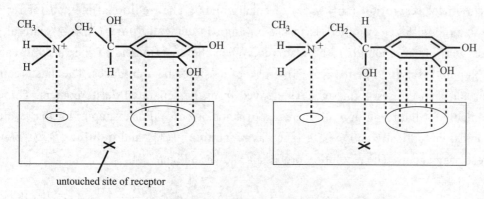

untouched site of receptor

Figure 3.3　Diagram of adrenaline binding to receptor

图3-3　肾上腺素与受体结合示意图

PPT

3.7　Separation of Enantiomers—Resolution (拆分)

Racemates (外消旋体) are composed of equal parts of a pair of enantiomers, which have the same physical properties except that they rotate light in opposite directions. Therefore, although racemates are a mixture of two compounds, the pairs of enantiomers cannot be separated by ordinary physical methods, such as **distillation** (蒸馏), **recrystallization** (重结晶), etc. Special methods must be used to separate them. The process of separating the enantiomers of racemates is usually called resolution. The common methods of resolution are as follows.

(1) **Mechanical resolution** (机械拆分) of enantiomers　By using the difference in the **crystal morphology** (晶体形态) of the enantiomers in the racemates, the two crystals are separated by identification with the naked eye or through a magnifying glass. In 1848, Pasteur first used this method to separate two crystals of **ammonium sodium tartrate** (酒石酸铵钠).Unfortunately, few racemic compounds crystallize as separate enantiomers, and other methods of resolution are required.

(2) **Microbial resolution** (微生物拆分) of enantiomers　The selective decomposition of an enantiomer by certain microorganisms or enzymes can be used to separate an enantiomer from the racemates. For example, add acetic anhydride to racemic α-amino acids for acyl reaction to prepare D, L-acetyl amino acids. Then, **acylase** (酰化酶) was added, and only L-amino acid was precipitated by the selective hydrolysis of acylate. Finally, D-acetyl amino acids are hydrolyzed to obtain D-amino acids. The disadvantage of this method is that only one isomer is utilized and the other isomer is consumed. In addition, it is difficult to find suitable enzymes.

(3) **Seed crystallization method** (晶种法) of enantiomers　When a certain amount of **levorotatory**

(左旋) or **dextrorotatory** (右旋) substance is added to the **supersaturated solution** (过饱和溶液) of the race mate as a **crystal seed** (晶种), the same isomer as the crystal seed will be precipitated first. After the crystal is filtered out, the racemic system will be added to the **mother liquor** (母液) to form supersaturated solution, and then another isomer in the solution will be crystallized. Hence the race mate can be completely resolved by repeated treatment.

(4) **Chemical resolution** (化学拆分) of enantiomers　The traditional method for resolving a racemic mixture into its enantiomers is to use an **enantiomerically pure** (光学纯) natural product that bonds with the compounds to be resolved. When the enantiomers of the racemic compound bond to the pure **resolving agent** (拆分剂), a pair of diastereomers produces. The diastereomers are separated, and then the resolving agent is cleaved from the separated diastereomers. This method is most suitable for the resolution of racemates of acids or bases. For example, for the resolution of racemic acids, the optically active bases such as **morphine** (吗啡) and **quinine** (金鸡纳碱) can be used as resolving agents. The resolution process is shown in Figure 3.4.

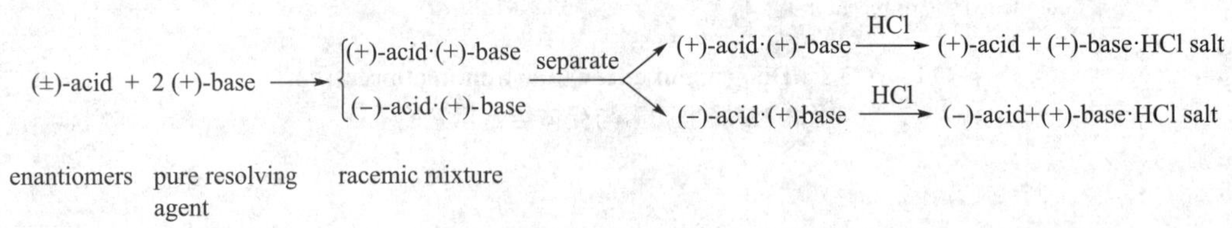

$$(\pm)\text{-acid} + 2\,(+)\text{-base} \longrightarrow \begin{cases} (+)\text{-acid}\cdot(+)\text{-base} \\ (-)\text{-acid}\cdot(+)\text{-base} \end{cases} \xrightarrow{\text{separate}} \begin{cases} (+)\text{-acid}\cdot(+)\text{-base} \xrightarrow{\text{HCl}} (+)\text{-acid} + (+)\text{-base}\cdot\text{HCl salt} \\ (-)\text{-acid}\cdot(+)\text{-base} \xrightarrow{\text{HCl}} (-)\text{-acid} + (+)\text{-base}\cdot\text{HCl salt} \end{cases}$$

enantiomers　pure resolving agent　racemic mixture

Figure 3.4　Resolution process of racemic acids
图3-4　外消旋酸的拆分工艺

For the resolution of racemic bases, the optically active acids such as **tartaric acid** (酒石酸) and **malic acid** (苹果酸) are often used.

(5) **Chromatographic resolution** (色谱拆分) of enantiomers　Chromatography is a powerful method for separating compounds. One type of chromatography involves passing a solution through a column containing particles whose surface tends to adsorb organic compounds. Enantiomers that are adsorbed strongly spend more time on the stationary particles. They come off the column later than less strongly adsorbed enantiomers, which spend more time in the mobile solvent phase.

重 点 小 结

一、异构现象介绍

1. 有机化合物普遍存在同分异构现象,即具有相同分子式但结构不同的现象。同分异构分为构造异构和立体异构两大类。

2. 构造异构是指具有相同分子组成,但分子中原子连接次序或方式不同所产生的同分异构现象,如碳架异构、位置异构、官能团异构和互变异构等。

3. 立体异构是指构造相同的分子,由于原子(团)在三维空间的排列方式不同所产生的同分异构现象,包括构型异构和构象异构。构型异构是指分子中原子(团)在空间的固定排列方式不同产生的同分异构现象,如对映异构和非对映异构(包括顺反异构体和其他非对映异构体)。构象异构是指具有一定构型的分子由于单键的旋转或扭曲使分子中原子（团）在空间产生不同排列的异构现象。

二、CIP顺序规则

CIP顺序规则用于命名构型异构体,例如Z-、E-构型标记和R-、S-构型标记。所谓"顺序规则"就是把原子（团）按照优先次序进行排序,方法如下：①把取代基按照原子序数由大到小排列,原子序数大的优先。当原子序数相同时,原子量大的优先。 ②比较取代基的优先次序时,先比较直接相连的第一个原子的原子序数。如果是相同原子,那就依次比较第二个、第三个 ……原子的原子序数。③当取代基为不饱和基团时,应把双键或叁键原子看作连有两个或三个相同的原子。 ④顺式优先于反式。⑤ R优先于S。

三、顺反异构体

1. 顺反异构体属于构型异构体,又称为几何异构体。具有刚体结构(如双键或环)的有机分子,由于其共价键的自由旋转受到阻碍,引起分子中原子（团）在空间的排列方式不同,从而有可能产生顺式和反式两种不同的构型,这种现象称为顺反异构。分子中两个相同原子（团）处在双键同侧的为顺式,处在双键异侧的为反式。采用顺-(cis-)、反-(trans-)构型标记法标明构型。

2. 当顺反异构体的双键碳原子上所连接的四个原子（团）各不相同时,采用Z-、E-构型标记法标明构型。根据CIP顺序规则,比较每一个双键碳原子上所连接的两个原子（团）的优先顺序,在双键的两端—优先的两个原子(团)处于双键同侧的为Z-构型,处于双键异侧的则为E-构型。

四、对映异构体

1. 两个互为实物和镜像而不重合的化合物称为对映异构体。一对对映体就像我们的左右手一样,相互对映而不重合,物质的这种性质称为手性,具有手性的分子称为手性分子。

2. 与四个不相同的原子（团）相连的碳原子称为手性碳原子或不对称碳原子,通常用C*表示。采用R-、S-构型标记法标明手性碳原子的构型。将手性碳原子所连接的四个原子（团）按照CIP顺序规则从大到小排序：a>b>c>d。让排序最小的原子（团）(d)远离观察者的视线,这时其他三个原子（团)(a、b、c)就指向观察者。观察a→b→c的顺序,如果是按顺时针方向排列的,此手性碳原子为R-构型;如果是按逆时针方向排列的,则为S-构型。

3. 费歇尔投影式的投影规则是一般将分子碳链竖立放置,编号最小的碳原子放在上方,编号最大的碳原子放在下方。费歇尔投影式"十"字交叉处通常为手性碳原子,竖键上的原子（团）指向纸平面的后方,横键上的原子（团）指向纸平面的前方,简称"横前竖后、十字交叉"原则。

4. 投影式的相互转化必须遵守下述基本操作法则:允许费歇尔投影式在纸平面上旋转180°;允许原子（团）相互交换偶数次;允许一个原子（团）不动,另外三个原子（团）按顺时针方向或逆时针方向旋转;以上操作不会改变原子（团）的空间排列关系,仍表示为同一构型的原化合物。费歇尔投影在纸平面上旋转90°、原子（团）交换奇数次、或者离开纸平面翻转,会改变原子（团）的空间排列关系,不再表示同一构型的原化合物。

5. 在费歇尔投影式中,当排序最小的原子（团）(d)在竖线上时,此时排序最小的原子（团)(d)远离观察者,如果a→b→c是按顺时针方向排列的为R-构型,按逆时针方向排列的则为S-构型。反之,当排序最小的原子（团）(d)在横线上时,此时排序最小的原子（团)(d)指向观察者,如果a→b→c是按顺时针方向排列的为S-构型,按逆时针方向排列的则为R-构型。

6．在相对构型表示法中,以甘油醛为标准,位号最大的手性碳原子具有与D-甘油醛相同构型的,则以D-构型进行标记;反之,具有与L-甘油醛相同构型的,以L-构型进行标记。

7．含有两个不同手性碳原子的化合物在空间有四种不同的排列方式,即有四个构型异构体,构成两对对映异构体。四个构型异构体之间不呈实物和镜像关系的称为非对映异构体。一对对映体等量混合组成外消旋体。外消旋体没有旋光性,可以拆分成左旋体和右旋体。

五、手性化合物的旋光性

手性化合物使偏振光的振动方向发生旋转,具有旋光性。其中使偏振光的振动方向向右(顺时针)旋转的物质称为右旋体,用(+)或d表示;使偏振光的振动方向向左(逆时针)旋转的物质称为左旋体,用(-)或l表示。旋光度的大小可以通过旋光仪来测定。

Problems
目 标 检 测

1. How many configurational isomers are there in
$$\begin{array}{c} CH_2OH \\ | \\ C=O \\ HO-\!|\!-H \\ H-\!|\!-OH \\ H-\!|\!-OH \\ | \\ CH_2OH \end{array}$$?

化合物 $\begin{array}{c} CH_2OH \\ | \\ C=O \\ HO-\!|\!-H \\ H-\!|\!-OH \\ H-\!|\!-OH \\ | \\ CH_2OH \end{array}$ 的构型异构体数目是（　　　　）。

A. 2   B. 4

C. 8   D. 16

2. Which of the following compounds is a Z isomer?
下列化合物中构型为 Z 型的是（　　　　）。

A. $\begin{array}{c} HO \\ CH_3 \end{array}\!C\!=\!C\!\begin{array}{c} H \\ CH_2OH \end{array}$ B. $\begin{array}{c} Br \\ H \end{array}\!C\!=\!C\!\begin{array}{c} C\!\equiv\!CH \\ CH\!=\!CH_2 \end{array}$

C. $\begin{array}{c} CH_3 \\ CH_3CH_2 \end{array}\!C\!=\!C\!\begin{array}{c} CH(CH_3)_2 \\ CH_2CH_3 \end{array}$ D. $\begin{array}{c} HOCH_2 \\ HOOC \end{array}\!C\!=\!C\!\begin{array}{c} CH_3 \\ H \end{array}$

3. Which of the following compounds do not exhibit optical activity?
下列化合物中没有旋光性的是（　　　　）。

A. $CH_3CH\!=\!C\!=\!CHCH_3$ B. $\begin{array}{c} CH_3-CH-COOH \\ | \\ Cl \end{array}$

C.

$$
\begin{array}{c}
CH_3 \\
H \text{——} Cl \\
Cl \text{——} H \\
H \text{——} Cl \\
CH_3
\end{array}
$$

D.

4. What is the configuration of the chiral carbon atom?

$$
\begin{array}{c}
CH_2Cl \\
H \text{——} Br \\
Br \text{——} H \\
CH_3
\end{array}
$$
的手性碳原子的构型为（　　）。

A. 2*R*, 3*R*　　　　　　　　　　B. 2*S*, 3*S*

C. 2*R*, 3*S*　　　　　　　　　　D. 2*S*, 3*R*

5. How to judge whether the compound has optical activity?

判断化合物是否具有旋光性的条件是（　　）。

　A. There are chiral carbon atoms in the molecule

　B. The molecule has polarity

　C. There are conformational isomers in the molecule

　D. The molecular is asymmetric

6. When a pair of enantiomers are mixed equally, the resulting mixture is called (　　).

将一对对映异构体等量混合，得到的混合物称为（　　）。

　A. low eutectic compound　　　　B. diastereomer

　C. racemates　　　　　　　　　　D. mesomers

7. Which of the following statements is correct?

下列说法正确的是（　　）。

　A. All compounds with *cis* configuration are *Z* configuration

　B. All compounds with *R* configuration are dextrorotatory

　C. Meso compound has no optical activity

　D. All compounds with chiral carbon atoms have optical activity

8. Which of the following structures is the enantiomer of ?

与化合物 是对映体的结构是（　　）。

A.

$$
\begin{array}{c}
CH_2CH_3 \\
H \text{——} Br \\
CH_3
\end{array}
$$

B.

C.

$$
\begin{array}{c}
Br \\
H \text{——} CH_3 \\
H_3C \text{——} H \\
H
\end{array}
$$

D.

9. Which of the following is the Fischer projection of $H_3C-\overset{H}{\underset{H}{C}}-\overset{H}{\underset{Cl}{C}}-CH_3$?

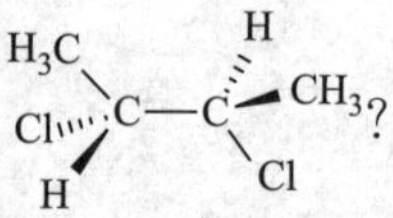

化合物 $\underset{H}{\overset{H_3C}{Cl}}C-\overset{H}{\underset{Cl}{C}}-CH_3$ 的Fischer投影式是（　　　）。

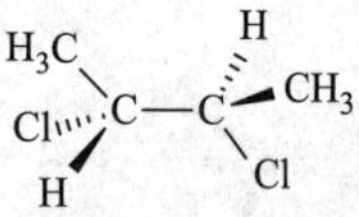

A. 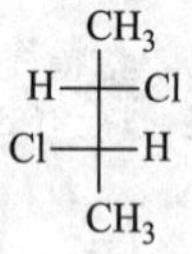
```
     CH₃
  H──┼──Cl
 Cl──┼──H
     CH₃
```

B. 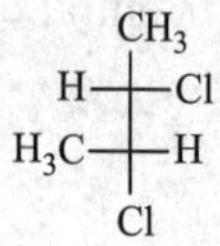
```
     CH₃
  H──┼──Cl
H₃C──┼──H
     Cl
```

C. 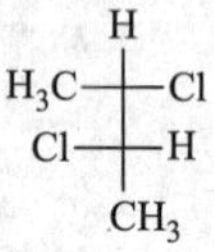
```
      H
 H₃C──┼──Cl
 Cl──┼──H
     CH₃
```

D. 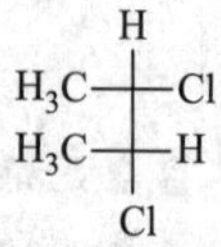
```
      H
 H₃C──┼──Cl
H₃C──┼──H
     Cl
```

10. Which of the following compounds exhibit *cis* and *trans* isomers?
下列化合物中具有顺反异构体的是（　　　）。

A. 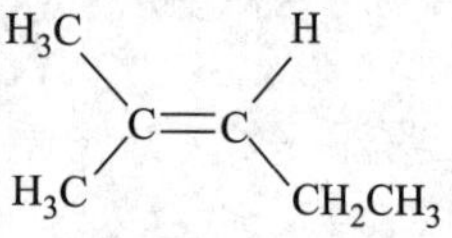 $\underset{H_3C}{\overset{H_3C}{}}C=\overset{H}{\underset{CH_2CH_3}{C}}$

B. $CH_3CH=C=CHCH_3$

C. $CH_3C\equiv CCH_3$

D. 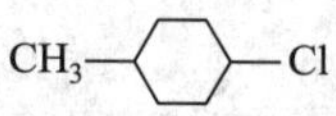 $CH_3-\!\!\bigcirc\!\!-Cl$

（张园园　杨淑珍）

第四章　烯　烃
Chapter 4　Alkenes

　学习目标

　　1.掌握　π键的形成以及π键的特性、烯烃的命名方法和烯烃的重要反应(亲电加成反应、氧化反应)

　　2.熟悉　次序规则及Z/E命名法、烯烃的亲电加成反应、马氏规则和过氧化物效应。

　　3.了解　烯烃的制备方法及其在医药领域的应用。

An unsaturated hydrocarbon contains at least one carbon-carbon double or triple bond, indicating that fewer hydrogen atoms are bonded to carbon than comparable alkanes with the same number of carbon atoms. Among the three most important classes of unsaturated hydrocarbons (*i.e.* alkenes, alkynes, and arenes), alkenes contain one or more carbon-carbon double bonds ($R_2C=CR_2$) with the general formula of C_nH_{2n}. Alkenes are sometimes called **olefins (烯烃)**, which are derived from *olefiant gas*. The simplest alkene is **ethene** (traditionally called **ethylene**), C_2H_4 or $CH_2=CH_2$. In this chapter, the structure, nomenclature, physical properties, reactions of alkenes as well as the typical application of alkenes in pharmacy area will be presented.

4.1　Structure of Alkenes

PPT

In the previous chapter, we can visualize the formation of both σ and π bonds in terms of the overlap of atomic orbitals. A carbon-carbon double bond consists of one σ bond and one π bond. Taking ethene for example (Figure 4.1), it has two sp^2-hybridized carbon atoms, which are σ-bonded to each other and to two hydrogen atoms each. The remaining unhybridized p orbitals on the two carbons form a π bond, resulting in its reactivity.

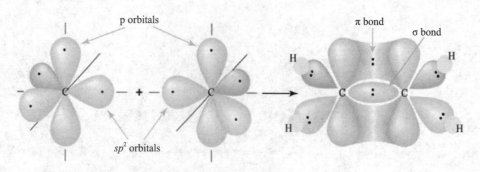

Figure 4.1   The formation process of ethene
图4-1   乙烯的形成过程

The sp^2 hybridization corresponds to the ideal **bond angles (键角)** of 120° due to the optimum separation of three atoms bonded to the carbon atom. For example, the observed H-C-C bond angle in ethene and C-C-C bond angle in propene is 121.7°and 124.8° respectively, which are typically within a few degrees of 120°. It is reasonable since the strain of electron clouds between the π bond and the bulky group.

One of the important structural characteristics is the potential *cis-trans* **isomerism (顺反异构)** only if each of the two sp^2-hybridized carbon atoms containing two different atoms or groups. Unlike single bonds with free rotation around carbon-carbon single bonds at room temperature, the two ends of carbon-carbon double bond cannot be twisted with each other without disrupting the π bond (Figure 4.2a), leading to remain in the same plane of the six atoms including the double-bonded carbon atoms and the four atoms bonded to them. Consequently, these two isomers are always different from each other, particularly in physical properties (*e.g.* melting point, boiling point). As shown in Figure 4.2b, the alkenes with the same groups located on the same side of a double bond are the *cis* isomer. Simultaneously, the molecules with same groups located on the opposite side of a double bond belong to the *trans* isomer. *Cis-trans* isomerism is not possible if one of sp^2-hybridized carbon atoms is attached to two identical groups.

(a)

cis-but-2-ene   rotation breaks π bond   *trans*-but-2-ene

(b)

cis (顺)   *trans* (反)

Figure 4.2   (a) Distinct *cis-trans* isomers of 2-but-2-ene; (b) *cis-trans* isomers of alkenes
图4-2   (a) 丁-2-烯的顺反异构; (b) 烯烃的顺反异构

The *cis-trans* system is not detailed enough to describe isomeric tri-substituted and tetra-substituted alkenes. For example, the following two compounds can not be designated as *cis* or *trans* isomers.

To deal with this problem, we can distinguish any double bond capable of geometric isomerism by the *E-Z* system. This system uses **sequence rules (顺序规则)** to assign relative priorities to each of two groups, which are bonded to sp^2-hybridized carbon atom of an alkene. Thus, we ought to separate the double bond into its two ends, followed by assigning the relative priorities to the two substituent groups on that end. The sequence rules are the same as the priority rules of the *R-S* system. If the higher priority groups locate at the same side of the double bond, the configuration of the alkene is *Z*. Similarly, if the higher priority groups locate at the opposite side of the double bond, the molecule is the *E* isomer. Please see the following examples.

4.2　How to Calculate the Index of Hydrogen Deficiency

PPT

If all the carbon atoms of a hydrocarbon have only single bonds and no rings are involved, the molecule would have the maximum number of hydrogen atoms. The presence of π bonds or the ring of a cyclic compound decreases the number of hydrogen atoms as compared to a corresponding **acyclic (非环状的)** alkane. In other words, the molecular **formula (分子式)** has two fewer hydrogens for each ring or π bond. We can perform to obtain this important information by calculating the **Index of Hydrogen Deficiency (IHD,缺氢指数)**. IHD, commonly called **Degrees of Unsaturation (不饱和度)**, is the sum of the number of rings and π bonds in a molecule. If molecular formula of a hydrocarbon is C_xH_y, the IHD can be deduced using the following calculation, providing the way to consider possible structures quickly.

$$IHD = \frac{2x+2-y}{2}$$

For example, the IHD of the C_4H_6 is 2 ($IHD = \dfrac{2 \times 4 + 2 - 6}{2} = 2$). It means that the molecule has either one double bond and a ring, or two double bonds, or two rings, or one triple bond.

C_4H_6 :

one double   two double bonds   two rings   one triple bond
bond and a ring

How about the compounds containing **heteroatoms (杂原子)**, which include any atoms other than carbon and hydrogen?

(1) **Halogens (卤素)**   A halogen atom (F, Cl, Br, I) is monovalent, which can be regarded as a replacement for hydrogen. Treating halogens as hydrogen atoms would not change the IHD. For example, the C_3H_7Br and C_3H_8 have the same IHD. In calculating the IHD, we can simply count halogens as hydrogen atoms.

(2) Divalent atoms of Group 6 elements (O, S, Se)   Taking oxygen atom for example, the number of hydrogens does not change if we rationally remove the oxygen atom from a structure. Ethanol and ethane have the same IHD. Thus, we can simply ignore the oxygen atoms in calculating the IHD.

(3) Trivalent Group 5 elements (N, P, As)   Taking nitrogen atom for example, nitrogen-containing compounds have one more hydrogen atom for per nitrogen atom than a hydrocarbon with an equal number of carbon atoms. In calculating the IHD, we can simply ignore the nitrogen atoms and must subtract the number of hydrogen atoms with an equal number of nitrogen atoms at the same time.

In a word, we can calculate the IHD of $C_xH_yO_zN_mCl_n$ using the following formula.

$$IHD = \frac{2x + 2 - y + m - n}{2}$$

4.3　Nomenclature of Alkenes

PPT

1. IUPAC Names of Alkenes　In terms of the nomenclature of alkenes, some groups derived from alkenes have their common names. The most often encountered three groups are the **vinyl (乙烯基)**, **allyl (烯丙基)**, and **isopropenyl (异丙烯基)** groups.

$$H_2C=CH-\qquad H_2C=CH-CH_2-\qquad \underset{\text{isopropenyl}}{\overset{\overset{\displaystyle CH_3}{|}}{H_2C=C-}}$$

vinyl　　　　　allyl

Naming alkenes are very similar to alkanes based on the IUPAC rules. The only difference is the correct position of the double bond. Please keep in mind *that* geometric isomers are possible when double bonds and/or rings are involved.

(1) Step 1: selecting the parent alkene　The parent alkene is the longest continuous chain of carbon atoms that contains as many double bonds as possible.

$$\boxed{H_3C-CH_2-CH_2-\overset{\overset{\displaystyle CH_2}{||}}{C}-CH_3}\qquad \times\quad \text{NOT containing the double bond}$$

$$H_3C-CH_2-CH_2-\overset{\overset{\displaystyle CH_2}{||}}{C}-CH_3 \qquad \sqrt{}\quad \begin{array}{l}\text{Containing most number of atoms and}\\ \text{double bond}\end{array}$$

(2) Step 2: Numbering and naming the parent alkene　Please number the carbon atoms in the selected parent alkene starting from the end of the chain that is nearer to the double bond. Meanwhile, the parent chain has the same name as the corresponding alkane only replacing the suffix *-ene* of the suffix *-ane*. The position of the double bond should be identified. Sometimes it is not given, indicating that the double bond is located between C-1 and C-2.

$$\overset{5}{CH_3}-\overset{4}{\underset{\underset{\displaystyle CH_3}{|}}{C}}\overset{}{\underset{3}{=}}\overset{}{\underset{}{CH_2}}-\overset{2}{\underset{}{CH}}-\overset{1}{CH_3} \qquad \times\quad \begin{array}{l}\text{NOT starting from the end of the chain}\\ \text{that is nearer to the double bond}\end{array}$$

$$\overset{1}{CH_3}-\overset{2}{\underset{\underset{\displaystyle CH_3}{|}}{C}}\overset{}{\underset{3}{=}}\overset{}{\underset{}{CH_2}}-\overset{4}{\underset{}{CH}}-\overset{5}{CH_3} \qquad \sqrt{}\quad \begin{array}{l}\text{Sarting from the end of the chain that is}\\ \text{nearer to the double bond}\end{array}$$

(3) Step 3: Naming the substituents　Please name all branched or substituted groups and their positions on the chain should be identified, which is in an identical pattern to that with alkanes.

$$\underset{2}{\overset{^1CH_2}{\underset{|}{C}}}H_3\text{-C=CH}_2\text{-CH-CH}_3 \quad \Longrightarrow \quad \text{2,4-dimethylpentene}$$

$$\underset{6}{H_3C}\text{-}\underset{5}{\overset{CH_3}{\underset{|}{C}}}\text{-}\underset{4}{CH}=\underset{3}{\overset{H}{C}}\text{-}\underset{2}{\overset{CH_3}{\underset{|}{C}}}H\text{-}\underset{1}{CH_3} \quad \times \quad \Longrightarrow \quad \text{2,5,5-trimethylhex-3-ene}$$

$$\underset{1}{H_3C}\text{-}\underset{2}{\overset{CH_3}{\underset{|}{C}}}\text{-}\underset{3}{CH}=\underset{4}{\overset{H}{C}}\text{-}\underset{5}{\overset{CH_3}{\underset{|}{C}}}H\text{-}\underset{6}{CH_3} \quad \checkmark \quad \Longrightarrow \quad \text{2,2,5-trimethylhex-3-ene}$$

Moreover, the following important items should be highlighted:

1) If a molecule contains more than one double bond, each double bond should be specified the number of its location. An appropriate prefix (*e.g. di, tri, tetra*...) to *-ene* implies the number of double bonds.

$$H_2C=CH-CH_2-\overset{CH_3}{\underset{H}{C}}-CH=CH_2 \qquad \text{3-methylhexa-1,5-diene}$$

$$H_2C=CH-CH=\overset{CH_3}{\underset{|}{C}}-CH=CH-CH_3 \quad \text{4-methylhepta-1,3,5-triene}$$

2) If a double bond can be assigned as an *E* or *Z* isomer, an appropriate prefix followed by a hyphen is placed within parentheses in front of its name. Morever, the locating number of the *E* or *Z* isomer is not necessary to label if there is only one double bond. However, if a molecule contains more than one double bond accompanied with *E* or *Z* isomer, we ought to specify each of the locating number for each *E* or *Z* isomer. Do not forget to assign the R/S configuration if it exists.

(2*E*)-3,5-dimethylhex-2-ene
or (*E*)-3,5-dimethylhex-2-ene

(2*E*, 5*E*)-3,4-dimethylhepta-2,5-diene

(2*E*, 4*S*, 5*E*)-3,4-dimethylhepta-2,5-diene

3) Naming a cycloalkene, we have to give the double-bonded carbon atoms of the numbers 1 and 2 so that the functional group can have the minimum number. Pease bear in mind that the first substituent on the ring need to receive the lower number.

3,4-dimethylcyclopentene

(×)   (√)

1,6-dimethylcyclohexene

(×)   (√)

2. Common Names   Some alkenes (particularly the molecules with low molecular weight) are known exclusively by their common names. For example:

$$H_2C=CH_2 \quad CH_2=CH-CH_3 \quad CH_2=\underset{\underset{CH_3}{|}}{C}-CH_3$$

| IUPAC name: | Ethene | Propene | 2-Methylpropene | Ethenylbenzene |
| Common name: | Ethylene | Propylene | Isobutylene | Styrene or vinylbenzene |

4.4　Physical Properties of Alkenes

PPT

The physical properties of the homologous series of alkenes (C_nH_{2n}) are quite similar to those of the homologous series of alkanes (C_nH_{2n+2}).

Alkenes are colorless and odorless in nature, but ethene having a faintly sweet odor is the exception that proves the rule. The first three alkenes exist as colorless gases at room temperature, and the next fourteen are liquids. Members of the 15 carbons or more are all solids.

Alkenes are lighter than water with the densities ranging from 0.6~0.8 g · cm^{-3}. Alkenes are either nonpolar or very slightly polar. Thus, they are virtually insoluble in water but soluble in most of the organic solvents.

Generally, the boiling points of alkenes increase with the increasing number of carbon atoms (Table 4.1). Similar to alkanes, straight-chain alkenes have higher boiling points than branched-alkyl alkenes.

Table 4.1　Physical properties of some alkenes
表 4-1　烯烃的物理性质

Name	Structural formula	Melting point /℃	Boiling point /℃
Ethylene	$H_2C = CH_2$	−169	−104
Propylene	$CH_2 = CH-CH_3$	−185	−47
But-1-ene	$CH_2 = CH-CH_2-CH_3$	−138	−6.4
trans-But-2-ene	$\underset{H_3C}{\overset{H}{\diagdown}}C=C\underset{H}{\overset{CH_3}{\diagup}}$	−105.8	0.9
cis-But-2-ene	$\underset{H_3C}{\overset{H}{\diagdown}}C=C\underset{CH_3}{\overset{H}{\diagup}}$	−138.9	3.7
Pent-1-ene	$CH_2 = CH-CH_2-CH_2-CH_3$	−165	30
Hex-1-ene	$CH_2 = CH-CH_2-CH_2-CH_2-CH_3$	−139.8	63.5

Specifically, geometric isomers always have different boiling points owing to their different polarities. For example, the boiling points of *cis*-but-2-ene (3.7℃) is a little higher than *trans*-but-2-ene (0.9℃). It is understandable since the *cis* isomer is polar and has a **net** dipole moment while the *trans* isomer is nonpolar and has no net dipole moment. In contrast, *cis*-but-2-ene (−139.8℃), packed in a

U-bending shape, has a lower melting point as compared to *trans*-but-2-ene (−105.8℃), whose structure is centrosymmetric.

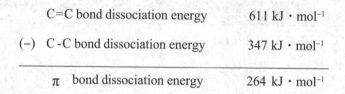

$\mu=0$ $\mu\neq0$

4.5 Reactions of Alkenes

PPT

4.5.1 Reactions of Alkenes–An Overview

The bonding energies of carbon-carbon single bond (σ bond) and carbon-carbon double bond are about $611\ kJ \cdot mol^{-1}$, $347\ kJ \cdot mol^{-1}$ respectively, from which the approximate bonding energy of π bond is estimated to be $264\ kJ \cdot mol^{-1}$. This value is much less than the σ bond energy ($347\ kJ \cdot mol^{-1}$), indicating that π bonds are more reactive than σ bonds. Consequently, carbon-carbon double bond is considered to be a functional group and alkenes are characterized by the reactions of their double bonds. By mastering the characteristic reactions arisen from the reactivity of the carbon-carbon double bond, we can have a prediction and better understanding the reactions of alkenes that we have never seen before.

	C=C bond dissociation energy	$611\ kJ \cdot mol^{-1}$
(−)	C - C bond dissociation energy	$347\ kJ \cdot mol^{-1}$
π	bond dissociation energy	$264\ kJ \cdot mol^{-1}$

The most characteristic reaction type of alkenes is addition. Two atoms or groups add to the carbon-carbon double bond, resulting in the fracture of π bond and formation of two new σ bonds. Table 4.2 summarizes several examples of addition reactions associated with the descriptive name(s). Some of reactions are treated separately in the part of *Oxidation and Reduction* (Section 4.5.5), but they are still included in this table as they belong to additions.

Addition $\underset{\pi \text{ bond broken}}{\diagup\!\!\!\diagup}$ + $\underset{\sigma \text{ bond broken}}{X\!-\!Y}$ $\longrightarrow$ $\underset{2\sigma \text{ bonds formed}}{\diagup\!\!\!\diagup}$

医药大学堂
WWW.YIYAODXT.COM

Table 4.2　Characteristic types of additions to alkenes

表 4-2　烯烃的典型加成反应

Reaction	Descriptive name(s)
>=< + H–H ⟶ ‖‖ H H	Hydrogenation (reduction)
>=< + X–X ⟶ ‖X‖X	Halogenation
>=< + H–X ⟶ ‖‖ H X	Hydrohalogenation
>=< + H–OH ⟶ ‖‖ H OH	Hydration (Oxidation)
>=< + H–OSO₃H ⟶ ‖‖ H OSO₃H	Alkyl hydrogen sulfates fromation
>=< + OsO₄ →H₂O₂→ ‖‖ HO OH	Diol formation (Oxidation)
>=< + H–BH₂ ⟶ ‖‖ H BH₂	Hydroboration
>=< + X–OH ⟶ ‖OH‖X	Halohydrin formation (Oxidation)
>=< + R–COOOH ⟶ epoxide	Epoxidation (Oxidation)

4.5.2　Electrophilic Additions

The **electrophilic addition (亲电加成)** requires two steps. Firstly, a strong **electrophile (亲电试剂, E⁺)** attracts the electrons of the π bond of an alkene, leading to the formation of a σ bond to one of the carbons of the double bond and generation of a **carbocation (碳正离子)** on the other carbon of the double bond. which is considered to be the **rate-determining step (决速步骤)**. Secondly, the carbocation acting as a strong electrophile reacts with a **nucleophile (亲核试剂, Nu⁻)** (often a weak nucleophile) to form another σ bond.

Step 1: Attack of the π bond on the electrophile forms a carbocation.

$$\text{>=<} \ + \ E^+ \ \xrightleftharpoons[]{\text{slow}} \ \text{‖\,+} \atop E$$

rate-determining step

Step 2: Attack by a nucleophile offers the addition product.

1. Electrophile Containing Acidic Hydrogen

(1) Addition of hydrogen halides (HX)   The hydrogen halides (HCl, HBr and HI) add to alkenes to give alkyl halides (haloalkanes). HCl has low reactivity compared to the other two acids.

$$H_3C-HC=CH-CH_3 \quad + \quad HBr \longrightarrow H_3C-H_2C-\underset{\underset{Br}{|}}{\overset{\overset{H}{|}}{C}}-CH_3$$

$$\underset{KI,\ H_3PO_4}{\longrightarrow}$$

(2) Addition of sulfuric acid (H_2SO_4)   **Concentrated** sulfuric acid can also add to the carbon-carbon bond of alkenes, resulting in forming **alkyl hydrogen sulfates (硫酸氢酯)**.

$$H_3C-HC=CH-CH_3 \quad + \quad H_2SO_4 \longrightarrow H_3C-H_2C-\underset{\underset{O_{\diagdown SO_3H}}{|}}{\overset{\overset{H}{|}}{C}}-CH_3$$

However, alkyl hydrogen sulfates could be easily converted to the corresponding alcohols upon heating them with water or steam, which is also called a **hydrolysis (水解)** reaction.

$$H_3C-H_2C-\underset{\underset{O_{\diagdown SO_3H}}{|}}{\overset{\overset{H}{|}}{C}}-CH_3 \quad \underset{\triangle}{\overset{H_2O}{\longrightarrow}} \quad H_3C-H_2C-\underset{\underset{OH}{|}}{\overset{\overset{H}{|}}{C}}-CH_3$$

(3) Hydration-addition of water (H_2O)   The addition of water to an alkene is carried out in the presence of a catalytic amount of dilute strong acid (*e.g.* H_2SO_4, H_3PO_4, HBF_4, TsOH), leading to form the corresponding alcohols.

$$H_3C-HC=CH-CH_3 \quad + \quad H_2O \quad \overset{H^+}{\longrightarrow} \quad H_3C-H_2C-\underset{\underset{OH}{|}}{\overset{\overset{H}{|}}{C}}-CH_3$$

$$+ \quad H_2O \quad \overset{H^+}{\longrightarrow}$$

Simultaneously, we can extend the acid-catalyzed electrophilic addition of an alkene to the reactants of alcohols and carboxylic acids, resulting in formation of the desired **ethers (醚)** and **esters (酯)**, respectively. These two types of electrophilic addition also undergo the carbocation as the reactive intermediate.

$$\begin{array}{l} \overset{H-OR,\ H^+}{\longrightarrow} \quad \text{ethers} \\ \\ \overset{O}{\overset{\|}{H-OCR}},\ H^+ \quad \text{esters} \end{array}$$

All the above-mentioned molecules involving electrophilic addition (HX, H_2SO_4, H_2O, ROH, RCOOH) are symmetrical alkene and only one possible product can be obtained. How about the unsymmetrical alkenes in which the two carbons of the double bond are not equivalently substituted?

Taking the addition of HBr to propene for example, in principle it ought to have two potential products (1-bromopropane and 2-bromopropane). Practically, only 2-bromopropane is the major product and 1-bromopropane is very limited. Thus, we say that addition of HBr to propene is a **regioselective (区域选择性的)** reaction.

$$H_3C-HC=CH_2 \ + \ H-Br \ \longrightarrow \ H_3C-\overset{\overset{H}{|}}{\underset{\underset{H}{|}}{C}}-CH_2 \ + \ H_3C-\overset{\overset{Br}{|}}{\underset{\underset{H}{|}}{C}}-CH_2$$

Propene ・ 1-Bromopropane (Not observed or very limited) ・ 2-Bromopropane (Major product)

Regioselectivity is the preference of one direction of chemical bond forming or breaking over all other possible directions. In 1870, the Russian chemist Vladimir Markovnikov firstly noticed the pattern that reagents add to unsymmetrical alkenes in a regiospecific way. Markovnikov stated:

Markovnikov's rule (马氏规则): When an unsymmetrically substituted alkene reacts with a proton acid, hydrogen adds to the carbon of the double bond that holds the greater number of hydrogen atoms.

Reactions that follow this rule are said to give the **Markovnikov product** (马氏产物) or follow **Markovnikovs orientation (马氏取向)**. Markovnikov's rule can also be extended to the electrophiles other than proton acids, provided that the electrophilic additions undergo the process of producing the most stable carbocation.

Markovnikov's rule (extended): *In an electrophilic addition to an unsymmetrically substituted alkene, the electrophile adds in such a way as to form the most stable carbocation.*

The Markovnikov's rule can be rationalized by the stability of carbocation intermediate formed in the rate-determining step. For example, the resulting carbocation intermediate I has more σ -p hyperconjugation in the addition of HBr to propene. Thus, the stability of carbocations I is more stable than II , leading to producing the major product of 2-bromopropane. In another words, the stable carbocation leads to major product.

In the electrophilic addition of an alkene, the importance of stability of a carbocation can be further demonstrated by the formation of rearrangement product. For example, in addition of HCl to 3,3-dimethyl-but-1-ene, the expected product is 2-chloro-3,3-dimethylbutane, which is obtained in a yield of 17%. In practice, the major product is 2-chloro-2,3-dimethylbutane, which has a change in connectivity of its atoms compared with that in the starting alkene. We ascribe the formation of 2-chloro-2,3-dimethylbutane to a carbocation **rearrangement (重排)**, which is extremely common in organic chemistry reactions. Carbocation rearrangement is defined as the movement of a carbocation from a relatively unstable state to a more stable state through group migration within a molecule.

Among the various types of carbocation rearrangement, the most common are 1,2-shifts that the migrating group moves to an adjacent position. Thus, the protonation of 3,3-dimethyl-1-butene gives a secondary carbocation, which is further subject to an adjacent alkyl shift to give a more stable tertiary carbocation. We call this rearrangement is 1,2-alkyl shift.

Not all carbocations have available alkyl groups located on adjacent carbon atoms for rearrangement. In this case, the reaction can undergo another mode of rearrangement known as 1,2-hyride shift. The major product of reaction of HCl with 3-methyl-1-butene is 2-chloro-2-methylbutane, which is subject to 1,2-hyride shift.

2. Addition of Halogens (X_2) to Alkenes　Normally, halogens react with alkenes by electrophilic addition, giving the products of **vicinal (邻位的)** dihalides. The commonly-used halogens are chlorine (Cl_2) and bromine (Br_2). The addition of fluorine (F_2) to alkenes is extremely fast and difficult to control, which is not a useful laboratory procedure. The addition of iodine (I_2) is not preparatively useful because the resulting vicinal diiodides are prone to lose I_2 and decompose to alkenes easily. Generally, halogenation with bromine or chlorine is performed in an inert solvent such as methylene chloride (CH_2Cl_2), chloroform ($CHCl_3$) and carbon tetrachloride (CCl_4). Meanwhile, a clear and deep red solution of bromine in carbon tetrachloride decolorizes quickly upon the addition of an alkene, which has been considered to be a simple and qualitative chemical test for efficient recognition of olefinic double bonds.

Alkene + Halogen ⟶ Vicinal dihalide

Importantly, the addition of chlorine or bromine to an alkene is a stereospecific *anti* (from the opposite side or face) addition, giving different stereoisomers of the product from starting materials with different stereoisomers. For example, addition of bromine to cyclohexene lead to *trans*-1,2-dibromocyclohexane. Two products (enantiomers) are formed from the *cis*-2-butene while only one obtained from the *trans* isomer

cyclohexene + Br₂ (CH₂Cl₂) ⟶ *trans*-1,2-dibromocyclohexane (a racemic mixture)

cis-but-2-ene + Br₂ ⟶ (±)-2,3-dibromobutane

trans-but-2-ene + Br₂ ⟶ *meso*-2,3-dibromobutane

The stereospecific anti addition of halogens to alkenes can be attributed to the generally accepted mechanism of bridged **halonium ion** (卤鎓离子) intermediates by the following two-step.

Step 1: The formation of bridged halonium ion intermediates.

Let us take the reaction of alkene with bromine for example. Both bromine and alkene are nonpolar molecules, however, they are polarizable. They are mutually attracted to each other through **induced-dipole/induced-dipole interaction** (诱导偶极间相互作用力), eventually causing the break of weak bromine-bromine bond and formation of bridged bromonium ion. The three-member cyclic bromonium intermediate, in which the positive charge locates on bromine rather than carbon, is a reasonable and acceptable Lewis structure because its atoms have octets of electrons. Consequently, the bridged bromonium ion is more stable than an alternative carbocation.

Step 2: Attack the bridged halonium ion intermediates by a halide anion in an anti coplanar orientation.

Bromide anion attacks **one of the carbons** of the bridged bromonium intermediate from the anti-coplanar side, leading to open the three-membered ring and offer the *anti* product.

3. Addition of HOX   In aqueous solution, halogens add to alkenes to form vicinal **halohydrins (卤代醇)**, compounds with a hydroxyl and a halogen group on adjacent carbon atoms.

This electrophilic addition is also *anti* stereoselective, whose mechanism involves a halonium ion. Thus, the hydroxyl group and halogen add to opposite side of the double bond. For example, the addition of bromine water to cyclohexene gives a couple of enantiomers *(trans*-2-bromocyclohexanol).

It should be highlighted that the addition of HOCl or HOBr is regioselective. That is to say, when water or hydroxyl attacks the bridged halonium ion, the electrophile (the chlorine or bromine atom) is bonded to the less substituted carbon of the double bond while the nucleophile (water or hydroxyl) is bonded to the more substituted carbon. For example, in the addition of bromine water to 1-methylcyclohexene, the two carbon atoms (C-1 and C-2) bonded to the halogen have partial positive charges. The C-1 with the more substituted has more stable transition state owing to its partial tertiary carbocation as compared to C-2 with a partial secondary carbocation. Thus, C-1 is prone to be attacked by nucleophile.

This mechanism can be further deduced to the electrophilic addition using similar reactants such as BrCl and IBr.

4.5.3   Hydroboration-Oxidation

Both the addition of water to an alkene (hydration) and addition of sulfuric acid followed by hydrolysis can convert an alkene to alcohol with Markovnikov orientation. However, hydration of an alkene with *anti*-Markovnikov orientation can be achieved through hydroboration-oxidation in alkaline aqueous solution, which is a very valuable laboratory method for the *stereoselective* and *regioselective* hydration of alkenes without rearrangement.

$$H_3C-HC=CH_2 \xrightarrow[H_2O_2, \; NaOH]{B_2H_6} H_3C-H_2C-CH_2-OH$$

This class of reactions proceeds in two steps.

Step 1: Hydroboration of the alkene

Hydroboration (硼氢化反应) is the addition of borane (BH_3) to an alkene to form a trialkylborane after three successive additions, which proceeds as a **concerted process (协同过程)** because bond formation and bond breaking occur at the same time. This simultaneous addition of boron and hydrogen to the double bond also accounts for the *syn* (from the same side) addition.

Borane is commonly used as a stable Lewis acid-base complex with tetrahydrofuran (THF), which is commercially available and easily measured and transferred.

$$2 \quad \text{[cyclopentane O]} \quad + \quad B_2H_6 \quad \longrightarrow \quad 2 \quad \text{[cyclopentane O}^+\text{-}\bar{B}\text{H}_3\text{]}$$

Importantly, as far as the unsymmetrical alkene is concerned, hydroboration of alkenes is **regioselective** as indicated by boron atom adding predominantly to the ***less hindered and less substituted*** carbon of the double bond. For example, in the hydroboration of propene, the C-1 of the double bond acquires a partial negative charge due to the electronic effects, which is more prone to be attacked by the B atom with a partial positive charge.

The importance of steric effects in controlling the regioselectivity of hydroboration can be further demonstrated by the following two examples.

Step 2: Oxidation of a trialkylborane to alcohol using alkaline hydrogen peroxide

Hydroperoxide anion acts as a nucleophile, attacking boron and causing the alkyl group migrate from boron to oxygen (1,2-shift) for three successive steps to obtain **trialkyl borate ester (三烷基硼酸酯)**. The alkyl group migrates with retention of its configuration. Finally, hydrolysis of the trialkyl borate ester gives the desired alcohol. Superficially, hydration of an alkene by hydroboration-oxidation does not follow the Markovnikov's rule. Actually, it still follows the essential reasoning behind Markovnikov's rule.

$$\text{H-O-O-H} \quad + \quad \text{OH}^- \quad \longrightarrow \quad {}^-\text{O-O-H} \quad + \quad \text{H}_2\text{O}$$

4.5.4　Free-Radical Addition of HBr-Peroxide Effect

As discussed before, HBr adds to unsymmetrical alkene *via* an ionic mechanism (with a carbocation intermediate) to give the product of Markovnikov's rule. However, when peroxides (ROOR) were present in the reaction mixture, HBr adds to unsymmetrical alkene in an *anti*-Markovnikov sense. This phenomenon of regioselectivity for addition of HBr is called **peroxide effect (过氧化物效应)**.

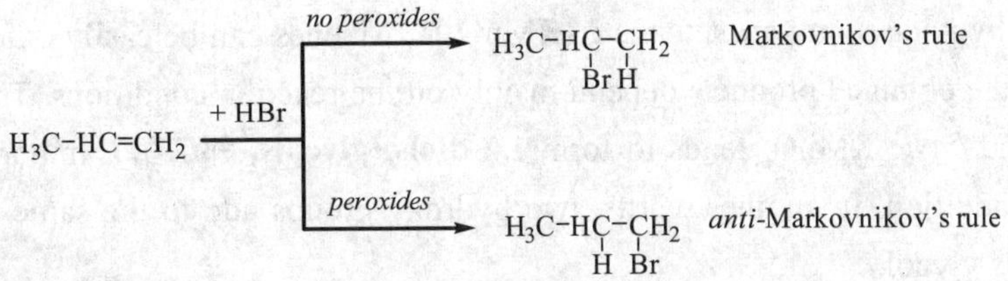

Importantly, the regioselectivity for other electrophiles (*e.g.* HCl, HI, H$_2$O) adding to unsymmetrical alkene is not altered even in the presence of peroxides. Peroxide effect is only specific to HBr, which undergoes the free-radical addition process. Peroxides are not incorporated into the product but act as initiators.

(a) Initiation

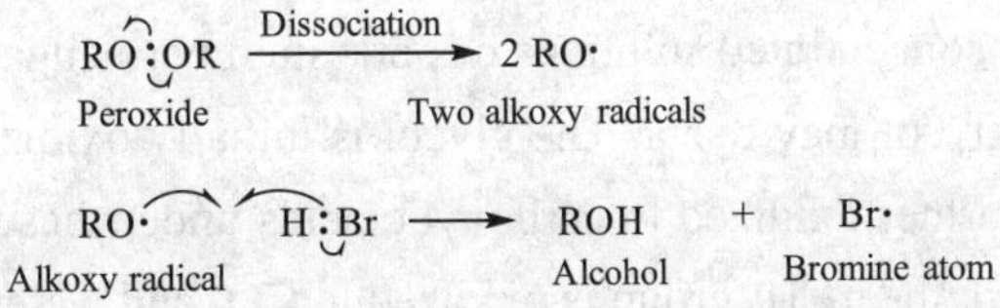

(b) Chain propagation

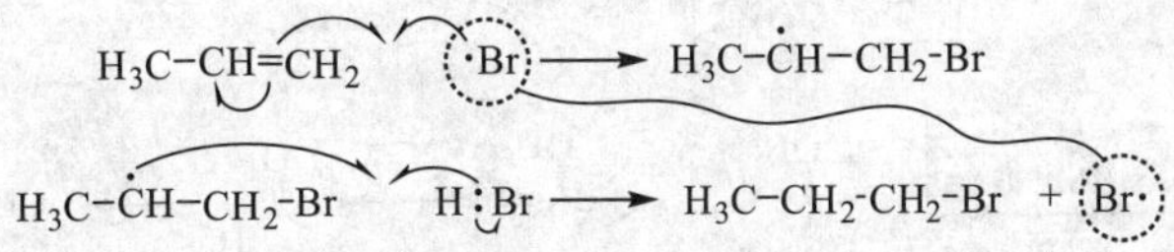

This free-radical addition mechanism can account for the peroxide effect with an *anti*-Markovnikov's product very well. Taking the reactant of propene for example, in the propagation steps, addition of a bromine atom to C-1 leads to a secondary alkyl radical, which is more stable than a primary radical derived from addition of a bromine atom to C-2. Thus, the regioselectivity of *anti*-Markovnikov addition is observed.

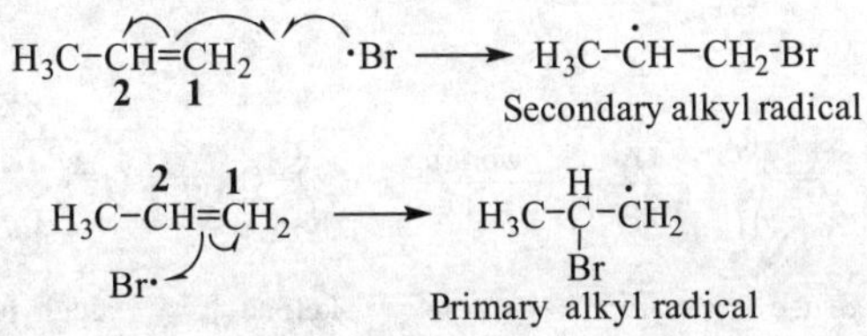

Furthermore, free-radical addition of HBr to unsymmetrical alkene can also be initiated photo chemically.

$$\text{cyclohexylidene=CH}_2 \;+\; \text{HBr} \xrightarrow{h\gamma} \text{cyclohexyl-CH}_2\text{Br}$$

4.5.5 Oxidation and Reduction

1. Oxidation of Alkene

(1) Oxidation by potassium permanganate (KMnO$_4$)   Alkenes can be easily oxidized by potassium permanganate and the obtained products depend mainly on the reaction conditions. Treatment of alkenes with *cold*, *dilute* and *basic* KMnO$_4$ tends to form **1,2-diols (glycols, 邻二醇)**. Importantly, the reaction proceeds with *syn* addition. In another words, two hydroxy groups add to the same side of the alkene, leading to form a *cis*-glycol.

$$\text{cyclohexene} \xrightarrow[\text{NaOH,cold}]{\textit{dil.}\text{KMnO}_4} \text{cyclohexane-1,2-diol (OH, OH)}$$

$$\text{butene} \xrightarrow[\text{NaOH,cold}]{\textit{dil.}\text{KMnO}_4} \quad \text{(OH, OH)} \;+\; \text{(OH, OH)}$$

When warm or acidic or concentrated solutions of potassium permanganate are employed, **oxidative cleavage (氧化断键)** of the glycol may occur. The glycol is initially oxidized to ketones and aldehydes. However, aldehydes can be further oxidized to carboxylic acids under these strong oxidizing conditions. If terminal olefins are involved, the =CH$_2$ group is oxidized to CO$_2$ and water eventually.

$$\text{cyclohexene} \xrightarrow[\text{H}^+,\text{ warm}]{\textit{concd.} \text{KMnO}_4} \left[\text{(OH, OH) } \textit{cleavage} \right] \longrightarrow \text{(CHO, CHO)} \longrightarrow \text{HOOC} \sim \text{COOH}$$

$$\text{butene} \xrightarrow[\text{H}^+,\text{ warm}]{\textit{concd.} \text{KMnO}_4} \left[\text{(OH, OH)} \; \textit{cleavage} \; + \; \text{(OH, OH)} \; \textit{cleavage} \right] \longrightarrow \text{(CHO)} \longrightarrow \text{CH}_3\text{COOH}$$

$$\text{terpene} \xrightarrow[\text{H}^+,\text{ warm}]{\textit{concd.} \text{KMnO}_4} \text{(COOH, COOH ketone)} \;+\; \text{CO}_2 \;+\; \text{H}_2\text{O}$$

(2) Oxidation by ozone (O$_3$)- ozonolysis   Treating an alkene with ozone (O$_3$) followed by a suitable **work-up (后处理)** can cleave the carbon-carbon double bond to give ketones and (or) aldehydes in its place.

$$\begin{array}{c} \text{R}_1 \quad \text{R}_3 \\ \diagup\!\!=\!\!\diagdown \\ \text{R}_2 \quad \text{H} \end{array} \xrightarrow{\text{O}_3} \xrightarrow{\text{work-up}} \begin{array}{c} \text{R}_1 \\ \diagdown \\ \text{R}_2 \end{array}\!\!=\!\!O \;+\; O\!\!=\!\!\begin{array}{c} \text{R}_3 \\ \diagup \\ \text{H} \end{array}$$

cleavage occurs here　　　　　ketone　　　aldehyde

The two-stage reaction sequence of cleaving double bonds by ozone followed by mild reducing is called **ozonolysis (臭氧化分解)**, which has been used for synthetic applications and determining the positions of double bonds in alkenes

(3) Epoxidation of alkenes by peroxy acid (RCOOOH)　Three-membered rings containing oxygen atom are called **epoxides** (also called **oxiranes, 环氧乙烷**). Epoxides are very easy to obtain *via* the reaction of an alkene with a **peroxy acid (过氧酸)**, a carboxylic acid that has an extra oxygen atom in a -O-O- (peroxy) linkage.

Epoxidation of alkenes with peroxy acids is a *syn* addition in concerted (one-step) electrophilic process that bonds are broken and formed at the same time. The epoxide retains whatever stereochemistry is present in the alkene. The *cis* substituents in an alkene are *cis* in epoxide and *vice versa*. The commonly-used peroxy acids are *m*-chloroperoxybenzoic acid and peroxyacetic acid (CH_3CO_2OH). The advantage of m-CPBA is its distinct solubility properties that the resulting formed acid precipitates out of solution. Meanwhile, a nonaqueous solvent such as chloroform, acetone or dioxane is necessary because water can attack the protonated epoxide in an acid aqueous medium, resulting in opening the three-membered ring and forming a 1,2-diol.

2. Reduction of Alkene-catalytic Hydrogenation　Alkenes can be reduced quantitatively or nearly so by molecular hydrogen (H₂) in the presence of transition metal as a catalyst to give alkanes, which is also called **catalytic hydrogenation (催化氢化)**. It can be observed that two C-H σ bonds of an alkane are formed at the expense of fracture with the π component of the alkene's double bond and the H-H σ bond of the H₂. Consequently, hydrogenation of a double bond is a thermodynamically favorable reaction due to a lower energy of product as compared to the energy of the reactant. Thus, hydrogenation is **exothermic (放热的)**. However, uncatalyzed addition of hydrogen to an alkene is very slow, meaning that any uncatalyzed mechanism must have a extremely high **activation energy (活化能)**. Among the various catalysts, platinum is the most often used.

$$\diagdown\!\!=\!\!\diagup \;+\; H_2 \;\xrightarrow[\text{(Pd, Pt, Ni)}]{\text{catalyst}}\; \underset{H \quad H}{\big| \; \big|}$$

In most case, hydrogenation takes place at the surface of the insoluble metal. Thus, the catalytic hydrogenation involves two phases (the solution and the metal), which is consider to be a **heterogeneous reaction (非均相反应)**. In the presence of a metal catalyst, both the hydrogen and the alkene are adsorbed on the metal surface. A hydrogen atom is then transferred to the alkene, forming a new C-H bond followed by transfering the second hydrogen to form another C-H bond. As a result, catalytic hydrogenation is a stereoselective reaction with the vast majority proceeding by *syn* addition of hydrogens to the carbon-carbon double bond. It should be noted that the alkene is adsorbed on the metal surface from the side with relatively small **steric hindrance (空间位阻)**.

Additionally, the heat evolved on hydrogenation of one mole of an alkene (*exothermic process*) is called **heat of hydrogenation (氢化热)**, which is a positive quantity equal to the value of $-\Delta H^{\ominus}$ for the reaction. *Heats of hydrogenation are used to estimate the relative stabilities of different alkenes*. As is known to all that lower energy molecules are more stable than higher energy molecules. More substituted alkenes ought to have a lower heat of hydrogenation, meaning that they are more stable than less substituted ones due to the σ-π hyperconjugation.

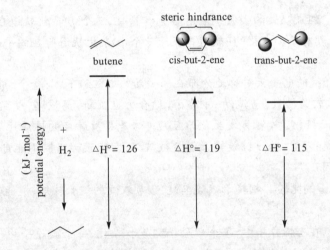

For example, catalytic hydrogenation of 1-butene, cis-but-2-ene or trans-but-2-ene results in the same product of butane. Both the *trans*-2-butene and *cis*-2-butene have lower heats of hydrogenation as compared to 1-butene, indicating that 2-butene is more stable than 1-butene. In terms of the two isomer of 2-butene, *trans* isomer of 2-butene is more stable (evolves less heat) than the corresponding *cis* isomer, which can be attributed to the *cis* isomer's steric hindrance from two alkyl groups that is located at the same side of the double bond.

4.6 Application of Alkenes in the Pharmacy Study

PPT

Alkenes are extremely important in pharmacy area. Plenty of medicines are closely related to the alkenes that act as irreplaceable building blocks in construction of specific medicines. In naturally occurring alkenes, **terpene (萜类)** compounds, whose carbon skeleton can be divided into two or more units that are identical with the carbon skeleton of isoprene, have widely demonstrated to have numerous biological activities. Typically, carbon 1 and carbon 4 of an isoprene unit are called the head and the tail, respectively. Thus, terpenes consist of bonding the tail of one isoprene unit to the head of another, which is called **isoprene rule (异戊二烯法则)**. Vitamin A is important for growth and development, for good vision, and for the maintenance of the immune system. Among them, retinol is one of the major forms of Vitamin A.

Vitamin A (retinol)

重点小结

一、烯烃的定义及通式

分子中含有碳碳双键的烃,称为烯烃。碳碳双键是烯烃的官能团。烯烃的通式为C_nH_{2n}。

二、烯烃的结构

1. **π键的形成**   组成双键的碳原子采用的是sp^2杂化,三个sp^2杂化轨道在同一个平面,轨道间对称轴的夹角为120°,没有参加杂化的P轨道垂直于三个sp^2杂化轨道所在的平面,两个p轨道侧面重叠,形成π键。

2. **π键的特点**   π键的电子云对称地分布在间平面的上方和下方;一个π键由于分成了两部分,相对于σ键来说,结构较为松散,易流动,重叠小,结合力也较弱,离核远,受到核的引力也小,容易受到其他基团的影响而极化,较容易发生化学反应(这是导致烯烃的性质比烷烃更活泼的主要原因);双键中有一个σ键和π键,若σ键旋转,π键会被破坏,所有碳碳双键不能自由旋转(这是导致烯烃存在顺反异构的原因)。

3. **碳碳双键与单键的比较**   相较于碳碳单键,碳碳双键的键长较短,键能较大;不能单独存在,只能与σ键共存。

4. **乙烯的结构**   在乙烯分子中,每个碳原子都是sp^2杂化,每个碳原子的一个sp^2杂化轨道与另一个碳原子的sp^2杂化轨道沿对称轴方向重叠形成σ键,另外两个sp^2杂化轨道分别与氢原子的s轨道沿对称轴方向重叠形成碳氢σ键,两个未参加杂化的p轨道侧面重叠,形成π键;乙烯分子中的两个碳原子之间,一个是碳碳σ键,一个碳碳π键;一个碳碳σ键和四个碳氢σ键共处同一个平面,乙烯是一个平面型的分子。

5. **烯烃的同分异构**   烯烃的同分异构现象比烷烃更复杂,异构体数目也较烷烃多,其异构体种类主要有碳架异构、官能团位置异构和顺反异构;碳架异构和官能团异构属于构造异构,顺反异构属于立体异构。 产生顺反异构体的条件是双键两端的同一个碳原子上不能连有相同的原子或基团。

三、烯烃的命名

1. **系统命名**   烯烃的系统命名法的命名步骤与烷烃的命名步骤相同,但由于烯烃有双键官能团,所以在命名规则上稍有不同。 ①等长碳链,选择含双键最长的碳链为主链。 ②离双键最近的一端开始编号。 ③将双键的位号(只写位号较小的一个)在母体名称的后缀之前。 ④其他与烷烃的命名相同。

2. 顺反异构体的命名　若双键两端连有相同的基团,则以此基团为准,相同基团在双键同侧为顺式,异侧为反式,并分别在其系统名称前加"顺-"("cis-")或"反-"("trans-")。双键两端若没有相同基团,则用Z/E法来命名。在次序规则中,较优的基团在双键的同侧为"Z",反之为"E"。

次序规则:①原子序数大的优先,同位素,质量数大的优先;②重建看作是相同的原子;③依次比较。

四、烯烃的化学性质

(一)加成反应

烯烃双键中的π键重叠小,结合力也较弱,离核远,受到核的引力也小,容易断开,在双键的两个碳原子上各加入一个原子或基团,形成两个新的σ键,即发生加成反应。

1. 催化加氢　在催化剂存在下,烯烃和氢发生加成反应生成烷烃,这一过程称为催化加氢。催化加氢常用的催化剂是Pt、Pd、Ni等。

2. 亲电加成反应　凡是具有亲电性的正离子或者缺电子化合物都叫亲电试剂,由亲电试剂进攻引起的加成反应叫亲电加成反应。

烯烃中双键的π键电子云突出在外,结构松散,易受到亲电试剂的进攻而发生加成反应;烯烃最典型的反应就是亲电加成反应。

(1)与卤化氢的加成　卤化氢的活性顺序为HI>HBr>HCl>HF;反应的取向遵循马氏规则:卤化氢中的氢总是优先加到含氢较多的双键碳原子上,经历碳正离子中间体历程;但在过氧化物存在下与溴化氢加成得到反马氏规则的加成产物,经历自由基历程。反应取向的本质是优先生成更稳定的中间体。

(2)与卤素的加成　烯烃与卤素加成生成邻二卤代烷,卤素的活性顺序为:$F_2>Cl_2>Br_2>I_2$。氟与烯烃反应十分剧烈,同时伴有多种副反应;烯烃与碘一般不反应。反应分两步进行,经历卤鎓离子中间体历程,反应具有立体选择性。

(3)与次卤酸的加成　反应的取向遵循马氏规则,反应经过卤鎓离子中间体历程,产物以反式加成为主。

3. 硼氢化-氧化反应　硼烷对π键的加成反应,称为硼氢化反应,是另一个可以由烯烃制备醇的反应,从反应结果看,是反马氏加成,但实质上是遵循马氏规则。

(二)氧化反应

π键键能较小($127kJ\cdot mol^{-1}$),易发生氧化反应,不同的氧化剂、催化剂和条件,可得到不同的产物。常见的氧化反应有高锰酸钾氧化和臭氧化还原反应。烯烃的氧化反应,可用于鉴别烯烃、推测原烯烃的结构、利用不同结构烯烃的氧化,制备有机酸、醛或酮。

1. 高锰酸钾氧化　在稀、冷的高锰酸钾碱性或中性溶液中,烯烃被氧化为邻二醇;加热、酸性条件下,碳碳键断裂,生成酮或羧酸。

2. 臭氧化还原反应　烯烃经臭氧化后,在水解时加入适量锌粉还原,得到醛或酮。烯烃的氧化反应,可用于鉴别烯烃、推测原烯烃的结构、利用不同结构烯烃的氧化,制备有机酸、醛或酮。

题库

Problems
目 标 检 测

1. Give a systematic (IUPAC) name for the following compounds.
请用系统命名法命名下列化合物。

$$H_3C \qquad \begin{array}{c} CH_3 \\ | \end{array} \qquad \qquad H_3C \qquad \begin{array}{c} CH_3 \\ | \end{array} \qquad \qquad \begin{array}{c} C_2H_5 \\ H \xrightarrow{} CH_3 \\ H_3C \xrightarrow{} H \end{array} \qquad \qquad \begin{array}{c} CH_3 \\ Et \\ H \xrightarrow{} Et \\ CH=CH_2 \end{array}$$

2. Give the structure of the reactant or principal organic product in each of the following reaction。
写出下列反应的反应物或主要产物。

$$\text{（cyclohexene with CH}_3\text{）} \xrightarrow[\text{②}H_2O_2, OH^-]{\text{①}B_2H_6}$$

$$\text{（decalin diene）} \xrightarrow[\text{②}Zn, H_2O]{\text{①}O_3}$$

$$\text{（decalin with CH}_3 \text{ and H）} + Br_2 \xrightarrow{CCl_4}$$

$$\text{（1-methyl-1-vinylcyclopentane）} + HBr \longrightarrow$$

$$A\ (C_8H_{12}) \xrightarrow[\text{②}Zn, H_2O]{\text{①}O_3} OHC-\text{（cyclohexane）}-CHO$$

$$\text{（4-vinylcyclohexene）} \xrightarrow{H^+, KMnO_4}$$

3. Determine the structure of α-pinene and of the reaction products A through E.
写出 α-蒎烯 和 A-E 的结构。

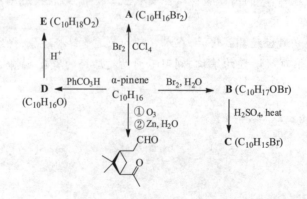

医药大学堂
WWW.YIYAODXT.COM

Discussion Topic

1. Whether the *cis-trans* isomer can be corresponding to the *Z/E* isomers respectively? Please give some examples.

2. Please give more examples of medications that are correlated to carbon-carbon double bond. You may consult any references and/or books you have.

（陈胡兰　方玉宇）

第五章 炔烃、二烯烃和周环反应
Chapter 5　Alkynes, Dienes and Pericyclic Reaction

 学习目标

1. **掌握**　炔烃和共轭二烯烃的结构、分类、命名及主要化学性质。
2. **熟悉**　炔烃和共轭二烯烃的异构，共轭效应及共轭二烯烃的稳定性。
3. **了解**　共振论，动（热）力学控制对共轭二烯烃反应产物的影响，炔烃的制法及其在医药领域的应用。

Alkynes (炔烃) are hydrocarbons that contain one **carbon–carbon triple bond (碳碳叁键). Dienes** (二烯烃) are hydrocarbons that contain two carbon–carbon double bonds. A triple bond gives an **alkyne** (炔烃) four fewer hydrogens than the corresponding alkane. Two double bonds also give a diene four fewer hydrogens than the corresponding alkane, so alkynes and dienes have the same general molecular formula C_nH_{2n-2}. Alkynes and dienes with the same molecular formula are isomers. They are common in nature and widely used in the field of traditional Chinese medicine. **Panaxynol (人参炔醇)**, isolated from ginseng, has anti-bacterial and anti-inflammatory effects. There are two carbon-carbon triple bonds in its molecule. **β-carotene (β-胡萝卜素)**, a **polyene (多烯烃),** with eleven carbon-carbon double bonds, is widely found in vegetables and fruits. It is known as the source of vitamin A and is an important active substance for human physiological function.

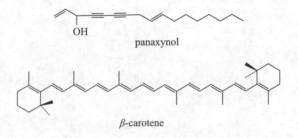

panaxynol

β-carotene

5.1　Alkynes

PPT

The functional group of an alkyne is a carbon-carbon triple bond. The simplest alkyne is ethyne, C_2H_2, more commonly named **acetylene (乙炔)**.

5.1.1　Structure of Alkynes

Acetylene is a linear molecule, and all the atoms are in a straight line and all bond angles are 180°. The carbon-carbon triple bond length in acetylene is 121 pm (1.21 Å), appreciably shorter than that of most double (1.34 Å) or single (1.54 Å) bonds (Figure 5.1). Accordingly, the carbon-carbon triple bond energy in acetylene is 837 kJ · mol^{-1}, appreciably stronger than those of most double (682 kJ · mol^{-1}) or single (368 kJ · mol^{-1}) bonds.

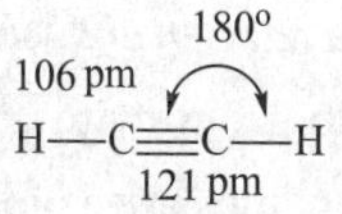

Figure 5.1　Bond lengths and bond angles in acetylene
图5-1　乙炔分子中的键长和键角

A triple bond is described in terms of the overlap of sp hybrid orbitals of adjacent carbon atoms to form an σ bond, the overlap of parallel 2p$_y$ orbitals to form one π bond, and the overlap of parallel 2p$_z$ orbitals to form a second π bond. In acetylene, each carbon-hydrogen bond is formed by the overlap of a 1s orbital of hydrogen with a sp orbital of carbon. The π electrons are cylindrical shape around carbon-carbon σ bond in alkynes (Figure 5.2). Due to more (50%) s-character in the hybrid orbitals of the triply bonded carbon atom, the acetylenic C-H bond is much stronger and more polar (535 kJ · mol^{-1}) than the C-H bonds in alkenes and alkynes (452 kJ · mol^{-1} and 410 kJ · mol^{-1}, respectively).

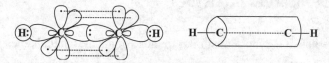

Figure 5.2　Bond formation in acetylene
图5-2　乙炔分子中键的形成过程

5.1.2　Isomerism and Nomenclature of Alkynes

The IUPAC rules for naming alkynes are similar to those for alkenes. The ending -yne is used to designate a carbon-carbon triple bond. When more than one triple bond is present, the ending should be -diyne, -triyne, and so on. Thus, HC≡CH is named ethyne, and CH$_3$C≡CH is named propyne. The IUPAC system retains the name acetylene. Therefore, there are two acceptable names for HC≡CH, ethyne and acetylene. Of these names, acetylene is used much more frequently.

There is no need to use a number to locate the position of the triple bond in ethyne and propyne,

113

because there is only one possible location for it in each compound. For larger molecules, number the longest carbon chain that contains the triple bond from the end that gives the triple bond carbons the lower numbers. Show the location of the triple bond by the number of its first carbon.

$$CH_3CH_2CH_2C\equiv CH \qquad CH_3CH_2C\equiv CCH_3 \qquad CH_3CHC\equiv CH$$
$$\qquad\qquad\qquad\qquad\qquad\qquad\qquad\qquad\qquad\qquad\qquad | $$
$$\qquad\qquad\qquad\qquad\qquad\qquad\qquad\qquad\qquad\qquad\qquad CH_3$$

pent-1-yne   pent-2-yne   3-methylbutyne

$$HC\equiv C-C\equiv CH \qquad\qquad HC\equiv C-C\equiv CCH_3$$

butadiyne   penta-1,3-diyne

Obviously, pent-1-yne and pent-2-yne differ in the position of a triple bond, while 3-methylbutyne has a branch. They are **constitutional isomers** (构造异构). Because the triple bond is linear, Alkynes don't have *cis-trans* **isomers** (顺反异构).

The group with one acetylenic hydrogen atom absence in alkyne is called an alkyne group. Their IUPAC names are as follows. These groups are usually used in common names.

$$HC\equiv C- \qquad CH_3C\equiv C- \qquad HC\equiv CCH_2-$$

ethynyl   1-propynyl   2-propynyl

If more than one multiple bond is present, number the chain from the end nearest the first multiple bond. If a double bond and a triple bond are equidistant from the end of the chain, the double bond receives the lowest numbers. To get the IUPAC name from the structural formula, put the suffix of alkene (en) before that of alkyne (yne), with the locating number of multiple bonds in front of the corresponding suffix.

$$H_3CHC=CHC\equiv CH \qquad HC\equiv CCH_2CH=CH_2 \qquad HC\equiv CC\equiv CCH=CH_2$$

pent-3-en-1-yne   pent-1-en-4-yne   hexa-1-en-3,5-diyne

5.1.3 Physical Properties of Alkynes

The physical properties of alkynes are similar to those of alkanes and alkenes with analogous carbon skeletons. The lower-molecular-weight alkynes are gases at room temperature. Those that are liquids at room temperature have densities less than $1.0\ g\cdot ml^{-1}$ (less dense than water). Listed in Table 5.1 are melting points, boiling points, and densities of several low-molecular-weight alkynes. Because alkynes, like alkanes and alkenes, are nonpolar compounds, they are insoluble in water and other polar solvents. They are soluble in each other and in other weak polar or nonpolar organic solvents as benzene and petroleum ether.

Table 5.1　Physical properties of alkynes

表5-1　炔烃的物理性质

Name	mp (°C)	bp (°C)	Density at 20°C (g · ml^{-1})
ethyne	−82	−83	(a gas)
propyne	−103	−23	(a gas)
but-1-yne	−126	9	(a gas)
but-2-yne	−32	27	0.69
pent-1-yne	−90	40	0.70
pent-2-yne	−101	55	0.71
3-methylbut-1-yne	−90	29	0.67
hex-1-yne	−132	71	0.72
hept-1-yne	−81	100	0.73
oct-1-yne	−79	126	0.75
non-1-yne	−50	151	0.76
dec-1-yne	−44	174	0.77

5.1.4　Chemical Properties of Alkynes

Many reactions of alkynes are similar to the corresponding reactions of alkenes because they both involve π bonds between two carbon atoms. Like the π bond of an alkene, the π bonds of an alkyne are electron-rich, and they readily undergo hydrogenation, electrophilic addition and oxidation reactions. Due to the difference of hybridization and electron distribution between alkyne and double bond, alkynes may also undergo nucleophilic addition. In addition, the hydrogen bonded to the triple bond carbon atom of a **terminal alkyne** (末端炔烃) shows weak acidity.

1. Acidity of Alkynes　The acidity of an **acetylenic hydrogen** (炔氢) stems from the nature of the sp hybrid ≡C—H bond. Table 5.2 shows how the acidity of a carbon-hydrogen bond varies with its hybridization, increasing with the increasing s character of the orbitals: $sp^3 < sp^2 < sp$ (Remember that a smaller value of pK_a corresponds to a stronger acid). The acetylenic proton is about 10^{19} times more acidic than a vinyl proton.

Acetylene is more acidic than ammonia and less acidic than alcohol. Protons are weakly dissociated in terminal alkyne, so acetylene cannot make **litmus paper** (石蕊试纸) turn red.

Table 5.2　Effect of hybridization on acidity

表5-2　杂化方式对酸性的影响

Compound	$CH_3\overset{H}{\underset{H}{C}}$—H	$H_2C=\overset{H}{C}$—H	H_2N—H	HC≡C—H	RO—H
Hybridization	sp^3	sp^2	-	sp	-
s Character	25%	33%	-	50%	-
pK_a	50	44	35	25	16~18

increasing acidity →

An acetylenic hydrogen is weakly acid and can be removed by a very strong base or an alkali metal. The reaction of a terminal alkyne with an alkali metal gives hydrogen gas and can be used as tests to distinguish terminal alkynes from the other alkynes.

$$RC{\equiv}C{-}H + Na^+ NH_2^- \xrightarrow{\text{liquid NH}_3} RC{\equiv}C^-Na^+ + NH_3$$

sodium amide   a sodium acetylide

$$2RC{\equiv}C{-}H + 2Na \xrightarrow{110\ ^{o}C} 2RC{\equiv}CNa + H_2\uparrow$$

$$HC{\equiv}CH + 2Na \xrightarrow{190{\sim}200\ ^{o}C} NaC{\equiv}CNa + H_2\uparrow$$

An alkyne with the *sp* hybrid ≡C—H bond can also react with **silver nitrate（AgNO₃，硝酸银）and cuprous chloride（CuCl，氯化亚铜）**. the products, the **silver acetylide** (炔化银) and **copper acetylide** (炔化亚铜) are white and red precipitates respectively. These reactions have sensitive phenomena and can be used as tests to distinguish terminal alkynes from the other alkynes.

$$RC{\equiv}CH + AgNO_3 \longrightarrow RC{\equiv}CAg\downarrow \text{white}$$

$$RC{\equiv}CH + CuCl \longrightarrow RC{\equiv}CCu\downarrow \text{red}$$

Metal acetylides are explosive when dry, so great care should be taken in their preparation. The metal acetylides can be destroyed when they are still wet by warming with dilute acid and then the parent alkynes will regenerate.

2. Addition Reaction

(1) Catalytic hydrogenation（催化加氢）

1) Catalytic hydrogenation to alkanes　In the presence of a suitable catalyst, hydrogen adds to an alkyne, reducing it to an alkane. Platinum, palladium, and nickel catalysts are commonly used in this reduction.

$$R{-}C{\equiv}C{-}R' + 2H_2 \xrightarrow{\text{Pt, Pd or Ni}} R{-}\underset{\underset{H}{|}}{\overset{\overset{H}{|}}{C}}{-}\underset{\underset{H}{|}}{\overset{\overset{H}{|}}{C}}{-}R'$$

2) Catalytic hydrogenation to *cis* alkenes　Hydrogenation of an alkyne can be stopped at the alkene stage by using a "poisoned" (partially deactivated) catalyst made by treating a good catalyst with a compound that makes the catalyst less effective. **Lindlar's catalyst** (林德拉催化剂) is a poisoned **palladium（Pd，钯）**catalyst, composed of powdered **barium sulfate（BaSO₄，硫酸钡）** coated with palladium, poisoned with **quinoline** (喹啉). The catalytic hydrogenation of alkynes is similar to the hydrogenation of alkenes, and both proceed with syn stereochemistry.

$$R{-}C{\equiv}C{-}R' + H_2 \xrightarrow[\text{quinoline}]{\text{Pd/BaSO}_4} \underset{H}{\overset{R}{>}}C{=}C\underset{H}{\overset{R'}{<}}$$

cis alkene

Nickel boride（Ni₂B，硼化镍) is a newer alternative to Lindlar's catalyst that is more easily made and often gives better yields.

$$CH_3C{\equiv}CCH_2CH_3 + H_2 \xrightarrow[\text{or Ni}_2\text{B}]{\text{Lindlar's}} \underset{H}{\overset{H_3C}{>}}C{=}C\underset{H}{\overset{CH_2CH_3}{<}}$$

3) Metal-Ammonia reduction to *trans* alkenes   To form a *trans* alkene, two hydrogens must be added to the alkyne with anti stereochemistry. Sodium metal in liquid ammonia (-33.5℃) reduces alkynes with anti stereochemistry, so this reduction is used to convert alkynes to *trans* alkenes.

$$R-C≡C-R' \xrightarrow[\text{NH}_3\ (\text{liquid})]{\text{Na}} \overset{R}{\underset{H}{>}}C≡C\overset{H}{\underset{R'}{<}}$$

$$\textit{trans} \text{ alkene}$$

(2) Electrophilic addition to alkynes   Analogous to alkenes, a characteristic reaction of alkynes is a π bond acting as a **nucleophile** (亲核试剂) to make a new bond with an electrophile. As a result, alkynes undergo many of the same electrophilic additions as alkenes. In this section, we illustrate the addition of bromine, hydrogen halides and water.

1) Addition of bromine and chlorine   Addition of 1 molecular Br_2 to an alkyne gives a dibromoalkene. This reaction is **stereoselective**. The major product corresponds to anti addition of the two bromine atoms. The addition activity of carbon-carbon triple bond is lower than that of double bond. Sometimes catalysts are used. Carrying out the bromination in acetic acid with added bromide ion (CH_3COOH, LiBr), increases the preference for anti addition significantly.

微课

$$R-C≡C-R' \xrightarrow[\text{CH}_3\text{COOH, LiBr}]{\text{Br}_2} \overset{R}{\underset{Br}{>}}C=C\overset{Br}{\underset{R'}{<}} \xrightarrow{\text{Br}_2} R-\overset{Br}{\underset{Br}{C}}-\overset{Br}{\underset{Br}{C}}-R'$$

$$\textit{trans} \text{ product}$$

Addition of a second molecular Br_2 gives a tetrabromoalkane. If 2 moles of halogen add to 1 mole of an alkyne, a tetrahalide results. After the first step addition, both of the double bond carbons are attached to electron-withdrawing bromine atoms. Sometimes it can keep the reaction from proceeding to the tetrahalide when we want it to stop at the dihalide. Alkynes similarly undergo addition of Cl_2, although less stereoselectively than with Br_2.

Addition of bromine to alkynes follows much the same type of mechanism as it does for addition to alkenes, namely, formation of a bridged bromonium ion intermediate, which is then attacked by bromide ion from the face opposite that occupied by the positively charged bromine atom.

A carbon-carbon triple bond is less active than a double bond. When both a double bond and a triple bond exist in one molecule, the halogens are first added to the double bond at low temperature.

$$H_2C=CHCH_2C≡CH + Cl_2 \longrightarrow H_2\overset{Cl}{\underset{|}{C}}-\overset{Cl}{\underset{|}{C}}HCH_2C≡CH$$

However, in the above reaction, the slight difference in activity between the double bond and the triple bond does not allow the halogen to distinguish alkynes from alkenes.

2) Addition of hydrogen halides (卤化氢)   Hydrogen halides add across the triple bond of an alkyne in much the same way they add across the alkene double bond. The initial product is a vinyl halide. When

a hydrogen halide adds to a terminal alkyne, the product has the orientation predicted by Markovnikov's rule. Because of the low activity of carbon-carbon triple bond, mercury halides are often used as catalysts. A second molecule of HX can add, usually with the same orientation as the first, the final product is a **geminal dihalide** (偕二卤代烷).

$$CH_3C{\equiv}CH \xrightarrow[HgCl_2]{HCl} CH_3C{=}CH_2 \xrightarrow{HCl} CH_3CCl_2CH_3$$
$$\underset{Cl}{|}$$

The mechanism is similar to the mechanism of hydrogen halide additions to alkene. The vinyl cation formed in the first step is more stable with the positive charge on the more highly substituted carbon atom. Attack by halide ion completes the reaction.

$$CH_3C{\equiv}CH \xrightarrow{H-Cl} CH_3\overset{+}{C}{=}CH_2 \xrightarrow{Cl^-} CH_3C{=}CH_2$$

vinyl cation   Markovnikov orientation

In the addition of the second mole of HX, the electron pair of the remaining π bond transfer to H$^+$ to form a carbocation. Of the two possible carbocations, the one with the positive charge on the carbon bearing the halogen is favored because of delocalization of the positive charge through *p-p* conjugation.

$$CH_3C{=}CH_2 \xrightarrow{H^+}$$

$$H_3C-\overset{Cl}{\underset{H}{C}}-\overset{+}{C}H_2$$

$$H_3C-\overset{+}{\underset{Cl}{C}}-CH_3 \xrightarrow{Cl^-} CH_3CCl_2CH_3$$

more stable

When a hydrogen halide adds to a non-terminal alkyne, the acetylenic carbon atoms are both substituted, and a mixture of products results.

$$CH_3CH_2C{\equiv}CCH_3 \xrightarrow{HBr} CH_3CH_2\overset{H}{C}{=}\overset{Br}{C}CH_3 + CH_3CH_2\overset{Br}{C}{=}\overset{H}{C}CH_3$$

3) Hydration of alkynes to ketones and aldehydes   **Mercuric Ion** (汞离子) **-Catalyzed Hydration** Alkynes undergo acid-catalyzed addition of water across the triple bond in the presence of mercuric ion as a catalyst. A mixture of mercuric sulfate in aqueous sulfuric acid is commonly used as the reagent. The hydration of alkynes is similar to the hydration of alkenes, and it also goes with Markovnikov orientation. The reaction gives a vinyl alcohol, called an **enol** (烯醇). Enols tend to be unstable and isomerize to the **ketone** (酮) form.

$$CH_3C{\equiv}CH + H_2O \xrightarrow[H_2SO_4]{HgSO_4} \left[CH_3\overset{OH}{C}{=}CH_2 \right] \rightleftharpoons CH_3\overset{O}{\overset{\|}{C}}CH_3$$

enol   acetone

This type of rapid equilibrium is called a **tautomerism** (互变异构). The one shown is the **keto–enol**

tautomerism (酮式-烯醇式互变异构), The keto form usually predominates.

$$-\overset{|}{C}=\overset{|}{C}- \rightleftharpoons -\overset{|}{C}-\overset{|}{C}-$$

enol   keto

According to the rules above，the hydration of acetylene gives acetaldehyde and other alkynes give ketones.

$$HC\equiv CH + H_2O \xrightarrow[H_2SO_4]{HgSO_4} \left[H_2C=\overset{OH}{\underset{|}{CH}} \right] \rightleftharpoons CH_3\overset{O}{\overset{||}{CH}}$$

acetaldehyde

(3) Hydroboration-oxidation (硼氢化–氧化)   In Chapter four we saw that hydroboration-oxidation adds water across the double bonds of alkenes with anti-Markovnikov orientation. A similar reaction takes place with alkynes. Hydroboration-oxidation of a terminal alkyne gives an aldehyde and other alkynes give ketones.

$$CH_3C\equiv CH \xrightarrow[2.\ H_2O_2,\ NaOH]{1.\ BH_3} \left[\begin{matrix} H & OH \\ H_3C & H \end{matrix} \right] \rightleftharpoons CH_3CH_2CHO$$

$$\xrightarrow[2.\ H_2O_2,\ NaOH]{1.\ BH_3} \left[\begin{matrix} H & OH \\ & \end{matrix} \right] \rightleftharpoons$$

(4) Free-radical addition (自 由 基 加 成)   In chapter four, we saw the effect of peroxides on the addition of HBr to alkenes. Peroxides catalyze a free-radical chain reaction that adds HBr across the double bond of an alkene in the anti-Markovnikov sense. A similar reaction occurs with alkynes, with HBr adding in anti-Markovnikov orientation.

$$CH_3C\equiv CH + HBr \xrightarrow{ROOR} CH_3CH=CHBr$$

(5) Nucleophilic addition   Terminal Alkynes can undergo nucleophilic addition reactions with nucleophiles such as HCN, ROH, NH_3 and RCOOH. The nucleophile uses its lone pair of electrons to form a new bond to the terminal acetylenic carbon. The π electrons are cylindrical shape around carbon-carbon σ bond in alkynes, so the nucleus of terminal acetylenic carbon is easy to exposed and attacked by nucleophiles.

$$HC\equiv CH + C_2H_5OH \xrightarrow[\substack{150\sim180\ ^oC \\ 0.1\sim0.5\ MPa}]{alkali} H_2C=CHOC_2H_5$$

ethyl vinyl ether

$$HC\equiv CH + HCN \xrightarrow[NH_4Cl]{Cu_2Cl_2} H_2C=CHCN$$

acrylonitrile

3. Oxidation of Alkynes   Permanganate Oxidations (高锰酸钾氧化)**:** Under mild conditions, potassium permanganate oxidizes alkenes to glycols, compounds with two groups on adjacent carbon atoms. If an alkyne is treated with cold, aqueous potassium permanganate under nearly neutral conditions, an α- diketone results. This oxidation involves adding two hydroxyl group to each end of the triple bond, then losing two molecules of water to give the diketone.

$$R-C \equiv C-R' \xrightarrow[\text{H}_2\text{O, neutral}]{\text{KMnO}_4} \left[\begin{array}{c} \text{OH OH} \\ R-C—C-R \\ \text{OH OH} \end{array} \right] \longrightarrow \begin{array}{c} \text{O O} \\ R-C-C-R \\ \text{diketone} \end{array}$$

For example, when pent-2-yne is treated with a cold, dilute solution of neutral permanganate, the product is pentane-2,3-dione.

$$CH_3C \equiv CCH_2CH_3 \xrightarrow[\text{H}_2\text{O, neutral}]{\text{KMnO}_4} H_3C-C-C-CH_2CH_3$$

pentane-2,3-dione

If the reaction mixture becomes warm or too basic, the diketone undergoes oxidative cleavage. The products are the **carboxylate salts** (羧酸盐) of **carboxylic acids** (羧酸), which can be converted to the free acids by adding dilute acid.

$$CH_3C \equiv CCH_2CH_3 \xrightarrow[\text{H}_2\text{O, heat}]{\text{KMnO}_4, \text{KOH}} CH_3COO^- + CH_3CH_2COO^- \xrightarrow{H^+} CH_3COOH + CH_3CH_2COOH$$

Terminal alkynes are cleaved similarly to give a carboxylate ion and **formate ion** (甲酸根). Under these oxidizing conditions, formate oxidizes further to **carbonate** (碳酸盐), which becomes CO_2 after protonation. The overall reaction is:

$$CH_3C \equiv CH \xrightarrow[\text{2. H}^+]{\text{1. KMnO}_4, \text{KOH, heat}} CH_3COOH + CO_2 \uparrow$$

Ozonolysis (臭氧化反应): Ozonolysis of an alkyne, followed by hydrolysis, cleaves the triple bond and gives two carboxylic acids. Either permanganate cleavage or ozonolysis can be used to determine the position of the triple bond in an unknown alkyne.

$$CH_3C \equiv CCH_2CH_3 \xrightarrow[\text{2. H}_2\text{O}]{\text{1. O}_3} CH_3COOH + CH_3CH_2COOH$$

4. Polymerization Acetylenes can be polymerized into chain or ring compounds selectively, but generally, it does not give polymers.

$$2HC \equiv CH \xrightarrow[\text{NH}_4\text{Cl}]{\text{Cu}_2\text{Cl}_2} H_2C=C-C \equiv CH \xrightarrow{HC \equiv CH} H_2C=C-C \equiv C-C=CH_2$$

$$3HC \equiv CH \xrightarrow{500 \text{ °C}} \bigcirc$$

$$4HC \equiv CH \xrightarrow[\text{50 °C, 1.5~2.0 MPa}]{\text{Ni(CN)}_2} \bigcirc$$

5.1.5   Synthesis of Alkynes

1. Alkylation of Acetylide Ions (炔负离子的烷基化)   An acetylide ion is a strong base and a powerful nucleophile. It can displace a halide ion from a suitable substrate, giving substituted acetylene.

$$RC \equiv C^-Na^+ + R'X \longrightarrow RC \equiv CR' + NaX$$
(R'X must be a methyl or primary alkyl halide)

If this reaction is to produce a good yield, the alkyl halide must be methyl or primary, with no bulky substituents or branches close to the reaction center. In the following examples, acetylide ions displace primary halides to form elongated alkynes.

$$HC \equiv CH \xrightarrow[\text{liquid } NH_3]{NaNH_2} HC \equiv CNa \xrightarrow{n\text{-}C_4H_9Br} CH_3CH_2CH_2CH_2C \equiv CH$$
$$89\%$$

$$HC \equiv CH \xrightarrow[\text{liquid } NH_3]{2\ NaNH_2} NaC \equiv CNa \xrightarrow{2\ n\text{-}C_3H_7Br} CH_3CH_2CH_2C \equiv CCH_2CH_2CH_3$$
$$60\% \sim 66\%$$

2. Dehydrohalogenation of Dihalide (二卤代烷脱卤化氢)   In some cases, we can generate a carbon-carbon triple bond by eliminating two molecules of HX from a dihalide. Under strongly basic conditions, dehydrohalogenation of a **geminal or vicinal dihalide** (偕二卤代烃或邻二卤代烃) gives an alkyne.

$$\begin{array}{c} \overset{Br}{} \ \overset{Br}{} \\ CH_3CH_2CHCHCH_3 \end{array} \xrightarrow[200\ ^\circ C]{KOH\ (fused)} CH_3CH_2C \equiv CCH_3$$
$$45\%$$

$$CH_3CH_2CH_2CH_2CHCl_2 \xrightarrow[150\ ^\circ C]{NaNH_2} CH_3CH_2CH_2C \equiv CNa \xrightarrow{H_2O} CH_3CH_2CH_2C \equiv CH$$
$$55\%$$

5.2   Dienes

PPT

Dienes are compounds that contain two carbon-carbon double bonds. Dienes can be divided into three groups: unconjugated, conjugated and cumulated. An **unconjugated diene** (非共轭二烯烃，孤立二烯烃) is one in which the double bonds are separated by two or more single bonds. A **conjugated diene** (共轭二烯烃) is one in which the double bonds are separated by one single bond. A **cumulated diene** (累积二烯烃) is one in which two double bonds share an *sp* hybridized carbon，The two π bonds are perpendicular to each other. Because of the geometry of this carbon, the 2*p* orbitals of the two double bonds do not overlap in a cumulated diene and are not conjugated.

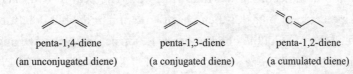

penta-1,4-diene   penta-1,3-diene   penta-1,2-diene

(an unconjugated diene)   (a conjugated diene)   (a cumulated diene)

5.2.1　Structure of Dienes

The double bonds in unconjugated dienes are independent, so their properties are similar to those of mono-olefins. Cumulated dienes are not stable because of two vertical π bonds on one *sp* hybridized carbon. Due to π-π conjugate between two double bonds, conjugated dienes are stable and have properties different from mono-olefins.

1. Structure of Buta-1,3-diene　In buta-1,3-diene, the C2-C3 bond (1.48 Å) is shorter than a carbon-carbon single bond in an alkane (1.54 Å) (Figure 5.3). This bond is shortened slightly by the increased *s* character of the sp^2 hybrid orbitals, but the most important cause of this short bond is its π bonding overlap and partial double-bond character. The planar conformation, with the *p* orbitals of the two double bonds aligned, allows overlap between the π bonds. In effect, the electrons in the double bonds are delocalized over the entire molecule, creating some π overlap and π bonding in the C2-C3 bond. The length of this bond is intermediate between the normal length of a single bond and that of a double bond.

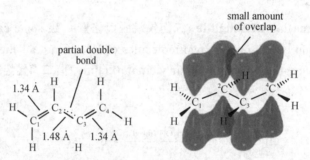

Figure 5.3　Structure of buta-1,3-diene in its most stable conformation.
图 5-3　丁 -1,3- 二烯的结构

Lewis structures are inadequate to represent delocalized molecules such as buta-1,3-diene. To represent the bonding in conjugated systems accurately, we must consider molecular orbitals that represent the entire conjugated π system, and not just one bond at a time.

The partial double-bond character between C2 and C3 in buta-1,3-diene explains why the molecule is most stable in a planar conformation. There are actually two planar conformations that allow overlap between C2 and C3. These conformations arise by rotation about the C2-C3 bond, and they are considered single-bond analogues of *trans* and *cis* isomers about a double bond. Thus, they are named *s-trans* ('single'- *trans*) and *s-cis* ('single'-*cis*) conformations.

2. Stability of Conjugated Dienes　Heats of hydrogenation for several alkenes and conjugated dienes are given in Table 5.3. By using these data, we can compare the relative stabilities of conjugated and unconjugated dienes.

The simplest conjugated diene is buta-1,3-diene. But because this molecule has only four carbon atoms, it has no unconjugated constitutional isomer. However, we can estimate the effect of conjugation of two double bonds in this molecule in the following way. The heat of hydrogenation of but-1-ene is -127 kJ · mol⁻¹. A molecule of buta-1,3-diene has two terminal double bonds, each with the same degree of substitution as the one double bond in penta-1,2-diene, therefore, we might predict that the heat of hydrogenation of buta-1,3-diene should be $2 \times (-127$ kJ $\cdot$ mol$^{-1})$ or -254 kJ · mol⁻¹. However,

Table 5.3　Heats of hydrogenation of several alkenes and conjugated dienes

表5-3　一些烯烃和共轭二烯烃的氢化热

Name	Structural formula	$\triangle H^{\circ}/\,(kJ \cdot mol^{-1})$
but-1-ene		−127(−30.3)
pent-1-ene		−126 (−30.1)
cis-but-2-ene		−120 (−28.6)
trans-but-2-ene		−115 (−27.6)
buta-1,3-diene		−237 (−56.5)
trans-penta-1,3-diene		−226 (−54.1)
penta-1,4-diene		−254 (−60.8)

the observed heat of hydrogenation of buta-1,3-diene is −237 kJ · mol^{-1}, a value 17 kJ · mol^{-1} less than estimated.

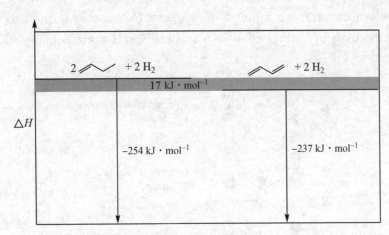

Both reactions are exothermic and give the same product, and the more stable compound releases less heat upon hydrogenation. The conclusion is that conjugation of two double bonds in buta-1,3-diene gives an extra stability to the molecule of approximately 17 kJ · mol^{-1}, these energy relationships are displayed graphically in Figure 5.4.

Figure 5.4　Conjugation of double bonds in butadiene gives the molecule an additional stability

图5-4　共轭烯烃的特殊稳定性

The extra stability in the conjugated molecule is called the **resonance energy** (共振能) of the system. Other terms favored by some chemists are **conjugation energy** (共轭能), **delocalization energy** (离域能), and **stabilization energy** (稳定能).

Calculations of this type for other conjugated and unconjugated dienes give similar results: conjugated dienes are more stable than isomeric unconjugated dienes by approximately 14.5~17 kJ · mol^{-1}. The effects of conjugation on stability are even more general. Compounds containing conjugated double bonds, not just those in dienes, are more stable than isomeric compounds containing unconjugated double bonds. For example, cyclohex-2-en-1-one is more stable than its isomer cyclohex-3-en-1-one.

cyclohex-2-en-1-one   cyclohex-3-en-1-one
(more stable)   (less stable)

The additional stability of conjugated dienes relative to unconjugated dienes arises from delocalization of electron density in the conjugated diene. In two unconjugated double bonds, each pair of π electrons is localized between two carbons. In a conjugated diene, however, the four π electrons are delocalized over the set of four parallel 2p orbitals. As we have seen many times before, delocalization leads to increased stability.

According to the molecular orbital model, the conjugated system of a diene is described as a set of four π molecular orbitals arising from combination of four 2p atomic orbitals. The key idea here is that in conjugated systems, the adjacent 2p orbitals overlap in space, even between the 2p orbitals on C2 and C3 in buta-1,3-diene.

As a result, they all combine to produce π molecular orbitals (MOs) that cover all the atoms of the conjugated system, in this case, the four carbon atoms. These MOs have zero, one, two, and three nodes, respectively, as illustrated in Figure 5.5. In the ground state, all four π electrons lie in π-bonding MOs. Because the lowest two MOs are at lower energies than that of two isolated p bonds, the net heat given off by filling these orbitals is more than would be the case for two isolated π bonds. Note that the electrons in these filled MOs are delocalized over the entire p orbital system. This π electron delocalization is the hallmark of conjugated systems and can be used to explain the spectroscopy and reactivity of conjugated molecules. Finally, it is worth pointing out that in order for

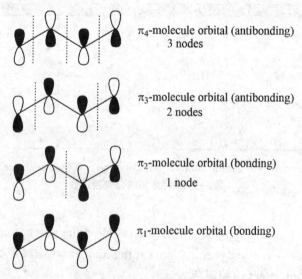

π_4-molecule orbital (antibonding)
3 nodes

π_3-molecule orbital (antibonding)
2 nodes

π_2-molecule orbital (bonding)
1 node

π_1-molecule orbital (bonding)

Figure 5.5   Structure of buta-1,3-diene molecular orbital model
图5-5   丁-1,3-二烯分子轨道结构

maximal orbital overlap to occur, the 2p orbitals must be parallel, restricting the four sp^2 hybridized atoms of conjugated systems to a planar geometry.

5.2.2   Nomenclature of Dienes

The IUPAC nomenclature of dienes is similar to that of alkenes, numbers are used to specify the locations of the double bonds. The chain is numbered starting from the end closest to the double bond, and the double bond is given the lower number of its two double-bonded carbon atoms. Cycloalkenes are assumed to have one of the double bonds in the number 1 position.

$H_2C=C-C=CH_2$ $H_2C=C-C=C-CH_3$

buta-1,3-diene   penta-1,3-diene   cyclohexa-1,3-diene   cyclohexa-1,4-diene

丁二烯   戊-1,3-二烯   环己-1,3-二烯   环己-1,4-二烯

Conformations of dienes arise by rotation about the C2-C3 bond in buta-1,3-diene, and they are considered single-bond analogues of *trans* and *cis* isomers about a double bond.

(2*Z*,4*E*)-3-methylhepta-2,4-diene   (2*E*,4*E*)-3-methylhepta-2,4-diene

s-cis-buta-1,3-diene   *s-trans*-buta-1,3-diene

5.2.3   Reactions of Dienes

1. Electrophilic Addition to Conjugated Dienes   Conjugated dienes undergo two-step electrophilic addition reactions just like simple alkenes. However, certain features are unique to the reactions of conjugated dienes.

(1) 1,2-Addition and 1,4-addition   Addition of one mole of HBr to buta-1,3-diene at −78°C gives a mixture of two constitutional isomers, 3-bromobut-1-ene and 1-bromobut-2-ene. The former results from Markovnikov addition across one of the double bonds, this process is called a 1,2-addition, and the latter double bond shifts to the C2-C3 position, such an addition is called a 1,4-addition.

$H_2C=C-C=CH_2 + HBr \longrightarrow H_2C-C-C=CH_2 + H_2C-C=C-CH_2$

	1,2-addition	1,4-addition
-80 °C	80%	20%
40 °C	20%	80%

(2) Mechanism of 1,2-addition and 1,4-addition   The mechanism is similar to other electrophilic additions to alkenes. The proton is the electrophile, adding to the alkene to give the most stable carbocation. Protonation of buta-1,3-diene gives an allylic cation, which is stabilized by resonance delocalization of the positive charge over two carbon atoms. Bromide can attack this resonance-stabilized intermediate at either of the two carbon atoms sharing the positive charge. Attack at the secondary carbon gives 1,2-addition; attack at the primary carbon gives 1,4-addition.

Step 1: Protonation of one of the double bonds forms a resonance-stabilized allylic cation.

allylic cation

Step 2: A nucleophile attacks at either electrophilic carbon atom.

1,2-addition   1,4-addition

The key to formation of these two products is the presence of a double bond in position to form a stabilized allylic cation. Molecules having such double bonds are likely to react via resonance-stabilized intermediates.

One of the interesting peculiarities of the reaction of buta-1,3-diene with HBr is the effect of temperature on the products. If the reagents are allowed to react briefly at -80℃, the 1,2-addition product predominates. If this reaction mixture is later allowed to warm to 40℃, however, or if the original reaction is carried out at 40℃, the composition favors the 1,4-addition product.

This variation in product composition reminds us that the most stable product is not always the major product. Of the two products, we expect 1-bromobut-2-ene (the 1,4-product) to be more stable, since it has the more substituted double bond. This prediction is supported by the fact that this isomer predominates when the reaction mixture is warmed to 40℃ and allowed to equilibrate.

A reaction-energy diagram for the second step of this reaction (Figure 5.6) helps to show why one product is favored at low temperatures and another at higher temperatures. The allylic cation is in the center of the diagram; it can react toward the left to give the 1,2-product or toward the right to give the 1,4-product. The initial product depends on where bromide attacks the resonance-stabilized allylic cation. Bromide can attack at either of the two carbon atoms that share the positive charge. Attack at the secondary carbon gives 1,2-addition, and attack at the primary carbon gives 1,4-addition.

$$H_2C=C-C=CH_2 + HBr \longrightarrow H_2C-C-C=CH_2 + H_2C-C=C-CH_2$$

	1,2-addition	1,4-addition
−80℃	80%	20%
40℃	20%	80%

(3) Kinetic control (动力学控制) and thermodynamic control (热力学控制)

1) Kinetic control at -80℃   The transition state for 1,2-addition has a lower energy than the transition state for 1,4-addition, giving the 1,2-addition a lower activation energy. This is not surprising, because 1,2-addition results from bromide attack at the more substituted secondary carbon, which bears more of the positive charge because it is better stabilized than the primary carbon. Because the 1,2-addition has a lower activation energy than the 1,4-addition, the 1,2-addition takes place faster (at all temperatures).

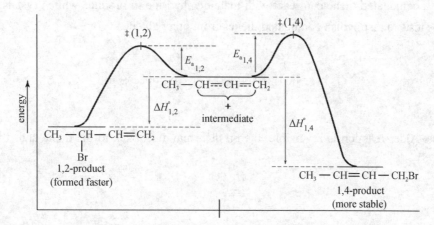

Figure 5.6   Reaction-energy diagram for the second step of the addition of HBr to buta-1,3-diene
图5-6   丁-1,3-二烯和HBr反应能量变化示意图

Attack by bromide on the allylic cation is a strongly exothermic process, so the reverse reaction has a large activation energy. At -80℃, few collisions take place with this much energy, and the rate of the reverse reaction is practically zero. Under these conditions, the product that is formed faster predominates. Because the kinetics of the reaction determine the results, this situation is called kinetic control of the reaction. The 1,2-product, favored under these conditions, is called the **kinetic product** (动力学产物).

2) Thermodynamic control at 40℃   At 40℃, a significant fraction of molecular collisions has enough energy for reverse reactions to occur. Notice that the activation energy for the reverse of the 1,2-addition is less than that for the reverse of the 1,4-addition. Although the 1,2-product is still formed faster, it also reverts to the allylic cation faster than the 1,4-product. At 40℃, an equilibrium is set up, and the relative energy of each species determines its concentration. The 1,4-product is the most stable species, and it predominates. Since thermodynamics determine the results, this situation is called thermodynamic control [or **equilibrium control** (平衡控制)] of the reaction. The 1,4-product, favored under these conditions, is called the **thermodynamic product** (热力学产物).

We will see many additional reactions whose products may be determined by kinetic control or by thermodynamic control, depending on the conditions. In general, reactions that do not reverse easily are kinetically controlled because no equilibrium is established. In kinetically controlled reactions, the product with the lowest-energy transition state predominates. Reactions that are easily reversible are thermodynamically controlled unless something happens to prevent equilibrium from being attained. In thermodynamically controlled reactions, the lowest-energy product predominates.

2. The Diels-Alder reaction（D-A 反应，狄尔斯 - 阿尔德反应）   In 1928, German chemists Otto Diels and Kurt Alder discovered that alkenes and alkynes with **electron-withdrawing groups** (吸电子基)

add to conjugated dienes to form six-membered rings. The Diels-Alder reaction has proven to be a useful synthetic tool, providing one of the best ways to make six-membered rings with diverse functionality and controlled stereochemistry. Diels and Alder were awarded the Nobel Prize in 1950 for their work.

Because a conjugated diene can react with maleic anhydride to produce white crystals, the reaction can be used as tests to distinguish conjugated dienes from other alkenes.

The Diels-Alder reaction is reversible, the product may turn back to the diene and the dienophile when heated.

5.3 Pericyclic Reaction (周环反应)

Pericyclic reaction is a reaction that takes place in a single step without radical or ionic intermediates. A concerted pericyclic reaction has a single cyclic transition state, whose activation energy may be supplied by heat or by ultraviolet light.

The Diels-Alder reaction is called a [4+2] **cycloaddition** (环加成反应), because a ring is formed by the interaction of four π electrons in the **diene** (双烯体) with two π electrons of the alkene or alkyne. Since the electron-poor alkene or alkyne is prone to react with a diene, it is called a **dienophile** (亲双烯体) ("lover of dienes"). In effect, the Diels-Alder reaction converts two π bonds into two σ bonds. We can symbolize the Diels-Alder reaction by using three arrows to show the movement of three pairs of electrons. This electron movement is concerted, with three pairs of electrons moving simultaneously.

diene dienophile   Diels-Alder product

1. Diene Must Be Able to Assume an *s-cis* Conformation The diene must be in the *s-cis* conformation to react. The *s-trans* conformation cannot undergo the Diels-Alder reaction. The energy barrier for interconversion of the s-*trans* to s-*cis* conformations for 1,3-butadiene is low, approximately 11.7 kJ

(2.8 kcal) /mol. Consequently, 1,3-butadiene can still be a reactive diene in Diels-Alder reactions.

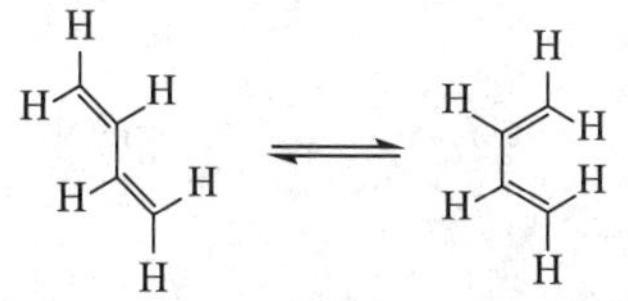

s-trans   s-cis

2. The Configuration of the Dienophile is Retained   If the dienophile is a *cis* isomer, then the substituents *cis* to each other in the dienophile are *cis* in the Diels-Alder adduct. Conversely, if the substituents that are *trans* in the dienophile are *trans* in the adduct.

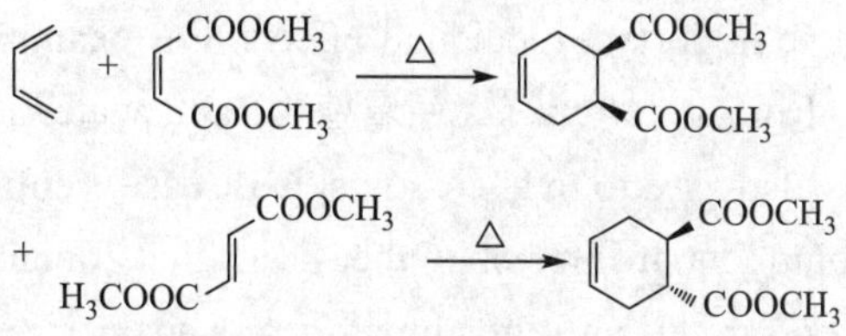

3. The Effect of Substituents on Rate   The Diels-Alder reaction gives excellent yields at moderate temperature if the diene has electron-donating groups or the dienophile has electron-withdrawing groups attached, as in the following examples.

The simplest example of a Diels-Alder reaction is that between 1,3-butadiene and ethylene, it is very slow and takes place only when the reactants are heated at 200°C under pressure.

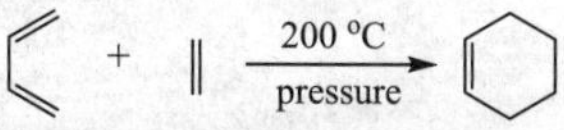

Placing a carbonyl group on the dienophile facilitates the reaction. To illustrate, 1,3-butadiene and 3-buten-2-one form a Diels-Alder adduct when heated at 140°C.

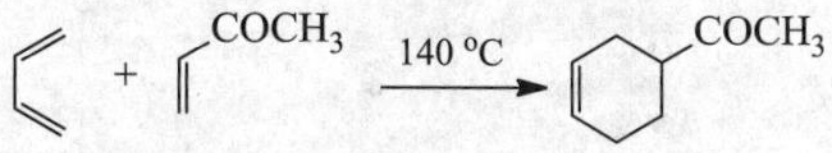

Placing electron-releasing methyl groups on the diene further facilitates reaction. 2,3-dimethyl-1,3-butadiene and 3-buten-2-one form a Diels-Alder adduct at 30°C.

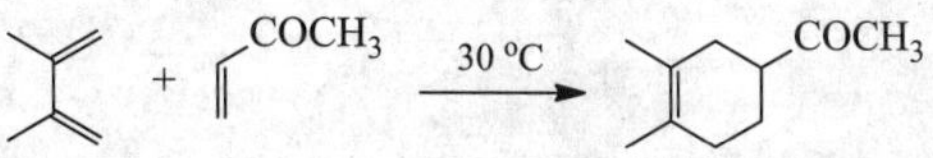

When the diene and dienophile are both unsymmetrically substituted, the Diels-Alder reaction usually gives a single product (or a major product) rather than a random mixture. The following examples show that an electron-donating substituent (-OCH$_3$) on the diene and an electron-withdrawing substituent (-CHO) on the dienophile usually show either a 1,2- or 1,4-relationship in the product.

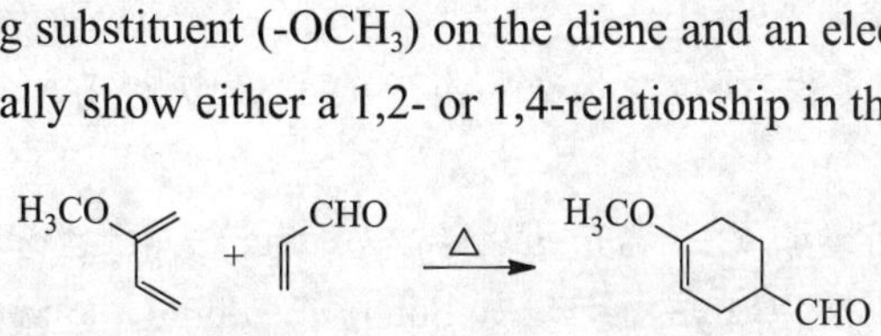

1,4-product

$$\text{OCH}_3 + \text{CHO} \xrightarrow{\Delta} \text{OCH}_3 \text{ CHO}$$

1,2-product

5.4　Application of Alkynes in the Pharmacy Study

Alkynes are not as common as alkenes in nature, and carbon-carbon triple bonds are not as common in medicine, but they also have important medicinal effects. For example, **cicutoxin** (毒芹素) is a kind of toxic compound isolated from **hemlock** (毒芹属植物) and water hemlock, which is used to treat osteomyelitis in folk, and it is also used to treat certain skin disease, ventilated, rheumatism, and neuralgia in European folk made into ointment or infusion. Other drugs that containing carbon-carbon triple bonds, such as **parsalmide** (帕莎米特), are the anti-inflammatory and analgesic, **ethynyl estradiol** (乙炔雌二醇) is used as the birth control pills, and Dynemicin A is an antibiotic with the antitumor effects.

cicutoxin

parsalmide　　　　ethynyl estradiol　　　　dynemicin A

重 点 小 结

一、炔烃

1. **炔烃的结构**　炔烃中的叁键碳原子(炔碳原子)为 sp 杂化。两个炔碳原子之间形成一个 σ 键,两个 π 键,π 电子云形成以 σ 键为对称轴的圆柱体形状。叁键碳原子与其相连的原子或基团位于同一直线上,键角为180°,因此乙炔为直线型分子。炔烃具有与烯烃相似的性质,能发生亲电加成、氧化和聚合反应。

2. **炔烃的异构和命名**　异构:炔烃有构造异构,如丁−1−炔和丁−2−炔。由于叁键是直线型的,炔烃没有顺反异构。

命名:①单炔——从靠近叁键的一端开始编号,以 −yne 结尾,如 pent−1−yne。②二炔——从最靠近末端的叁键一端开始编号,以 diyne 结尾,如 buta−1,3−diyne。③烯炔——从最靠近叁键或双键的一端开始编号,同等条件下,双键优先编号。

3. 炔烃的物理性质　同系列炔烃,沸点随相对分子质量升高而升高。与烷烃、烯烃相似,炔烃不溶于水,易溶于有机溶剂。

4. 炔烃的化学性质　与烯烃相似点,能发生加氢、亲电加成和氧化反应。与烯烃不同点,能发生亲核加成,炔氢显示出酸性。

(1) 炔氢的酸性　弱酸性,能和强碱如氨基钠或碱金属反应;能和硝酸银或炔化亚酮反应生成沉淀,可用以此鉴别末端炔烃。

(2) 催化加氢　可催化加氢得烷烃;可以在林德拉催化剂催化下加氢得顺式烯烃,或在钠–液氨体系中还原得反式烯烃。

(3) 亲电加成　+X_2,得到四卤代烷烃,反应可停留在二卤代烯烃阶段。若双键、叁键并存,双键优先加卤素。+HX,加成符合马氏规则（本质:中间体碳正离子的稳定性决定主产物）。+H_2O,Hg^{2+}催化,加成符合马氏规则,产物烯醇经过烯醇互变得酮,仅乙炔和水反应得乙醛。

(4) 硼氢化–氧化　末端炔烃经硼氢化–氧化后得到醛,其余炔烃得到酮,加成符合反马氏规则。

(5) 自由基加成　在过氧化物催化下,炔烃加 HBr 得到反马氏规则的产物。

(6) 亲核加成　叁键的 π 电子围绕碳碳 σ 键呈圆筒状分布,因此末端炔烃的炔碳原子核相对比较暴露,易受亲核试剂的进攻而发生亲核加成反应。

(7) 氧化反应　高锰酸钾氧化炔烃,在中性条件下得邻二酮;在碱性或酸性条件下叁键断裂,碱性条件下得羧酸盐,酸性条件下得羧酸,末端炔碳被氧化成 CO_2。臭氧氧化后水解,叁键断裂得羧酸,末端炔碳被氧化成 CO_2。

(8) 聚合反应　在不同催化剂和反应条件下,乙炔可选择性地聚合成链状或环状化合物。与烯烃不同,它一般不聚合成高聚物。

5. 炔烃的制备　通过炔负离子的烷基化反应制备。炔钠与伯卤代烷发生亲核取代反应生成高级炔烃,这是有机合成中增长碳链的常用方法之一。

二卤代烷在强碱条件下脱卤化氢也可制备炔烃。

二、二烯烃

1. 二烯烃的分类　根据双键的的位置不同,二烯烃分为共轭二烯烃、累积二烯烃和孤立二烯烃。

2. 二烯烃的结构　重点是共轭二烯的结构。在丁–1,3–二烯中,所有原子共平面,碳原子均为 sp^2 杂化,4 个 p 轨道互相平行且共平面,为典型的 π–π 共轭体系。体系稳定,存在较大的共轭能。

3. 二烯烃的命名　与单烯烃类似。以-diene结尾,标注双键的位次,例如buta-1,3-diene,环二烯的命名,选择其中一个双键标注1号位。

4. 二烯烃的化学性质

(1) 亲电加成反应　类似烯烃的亲电加成反应,分为 1,2-加成反应和 1,4-加成反应两类。1,2-加成为动力学控制步骤,1,2-加成产物为动力学产物;1,4-加成为热力学控制步骤,1,4-加成产物为热力学产物,又称平衡控制产物。

(2) Diels-Alder 反应　简称为 D-A 反应,又称双烯合成反应。丁–1,3–二烯和顺丁烯二酸酐反应生成白色沉淀,此反应可以作为共轭二烯烃的鉴别反应。

狄尔斯–阿尔德反应是可逆的,加成产物在较高温度下又转变为双烯体和亲双烯体。

共轭二烯烃在该反应中称为双烯体,烯烃称为亲双烯体,双烯体连有供电子基,亲双烯体连有吸电子基,有利于反应。不对称的双烯体和亲双烯体发生 D-A 反应时,遵循邻对位规律。

三、周环反应

在加热或光照条件下,反应通过一个环状过渡态一步完成,不产生离子或自由基中间体。

共轭二烯烃及其衍生物（双烯体）与含有碳碳双键、叁键等不饱和化合物（亲双烯体）发生 1,4-加成反应，生成六元环状化合物，称为双烯合成，又称 [4+2] 环加成反应。

双烯体上连有供电子基或亲双烯体上连有吸电子基时，反应更易进行。

不对称的双烯体和亲双烯体发生 D-A 反应时，易生成取代基处于邻位（1,2-）或对位（1,4-）的产物，遵循邻对位规律。

四、炔烃在药学中的应用

在药学上，碳碳叁键和炔烃不及碳碳双键和烯烃普遍，但也有重要的药用价值。 例如毒芹素、帕莎米特、乙炔基雌二醇、达内霉素。

Problems
目 标 检 测

1. Name $CH_3(CH_2)_3C{\equiv}CCH_3$ by the IUPAC system.
用IUPAC命名化合物$CH_3(CH_2)_3C{\equiv}CCH_3$是（　　　）。

    A. hex-2-yne   B. hept-2-yne

    C. hexa-2-yne   D. hepta-3-yne

2. Name $CH_3C{\equiv}CCH{=}CH_2$ by the IUPAC system.
用IUPAC命名化合物$CH_3C{\equiv}CCH{=}CH_2$是（　　　）。

    A. hept-3-en-1-yne   B. pent-3-en-1-yne

    C. pent-1-en-3-yne   D. hept-1-en-3-yne

3. Write a structural formula for 4,5-dimethylhept-2-yne.
写出4,5-二甲基庚-2-炔的结构式是（　　　）。

    A.   B.

    C.   D.

4. Write the product for the reaction: pent-1-yne + H_2O (Hg^{2+}, H^+).
写出反应产物：戊-1-炔 + H_2O （ Hg^{2+}, H^+ ）→（　　　）。

    A.   B.   C.   D.

5. Write the product for the reaction: but-2-yne + H_2 （1mol, Lindlar's catalyst）.
写出反应产物：丁-2-炔 + H_2 （1mol, Lindlar's catalyst）→（　　　）。

    A.   B.   C.   D.

6. How to distinguish the following compounds by chemical method?
如何用化学方法鉴别下列化合物？

but-1-yne，but-1-ene and butane

7. Propose a mechanism for the entire reaction of but-1-yne with 2 moles of HBr. Show why Markovnikov's rule should be observed in both the first and second additions of HBr.
写出丁-1-炔和2摩尔HBr加成的机理，解释第一步和第二步反应中的马尔科夫尼科夫规则。

8. Synthesis *cis*-pent-2-ene by using propyne and ethane. Write equations for all steps.

以丙炔和乙烷为原料合成顺戊-2-烯，写出每一步的化学方程式。

9. Synthesis adipic acid (HOOC ⌇⌇ COOH) from acetylene. Write equations for all steps.

以乙炔为原料合成己二酸，写出每一步的化学方程式。

10. Using data from Table 5.3, estimate the extra stability that results from the conjugation of double bonds in *trans*-penta-1,3-diene。

利用表5-3中数据，估算反戊-1,3-二烯的共轭能。

11. Predict the product(s) formed by addition of one mole of Br_2 to hexa-2,4-diene.

预测1mol Br_2与己-2,4-二烯的加成产物。

12. Propose a mechanism for the reaction, showing explicitly how the observed mixtures of products are formed.

提出一种反应机理，明确说明观察到的产物混合物是如何形成的。

$$HO \diagdown \diagup + HBr \longrightarrow Br \diagdown \diagup + \diagdown \diagup Br$$

13. Predict the products of the following Diels–Alder reactions.

预测下列D-A反应的产物。

a.

b.

14. The addition of *cis*-maleic anhydride to hexa-2,4-diene is a Diels-Alder reaction. The product is a white crystal, so this reactions can be used as a test to distinguish conjugated dienes from the other alkenes. Please write the equation.

顺-丁烯二酸酐和己-2,4-二烯的加成反应是狄尔斯-阿尔德反应，其产物是白色的晶体，因此可作为共轭二烯烃的特征鉴别。请写出该反应的方程式。

Discussion Topic

Hemlock belongs to Apiaceae (伞形科) and likes growing in moist areas on the edges of lakes and marshes. It is a native plant of Asia, distributed in central China, north China and northeast China. Cicutoxin is an poisonous organic compound found in hemlock which can strongly act on the central nervous system, leading to tremors, convulsions, and even death. It is used to treat osteomyelitis in folk and skin disease, ventilated, rheumatism, and neuralgia in European folk made into ointment or infusion. It is only for external use because of its poison.

cicutoxin

How many stereoisomers are possible for cicutoxin?

（徐春蕾　盛文兵）

第六章　芳香化合物
Chapter 6　Aromatic Compound

 学习目标

1. **掌握**　休克尔规则与化合物的芳香性；芳香环的亲电取代反应和机理；芳香环上的取代基定位效应及其在合成上的应用。
2. **熟悉**　芳香化合物的合成。
3. **了解**　苯和共轭环状体系的分子轨道及其在芳香性判断中的应用。

6.1　Introduction to Aromatic Compounds

PPT

In 1825, Michael Faraday isolated a pure compound of boiling point 80°C from the oily mixture that condensed from illuminating gas, the fuel burned in gas lights. Elemental analysis showed an unusually small hydrogen-to-carbon ratio of 1:1, corresponding to an empirical formula of CH. Faraday named the new compound "bicarburet of hydrogen." Eilhard Mitscherlich synthesized the same compound in 1834 by heating benzoic acid, isolated from gum benzoin, in the presence of lime. Like Faraday, Mitscherlich found that the empirical formula was CH. He also used a vapor-density measurement to determine a molecular weight of about 78, for a molecular formula of CH. Since the new compound was derived from gum benzoin, he named it benzin, now called **benzene** (苯).

Many other compounds discovered in the nineteenth century seemed to be related to benzene. These compounds also had low hydrogen-to-carbon ratios as well as pleasant aromas, and they could be converted to benzene or related compounds. This group of compounds was called **aromatic** (芳香族) because of their pleasant odors. Other organic compounds without these properties were called **aliphatic** (脂肪族), meaning "fatlike." As the unusual stability of aromatic compounds was investigated, the term *aromatic* came to be applied to compounds with this stability, regardless of their odors.

1. The Kekulé Structure (凯库勒结构式)　In 1866, Friedrich Kekulé proposed a cyclic structure for benzene with three double bonds. Considering that multiple bonds had been proposed only recently (1859), the cyclic structure with alternating single and double bonds was considered somewhat bizarre. The Kekulé structure has its shortcomings, however. For example, it predicts two different 1,2-dichlorobenzenes, but only one is known to exist.

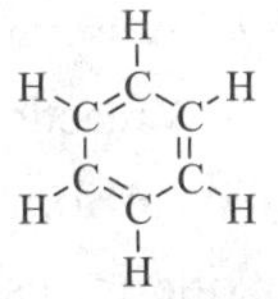

Kekulé structure of benzene

2. The Resonance Representation (共振论)   The resonance picture of benzene is a natural extension of Kekulé's hypothesis. In a Kekulé structure, the single bonds would be longer than the double bonds. Spectroscopic methods have shown that the benzene ring is planar and all the bonds are the same length (1.397 Å). Because the ring is planar and the carbon nuclei are positioned at equal distances, the two Kekulé structures must differ only in the positioning of the π electrons.

Benzene is actually a resonance hybrid of the two Kekulé structures. This representation implies that the π electrons are delocalized. The carbon–carbon bond lengths in benzene are shorter than typical single-bond lengths, yet longer than typical double-bond lengths.

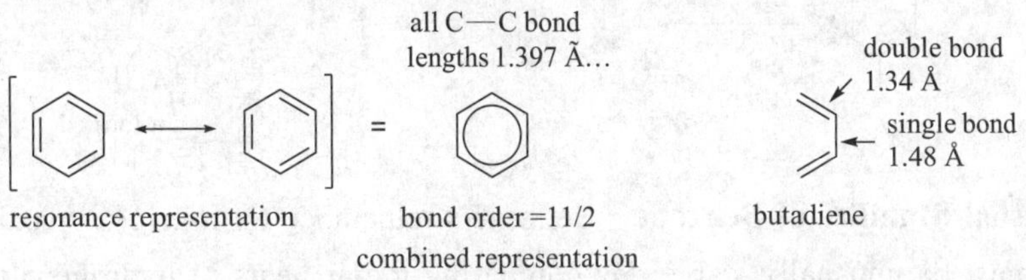

all C—C bond
lengths 1.397 Å…

double bond
1.34 Å
single bond
1.48 Å

resonance representation

bond order =11/2
combined representation

butadiene

The resonance-delocalized picture explains most of the structural properties of benzene and its derivatives—the **benzenoid** (苯类) aromatic compounds. Because the π bonds are delocalized over the ring, we often inscribe a circle in the hexagon rather than draw three localized double bonds. This representation helps us remember there are no localized single or double bonds, and it prevents us from trying to draw supposedly different isomers that differ only in the placement of double bonds in the ring. We often use Kekulé structures in drawing reaction mechanisms, however, to show the movement of individual pairs of electrons.

Using this resonance picture, we can draw a more realistic representation of benzene (Figure 6.1). Benzene is a ring of six sp^2 carbon atoms, each bonded to one hydrogen atom. All the carbon–carbon bonds are the same length, and all the bond angles are exactly 120°. Each sp^2 carbon atom has an unhybridized p orbital perpendicular to the plane of the ring, and six electrons occupy this circle of p orbitals.

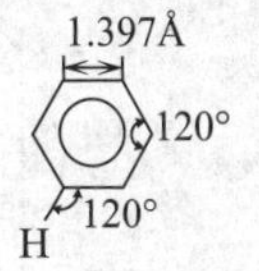

1.397Å
120°
120°
H

Figure 6.1   Benzene is a flat ring of sp^2 hybridcarbon atoms with their unhybridized p orbitals all aligned and overlapping
图6-1   苯是sp²杂化碳原子构成的环平面，其未杂化的p轨道均对齐且重叠

At this point, we can define an aromatic compound to be a cyclic compound containing some number of conjugated double bonds and having unusually large resonance energy. Using benzene as the example, we will consider how aromatic compounds differ from aliphatic compounds. Then we will discuss why an aromatic structure confers extra stability and how we can predict aromaticity in some interesting and unusual compounds.

3. The Unusual Reactions of Benzene   Benzene is actually much more stable than we would expect from the simple resonance-delocalized picture. Both the Kekulé structure and the resonance-delocalized picture show that benzene is a cyclic conjugated triene. We might expect benzene to undergo the typical reactions of polyenes. In fact, its reactions are quite unusual. For example, an alkene decolorizes potassium permanganate by reacting to form a glycol. The purple permanganate color disappears, and a precipitate of manganese dioxide forms. When permanganate is added to benzene, however, no reaction occurs.

Most alkenes decolorize solutions of bromine in carbon tetrachloride.The red bromine color disappears as bromine adds across the double bond. When bromine is added to benzene, no reaction occurs, and the red bromine color remains.

Addition of a catalyst such as ferric bromide to the mixture of bromine and benzene causes the bromine color to disappear slowly. HBr gas is evolved as a by-product, but the expected addition of Br₂ does not take place. Instead, the organic product results from substitution of a bromine atom for ahydrogen, and all three double bonds are retained.

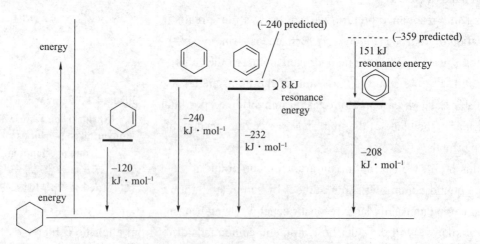

4. The Unusual Stability of Benzene   Benzene's reluctance to undergo typical alkene reactions suggests that it must be unusually stable. By comparing molar heats of hydrogenation, we can get a quantitative idea of its stability. Benzene, cyclohexene, and the cyclohexadienes all hydrogenate to form cyclohexane. Figure 6.2 shows the experimentally determined heats of hydrogenation and they are used to compute the **resonance energies** (共振能量):

(1) Hydrogenation of cyclohexene is exothermic by 120 kJ · mol⁻¹.

(2) Hydrogenation of cyclohexa-1,4-diene is exothermic by 240 kJ · mol⁻¹, about twice the heat of hydrogenation of cyclohexene. The resonance energy of the isolated double bonds in cyclohexa-1,4-diene is about zero.

Figure 6.2   The molar heats of hydrogenation and the relative energies of cyclohexene, cyclohexa-1,4-diene, cyclohexa-1,3-diene, and benzene. The dashed lines represent the energies that would be predicted if every double bond had the same energy as the double bond in cyclohexene

图6-2   环己烯，环己-1,4-二烯，环己-1,3-二烯和苯的相对能量和摩尔氢化热。虚线表示如果每个双键都具有与环己烯中的双键相同的能量时可以预测的能量

(3) Hydrogenation of cyclohexa-1,3-diene is exothermic by 232 kJ · mol^{-1}, about 8 kJ less than twice the value for cyclohexene. A resonance energy of 8 kJ is typical for a conjugated diene.

(4) Hydrogenation of benzene requires higher pressures of hydrogen and a more active catalyst. This hydrogenation is exothermic by 208 kJ · mol^{-1}, about 151 kJ less than 3 times the value for cyclohexene.

$$\bigcirc + 3H_2 \xrightarrow[\text{high pressure}]{\text{catalyst}} \bigcirc$$

$$\triangle H° = 208 \text{ kJ} \cdot \text{mol}^{-1}$$
$$3 \times \text{cyclohexene} = 359 \text{ kJ} \cdot \text{mol}^{-1}$$
$$\overline{\text{resonance energy} = 151 \text{ kJ} \cdot \text{mol}^{-1}}$$

The huge 151 kJ · mol^{-1} resonance energy of benzene cannot be explained by conjugation effects alone. The heat of hydrogenation for benzene is actually smaller than that for cyclohexa-1,3-diene. The hydrogenation of the first double bond of benzene is endothermic, the first endothermic hydrogenation we have encountered. In practice, this reaction is difficult to stop after the addition of 1 mole of because the product (cyclohexa-1,3-diene) hydrogenates more easily than benzene itself. Clearly,the benzene ring is exceptionally unreactive.

$$\bigcirc + H_2 \xrightarrow{\text{catalyst}} \bigcirc$$

$$\triangle H°\text{hydrogenation}$$
$$\text{benzene: } -208\text{kJ } (-49.8\text{kcal})$$
$$\text{cyclohexa-1,3-diene: } -232\text{kJ } (-55.4\text{kcal})$$
$$\overline{\triangle H° = +24 \text{ kJ } (+5.6 \text{ kcal})}$$

5. Failures of the Resonance Picture   For many years, chemists assumed that benzene'slarge resonance energy resulted from having two identical, stable resonance structures.They thought that other hydrocarbons with analogous conjugated systems of alternating single and double bonds would show similar stability. These cyclic hydrocarbons with alternating single and double bonds are called **annulenes (轮烯).** For example, benzene is the six-membered annulene, so it can be named [6]annulene. Cyclobutadiene is [4]annulene, cyclooctatetraene is [8]annulene, and larger annulenes are named similarly.

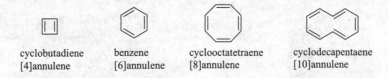

cyclobutadiene　　benzene　　cyclooctatetraene　　cyclodecapentaene
[4]annulene　　[6]annulene　　[8]annulene　　[10]annulene

For the double bonds to be completely conjugated, the annulene must be planar so the *p* orbitals of the pi bonds can overlap. As long as an annulene is assumed to be planar, we can draw two Kekulé-like structures that seem to show a benzene-like resonance. Figure 6.3 shows proposed benzene-like resonance forms for cyclobutadiene and cyclooctatetraene. Although these resonance structures suggest that the [4] and [8]annulenes should be unusually stable (like benzene), experiments have shown that cyclobutadiene and cyclooctatetraene are not unusually stable. These results imply that the simple resonance picture is incorrect.

Cyclobutadiene has never been isolated and purified. It undergoes an extremely fast Diels–Alder dimerization. In 1911, Richard Willstäer synthesized cyclooctatetraene and found that it reacts like a normal polyene. Bromine adds readily to cyclooctatetraene, and permanganate oxidizes its double bonds. This evidence shows that cyclooctatetraene is much less stable than benzene. In fact, structural studies have shown that cyclooctatetraene is not planar. It is most stable in a "tub" conformation, with poor overlap between adjacent pi bonds.

Visualizing benzene as a resonance hybrid of two Kekulé structures cannot fully explain the unusual stability of the aromatic ring. As we have seen with other conjugated systems, molecular orbital theory provides the key to understanding aromaticity and predicting which compounds will have the stability of an aromatic system.

6. The Energy Diagram of Benzene   Figure 6.4 shows the six molecular orbitals of benzene as viewed from above, exhibiting the sign of the top lobe of each *p* orbital.

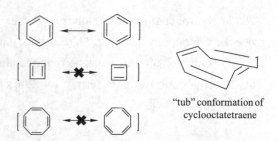

"tub" conformation of cyclooctatetraene

Figure 6.3   Cyclobutadiene and cyclooctatetraene have alternating single and double bonds similar to those of benzene. These compounds were mistakenly expected to be aromatic

图6-3　环丁二烯和环辛酸酯具有与苯类似的交替单键和双键。这些化合物被误认为是芳香族的

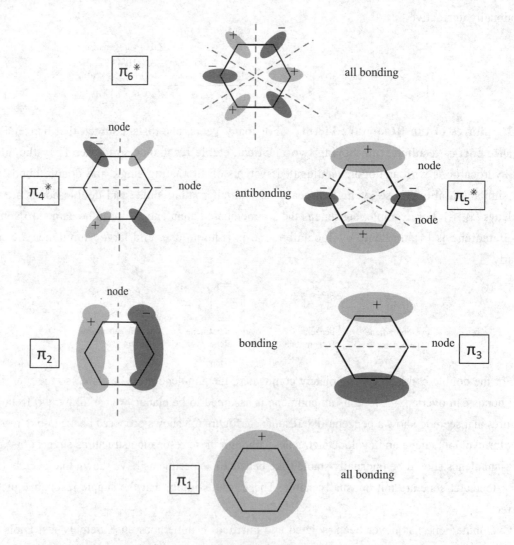

Figure 6.4   The six π molecular orbitals of benzene, viewed from above. The number of nodal planes increases with energy, and there are two degenerate MOs at each intermediate energy level

图6-4　苯的六个π分子轨道。节点平面的数量随能量的增加而增加，并且在每个中间能级上都有两个简并的MO

The energy diagram of the benzene MOs (Figure 6.5) shows them to be symmetrically distributed above and below the nonbonding line (the energy of an isolated p orbital). The all-bonding and all-antibonding orbitals (π_1 and π^*_6)are lowest and highest in energy, respectively. The degenerate bonding orbitals (π_2 and π_3) are higher in energy thanπ_1, but still bonding.

The Kekulé structure for benzene shows three π bonds, representing six electrons (three pairs) involved in π bonding. Six electrons fill the three bonding MOs of the benzene system. This electronic configuration explains the unusual stability of benzene. The first MO is all-bonding and is particularly stable. The second and third (degenerate) MOs are still strongly bonding, and all three of these bonding MOs delocalize the electrons over several nuclei. This configuration, with all the bonding MOs filled (a "closed bonding shell"), is energetically very favorable.

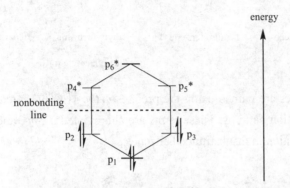

Figure 6.5　Energy diagram of the molecular orbitals of benzene. Benzene's six electrons fill the three bonding orbitals, leaving the antibonding orbitals vacant
图6-5　苯的分子轨道能量图。苯的六个电子充满三个键合轨道，反键轨道空着

6.2　Nomenclature of Aromatic Compounds

Benzene derivatives have been isolated and used as industrial reagents for well over 100 years. Many of their names are rooted in the historical traditions of chemistry. The following compounds are usually called by their historical common names, and almost never by the systematic IUPAC names:

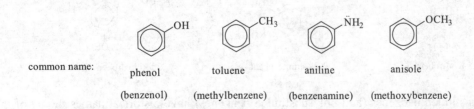

common name:　　phenol　　　toluene　　　aniline　　　anisole

（benzenol）　（methylbenzene）　（benzenamine）　（methoxybenzene）

微课

common name: styrene   acetophenone   benzaldehyde   benzoic acid

vinylbenzene   (methyl phenyl ketone)

Many compounds are named as derivatives of benzene, with their substituents named just as though they were attached to an alkane.

tert-butylbenzene   nitrobenzene   ethynylbenzene   benzenesulfonic acid

(phenylacetylene)

Disubstituted benzenes are named using the prefixes *ortho-*（邻位）, *meta-*（间位）, and *para-*（对位） to specify the substitution patterns. These terms are abbreviated *o-*, *m-*, and *p-*. Numbers can also be used to specify the substitution in disubstituted benzenes.

1,2 or *ortho*   1,3 or *meta*   1,4 or *para*

common name   *o*-dichlorobenzene   *m*-chloroperoxybenzoic acid   *p*-nitrophenol

IUPAC name   1,2-dichlorobenzene   3-chloroperoxybenzoic acid   4-nitrophenol

With three or more substituents on the benzene ring, numbers are used to indicate their positions. Assign the numbers as you would with a substituted cyclohexane, to give the lowest possible numbers to the substituents. The carbon atom bearing the functional group that defines the base name (as in phenol or benzoic acid) is assumed to be C1.

1,3,5-trinitrobenzene   2,4-dinitrophenol   3,5-dihydroxybenzoic acid

Many disubstituted benzenes (and polysubstituted benzenes) have historical names. Some of these are obscure, with no obvious connection to the structure of the molecule.

医药大学堂
WWW.YIYAODXT.COM

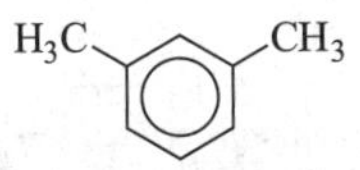

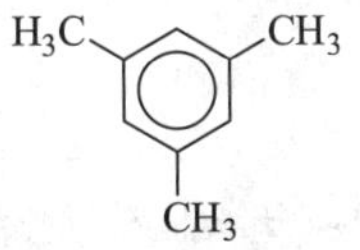

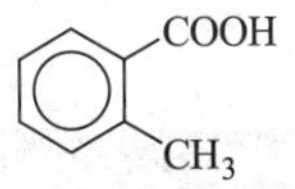

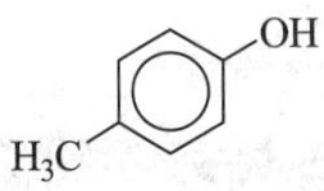

common name	m-xylene	mesitylene	o-toluic acid	p-cresol

| IUPAC name | 1,3-dimethylbenzene | 1,3,5-trimethylbenzene | 2-methylbenzoic acid | 4-methylphenol |

When the benzene ring is named as a substituent on another molecule, it is called a **phenyl group** (苯基). The phenyl group is used in the name just like the name of an alkyl group, and it is often abbreviated Ph (or Φ) in drawing a complex structure.

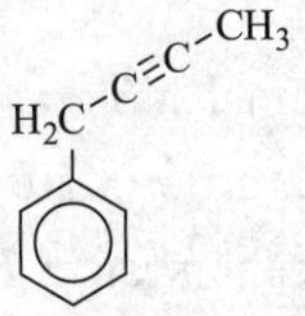

or Ph—$\overset{H_2}{C}$—C≡C—CH$_3$

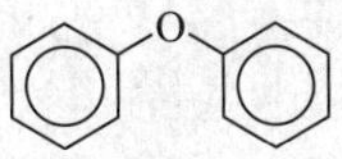

or Ph$_2$O

diphenyl ether

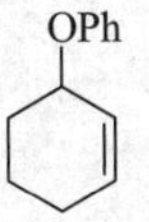

3-phenoxycyclohexene

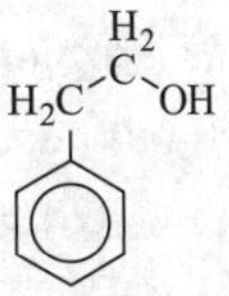

or PhCH$_2$CH$_2$OH

2-phenylethanol

The seven-carbon unit consisting of a benzene ring and a methylene ($-CH_2-$) group is often named as a **benzyl group** (苄基). Be careful not to confuse the benzyl group (seven carbons) with the phenyl group (six carbons).

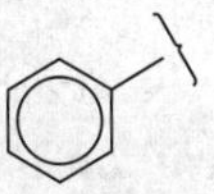

a pheny group

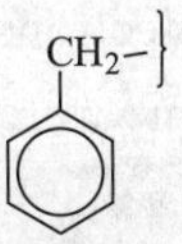

a benzyl group

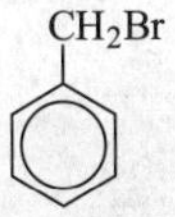

benzyl bromide

(a-bromotoluene)

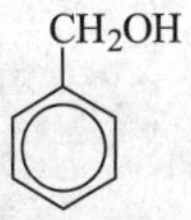

benzyl alcohol

Aromatic hydrocarbons are sometimes called **arenes** (芳烃). An **aryl group** (芳基), abbreviated Ar, is the aromatic group that remains after the removal of a hydrogen atom from an aromatic ring. The phenyl group, Ph, is the simplest aryl group. The genericaryl group (Ar) is the aromatic relative of the generic alkyl group, which we symbolize by R.

Examples of aryl groups

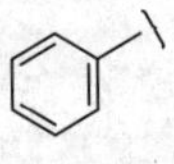

the phenyl group

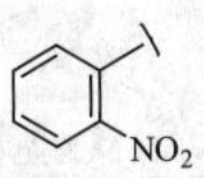

the o-nitrophenyl group

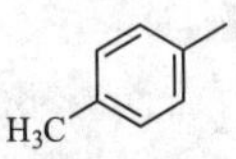

the p-tolyl group

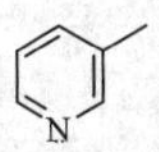

the 3-pyridyl group

Examples of the use of a generic aryl group

Ar—MgBr   Ar$_2$O or Ar—O—Ar   Ar—NH$_2$ Ar-SO$_3$H

an arylmagnesium bromide   a diaryl ether   an arylamine   a aryllsulfonic acid

6.3   Structure of Aromatic Compound of Hückel's Rule

6.3.1   Hückel's Rule

Erich Hückel developed a shortcut for predicting which of the annulenes and related compounds are aromatic and which are antiaromatic. In using Hückel's rule, we must be certain that the compound under consideration meets the criteria for an aromatic or antiaromatic system.

To qualify as aromatic or antiaromatic, a cyclic compound must have a continuous ring of overlapping p orbitals, usually in a planar conformation. Once these criteria are met, Hückel's rule applies: If the number of π electrons in the cyclic system is: (4N+2), the system is aromatic.(4N) the system is antiaromatic. N is an integer, commonly 0, 1, 2, or 3.

Benzene is [6]annulene, cyclic, with a continuous ring of overlapping p orbitals.There are six π electrons in benzene (three double bonds in the classical structure), soit is a (4N+2) system, with Hückel's rule predicts benzene to be aromatic.

Like benzene, cyclobutadiene ([4] annulene) has a continuous ring of overlapping p orbitals. But it has four π electrons (two double bonds in the classical structure), whichis a (4N) system with N = 1. Hückel's rule predicts cyclobutadiene to be antiaromatic.

Cyclooctatetraene is [8] annulene, with eight π electrons (four double bonds) in the classical structure. It is a (4N) system, with N = 2. If Hückel's rule were applied to cyclooctatetraene, it would predict antiaromaticity. However, cyclooctatetraene is a stable hydrocarbon with a boiling point of 153℃. It does not show the high reactivity associated with antiaromaticity, yet it is not aromatic either. Its reactions are typical of alkenes. Remember that Hückel's rule applies to a compound only if there is a continuous ring of overlapping p orbitals, usually in a planar system. Cyclooctatetraene is more flexible than cyclobutadiene, and it assumes a nonplanar "tub" conformation that avoids most of the overlap between adjacent π bonds. Hückel's rule simply does not apply.

Larger annulenes with (4N) systems do not show antiaromaticity because they have the flexibility to adopt nonplanar conformations. Even though [12]annulene, [16]annulene, and [20]annulene are (4N) systems (with N = 3,4, and 5, respectively), they all react as partially conjugated polyenes.

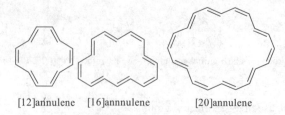

[12]annulene   [16]annnulene   [20]annulene

Aromaticity in the larger (4N+2) annulenes depends on whether the molecule can adopt the necessary planar conformation. In the all-cis [10]annulene, the planar conformation requires an excessive amount of angle strain. The [10]annulene isomer with two trans double bonds cannot adopt a planar conformation either, because two hydrogenatoms interfere with each other. Neither of these [10]annulene isomers is aromatic, even though each has ($4N$+2) π electrons, with N = 2. If the interfering hydrogen atoms in the partially trans isomer are removed, the molecule can be planar. When these hydrogen atoms are replaced with a bond, the aromatic compound naphthalene results.

all-*cis*   two *trans*   naphthalene
nonaromatic   nonaromatic   aromatic

6.3.2 Molecular Orbital Derivation of Hückel's Rule

Benzene is aromatic because it has a filled shell of equal-energy orbitals. The degenerate orbitals π_2 and π_3 are filled, and all the electrons are paired. Cyclobutadiene, by contrast, has an open shell of electrons. There are two half-filled orbitals easily capable of donating or accepting electrons. To derive Hückel's rule, we must show under what general conditions there is a filled shell of orbitals.

Recall the pattern of MOs in a cyclic conjugated system. There is one all-bonding, lowest-lying MO, followed by degenerate pairs of bonding MOs. (There is no need to worry about the antibonding MOs because they are vacant in the ground state.) The lowest-lying MO is always filled (two electrons). Each additional shell consists of two degenerate MOs, requiring four electrons to fill a shell.

A compound has a filled shell of orbitals if it has two electrons for the lowest-lying orbital, plus ($4N$) electrons, where N is the number of filled pairs of degenerate orbitals. The total number of π electrons in this case is ($4N$+2). If the system has a total of only ($4N$) electrons, it is two electrons short of filling N pairs of degenerate orbitals. There are only two electrons in the Nth pair of degenerate orbitals. This is a half-filled shell, and Hund's rule predicts these electrons will be unpaired (a diradical).

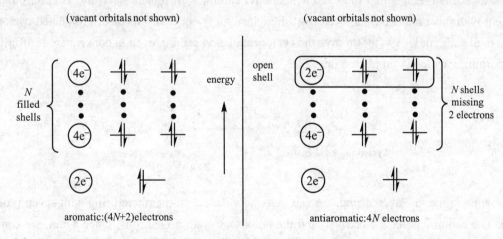

Figure 6.6   Pattern of molecular orbitals in a cyclic conjugated system. In a cyclic conjugated system, the lowest-lying MO is filled with two electrons. Each of the additional shells consists of two degenerate MOs, with space for four electrons. If a molecule has $4N$ pi electrons, it will have a filled shell. If it has ($4N$+2) electrons, there will be two unpaired electrons in two degenerate orbitals

图6–6   环状共轭体系中分子轨道的模式。在一个环状共轭体系中，能量最低的MO填充了两个电子。下面的每一个能量对应的核外的电子层由两个简并的分子轨道(MOs)组成，每层空间里有四个电子。如果一个分子有$4N$个电子，它的电子层就会被填满。如果它有($4N$+2)个电子，两个简并轨道中就会有两个不成对的电子

6.3.3　Aromatic Ions

Up to this point, we have discussed aromaticity using the annulenes as examples. Annulenes are uncharged molecules having even numbers of carbon atoms with alternating single and double bonds. Hückel's rule also applies to systems having odd numbers of carbon atoms and bearing positive or negative charges. We now consider some common aromatic ions and their antiaromatic counterparts.

We can draw a five-membered ring of sp^2 carbon atoms with all the unhybridized p orbitals lined up to form a continuous ring. With five pi electrons, this system would be neutral, but it would be a radical because an odd number of electrons cannot all be paired. With four pi electrons (a cation), Hückel's rule predicts this system to be antiaromatic. With six π electrons (an anion), Hückel's rule predicts aromaticity.

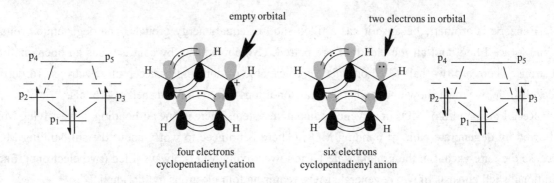

four electrons
cyclopentadienyl cation

six electrons
cyclopentadienyl anion

Because the cyclopentadienyl anion (six pi electrons) is aromatic, it is unusually stable compared with other carbanions. It can be formed by abstracting a proton from cyclopentadiene, which is unusually acidic for an alkene. Cyclopentadiene has a pK_a of 16, compared with a pK_a of 46 for cyclohexene. In fact, cyclopentadiene is nearly as acidic as water and more acidic than many alcohols. It is entirely ionized by potassium *tert*-butoxide.

Hückel's rule predicts that the cyclopentadienyl cation, with four π electrons, is antiaromatic. In agreement with this prediction, the cyclopentadienyl cation is not easily formed. Protonated cyclopenta-2,4-dien-1-ol does not lose water (to give the cyclopentadienyl cation), even in concentrated sulfuric acid. The antiaromatic cation is simply too unstable.

Cyclopenta-2,4-dien-1-ol　　$\xrightarrow{H_2SO_4}$　(does not occur)　not formed (four π electrons)　$+ H_2\overset{..}{O}:$

As with the five-membered ring, we can imagine a flat seven-membered ring with seven p orbitals aligned. The cation has six π electrons, and the anion has eight π electrons. Once again, we can draw resonance forms that seem to show either the positive charge of the cation or the negative charge of the anion delocalized over all seven atoms of the ring. By now, however, we know that the six-electron system is aromatic and the eight-electron system is antiaromatic (if it remains planar).

cycloheptatrienyl cation (tropylium ion): six π electrons, aromatic

cycloheptatrienyl anion: eight π electrons, antiaromatic (if planar)
The resonance picture gives a misleading suggestin of stability

The cycloheptatrienyl cation is easily formed by treating the corresponding alcohol with dilute (0.01 molar) aqueous sulfuric acid. This is our first example of a hydrocarbon cation that is stable in aqueous solution.

tropylium ion, six π electrons

The cycloheptatrienyl cation is called the **tropylium ion** (对离子). This aromatic ion is much less reactive than most carbocations. Some tropylium salts can be isolated and stored for months without decomposing.

We have seen that aromatic stabilization leads to unusually stable hydrocarbon anions such as the cyclopentadienyl anion. Dianions of hydrocarbons are rare and are usually much more difficult to form. Cyclooctatetraene reacts with potassium metal, however, to form an aromatic dianion.

ten pi electrons

The cyclooctatetraene dianion has a planar, regular octagonal structure with C−C bond lengths of 1.40 Å close to the 1.397 Å bond lengths in benzene. Cyclooctatetraene itself has eight π electrons, so the dianion has ten: ($4N+2$), with $N = 2$ the cyclooctatetraene dianion is easily prepared because it is aromatic.

6.3.4   Polynuclear Aromatic Hydrocarbons

The polynuclear aromatic hydrocarbons (abbreviated PAHs or PNAs) are composed of two or more fused benzene rings. **Fused rings** (稠环) share two carbon atoms and the bond between them.

1. Naphthalene (萘)   Naphthalene($C_{10}H_8$) is the simplest fused aromatic compound, consisting of two fused benzene rings. We represent naphthalene by using one of the three Kekulé resonance structures or using the circle notation for the aromatic rings.

The two aromatic rings in naphthalene contain a total of 10 π electrons. Two isolated aromatic rings would contain 6 π electrons in each aromatic system, for a total of 12. The smaller amount of electron density gives naphthalene less than twice there sonance energy of benzene: 252 kJ · mol⁻¹, or 126 kJ, per aromatic ring, compared with benzene's resonance energy of 151 kJ · mol⁻¹.

2. Anthracene and Phenanthrene (蒽和菲） As the number of fused aromatic rings increases, the resonance energy per ring continues to decrease and the compounds become more reactive. Tricyclic anthracene has a resonance energy of 351 kJ · mol⁻¹, or 117 kJ, per aromatic ring. Phenanthrene has a slightly higher resonance energy of 381 kJ · mol⁻¹, or about 127 kJ, per aromatic ring. Each of these compounds has only 14 π electrons in its three aromatic rings, compared with 18 electrons for three separate benzene rings.

anthracene   phenanthrene

6.4 Properties of Aromatic Compounds

PPT

6.4.1 Electrophilic Aromatic Substitution

Aromatic compounds undergo many reactions, but relatively few reactions that affect the bonds to the aromatic ring itself. Most of these reactions are unique to aromatic compounds. A large part of this chapter is devoted to **electrophilic aromatic substitution** (亲电芳香取代), the most important mechanism involved in the reactions of aromatic compounds. Many reactions of benzene and its derivatives are explained by minor variations of electrophilicaromatic substitution. We will study several of these reactions and then consider how substituents on the ring influence its reactivity toward electrophilic aromatic substitution and the regiochemistry seen in the products. We will also study other reactions of aromatic compounds, including nucleophilic aromatic substitution, addition reactions,

reactions of side chains, and special reactions of phenols.

Like an alkene, benzene has clouds of pi electrons above and below its sigma bond framework. Although benzene's pi electrons are in a stable aromatic system, they are available to attack a strong electrophile to give a carbocation. This resonance-stabilized carbocation is called a **sigma complex** because the electrophile is joined to the benzene ring by a new sigma bond.

The sigma complex (also called an *arenium ion*) is not aromatic because the sp^3 hybrid carbon atom interrupts the ring of p orbitals. Loss of aromaticity contributes to the highly endothermic nature of this first step. The sigma complex regains aromaticity either by a reversal of the first step (returning to the reactants) or by loss of the proton on the tetrahedral carbon atom, leading to the aromatic substitution product.

The overall reaction is the *substitution* of an electrophile (E^+) for a proton (H^+) on the aromatic ring: electrophilic aromatic substitution. This class of reactions includes substitutions by a wide variety of electrophilic reagents. Because it enables us to introduce functional groups directly onto the aromatic ring, electrophilic aromatic substitution is the most important method for synthesis of substituted aromatic compounds.

Electrophilic Aromatic Substitution

Step 1: Attack on the electrophile forms the sigma complex.

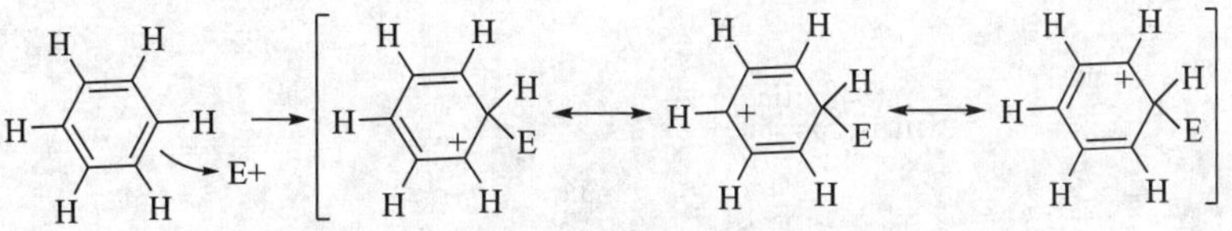

sigma complex (arenium ion)

Step 2: Loss of a proton regains aromaticity and gives the substitution product.

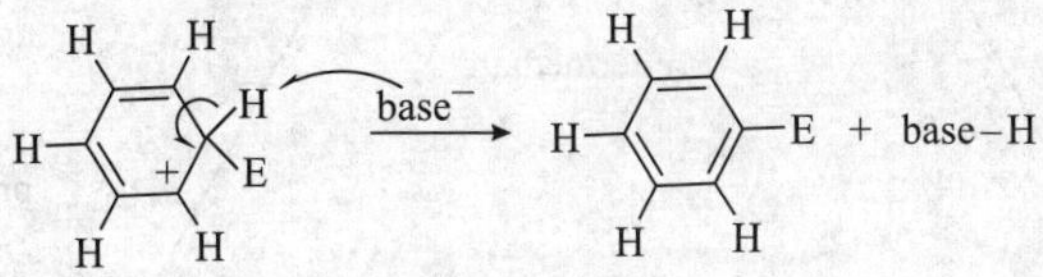

1. Halogenation of Benzene

(1) Bromination of benzene (苯的溴代)　As shown next, bromination follows the general mechanism for electrophilic aromatic substitution. Bromine itself is not sufficiently electrophilic to react with benzene, and the formation of bromine cation is difficult. A strong Lewis acid catalyzes the reaction, by forming a stronger electrophile with a weakened bond and a partial positive charge on one of the bromine atoms.

Step 1: Formation of a stronger electrophile.

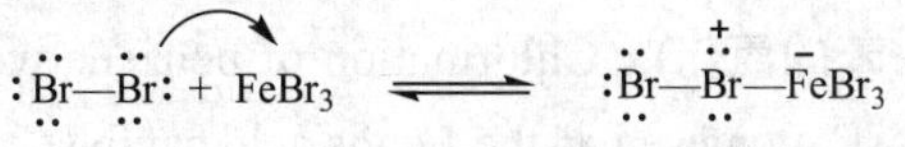

$Br_2 \cdot FeBr_3$ intermediate
(a stronger electrophile than Br_2)

Step 2: Electrophilic attack and formation of the sigma complex.

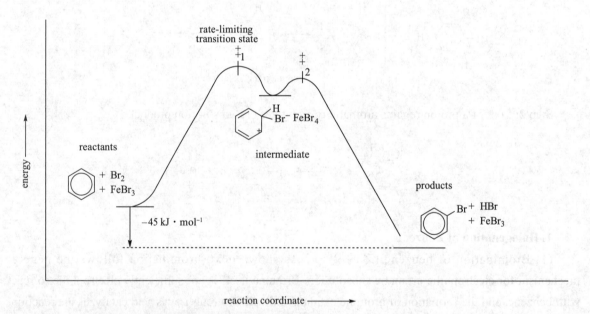

Step 3: Loss of a proton gives the products.

Formation of the sigma complex is rate-limiting, and the transition state leading to it occupies the highest-energy point on the energy diagram (Figure 6.7). This step is strongly endothermic because it forms a nonaromatic carbocation. The second step is exothermic because aromaticity is regained and a molecule of HBr is evolved. The overall reaction is exothermic by 45 kJ · mol^{-1}.

Figure 6.7 **The energy diagram for the bromination of benzene shows that the first step is endothermic and rate limiting and the second step is strongly exothermic**

图6-7　苯的溴代反应能量图,第一步为吸热和决速步骤,第二步为强效热反应

(2) Chlorination of benzene (苯的氯化)　Chlorination of benzene works much like bromination, except that aluminum chloride is most often used as the Lewis acid catalyst.

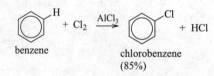

(3) Iodination of benzene (苯的碘化)　Iodination of benzene requires an acidic oxidizing agent, such as nitric acid. Nitric acid is consumed in the reaction, so it is a reagent (anoxidant) rather than a catalyst.

$$\text{benzene} + 1/2 I_2 + HNO_3 \longrightarrow \text{iodobenzene (85\%)} + NO_2 + H_2O$$

Iodination probably involves an electrophilic aromatic substitution with the iodinecation (I^+) acting as the electrophile. The iodine cation results from oxidation of iodine bynitric acid.

$$H^+ + HNO_3 + 1/2 I_2 \longrightarrow I^+ + NO_2 + H_2O$$
$$\text{iodine cation}$$

2. Nitration of Benzene　Benzene reacts with hot, concentrated nitric acid to give nitrobenzene. This sluggish reaction is hazardous because a hot mixture of concentrated nitric acid with any oxidizable material might explode. A safer and more convenient procedure uses a mixture of nitric acid and sulfuric acid. Sulfuric acid is a catalyst, allowing nitration to take place more rapidly and at lower temperatures.

$$\text{benzene} + HNO_3 \xrightarrow{H_2SO_4} \text{nitrobenzene(85\%)} + H_2O$$

The mechanism is shown next. Sulfuric acid reacts with nitric acid to form the nitronium ion, which is a powerful electrophile. The mechanism is similar to other sulfuricacid–catalyzed dehydrations. Sulfuric acid protonates the hydroxyl group of nitric acid, allowing it to leave as water and form a nitronium ion. The nitronium ion reacts with benzene to form a sigma complex. Loss of a proton from the sigma complex gives nitrobenzene.

Preliminary steps: Formation of the nitronium ion, NO_2^+.

Nitric acid has a hydroxyl group that can become protonated and leave as water, similar to the dehydration of an alcohol.

$$H-O-N=O + H-O-S-O-H \rightleftharpoons H-O-N=O + HSO_4^- \rightleftharpoons O=N=O + H_2O$$

Electrophilic aromatic substitution by the nitronium ion gives nitrobenzene.

Step 1: Attack on the electrophile forms the sigma complex.

benzene　　nitroium ion　　　　　　　　　　　　　sigma complex

Step 2: Loss of a proton gives nitrobenzene.

sigma complex
resonance-delocalized

nitrobenzene

Aromatic nitro groups are easily reduced to amino groups by treatment with an active metal such as tin, zinc, or iron in dilute acid. Nitration followed by reduction is often the best method for adding an amino group to an aromatic ring.

an alkylbenzene   a nitrated alkylbenzene   a substituted aniline

3. Sulfonation of Benzene   Aryl sulfonic acids are easily synthesized by sulfonation of benzene derivatives, an electrophilic aromatic substitution using sulfur trioxide as the electrophile.

benzene   sulfur trioxide   benzenesulfonic acid(95%)

"Fuming sulfuric acid" is the common name for a solution of 7% SO_3 in H_2SO_4 Sulfur trioxide is the **anhydride (酐)** of sulfuric acid, meaning that the addition of water to SO_3 gives H_2SO_4. Although it is uncharged, sulfur trioxide is a strong electrophile,with three sulfonyl (S=O) bonds drawing electron density away from the sulfur atom. Benzene attacks sulfur trioxide, forming a sigma complex. Loss of a proton on the tetrahedral carbon and reprotonation on oxygen gives benzenesulfonic acid.

sulfur trioxide, a powerful electrophile

Sulfur trioxide is a powerful electrophile.
Step 1: Attack on the electropile forms the sigma complex.

benzene   sulfur trioxide   sigma complex
(resonance-delocalized)

Step 2: Loss of a proton regenerates an aromatic ring.

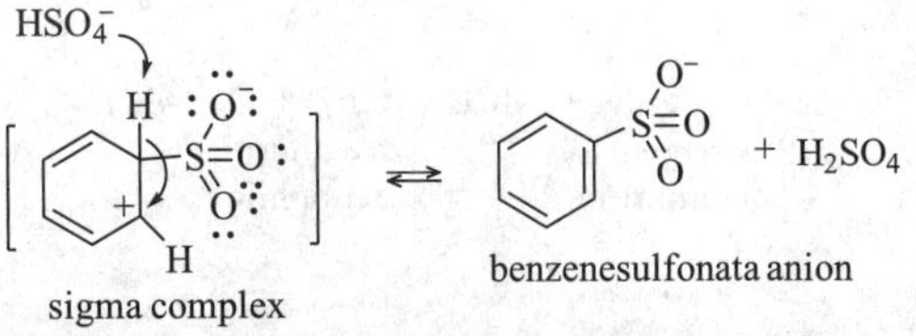

sigma complex　　　　　　benzenesulfonata anion

Step 3: The sulfonate group may become protonated in strong acid.

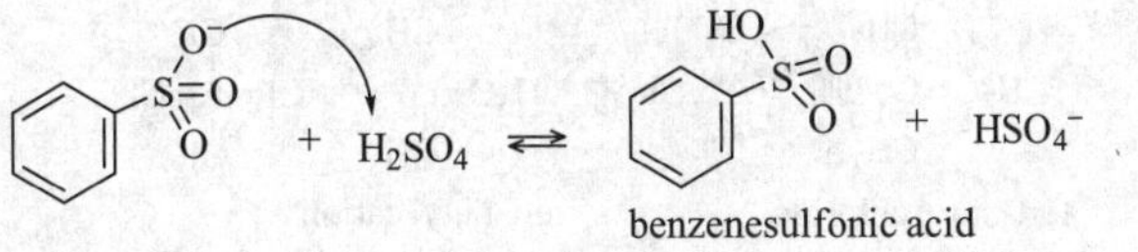

benzenesulfonic acid

Sulfonation is reversible, and a sulfonic acid group may be removed from an aromatic ring by heating in dilute sulfuric acid. In practice, steam is often used as a source of both water and heat for **desulfonation (脱磺化).**

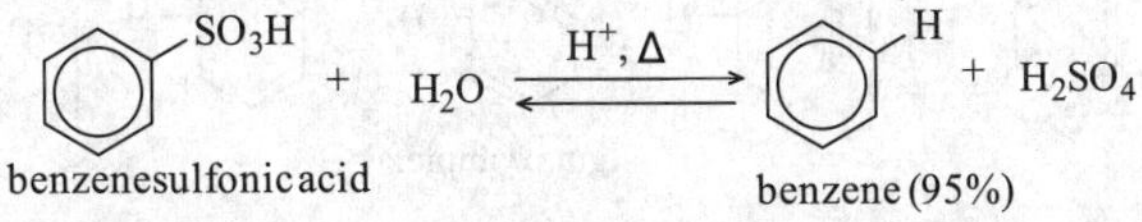

benzenesulfonic acid　　　　　　　benzene (95%)

4. The Friedel-Crafts Alkylation　Carbocations are perhaps the most important electrophiles capable of substituting onto aromatic rings, because this substitution forms a new carbon–carbon bond. Reactions of carbocations with aromatic compounds were first studied in 1877 by the French chemist Charles Friedel and his American partner, James Crafts. In the presence of Lewis acid catalysts such as aluminum chloride ($AlCl_3$) or ferric chloride ($FeCl_3$) alkylhalides were found to alkylate benzene to give alkyl benzenes. This useful reaction is called the **Friedel–Crafts alkylation (付氏烷基化).**

Friedel-Ceafts alkylation

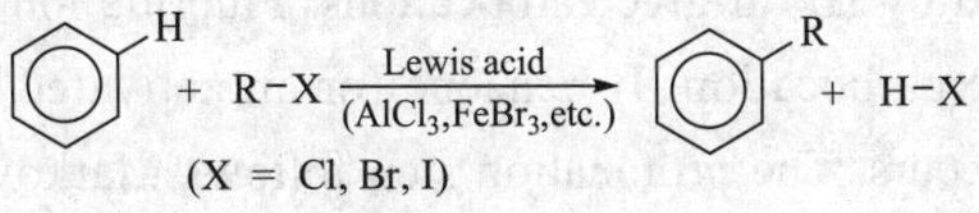

(X = Cl, Br, I)

For example, aluminum chloride catalyzes the alkylation of benzene by *tert*-butylchloride. HCl gas is evolved.

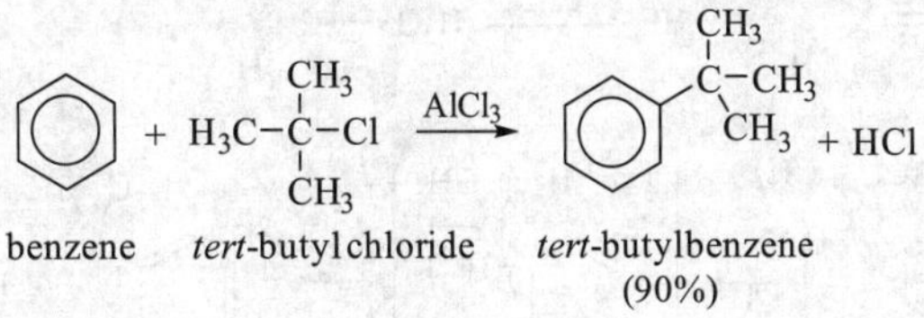

benzene　　*tert*-butyl chloride　　*tert*-butylbenzene
(90%)

This alkylation is a typical electrophilic aromatic substitution, with the *tert*-butyl cation acting as the electrophile. The *tert*-butyl cation is formed by reaction of *tert*-butyl chloride with the catalyst, aluminum chloride. The *tert*-butyl cation reacts with benzene to form asigma complex. Loss of a proton gives the product, *tert*-butylbenzene. The aluminumchloride catalyst is regenerated in the final step.

Friedel–Crafts alkylations are used with a wide variety of primary, secondary, andtertiary alkyl halides. With secondary and tertiary halides, the reacting electrophile is probably the carbocation.

$$R-X + AlCl_3 \rightleftharpoons R^+ + X-\bar{A}lCl_3$$

(R is secondary or tertiary) reacting electrophile

Friedel–Crafts Alkylation

Step 1: Formation of a carbocation.

H$_3$C-C-Cl : + Al-Cl $\rightleftharpoons$ H$_3$C-C$^+$ + Cl-Al$^-$Cl

tert-butyl chloride   tert-butyl cation

Step 2: Electrophilic attack forms a sigma comples.

sigma complex

Step 3: Loss of a proton regenerates the aromaticring and gives the alkylated prouct.

+AlCl$_3$

+HCl

We have seen several ways of generating carbocations, and most of these can be used for Friedel–Crafts alkylations. Two common methods are protonation of alkenes and treatment of alcohols with BF$_3$. Alkenes are protonated by HF to give carbocations. Fluoride ion is a weak nucleophile and does not immediately attack the carbocation. If benzene (or an activated benzene derivative) is present, electrophilic substitution occurs. The protonation step follows Markovnikov's rule, forming the more stable carbocation, which alkylates the aromatic ring.

H$_2$C=C + HF $\rightleftharpoons$ H$_3$C-C + F$^-$

+ HF

Alcohols are another source of carbocations for Friedel–Crafts alkylations. Alcohols commonly form carbocations when treated with Lewis acids such as boron trifluoride (BF$_3$). If benzene (or an activated benzene derivative) is present, substitution may occur.

Although the Friedel–Crafts alkylation looks good in principle, it has three major limitations that severely restrict its use.

Friedel–Crafts reactions work only with benzene, activated benzene derivatives, and halobenzenes. They fail with strongly deactivated systems such as nitrobenzene, benzenesulfonic acid, and phenyl ketones.

Like other carbocation reactions, the Friedel–Crafts alkylation is susceptible to carbocation rearrangements. As a result, it is impossible to make *n*-propylbenzene by the Friedel–Crafts alkylation.

Because alkyl groups are activating substituents, the product of the Friedel–Crafts alkylation is more reactive than the starting material. Multiple alkylations are hard to avoid.

5. The Friedel-Crafts Acylation   An **acyl group** (乙酰基) is a carbonyl group with an alkyl group attached. An **acyl chloride** is an acyl group bonded to a chlorine atom.

In the presence of aluminum chloride, an acyl chloride reacts with benzene (or an activated benzene derivative) to give a phenyl ketone: an acylbenzene. The **Friedel–Crafts acylation** (付氏酰基化) is analogous to the Friedel–Crafts alkylation, except that the reagent is an acyl chloride instead of an alkyl halide and the product is an acylbenzene (a "phenone") instead of an alkylbenzene.

Friedel–Crafts acylation

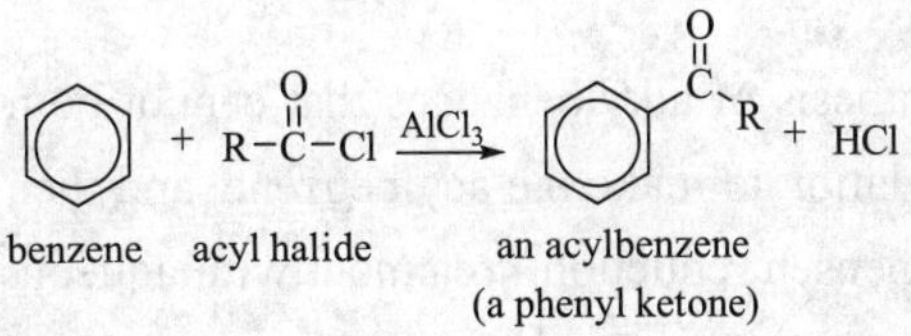

benzene   acyl halide   an acylbenzene
                            (a phenyl ketone)

Example

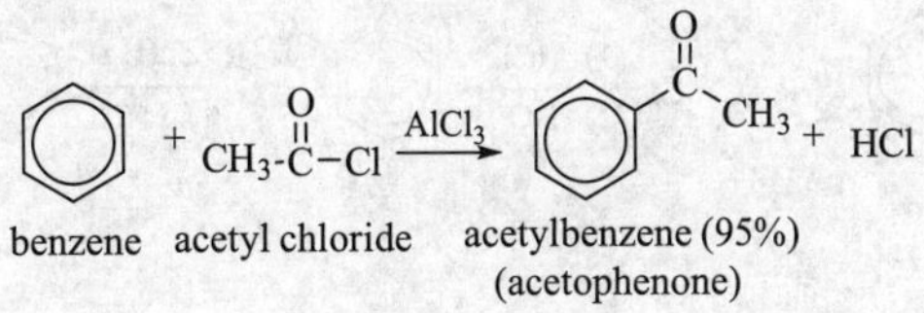

benzene   acetyl chloride   acetylbenzene (95%)
                            (acetophenone)

The mechanism of Friedel–Crafts acylation resembles that for alkylation, except that the electrophile is a resonance-stabilized acylium ion.

Friedel–Crafts Acylation

Step 1: Formation of an acylium ion.

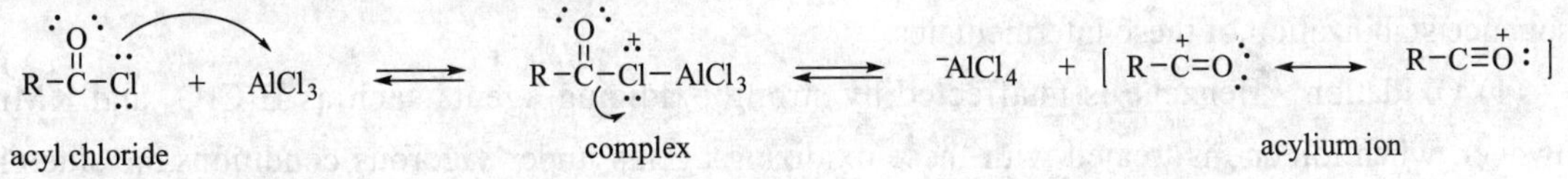

acyl chloride   complex   acylium ion

Steps 2 and 3: Electrophilic attack forms a sigma complex, and loss of a proton regenerates the aromatic system.

sigma complex   sigma complex   acylbenzene

Step 4: Complexation of the product. The product complex must be hydrolyzed (by water) to release the free acylbenzene.

acylbenzene   product complex   free acylbenzene

Friedel–Crafts acylation overcomes two of the three limitations of the alkylation. The acylium ion is resonance-stabilized, so that no rearrangements occur; and the acylbenzene product is deactivated, so that no further reaction occurs. Like the alkylation, however, the acylation fails with strongly deactivated aromatic rings.

Following shows the synthesis of alkylbenzenes that cannot be made by Friedel–Crafts alkylation. We use the Friedel–Crafts acylation to make the acylbenzene, and then we reduce the acylbenzene to the alkylbenzene using the Clemmensen reduction: treatment with aqueous HCl and amalgamated zinc (zinc treated with mercury salts).

6.4.2 Reactions at Benzylic Position

The alkyl substituted aromatic hydrocarbons may have reactions occuring at the benzylic position. Reactions involving alkyl side chains of aromatic compounds occur preferentially at the benzylic position for two reasons. First, the benzene ring is especially resistant to reaction with many of the reagents that normally attack alkanes. Second, benzyliccations and benzylic radicals are easily formed because of resonance stabilization of these intermediates.

(1) Oxidation Benzene is unaffected by strong oxidizing agents such as H_2CrO_4 and $KMnO_4$. However, when toluene is treated with these oxidizing agents under vigorous conditions,the side-chain methyl group is oxidized to a carboxyl group to give benzoic acid.

Toluene   Benzoic acid

Ethyl and isopropyl side chains are also oxidized to carboxyl groups. The side chain of tert-butylbenzene, however, is not oxidized.

$$H_3C-\overset{\overset{\displaystyle CH_3}{|}}{\underset{\underset{\displaystyle}{|}}{C}}-CH_3$$

tert-Butylbenzene $\xrightarrow{H_2CrO_4}$ No oxidation

From these observations, we conclude that if a benzylic hydrogen exists, then the benzylic carbon is oxidized to a carboxyl group and all other carbons of the side chain are removed as CO_2. If no benzylic hydrogen exists, as in the case of tert-butylbenzene, no oxidation of the side chain occurs.

(2) Halogenation   Reaction of toluene with chlorine in the presence of heat or light results in formation of chloromethylbenzene and HCl.

CH_3 + Cl_2 $\xrightarrow[\text{or light}]{\text{heat}}$ CH_2Cl + HCl

Toluene   Chloromethylbenzene
                           (Benzyl chloride)

Bromination is easily accomplished by using N-bromosuccinimide (NBS) in the presence of a peroxide catalyst. Halogenation of a larger alkyl side chain is highly regioselective, as illustrated by the halogenation of ethylbenzene. When treated with NBS, the only monobromo organic product formed is 1-bromo-1-phenylethane. This regioselectivity is dictated by the resonance stabilization of the benzylic radical intermediate.

$\xrightarrow[\text{(phCO}_2)_2, \text{CCl}_4]{\text{NBS}}$

Ethylbenzene   1-Bromo-1-phenylethane

6.4.3   The Effect of Substitution

When we introduce the limitation of Friedel-Crafts alkylations, we find that it is difficult to stop at monoalkylation unless reaction conditions are very carefully controlled. When a first alkyl group is introduced onto an aromatic ring, the ring is activated toward further alkylation. Illuminated by this kind of phenomenon, scientists fartherly studied the substitution effect and made the following generalizations about the manner in which existing groups influence further substitution reactions.

1. Substituents affect the orientation of new groups. Certain substituents (e.g., OCH_3) direct an incoming group preferentially to the *ortho* and *para* positions; other substituents (e.g., NO_2) direct it preferentially to the *meta* position. In other words, substituents on a benzene ring can be classified as *ortho-para* directing or as *meta* directing.

2. Substituents affect the rate of further substitution. Certain substituents cause the rate of a second substitution to be greater than that for benzene itself, whereas other substituents cause the rate of a second

substitution to be lower than that for benzene. In other words, groups on a benzene ring can be classified as activating or deactivating toward further substitution.

By combining the effect on orientation and reaction rate, we can classify the substituent into three groups: (Note, the groups in the same class are listed by the order of strength of the activity)

(1) Activating, *Ortho*, *Para*-Directors

$$-O^->-NR_2>-NHR >-NH_2>-OH, -OR >-NHCOR >-OCOR >-R >- CH_3$$

(2) Deactivating, *Meta*-Directors

$$-N^+R_3>-NO_2>-CF_3>-CCl_3>-CN>-SO_3H>-CHO >-COR >-COOR>-CONH_2$$

(3) Deactivating, *Ortho*, *Para*-Directors (halogen group)

From the above summary, we may find that the **resonance-donating (共轭给电子)** groups, like alkyl, amino, alkoxyl (or hydroxy) and halogen, which donate electron density through σ-π hyperconjugation or p-π conjugation are *ortho* and *para* directing.

For example, resonance forms show that the methoxy group effectively stabilizes the sigma complex if it is *ortho* or *para* to the site of substitution, but not if it is *meta*. Resonance stabilization is provided by a pi bond between the substituent and the ring.

Ortho attack

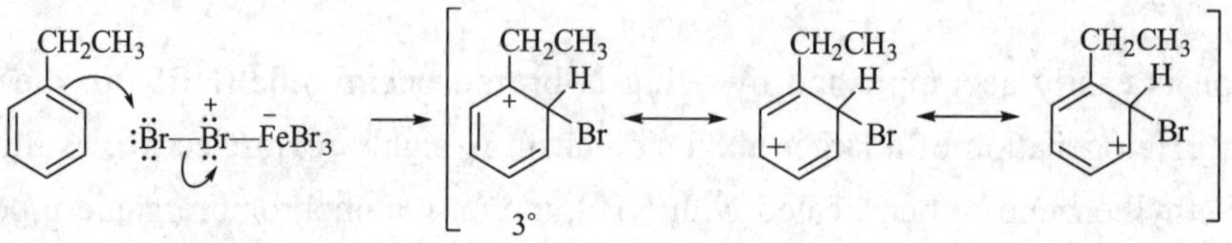

Meta attack

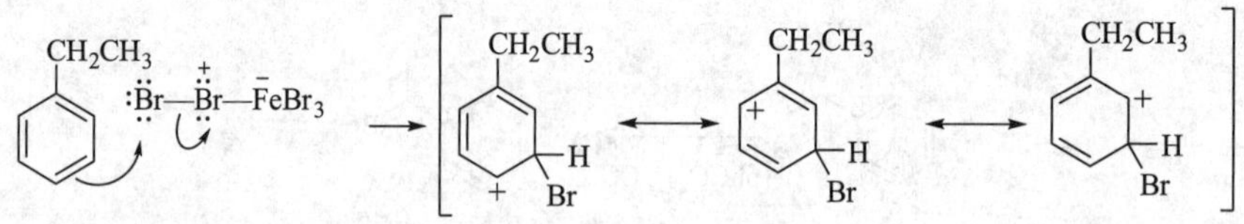

Para attack

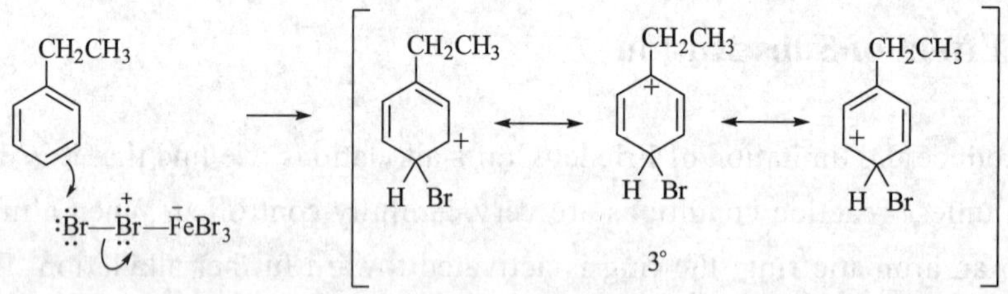

Similarly, we may find that the **resonance-withdrawing** groups, like nitro and carbonyl, which withdraw electrondensity through p-π conjugation are *meta* directing. *Meta*-directors deactivate the *meta* position less than the *ortho* and *para* positions, allowing *meta* substitution.

For example, in the sigma complex of nitrobenzene for *meta* substitution, the carbon bonded to the nitro group does not share the positive charge of the ring. This is a more stable situation because the positive charges are farther apart.

Ortho attack

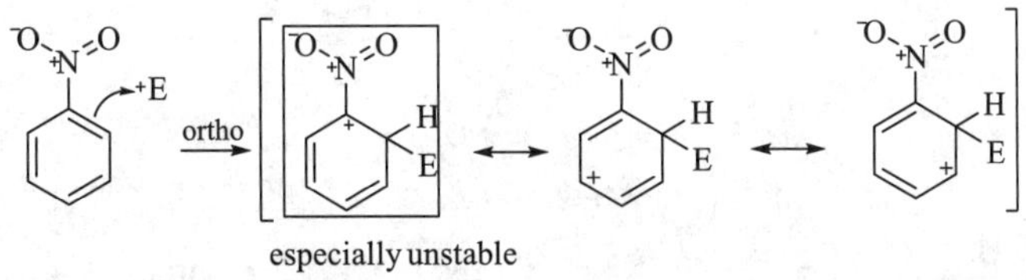

ortho

especially unstable

Meta attack

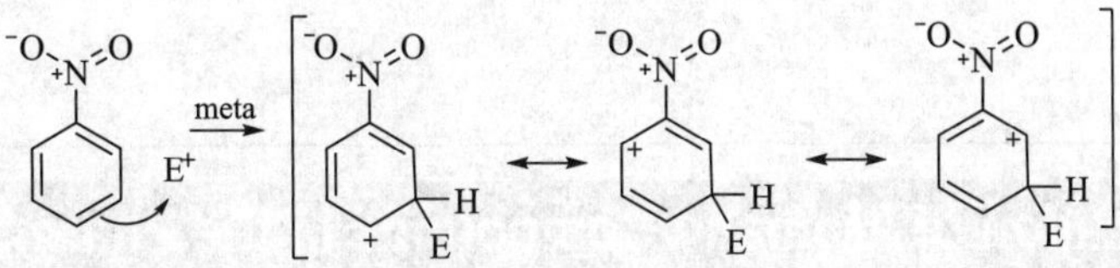

meta

Para attack

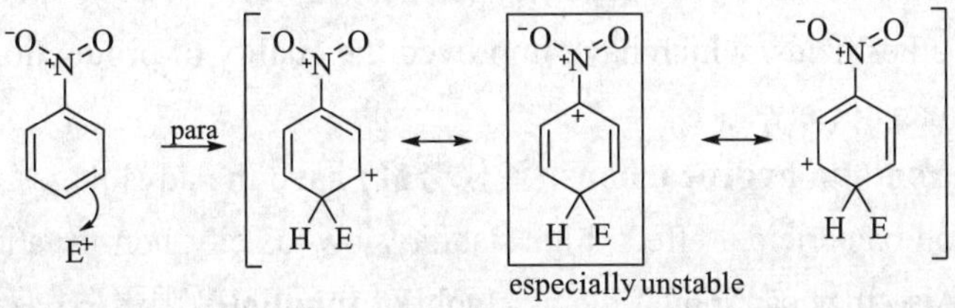

para

especially unstable

The effect on the rate of the reaction depends on the electronic effect. The electron-donating group activates the reaction, while the electron-withdrawing group deactivates the reaction. This is pretty easy to understand, since the rate-determining step is electrophilic. And the electronic effect can be obtained through the comprehensive analysis of the inductive effect and conjugation effect, which we have learned in Chapter 1. For example, an alkoxyl group has an oxygen atom which has an electron withdrawing inductive effect, however it also has a nonbonding pair of electrons serving as a powerful electron donating conjugation group. And when the two effects add up, alkoxyl group is a medium activating group.

We can illustrate the usefulness of these generalizations about the substitution effect by considering the synthesis of disubstituted derivatives of benzene. Suppose we want to prepare p-nitrobenzoic acid from toluene. The nitro group can be introduced with a nitrating mixture of nitric and sulfuric acids. The carboxyl group can be produced by oxidation of the methyl group of toluene. Nitration of toluene yields a product with the two substituents in the desired *para* relationship. Nitration of benzoic acid, on the other hand, yields a product with the substituents *meta* to each other. We see that the order in which the reactions are performed is critical.

$$\text{CH}_3 \xrightarrow[\text{H}_2\text{SO}_4]{\text{HNO}_3} \text{CH}_3\text{-NO}_2 \xrightarrow[\text{H}_2\text{SO}_4]{\text{K}_2\text{Cr}_2\text{O}_7} \text{COOH-NO}_2$$

p-Nitrobenzoic acid

$$\xrightarrow[\text{H}_2\text{SO}_4]{\text{K}_2\text{Cr}_2\text{O}_7} \text{COOH} \xrightarrow[\text{H}_2\text{SO}_4]{\text{HNO}_3} \text{COOH-NO}_2$$

m-Nitrobenzoic acid

6.5 Application of Aromatic Compounds in the Pharmacy Study

PPT

Halogenated aromatic hydrocarbons (卤代芳烃) are widely used in pesticides, medicine, synthetic dyes and other fields, which have improved the quality of production and life and promoted the development of human society.

Brominated aromatic hydrocarbons (溴代芳烃) have the advantages of **strong bacteriostati (抑菌作用)** ability, good bactericidal effect, high stability, low toxicity, non-irritating to the skin, non-allergenic and non-corrosive. Also they can inhibit the assembly of **tubulin(微观蛋白)** associated with carcinogenesis, aldose reductase related to diabetic complications and fat conversion enzymes related to trachea. Bromhexine hydrochloride can be used as the **anti-gout (抗痛风)** drug and benzbromarone is the spectrum anti-infective drug.

Bromhexine Hydrochloride

Benzbromarone

And the electrophilic aromatic reaction we learned in this chapter can be applied in the synthesis of many kinds of medicinal medicine. For example, p-nitrotoluene is the starting point for the syntheses of benzocaine and procaine, two compounds used as local anesthetics.

Benzocaine(R=H)
Procaine(R=NEt₂)

重 点 小 结

一、芳香化合物命名

IUPAC体系保留了几种较简单的单取代烷基苯的通用名称。从苯中失去一个H而得到的取代基是一个苯基，缩写为Ph–，由甲苯的甲基失去一个H而得到的是苄基，简写为Bn–。在含有其他官能团的分子中，苯基及其衍生物被称为取代基。

当苯环上有两个取代基时，可以通过给环上的原子编号或使用邻位、间位和对位来定位取代基命名。当环上有三个或三个以上取代基时，它们的位置用数字表示。如果其中一个取代基起了一个特殊的名字，那么这个化合物就被命名为这个分子的衍生物。如果没有一个取代基有一个特殊的名字，取代基就会被编号以给出最小的一组数字，并在苯的前面按字母顺序排列。

多环芳烃含有两个或两个以上的苯环，每对苯环共用两个环碳原子。萘、蒽和菲是最常见的多环芳烃。

二、芳香性的判断

Hückel's rule 导致苯的稳定性及其独特的反应性。如果分子是芳香族的，则必须满足几个通用标准。

1. 芳香分子必须是环状的。
2. 芳香分子必须是平面的。
3. 芳香环必须仅包含可形成 π 分子轨道离域系统的 sp^{2-} 杂化原子。
4. 离域 π 系统中 π 电子的个数必须等于 $4n + 2$，其中 n 是整数。

三、芳香化合物的化学性质

1. 亲电芳香取代

> 卤代反应：在催化剂存在的情况下，苯较容易和氯或者溴作用，生成氯苯或溴苯

> 硝化反应：苯与浓硫酸和浓硝酸共热，苯环上一个氢原子被硝基取代，生成硝基苯
> 烷基苯硝化能力大于苯

> 磺化：苯与浓硫酸或发烟硝酸反应，苯环上的氢被磺酸基取代生成苯磺酸
> 在合成中，常用磺化反应来保护对位

> 傅–克反应：芳烃中的氢被烷基取代的反应统称为烷基化反应，被酰基取代的反应称为酰基化反应，统称为傅–克反应

2. 加成反应　与烯烃相比不易发生，但在特殊条件下也可发生。在加热、加压、催化剂的条件下，可与 H_2 加成。在紫外线照射下可与氯气发生加成反应。

3. 烷基苯的侧链反应

（1）氧化反应　含 α–H 的烷基苯可被氧化剂氧化成苯甲酸。

（2）卤代反应　α–H 受苯环的影响而活化,烷基苯在光照或过氧化物等自由基引发剂的作用下与卤素发生自由基取代反应,与苯环直接相连的碳原子上的氢被取代(一般情况下,α–H 比 β–H 活性大)。

4. 定位基效应

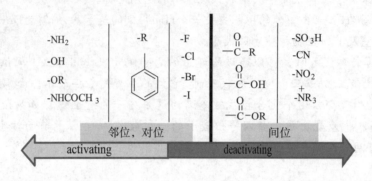

四、芳香化合物在医药领域的应用

在医药领域中主要的芳香化合物是卤代芳烃。常见的药物类型为溴代芳烃。而在农药领域中主要是多溴代芳烃。此外,它还可应用于阻燃剂、合成染料及荧光材料等。

题库

Problems
目 标 检 测

1. Which of the following does not belong to *ortho-para* directing group?

不属于邻对位定位基的是（　　　）

  A. Acetyl amino group   B. Methoxyl group

  C. Aminacetyl group   D. —Cl

2. Which of the following is aromatic?

具有芳香性的是（　　　）。

  A. Cyclopropene anion

  B. Cyclobutadiene

  C. γ-pyran

  D. Cyclopentadiene Anion

3. Which is the main reason that amino ($-NH_2$) has both *ortho* and *para* positioning effects and can activate benzene rings?

氨基（—NH_2）既有邻、对位定位效应又可使苯环致活的主要原因是（　　　）。

  A. +I 效应

  B. 供电子p-π共轭效应（+C ）

  C. electron-withdrawing p-π conjugate effec（-C ）

  D. π-π conjugate effect

4. Which of the following compounds is more reactive than benzene substitution?

医药大学堂
WWW.YIYAODXT.COM

下列化合物中比苯取代反应活性强的是（   ）。

    A. Benzenesulfonic acid   B. Chlorobenzene

    C. Phenol   D. Nitrobenzene

5. Which of the following carbocations is the most stable?

下列碳正离子最稳定的是（   ）。

    A. $C_6H_5\overset{+}{C}HCH_3$ B. $(C_6H_5)_2\overset{+}{C}CH_3$

    C. $(C_6H_5)_3\overset{+}{C}$ D. $CH_3\overset{+}{C}HCH_3$

6. Which of the following is not aromatic.

下列不具有芳香性的是（   ）。

    A.   B.

    C.   D.

7. The activity sequence of the following aromatic hydrocarbon electrophilic substitution reactions is (   ).

下列芳香烃亲电取代反应的活性顺序是（   ）。

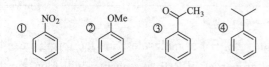

    A. ① > ④ > ③ > ②   B. ② > ④ > ③ > ①

    C. ② > ① > ③ > ④   D. ③ > ④ > ② > ①

8. The electrical effect present in bromobenzene is (   ).

溴苯中存在的电效应是（   ）。

    A. σ-p hyperconjugation

    B. π-π conjugate

    C. p-π conjugate

    D. σ-π hyperconjugation

9. Draw a detailed mechanism for the reaction of ethylbenzene with bromine, and show why the sigma complex (and the transition state leading to it) is lower in energy for substitution at the *ortho* and *para* positions than it is for substitution at the *meta* position.

画出乙苯与溴反应的详细机理，并说明为什么在邻位和对位取代时，σ络合物（及其导致的过渡态）的能量比在间位取代的能量低的原因。

Discussion Topics

Explain why *m*-xylene undergoes nitration 100 times faster than *p*-xylene.

（朱　静　刘晓芳）

第七章 卤 代 烃
Chapter 7　Alkyl Halides

学习目标

1. **掌握** 卤代烃的结构、命名和化学性质,双键位置对卤烃活泼性的影响和不同结构卤代烃的鉴别。

2. **熟悉** 卤代烃亲核取代反应、消除反应的历程,卤代烃的制备和物理性质。

3. **了解** 了解重要的卤烃及其在医药领域的应用。

Alkyl halides (卤代烃) contain the **halogen atom** (卤原子) in molecules, and the formula can be expressed as R−X (X=F, Cl, Br, I). Due to the special properties of R-F, this chapter will focus on the structure and properties of R−Cl, R−Br, R−I. The halogen atom is the functional groups of alkyl halides.

There are not many kinds of alkyl halides in nature, most of which are synthetic. Halogen atoms of alkyl halides can be transformed into other functional groups. It is active and can prepare a variety of compounds, which play an important bridge role in organic synthesis. Some alkyl halides can be used as solvents, **pesticides** (农药), **refrigerants** (制冷剂), **anesthetics** (麻醉剂), **preservatives** (防腐剂), **extinguishing agents** (灭火剂), etc. Alkyl halides are important compounds and play an important role in organic chemistry. In many drug molecules, the introduction of halogen has obvious changes in **pharmacological activity** (药理活性) and **pharmacokinetic** (药物动力学) characteristics, so alkyl halides are often used as intermediates for drug synthesis.

7.1　Structure of Alkyl Halides

PPT

In an alkyl halide, the halogen atom is bonded to a carbon atom. The halogen is more electronegative than carbon, and the bond is polarized with a partial positive charge on carbon and a partial negative charge on the halogen.

The **electronegativities** (电负性) of the halogens increase in the order: F>Cl>Br>I. But the C−X bond (Figure 7.1) lengths increase as the halogen atoms become bigger (larger atomic

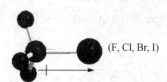
(F, Cl, Br, I)

Figure 7.1　Schematic diagram of polarity of C−X bonds in halogenated hydrocarbon molecules

图7-1　卤代烃分子中C−X键极性示意图

radii) in the order: C−I>C−Br>C−Cl>C−F. The **dipole moment** (偶极矩), bond length and bond energy of C−X are shown in Table 7.1.

Table 7.1   The dipole moment, bond length and bond energy of CH₃—X

表7-1　四种CH₃—X的键长、偶极矩和键能

Compound	Bond length/pm	Polarity/(μ/D)	Bond energy/(kJ · mol⁻¹)
CH_3F	138.2	1.85	485.6
CH_3Cl	178.1	1.87	339.1
CH_3Br	193.9	1.81	284.6
CH_3I	213.9	1.62	217.8

From Table 7.1, it can be seen that the reactivity of the alkyl halides is RI > RBr > RCl. As the electronegativity of the halogen atom decreases with increasing of the **atomic radius** (原子半径), the binding force on the outer electrons is weakened, and the outer electrons are prone to flow. The reactivity is higher.

7.2 Nomenclature of Alkyl Halides

7.2.1 Common Nomenclature of Alkyl Halides

Common names are useful only for simple alkyl halides, and are constructed by naming the alkyl group and then the halide.

$CH_3CH_2CH_2CH_2Br$　　$CH_2{=}CHBr$　　$CH_2{=}CHCH_2Cl$　　〇—Cl　　〇—CH₂Br

n-butyl bromide　　vinyl bromide　　allyl chlroride　　chlorobenzene　　benzyl bromide

Polysubstituted halides are named using **trivial name** (俗名).

$CHCl_3$　　　　　CHI_3

chloroform (trichloromethane)　　iodoform (triiodomethane)

7.2.2 Systematic Nomenclature of Alkyl Halides

The systematic (IUPAC) nomenclature treats an alkyl halide as an alkane with a halo- substituent: Fluorine is fluoro-, chlorine is chloro-, bromine is bromo-, and iodine is iodo-. The result is a systematic alkyl halide name, as in halo-alkane.

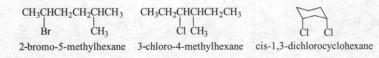

2-bromo-5-methylhexane　　3-chloro-4-methylhexane　　cis-1,3-dichlorocyclohexane

A vinyl halide is often treated as an alkene with a halo-substituent.

$$CH_3CH_2\underset{H}{\overset{}{C}}=\underset{Cl}{\overset{CH(CH_3)_2}{C}}$$

(E)-3-chloro-2-methylhex-3-ene

3-bromocyclohexene

An aryl halide is often treated as substituted benzenes.

2-chloronaphthalene

1,2-dibromobenzene

7.3 Physical Properties of Alkyl Halides

PPT

At room temperature, **alkyl fluorides (氟代烃)** below four carbon atoms and **alkyl chlorides (氯代烃)** below two carbon atoms are gases, other common alkyl halides are generally liquid, and alkyl halides with more than 15 carbon atoms are solid.

Due to the substitution of halogen atoms, the polarity of alkyl halides increases, and the boiling point of alkyl halide is higher than that of alkane with similar molecular weight. Molecules with higher molecular weights generally have higher boiling points because they are heavier (and therefore slower moving), and they have greater surface area. The surface areas of the alkyl halides vary with the surface areas of halogens. Notice that compounds with branched, more spherical shapes have lower boiling points as a result of their smaller surface areas. This effect is similar to the one we saw with alkanes.

Like their boiling points, their densities follow a predictable trend. Alkyl fluorides and alkyl chlorides (those with just one chlorine atom) are less dense than water ($1.00 \text{ g} \cdot \text{ml}^{-1}$). Alkyl chlorides with two or more chlorine atoms are denser than water, and all alkyl bromides and alkyl iodides are denser than water. Table 7.2 lists the boiling points and densities of some simple alkyl halides.

Table 7.2 Physical properties of alkyl halides
表7-2 卤代烃的物理性质

Compound	Boiling point / ℃	Density / (g · ml^{-1})
CH_3Cl	−24	0.92
CH_3Br	4	1.73
CH_3I	42	2.28
$CHCl_3$	61	1.50
$CHBr_3$	150	2.89
CHI_3	−	4.01
CH_3CH_2Cl	13	0.92
CH_3CH_2Br	38	1.461

Continued

Compound	Boiling point / ℃	Density / (g · ml⁻¹)
CH_3CH_2I	72	1.93
$CH_3CH_2CH_2Cl$	46	0.89
$CH_3CH_2CH_2Br$	71	1.35
$CH_3CHClCH_3$	35	0.86
$CH_3CHBrCH_3$	60	1.31
$CH_2=CHCH_2Cl$	45	0.94
$CH_2=CHCH_2Br$	70	1.43
C_6H_5Cl	132	1.11
C_6H_5Br	156	1.50
o-CH_3-C_6H_4Cl	159	1.08
o-CH_3-C_6H_4Br	182	1.42
p-CH_3-C_6H_4Cl	162	1.07
p-CH_3-C_6H_4Br	184	1.39
$C_6H_5CH_2Cl$	179	1.10
$C_6H_5CH_2Br$	198	1.44

The alkyl halides are toxic, in particular to the liver. The vapor of the chlorinated hydrocarbon and the iodine-producing hydrocarbon can cause damage to the human body through the absorption of the skin, and special attention should be paid to the use of such a solvent.

7.4　Reactions of Alkyl Halides

PPT

In alkyl halides, the electronegativity of halogen atoms is larger than that of carbon atoms, and the electron cloud in C–X is biased towards halogen atoms. The halogen atom can leave with its bonding pair of electrons to form a stable halide ion. When another group replaces the halide ion, the reaction is a **substitution** (取代). Alkyl halides are easily converted to many other functional groups, such as OH⁻, NH_2^-, CN⁻, et al. Notice that halide ions and silver ions can produce precipitation; this reaction can be used to identify the alkyl halides with different structures.

Nucleophilic substitution：

$$\overset{\delta^+}{R}-\overset{\delta^-}{X} + Nu^- \longrightarrow RNu + X^-$$

When the halide ion leaves with another atom or ion (often) and forms a new π bond, the reaction is an **elimination** (消除). In many eliminations, a molecule of is lost from the alkyl halide to give an alkene. These eliminations are called **dehydrohalogenations** (脱卤化氢) because a hydrogen halide has been removed from the alkyl halide. In the elimination, the reagent reacts as a base, abstracting a proton from the alkyl halide.

Elimination：

$$\overset{\underset{|}{H}}{-\underset{|}{C}}-\overset{\underset{|}{X}}{\underset{|}{C}}- \xrightarrow{B^-} \ >C{=}C< \ + \ HB + X^-$$

Substitution and elimination reactions often compete with each other. Most nucleophiles are also basic and can engage in either substitution or elimination, depending on the alkyl halide and the reaction conditions. Besides alkyl halides, many other types of compounds undergo substitution and elimination reactions. Substitutions and eliminations are introduced in this chapter using the alkyl halides as examples.

7.4.1　Nucleophilic Substitution in Alkyl Halides

In substitution, because the carbon atoms often have positive ions, the carbon atoms should be attacked by the nucleophilic reagents which have negative ions or be rich in electrons. This kind reaction is **nucleophilic substitution** (亲核取代反应, S_N).

A nucleophilic substitution has the general form.

$$\overset{\delta^+}{R}{-}\overset{\delta^-}{X} + Nu^- \longrightarrow RNu + X^-$$

where Nu^- is the nucleophile，X^- is the **leaving group (离去基团)**, RCH_2X is the **substrate (底物)**, and RNu is the product.

1. Different Kinds of Nucleophilic Substitution in Alkyl Halides

(1) React with H_2O　When alkyl halides and water are mixed and heated, hydroxy group replaces the halide atom and produces alcohols. The reaction is a hydrolytic reaction.

$$R{-}X + H_2O \underset{\triangle}{\overset{}{\rightleftharpoons}} ROH + HX$$

This is a **reversible reaction (可逆反应)**. Due to accelerate reaction velocity and increase the yield of alcohols, NaOH or KOH solution is often used to replace water.

$$R{-}X + NaOH \xrightarrow{H_2O} ROH + NaX$$

(2) React with ROH　When alkyl halides and ROH react, alkoxy group replaces the halide atom and produces ethers. This is also a reversible reaction. To promote the reaction NaOR is often used to replace ROH.

$$R{-}X + NaOR' \longrightarrow ROR' + NaX$$

This is a main method to produce ethers, named as Williamson ether synthesis. This method may produce **simple ethers (简单醚)** or **mixed ethers (混合醚)**. Using this method, it is the best choice to select the **primary alkyl halide (一级卤代烷)**, because the **tertiary alkyl halide (三级卤代烷)** would mainly produce alkenes instead of ethers in base conditions.

$$CH_3CH_2Cl + (CH_3)_3CONa \longrightarrow CH_3CH_2OC(CH_3)_3 + NaCl$$

(3) React with NaCN   When alkyl halides and NaCN react in alcohols solution, **cyano group (氰基）** replaces the halide atom and produces **nitriles (腈)**. The nitrile can be fourthly transferred to carboxylic acid, **amides (酰胺)**, **amines (胺)** etc.

$$R-X + NaCN \longrightarrow RCN + NaX$$

This reaction is a common method for the synthesis of **carboxylic acids (羧酸)**. But NaCN is highly poison, we must take special caution when using it.

$$\langle \rangle-CH_2Cl + NaCN \longrightarrow \langle \rangle-CH_2CN \xrightarrow[H^+]{H_2O} \langle \rangle-CH_2COOH$$

(4) React with NaC≡CR   Alkyl halides and NaC≡CR react to yield alkynes, and this is the important method to synthesis long-chain alkynes from short-chain alkynes.

$$R-X + R'C≡CNa \longrightarrow RC≡CR' + NaX$$

Notice that **sodium alkynide (炔基钠)** is a strong base, it only reacts with the primary halide to produce alkynes. This method can not be used with the **secondary alkyl halide (二级卤代烷)** and tertiary alkyl halide, because they are prone to elimination reactions.

(5) React with NH_3 Alkyl halides and NH_3 react to yield **ammonium salts (铵盐)**.

$$R-X + NH_3 \longrightarrow R\overset{+}{N}H_3 \, X^-$$

Ammonium salt can produce amine with strong base, such as sodium hydroxide, potassium hydroxide et al. Amine can react fatherly with alkyl halides to yield finally the mixture of amines.

$$R\overset{+}{N}H_3 \, X^- + NaOH \longrightarrow RNH_2 + H_2O + NaX$$
$$\xrightarrow{RX} R_2NH + HX$$
$$\xrightarrow{RX} R_3N + HX$$

(6) React with $AgNO_3$ Alkyl halides and $AgNO_3$ react to yield nitric ether and **silver halide precipitation (卤化银沉淀)** in alcohols solution.

$$R-X + AgNO_3 \xrightarrow{EtOH} RONO_2 + AgX\downarrow$$

Because halide ions and silver ions can produce precipitation, this reaction can be used to identify the alkyl halides with different structures.

When the structure of alkyl is same, the reactivity of alkyl halides is different, such as RI>RBr>RCl. When the kind of halide atom is same, the reactivity of alkyl halides is different, such as follows.

$$\begin{matrix} CH_2{=}CHCH_2X \\ PhCH_2X \end{matrix} > R_3CX > R_2CHX > RCH_2X > \begin{matrix} CH_2{=}CHX \\ PhX \end{matrix}$$

Allyl halides (烯丙型卤烃), **benzyl halides (苄基型卤烃)**, tertiary alkyl halides, and **alkyl iodide (碘代烷)** will immediately produce precipitation in room temperature. Primary alkyl halides and secondary alkyl halides will produce precipitation by heating in some minutes. Vinyl halides and phenyl halides will be hard to produce precipitation even by heating.

(7) React with NaI   When chloroalkane or bromoalkane react with NaI in acetone solution, iodine ion replaces the chlorine or bromine ions. The reaction is a **halogen exchange reaction** (卤素交换反应).

$$\begin{matrix} RCl \\ RBr \end{matrix} + NaI \xrightarrow{\text{acetone}} RI + \begin{matrix} NaCl \\ NaBr \end{matrix} \downarrow$$

This is a reversible reaction. To increase yield, acetone is used as solvent because NaCl and NaBr can precipitate in it and thus be removed from the reaction mixture.

2.Mechanism of Nucleophilic Substitution in Alkyl Halides   The nucleophilic substitution reaction (S_N) can be divided into two different processes **single molecule nucleophilic substitution reaction mechanism** (单分子亲核取代反应机理, S_N1) and **bimolecular nucleophilic substitution reaction mechanism** (双分子亲核取代反应机理, S_N2).

(1) Bimolecule nucleophilic substitution reaction mechanism (S_N2)   The reaction rate of bromomethane in 80% alcoholic solution is very slow, but the reaction rate increases with adding sodium hydrate solution. The reaction is therefore first order in each of the reactants and second order for the overall reaction. The rate equation has the following form:

Rate $\nu = k\,[CH_3Br]\,[OH^-]$

Mechanism as follows:

$$OH^- + C{-}Br \xrightarrow{\text{slow}} \left[HO{\cdots}C{\cdots}Br \right] \longrightarrow HO{-}C{\cdots}H + Br^-$$

nucleophile   substrate   transition state   product   leaving group

The reaction of **bromomethane** (溴甲烷) with hydroxide ion is a concerted reaction, taking place in a single step with bonds breaking and forming at the same time. This one-step nucleophilic substitution is a characteristic of the S_N2 mechanism. Hydroxide ion attacks the back side of the **substrate** carbon atom, donating a pair of electrons to form a new bond. Notice that curved arrows are used to show the movement of electron pairs, from the electron-rich nucleophile to the electron-poor carbon atom of the substrate. Then the hybrid state of the carbon atom transforms from sp^3 to sp^2. **In transition state** (过渡态), the hybrid state of the carbon atom is sp^2. The carbon atom and the three hydrogen atoms are in the same plane, and hydroxyl and bromine are on the opposite side. Carbon can accommodate only eight electrons in its **valence shell** (价电子层), so the C–Br bond must begin to break as the C–O bond begins to form. Bromine ion is the **leaving group** (离去基团). It leaves with the pair of electrons that once bonded it to the carbon atom. Moreover, the hybrid state of the carbon atom recovers sp^3, which is **tetrahedral configuration** (四面体构型).

The reaction-energy diagram for this substitution is shown as Figure 7.2.

With the variation of the substrate structure, the **system energy** (体系能量) will constantly change. Hydroxide ion attacks the back side of the substrate carbon atom, and three C–H bonds are squashed together on some plane. In order to overcome the resistance of hydrogen atom, the system energy is rising. The transition state is the highest in energy because it involves a five coordinate carbon atom with two partial bonds. When the bromine is leaving, the system energy is gradually decreasing. As this reaction is consistent with a mechanism that requires a collision between a molecule of methyl bromine and a hydroxide ion, and the middle structure is a transition state rather than an intermediate. This substitution is called bimolecular nucleophilic reaction（S_N2）.

In the S_N2 mechanism, the reaction requires attack by a nucleophile on the back side of a substrate carbon atom. A carbon atom can have only four **filled bonding orbitals (成键轨道)**, so the leaving group must leave as the nucleophile bonds to the carbon atom. The electrons of the nucleophile insert into the back lobe of carbon's hybrid orbital in its antibonding combination

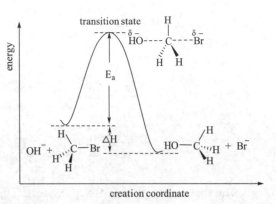

Figure 7.2　The reaction-energy diagram for the S_N2 reaction

图7-2　S_N2反应能级变化示意图

with the orbital of the leaving group. The transition state shows partial bonding to both the nucleophile and the leaving group. Back-side attack literally turns the tetrahedron of the carbon atom inside out, like an umbrella caught by the wind. In the product, the nucleophile assumes a **stereochemical (立体化学的)** position opposite the position the leaving group originally occupied. We call this result **an inversion of configuration at the carbon atom (碳原子的构型翻转)** or **Walden inversion (瓦尔登转化)**.

In the S_N2 mechanism, back-side attack (Figure 7.3) turns the tetrahedron of the carbon atom inside out, like an umbrella caught by the wind. In the product, the nucleophile assumes a stereochemical position opposite to the position the leaving group originally occupied. We call this result an **inversion of configuration at the carbon atom (碳原子的构型翻转)** or **Walden inversion (瓦尔登转化)**.

back-side attack on C-Br sp³ orbital　　transition state　　　　products

Figure 7.3　Back-side attack in the reaction

图7-3　反应背面进攻示意图

For example, when (S)-2-bromobutane undergoes displacement by hydroxide ion, inversion of configuration gives (R)-butan-2-ol.

(S)-2-bromobutane　　　　　　　　　　　(R)-butan-2-ol

(2) single nucleophilic substitution reaction mechanism（S_N1）The reaction rate of tert-butyl bromide in basic solution is found to be proportion to the concentration of alkyl halides, however independent on the concentration of the nucleophile (OH^-). The reaction is therefore first order overall. The rate equation has the following form.

Rate $v = k \, [(CH_3)_3CBr]$

Mechanism as follows:

Step 1: $(CH_3)_3C—Br \xrightarrow{slow} \left[(CH_3)_3\overset{\delta+}{C} \cdots \overset{\delta-}{Br} \right]^{\ddagger} \longrightarrow (CH_3)_3C^+ + Br^-$

     the substrate   transition state 1 the carbocation intermediate

Step 2: $(CH_3)_3C^+ + OH^- \xrightarrow{fast} \left[(CH_3)_3\overset{\delta+}{C} \cdots \overset{\delta-}{OH} \right]^{\ddagger} \longrightarrow (CH_3)_3C—OH$

     the nucleophile transition state 2   the product

The mechanism is a two-step process. The first step is that tert-butyl bromide in solution ionizes a tert-butyl carbocation and bromide negative ions intermediate. Because no nucleophile is assisting the departure of the halide anion, this is the relatively slow, **rate-determining step (决速步骤)** of the reaction. The second step is a fast attack on the tert-butyl carbocation by a nucleophile（OH^-）, then produce tert-butanol.

The reaction-energy diagram for this substitution is showed as Figure 7.4.

The carbocation forming in the first step is the intermediate. It is at the bottom of the two peaks of transition state and is highly active.

As shown in Figure 7.5, in the S_N1 reaction, the first step forms the carbocation intermediate, which is sp^2 hybridized and planar. The **central carbon atom (中心碳原子)** has an empty p-orbit distributed on both sides of the plane, and when the nucleophilic reagent is combined with the positive ion of the carbon, both sides of the plane may be attacked with equal probability. Such a process, giving both **enantiomers (对映体)** of the product in equal amounts, is called **racemization (外消旋化)**.

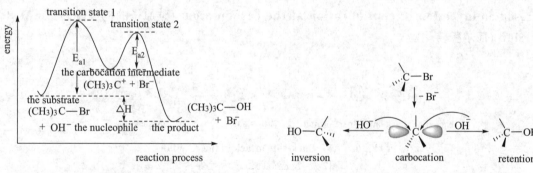

Figure 7.4 **Reaction-energy diagrams of the S_N1 reactions**

图7-4 **S_N1反应的能级变化示意图**

Figure 7.5 **Schematic diagram of racemization of S_N1 reaction mechanism**

图7-5 **S_N1反应机制外消旋化示意图**

In S_N1 reaction，the product is either racemic or at least less optically pure than the starting material. If a nucleophile attacks the carbocation from the front side (the side the leaving group left), the product molecule shows retention of configuration. Attack from the back side gives a product showing inversion of configuration. Racemization is simply a combination of retention and inversion. When racemization occurs, the product is rarely completely racemic, however, there is often more inversion than **retention of configuration (构型保持)**. As the leaving group leaves, it partially blocks the front side of the carbocation. The back side is unhindered, so attack is more likely there. For example: (R)-2,6-dimethyl-6-chlorooctane reaction in the 80% acetone solution produces 60.5% the inversion product and 39.5% the retention product.

(R)-6-chloro-2,6-dimethyloctane	(R)-3,7-dimethyloctan-3-ol	(S)-3,7-dimethyloctan-3-ol
	39.5%	60.5%

In S_N1 reaction mechanism, the carbocation frequently undergoes structural changes, called **rearrangements** (重排), to form more stable ions. It is also called **Wangner-Meerwein rearrangements** (瓦格涅尔-麦尔外因重排).

This is a characteristic property of the carbocation, and can be divided into two rearrangement types: a hydride shift or an alkyl shift. No matter a hydride shift or an alkyl shift is involved in the rearrangement, the ultimate driving force is to form the most stable carbocation. Most rearrangements convert $2°$ (or incipient $1°$) carbocation to $3°$ or **resonance-stabilized carbocation** (共振稳定碳正离子).

3. Comparison of S_N1 and S_N2 Mechanism

(1) Reaction Characteristics　It is shown in Table 7.3.

Table 7.3　Comparison of S_N1 and S_N2 reaction characteristics
表7-3　两种反应机制的反应特点比较

Reaction type	S_N1	S_N2
Reaction rate	$v= k\,[RX]$	$v= k\,[RX]\,[Nu^-]$
Reaction process	Two steps	One step
Intermediate	Carbocation and rearrangement	No intermediate
Stereochemistry	racemization	inversion of configuration

(2) Effect of substrate　In S_N1 and S_N2 reaction, the effect of substrate is different, because the reaction process is different. In general, the reaction activity shows as follow.

S_N1 mechanism: $\begin{array}{c}CH_2=CHCH_2X\\ phCH_2X\end{array} > R_3CX > R_2CHX > RCH_2X > \begin{array}{c}CH_2=CHX\\ phX\end{array}$

S_N2 mechanism: $\begin{array}{c}CH_2=CHCH_2X\\ phCH_2X\end{array} > RCH_2X > R_2CHX > R_3CX > \begin{array}{c}CH_2=CHX\\ phX\end{array}$

1) Allylic halides (烯丙型卤烃) and benzylic halides (苄基型卤烃)

$$CH_2=CHCH_2X \qquad PhCH_2X$$
Allyl halides   benzyl halides

The structures of allylic halides and benzyl halides have the same characteristic: halogen atom and sp^2 hybrid-carbon is separated by one carbon. Both of them have high reactivity of the nucleophilic substitution reaction. As shown in Figure 7.6, in the S_N1 reaction, the carbocation formed by breaking the bond of C−X can forms **p-π conjugate effect（p–π共轭效应）**, and the system energy is decreasing, so the reaction rate is fast. In S_N2 reaction, the transition state energy of this kind of halide is very low, and the reaction rate is also fast. For example, the S_N2 reaction rate of 3-chloropropene with iodine anion was 73 times higher than that of 1-chloropropane.

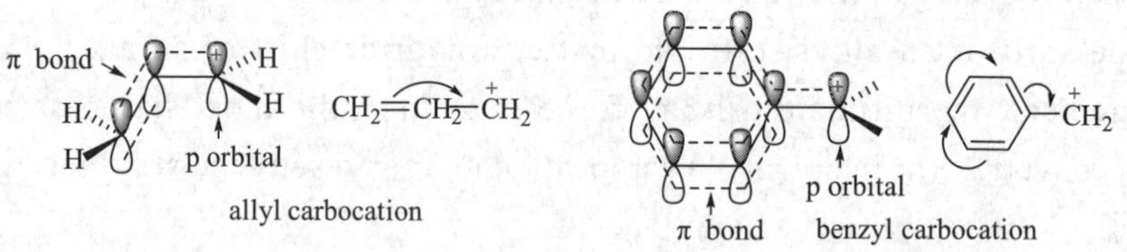

Figure 7.6 Schematic diagram of electronic delocalization in alkyl halide and benzyl carbocation
图7-6 烯丙型和苄基型碳正离子的电子离域示意图

2) Alkyl halides   For S_N1 reaction, the reactivity is mainly affected by electronic effect. In the process of reaction, the step of forming **carbocation intermediate (碳正离子中间体)** determines the reaction rate, so the stability of carbocation determines the speed of reaction rate (Table 7.4). The stability of carbocation is as follows: $R_3C^+>R_2CH^+>RCH_2^+>CH_3^+$. It shows the more stable the carbocation, the smaller the activation energy required for the reaction, and the faster the reaction rate. Therefore, the order of S_N1 reaction activities of alkyl halides are as follows: $R_3CX>R_2CHX>RCH_2X>CH_3X$.

Table 7.4 Reaction rates for hydrolysis of several bromoalkane S_N1 reaction mechanism
表7-4 几种溴代烷S_N1反应机制进行水解的反应速率

RBr	CH_3Br	CH_3CH_2Br	$(CH_3)_2CHBr$	$(CH_3)_3CBr$
Relative rate	1.0	1.7	45	10^8

For S_N2 reaction, the reactivity is mainly affected by space factors. In the process of reaction, the nucleophilic reagent always attacks the back side of the substrate, so the steric resistance is smaller, and the reaction rate is faster.

From Table 7.5, we can see that the order of S_N2 reactivity of halides is as follows: $CH_3X>RCH_2X>R_2CHX>R_3CX$. In primary alkyl halides, the reaction rate is slower with more branched chains of β-carbon.

Table 7.5   Reaction rates for hydrolysis of several bromoalkane S_N2 reaction mechanisms

表7-5   几种溴代烷S_N2反应机制进行碘交换的反应速率

RBr	Relative rate	RBr	Relative rate
CH_3Br	30	CH_3CH_2Br	1
CH_3CH_2Br	1	$CH_3CH_2CH_2Br$	0.82
$(CH_3)_2CHBr$	0.02	$(CH_3)_2CHCH_2Br$	0.036
$(CH_3)_3CBr$	≈ 0	$(CH_3)_3CCH_2Br$	0.000012

In conclusion, the reactivity of nucleophilic substitution reaction of alkyl halides is as follows.

Increasing reactivity of S_N1 →

CH_3X CH_3CH_2X $(CH_3)_2CHX$ $(CH_3)_3CX$

← Increasing reactivity of S_N2

In general, the structure of the substrate (the alkyl halide) is an important factor in determining which of these substitution mechanisms might operate. Most methyl halides and primary alkyl halides are poor substrates for S_N1 mechanism, because they cannot easily ionize to high-energy methyl and primary carbocation, so they are good substrates for S_N2 mechanism. Tertiary alkyl halides are too hindered to undergo displacement, but they can ionize to form tertiary carbocation, so they undergo substitution exclusively through the S_N1 mechanism. Secondary alkyl halides can undergo substitution by either mechanism, depending on the other conditions.

Notice that the nucleophilic substitution can not occur by either mechanism for alkyl halides in which halogen atom links to bridge-head carbon atoms. With the S_N1 mechanism, the substrate needs to form planar carbocation intermediate, but the bridge-head carbon is limited by the ring and cannot be extended into planar structure. With the S_N2 mechanism, the nucleophilic reagent needs to attack the central carbon atom from the back of the leaving group. But the ring hinders the attack of the nucleophilic reagent, so it is also difficult to react.

It is difficult to form carbocation.   The ring hinders the attack of the nucleophilic reagent.

3) Vinyl halides (乙烯型卤烃) and phenyl halides (苯基型卤烃)

$CH_2=CHX$
vinyl halides

phenyl halides

In the vinyl halide and phenyl halide halogen atom is linked to sp^2 hybrid-carbon. This kind of halide has the lowest reactivity of the nucleophilic substitution reaction. Because halogen atom linked to sp^2 hybrid-carbon forms **p-π conjugation system** (**p –π共轭体系**) and the bond energy of C−X get higher, which lower the reactivity of the bond (Figure 7.7).

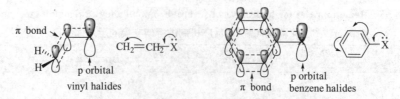

Figure 7.7   Schematic diagram of electronic delocalization in vinyl halides and benzene halides
图7-7   乙烯型和卤苯型卤代烃中p-π共轭效应示意图

(3) Effect of the Leaving Group   In S_N1 and S_N2 mechanism, the effect of the leaving group is same, as that the higher the leaving ability is, the easier the reaction is.

In the transition state for nucleophilic substitution of a halide, the leaving group develops a partial negative charge in both S_N1 and S_N2 reactions. So the ability of a group to function as a leaving group is related to how stable it is as an anion. Good leaving groups are the weak conjugate bases of strong acids. The leaving ability order of common groups is shown here.

reactivity as a leaving group   rarely function as leaving groups

$$I^- > Br^- > Cl^- \sim H_2O >> F^- > CH_3COO^- > HO^- > CH_3O^- > NH_2^-$$

stability of anions; stength of conjugate acid

(4) Effect of the Nucleophile   In the S_N1 mechanism, the reaction rate is only proportion to the concentration of alkyl halides, but not to the nucleophilic reagent. So the nucleophilic reagent rarely affect the reaction rate. In the S_N2 mechanism, the reaction rate is proportion to the concentration of both the halides and nucleophile.

Nucleophilic reagents are negative ions or neutral molecules that can provide electrons. Nucleophilicity are related to the ability to provide electrons, polarizability, solvent and so on. Because all nucleophiles are bases as well, we also study correlations between nucleophilicity and basicity. Basicity and nucleophilicity are often related because they both involve donating a lone pair of electrons to another group. In the case of a base, the lone pair makes a bond to a proton, while with a nucleophile, the one pair most commonly creates a bond to an **electrophilic** (亲电) carbon. In general, sterically unhindered strong bases are good nucleophiles. Polarizability refers to the degree of electron cloud deformation under the external electric field. When the atom is larger and the electronegativity is smaller, the binding force on the outer electron is weaker. The reagent is easy to polarize, and the nucleophilicity of the reagent is stronger.

The general rules of the nucleophilic reagent are as follows: ① The **nucleophilicity** (亲核性) of reagent, which has the same element, is same as the order of basicity. The nucleophilicity and basicity: $RO^- > HO^- > PhO^- > RCOO^- > NO_3^- > ROH > H_2O$. ② The nucleophilicity of reagent, which has the **same period element** (同周期元素), is same as the order of **basicity** (碱性). The nucleophilicity and basicity: $R_3C^- > R_2N^- > RO^- > F^-$; $RS^- > Cl^-$. ③ The nucleophilic reagent, which is **congeners** (同族元素), is inverse with the order of basicity. The nucleophilicity: $I^- > Br^- > Cl^- > F^-$; $HS^- > HO^-$. The basicity: $I^- < Br^- < Cl^- < F^-$; $HS^- < HO^-$.

(5) Effect of the Solvent   The solvent can affect the halides and nucleophilic reagent. It can be divided into protonic solvent, dipole solvent and non-polar solvent according to the presence of active hydrogen and polarity.

1) Protonic solvent (质子型溶剂)   The molecule of protonic solvent has the hydrogen atom, which can form hydrogen bond, for example H_2O, ROH, RCOOH, et al. In S_N1 reaction, because the step of forming carbocation is the rate-determining step, the protonic solvent is beneficial to highly polarize and thoroughly crack on the bond of C−X. Meanwhile it forms the stable carbocation and halogen anion, the protonic solvent can promote their electric charge to further decentralize, and the system is more stable. So, the protonic solvent favors the S_N1 reaction.

The reaction rate of S_N2 depends on the contact of alkyl halides and nucleophilic reagent, the nucleophilicity of nucleophilic reagent, etc. The protonic solvent can wrap the nucleophilic reagent by solvation, which greatly reduces the nucleophilicity. Therefore, the protonic solvent disfavors the S_N2 reaction.

2) Dipole solvent (偶极溶剂)   Protonic solvent does not have the hydrogen atom, its polarity comes from the dipole in the structure. For example chloroform, acetone, **dimethyl sulfoxide (二甲亚砜, DMSO）**, **N, N-dimethyl formamide（*N,N*-二甲基甲酰胺, DMF）**, **tetrahydrofuran (四氢呋喃, THF）**, et al.

$$\underset{\text{acetone}}{\overset{\displaystyle O\delta-}{\underset{CH_3 \qquad CH_3}{C\delta+}}} \qquad \underset{\text{dimethyl sulfoxide (DMSO)}}{\overset{\displaystyle O\delta-}{\underset{CH_3 \qquad CH_3}{S\delta+}}} \qquad \underset{N, N\text{-dimethyl formamide (DMF)}}{\overset{\displaystyle O\delta-}{\underset{H \qquad N(CH_3)_2}{C\delta+}}}$$

Because the positive pole of the dipole solvent is buried inside the molecule, it cannot solvate the negative ion; while the negative pole is at the outer layer and can solvate the positive ion. The nucleophilicity will increase by the solvation of nucleophile with dipole solvent. Therefore, the dipole solvent is in favor of the S_N2 reaction.

3) Nonpolar solvent (非极性溶剂)   The molecule of nonpolar solvent hasn't the hydrogen atom, and the dipole moment is less than $6.67×10^{-30}$, such as hexane, benzene, ether, et al. In the nonpolar solvent, the polar molecules are not easy to dissolve, and the molecules are in the state of association, so the reactivity is reduced.

7.4.2   β-Elimination Reaction

The alkyl halides with strong base in boiling alcohol solution eliminate a molecule of halogenated and form an alkene, called as **elimination reaction (消除反应, E)**. It is a common method for the preparation of the alkene. The general formula is as follows.

$$\overset{\beta \qquad \alpha}{\underset{H \qquad X}{\diagdown\diagup}} \xrightarrow[\text{alcohol, } \triangle]{\text{strong base}} \diagdown\!\!=\!\!\diagup + HX$$

Because this elimination reation removes the hydrogen of *β*-carbon, it is also called *β*-H elimination reation.

The elimination of alkyl halides may also be used to prepare the alkyne.

$$RC \underset{\underset{H}{|}}{\overset{\overset{H}{|}}{C}} - \underset{\underset{X}{|}}{\overset{\overset{X}{|}}{C}}R \xrightarrow[\text{alcohol, }\triangle]{\text{strong base}} RC \equiv CR + 2HX$$

1. β-Elimination of Saturated Alkyl Halides　In the elimination reaction of primary alkyl halides, there is only one kind of *β*-H in the molecule, which forms a single elimination product. For example, 1-bromobutane is treated with **sodium ethoxide (醇钠)** in boiling ethanol, and forms but-1-ene.

$$CH_3CH_2\overset{\beta}{C}H - \overset{\alpha}{C}H_2 \xrightarrow[\text{alcohol, }\triangle]{\text{sodium ethoxide}} CH_3CH_2CH = CH_2 + HBr$$

In the elimination reaction of secondary and tertiary alkyl halides, there is multiple kind of *β*-H in the molecule; therefore, formation of more than one kind of alkenes is possible. It is the problem of regioselectivity.

$$CH_3\overset{\beta}{C}H - \overset{\alpha}{C}H - \overset{\beta}{C}H_2 \xrightarrow[\text{alcohol, }\triangle]{\text{sodium ethoxide}} CH_3CH = CHCH_3 + CH_3CH_2CH = CH_2$$

but-2-ene (81%)　　　but-1-ene (19%)

$$CH_3\overset{\beta}{C}H - \overset{\overset{CH_3}{|}}{C} - \overset{\beta}{C}H_2 \xrightarrow[\text{alcohol, }\triangle]{\text{sodium ethoxide}} CH_3CH = CCH_3 + CH_3CH_2C = CH_2$$

2-methylbut-2-ene (71%)　2-methyl-but-1-ene (29%)

Based on many of the experimental results, the Russian chemist, Zaitsev concluded a rule in elimination reaction: the hydrogen connected to the carbon with less hydrogen is eliminated and the more substituted (and therefore the more stable) alkene should be formed. Because the C−H bond of alkyl can form σ-π hyper conjugation with the π bond of the alkene. The order of the stability of alkene is that:

$$R_2C = CR_2 > R_2C = CHR > R_2C = CH_2 > RCH = CHR > RCH = CH_2 > CH_2 = CH_2$$

2. β-Elimination of Unsaturated Alkyl Halides　In the elimination reaction of unsaturated halides, it is easy to eliminate the hydrogen next to the carbon of double bond, because the hydrogen of this carbon is more acidic, and the products (alkene) is more stable.

$$CH_2 = CHCH\overset{\beta}{C}H_3 \xrightarrow[\text{alcohol, }\triangle]{\text{sodium ethoxide}} CH_2 = CHCH = CH_2 + HCl$$

（下同）

major product

major product

3. E1 Mechanism and E2 Mechanism　The elimination reaction of alkyl halides is similar to the nucleophilic substitution reaction. A fundamental difference between them is the timing of the bond-breaking and bond-forming steps. According to that, β-Elimination might be a **first-order (unimolecular elimination, E1,单分子消除)** and **second-order (bimolecular elimination, E2，双分子消除)** process.

(1) E2 mechanism　The E2 reaction is similar with S_N2, and the process has a single step. With the β-C−H and α-C−X bond breaking, the α-carbon and β-carbon double bond forms.

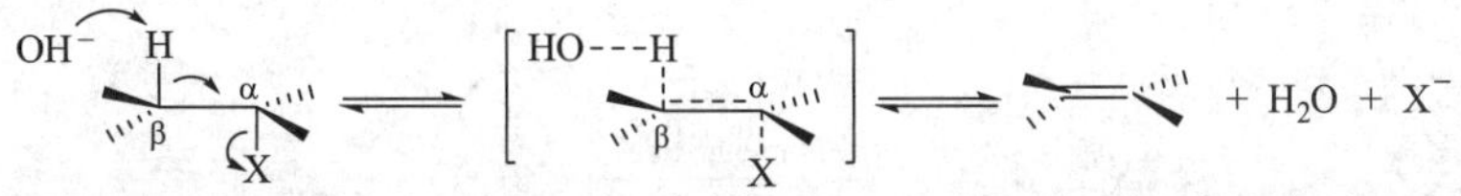

The rate of E2 elimination is proportional to the concentrations of both the alkyl halide and the base, giving a second-order rate equation. Because formation of the carbon-carbon double bond, and ejection of the halogen ion occur simultaneously; all bond-breaking and bond-forming steps are concerted. The reactivity order reflects the greater stability of highly substituted double bonds. The stabilities of the alkene products are reflected in the transition states (Figure 7.8), giving lower activation energies and higher rates for elimination of alkyl halides that lead to highly substituted alkenes. Therefore, the order of E2 creation is $R_3CX>R_2CHX>RCH_2X$.

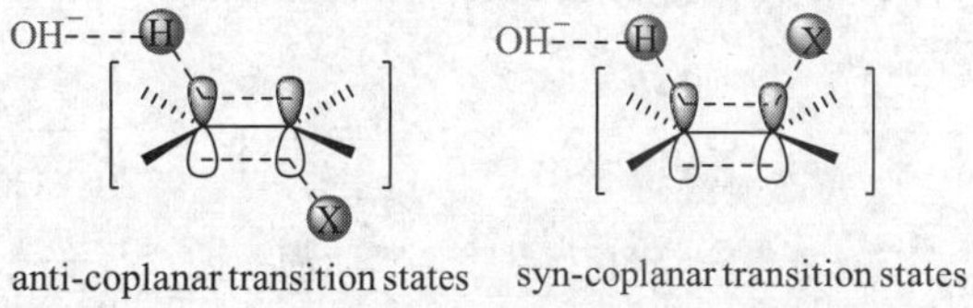

anti-coplanar transition states　　syn-coplanar transition states

Figure 7.8　Concerted transition states of the E2 reaction
图7-8　E2过渡态中轨道结合示意图

As all bond-breaking and bond-forming steps occur simultaneously, E2 mechanism has strict requirements for the spatial position. The leaving group and β-hydrogen must be in **anti-coplanar conformation** (共面反位构象). When the leaving group and β-hydrogen are in a coplanar, the orbitals of α-C and β-C can be parallel overlap to form the new π bond. When the hydrogen and the halogen are anti to each other their orbitals are aligned. The conformational structure is cross type, and the activation energy needed to form the transition state is lower than that of the full overlapping conformational state, and the reaction is more likely to take place.

The E2 is a stereospecific reaction, because different stereoisomers of the starting material react to give different stereoisomers of the product. This **stereospecificity** (立体定向性) results from the anti-coplanar transition state that is usually involved in the E2.

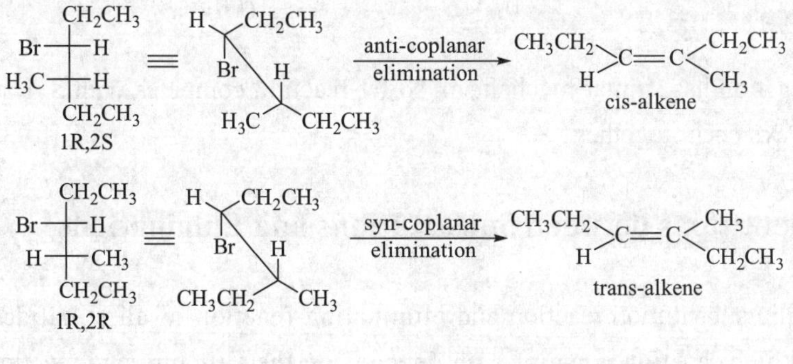

Since the C—C bond can rotate freely, the stereospecificity of the E2 reaction affects the configuration of the formed alkene. When the reactants are cyclic-halides, because of the rigid structure of the ring, the substituents cannot be freely flipped, the product must follow the anti-coplanar conformation of the E2 reaction; otherwise it cannot react. When there are two β-hydrogens in the anti-coplanar conformation, the reaction should follow the Zaitsev's rule and the stereochemistry at the same time.

$$\text{(structure with } C(CH_3)_3 \text{)} \xrightarrow[\text{alcohol, } \triangle]{\text{sodium ethoxide}} \text{(cyclohexene)}-C(CH_3)_2$$

$$\text{(structure with } C(CH_3)_3 \text{)} \xrightarrow[\text{alcohol, } \triangle]{\text{sodium ethoxide}} \text{(cyclohexene with } C(CH_3)_3 \text{)}$$

(2) E1 mechanism   The E1 reaction is similar with S_N1, and the process is two steps. At first, the $C-X$ bond of alkyl halides breaks to give a carbocation. Secondary, the base attacks β-hydrogen and the carbon-carbon double bond forms.

first step $\xrightarrow[\text{}]{\text{slow}}$ $+ X^-$

second step $\xrightarrow[\text{}]{\text{fast}}$ $+ H_2O$

Formation of the carbocation intermediate in Step 1 crosses the higher energy barrier and is the rate-determining step. The reaction rate is proportional to the concentration of halides, so this mechanism is designated an E1 reaction, where E stands for Elimination and 1 stands for unimolecular.

In an E1 mechanism, one transition state exists for the formation of the carbocation. The stability of carbocation determines the reaction rate, so the order of the activity of alkyl halides is that: $R_3CX>R_2CHX>RCH_2X$. As the carbocation is ease to rearrange to form the more stable intermediate, there often is the rearrangement product formed as the major product. For example, 2-bromo-3,3-dimethylbutane occurs the elimination and the major product is 2,3-dimethylbut-2-ene, which is the rearrangement product.

$$CH_3\overset{\underset{\displaystyle CH_3}{|}}{\underset{\underset{\displaystyle CH_3Br}{|}}{C}}-CHCH_3 \xrightarrow{-Br^-} CH_3\overset{\underset{\displaystyle CH_3}{|}}{C}-\overset{+}{C}HCH_3 \longrightarrow CH_3\overset{+}{C}-\overset{\underset{\displaystyle CH_3}{|}}{C}CH_3 \longrightarrow CH_3\overset{\underset{\displaystyle CH_3}{|}}{C}=\overset{\underset{\displaystyle CH_3}{|}}{C}CH_3$$

S_N1 and E1 reaction has similar mechanism, so E1 reaction competes with S_N1 substitution, and E1 and S_N1 almost always occur together.

7.4.3   Competitions between Substitutions and Eliminations

The nucleophilic substitution reaction and elimination reaction of alkyl halides usually compete with each other. Which reaction is major path depends on the structure and reaction conditions of the reactants.

1. Effect of Substrate   The substitution reaction of primary alkyl halides is easy to occur, and the elimination reaction is the main one under the condition of strong base and weak polar solvent or heating. The reaction is usually carried out according to the mechanism of bimolecule reaction (S_N2 or E2).

$$CH_3CH_2CH_2Cl \xrightarrow[\text{EtOH}]{\text{EtONa}} CH_3CH_2CH_2OCH_2CH_3 \quad \text{substitution reaction}$$

$$CH_3CH_2CH_2Cl \xrightarrow[\text{EtOH, } \triangle]{\text{EtONa}} CH_3CH{=}CH_2 \quad \text{elimination reaction}$$

When there is a branched chain on β-carbon, the steric hindrance is large, which hinders the nucleophilic reagent attack on α-carbon, and increases the attack probability on β-hydrogen, resulting in the increase of the elimination reaction product.

$$\underset{\underset{CH_3}{|}}{CH_3CHCH_2Br} + CH_3CH_2ONa \xrightarrow{\text{EtOH}} \underset{\underset{CH_3}{|}}{CH_3CHCH_2OCH_2CH_3} + \underset{\underset{CH_3}{|}}{CH_3C{=}CH_2}$$
$$\qquad\qquad\qquad\qquad\qquad\qquad\qquad\qquad 40.4\% \qquad\qquad 59.5\%$$

When there are benzene rings and vinyl groups on the β-carbon, the products formed by the elimination reaction have π-π **conjugate system** (共轭体系), and the stability is strong, which is beneficial to the E2 reaction, and the elimination product is the major one.

$$\text{Ph}{-}CH_2CH_2Br \xrightarrow[\text{EtOH}]{\text{EtONa}} \text{Ph}{-}CH{=}CH_2$$

The tertiary alkyl halides are prone to elimination reactions, and only in pure water or ethanol can they be mainly substituted.

$$\underset{\underset{Cl}{|}}{\overset{\overset{CH_3}{|}}{CH_3CH_2CCH_2CH_3}} \xrightarrow[\text{EtOH}]{\text{EtONa}} \underset{\overset{CH_3}{|}}{CH_3CH_2C}{=}CHCH_3$$

$$\underset{\underset{Cl}{|}}{\overset{\overset{CH_3}{|}}{CH_3CH_2CCH_2CH_3}} \xrightarrow[\triangle]{H_2O} \underset{\underset{OH}{|}}{\overset{\overset{CH_3}{|}}{CH_3CH_2CCH_2CH_3}}$$

With the increase of substituents on β-carbon, tertiary alkyl halides are beneficial to E1 and is unfavorable to S_N1. In the S_N1 reaction, the structure of the central carbon atom is from tetrahedron → plane → tetrahedron, and the space obstruction is not conducive to the reaction. In E1 reaction, the configuration of central carbon atom is more stable from tetrahedral → plane, and more substituted alkenes are more stable, which is beneficial to the reaction.

The secondary alkyl halides may react with both paths. With the reagent basicity increased, the reaction temperature increased and the branches on β -carbon increased, the proportion of elimination reaction increases.

$$\underset{\underset{Cl}{|}}{CH_3CHCH_3} + CH_3CH_2ONa \xrightarrow{\text{EtOH}}{\triangle} \underset{\underset{OCH_2CH_3}{|}}{CH_3CHCH_3} + CH_3CH{=}CH_2$$
$$\qquad\qquad\qquad\qquad\qquad\qquad\qquad 21\% \qquad\qquad 79\%$$

In conclusion, for the different structure of alkyl halides, the order of preference for substitution and elimination is as follows.

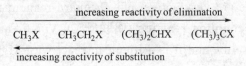

increasing reactivity of elimination

$$CH_3X \qquad CH_3CH_2X \qquad (CH_3)_2CHX \qquad (CH_3)_3CX$$

increasing reactivity of substitution

2. Effect of the Nucleophile   The nucleophilicity and basicity of the reagent are different. Strong nucleophile favors the substitution reaction; while strong base favors elimination reaction.

$$CH_3CH_2\underset{\underset{Cl}{|}}{\overset{\overset{CH_2CH_3}{|}}{C}}CH_2CH_3 + CH_3CH_2ONa \xrightarrow{EtOH} CH_3CH_2\underset{\underset{OCH_2CH_3}{|}}{\overset{\overset{CH_2CH_3}{|}}{C}}CH_2CH_3 + CH_3CH_2\overset{\overset{CH_2CH_3}{|}}{C}{=}CHCH_3$$
major product

In the S_N1 and E1 reaction, the reaction rate is independent of the nucleophile, it has little effect. The strong nucleophile is favorable to S_N2, the strong base is favorable to E2, and the increase of reagent concentration is beneficial to both.

3. Effect of the Solvent   Increasing the polarity of the solvent is more beneficial to the substitution reaction than elimination reaction. In general, the weak base and strong polar solvent is used for substitution reaction; the strong base and weak polar solvent is used for elimination reaction.

4. Effect of the Temperature   The increase of reaction temperature is beneficial to both the substitution reaction and the elimination reaction, but it is more beneficial to the elimination reaction. Because the activation energy of breaking C—H bond is high, the increase of temperature is more beneficial to the elimination reaction.

7.4.4   Reaction to Form Organometallic Compounds

Alkyl halides can combine with magnesium, lithium, potassium, sodium and other metals to form organometallic compounds, which is important in organic synthesis.

1. Reaction with Mg   Alkyl halides and Mg form the organo magnesium halide which is also known as **Grignard reagent** (格氏试剂).

$$RX + Mg \xrightarrow{C_2H_5OC_2H_5} RMgX \quad \text{organomagnesium halide}$$

The preparation of Grignard reagent needs to be completed under the condition of anhydrous ethyl ether, because it is not only solvent, but can also form complex with Grignard reagent to stabilize it.

$$\underset{C_2H_5}{\overset{C_2H_5}{}}O:{\rightarrow}\underset{\underset{R}{|}}{\overset{\overset{X}{|}}{Mg}}{\leftarrow}:O\underset{C_2H_5}{\overset{C_2H_5}{}}$$

The reactivity of Grignard reagent generation is related to the alkyl structure and halogen species in halides. When the alkyl structures are same and the different halogen, the reaction order is that: RI>RBr>RCl.

Because of the high price of iodides and the poor reaction selectivity of chlorides, bromides are commonly used to form Grignard reagent. When there are the different alkyl structure and the same halogen, the reaction order is $RCH_2X>R_2CHX>R_3CX>PhX$ and $CH_2{=}CHX$.

Allyl and benzyl halides are very active, but Grignard reagent formed with the reaction is quickly decomposed by **the coupling reaction** (偶联反应). Vinylic and benzenic halides (乙烯型和苯基型卤烃) have low activity, and it is difficult to react with Mg under ether condition. It is necessary to use tetrahydrofuran (THF) as solvent and increase reaction temperature in order to form Grignard reagent.

$$\text{C}_6\text{H}_5\text{-Br} + \text{Mg} \xrightarrow{\text{THF}} \text{C}_6\text{H}_5\text{-MgBr}$$

Due to the high electropositivity of magnesium in Grignard reagent, the polarity of C—Mg bond is large, with partial positive charge on magnesium and partial negative charge on carbon. Grignard reagent reacts as strong nucleophilic reagent. Due to the high reactivity of Grignard reagent, when encounters the compounds with active hydrogen (for example: alcohol, carboxylic acid, amine, etc.), it will decompose and form alkane rapidly. Therefore, Grignard reagent must be prepared with anhydrous operation.

$$\text{CH}_3\text{CH}_2\text{MgX} + \begin{cases} \text{H}_2\text{O} \\ \text{ROH} \\ \text{RCOOH} \\ \text{NH}_3 \\ \text{RNH}_2 \\ \text{R}_2\text{NH} \\ \text{RC}\equiv\text{CH} \end{cases} \longrightarrow \text{CH}_3\text{CH}_3 + \begin{cases} \text{HOMgX} \\ \text{ROMgX} \\ \text{RCOOMgX} \\ \text{NH}_2\text{MgX} \\ \text{RNHMgX} \\ \text{R}_2\text{NMgX} \\ \text{RC}\equiv\text{CMgX} \end{cases}$$

Grignard reagent often reacts as a strong nucleophile to attack the substrate with positively charged carbon, such as alkyl halides, carbon dioxide, carbonyl, and nucleophilic substitution or nucleophilic addition reaction may happen.

$$\text{CH}_2=\text{CH}\overset{\delta+}{\text{CH}_2}\overset{\delta-}{\text{Cl}} + \overset{\delta-}{\text{CH}_3}\overset{\delta+}{\text{CH}_2}\text{MgBr} \longrightarrow \text{CH}_2=\text{CHCH}_2\text{CH}_2\text{CH}_3 + \text{MgClBr}$$

$$\text{C}_6\text{H}_5\overset{\delta+}{\text{MgBr}} + \overset{\delta-}{\text{O}}=\overset{\delta+}{\text{C}}=\overset{\delta-}{\text{O}} \longrightarrow \text{C}_6\text{H}_5\text{C(=O)-OMgBr} \xrightarrow{\text{H}_3\text{O}^+} \text{C}_6\text{H}_5\text{C(=O)-OH}$$

$$\text{CH}_3\text{CH}_2\overset{\delta+}{\text{CH}}=\overset{\delta-}{\text{O}} + \overset{\delta-}{\text{CH}_3}\overset{\delta+}{\text{CH}_2}\text{MgBr} \longrightarrow \text{CH}_3\text{CH}_2\text{CH(OMgBr)CH}_2\text{CH}_3 \xrightarrow{\text{H}_3\text{O}^+} \text{CH}_3\text{CH}_2\text{CH(OH)CH}_2\text{CH}_3$$

2. Reaction with Li Alkyl halides and Li form the **organolithium** (有机锂化物) in non-polar solvent.

$$\text{RX} + 2\text{Li} \xrightarrow[\text{organolithium}]{\text{benzene}} \text{RLi} + \text{LiX}$$

Like organomagnesium, lithium can react with alkyl halides, vinyl halides, and **aryl halides (芳基卤烃)** to form organometallic compounds. Organolithium is more active and expensive than Grignard reagents. For example, organolithium can react with **cuprous iodide (碘化亚铜)** to form **dialkyl copper-lithium (二烷基铜铝)**.

$$2RLi + CuI \longrightarrow R_2CuLi + LiI$$
$$\text{dialkyl copper lithium}$$

The alkyl groups of dialkyl copper-lithium can be phenyl, alky, alkyl, and dialkyl copper-lithium is an important alkylation reagent in organic synthesis. It can react with halides to synthesize alkane, which is a common synthetic method for the synthesis of asymmetric alkane. This is also called **Corey-House synthesis** (科瑞-郝思合成法).

$$RX \xrightarrow[\text{benzene}]{Li} RLi \xrightarrow{CuI} R_2CuLi \xrightarrow{R'X} RR'$$

The primary alkyl halide is the best reactant. The reaction is not affected by other groups, and the reaction yield is high, so it is widely used in organic synthesis.

$$CH_2{=}CHBr \xrightarrow[\text{benzene}]{Li} CH_2{=}CHLiBr \xrightarrow{CuI} (CH_2{=}CH)_2CuI \xrightarrow{CH_3CH_2Cl} CH_2{=}CHCH_2CH_3$$

3. Reaction with Na   Halides react with sodium metal, the two part alkyl of halides are coupled. This reaction is used to synthesize alkane with symmetrical structure, called **Wurtz synthetic method** (武兹合成法).

$$2RX + 2Na \longrightarrow R{-}R + 2NaX$$

In this reaction, the alkyl chain has doubled，but it can only be used to synthesize **alkanes with even carbon atoms** (偶数个碳原子的烃类) or **high-grade alkane with symmetrical structure** (结构对称的高级烃类). Because Na is a base, the primary alkyl halide is the best reactant.

7.4.5 Reduction Reaction of Alkyl Halides

In alkyl halides，halogen atoms can be replaced by active hydrogen to form alkanes. The commonly used reducing agents are catalytic hydrogenation, zinc and hydrochloric acid, sodium borohydride ($NaBH_4$), and lithium aluminum hydride ($LiAlH_4$), etc.

$$CH_3CH_2CH_2CH_2Br \xrightarrow{Zn+HCl} CH_3CH_2CH_2CH_3$$

Lithium aluminum hydride and sodium borohydride don't reduce the π bond of C—C bond. Lithium aluminum hydride is a strong reducing agent, and can reduce other functional group, such as —COOH, —COOR, —CN, etc. Sodium borohydride is less reactive than lithium aluminum hydride.

PPT

7.5　Application of Alkyl Halides in the Pharmacy Study

1. Trichlormethane (三氯甲烷)　Trichlormethane ($CHCl_3$) is also called **chloroform** (氯仿), which is a colorless liquid with slightly sweet, and its boiling point is 61.3℃. It is unburnable and insoluble in water, also an important insolvent. Chloroform was found to produce general **anesthesia** (麻醉), opening new possibilities for careful surgery with a patient who is unconscious and relaxed. Chloroform is toxic and **carcinogenic** (致癌), however, and it was soon abandoned in favor of safer **anesthetics** (麻醉剂), **such as diethyl ether** (乙醚).

2. Carbon Tetrachloride (四氯化碳)　Carbon tetrachloride is also called **tertachlormethane** (四氯甲烷) or **chlorane** (氯烷), and is a colorless liquid in room temperature. Its boiling point is 76.8℃. It is slightly insoluble in water, and is an important insolvent. Because its steam is heavier than air and can insulate the **combustion** (燃烧) from the air, it isn't help to burn, so can be used as **extinguishing agent** (灭火剂). It has a pleasant smell, but its vapor is toxic, and we should prevent inhalation using it. Because of its toxicity, it is limited to the use of **veterinary drugs** (兽药).

3. Polyvidone Iodine (聚维酮碘)　Polyvidone iodine, also called **active iodine** (活力碘), is a new type of **skin disinfectant** (皮肤消毒剂) which replaces **iodine tincture** (碘酊). It has the characteristics of non-toxic, **non-irritating** (无刺激), no corrosion, no odor and stable performance. It can overcome the characteristics of iodine in iodine tincture, such as easy **sublimation** (升华) and **irritation** (刺激), and its **bactericidal effect** (杀菌效果) is higher than that in iodine tincture.

4. Ofloxacin (氧氟沙星)　Ofloxacin is a **broad-spectrum antibiotic** (广谱抗菌素), and has strong **antibacterial action** (抗菌活性)**,** which belongs to **fluoroquinolones** (氟喹诺酮类). It mainly acts as the dugs of the anti-infection, which caused by Gram-negative bacteria in the **respiratory tract** (呼吸道), **urinary tract** (泌尿道), skin, **intestinal tract** (肠道) and other parts. **Levofloxacin** (左氧氟沙星) is the **l-isomer** (左旋体) of ofloxacin. It's antibacterial action is about twice that ofloxacin in vitro and the side effects are lower than most fluoroquinolones.

5. Chloroamphenicol (氯霉素)　Chloroamphenicol，also called **chloramphenicol** (氯胺苯醇), is an **amidol** (酰胺醇) antibiotic with **optical activity** (光学活性). It is mainly used for **typhoid** (伤寒), **paratyphoid** (副伤寒) and other **Salmonella** (沙门菌), Bacillus fragile infection. It has the inhibitory effect on **gram-positive and negative bacteria** (革兰阳性、阴性细菌), and has strong effect on the latter. Due to the serious impairment in the **hematopoietic** (造血) system, it should be used carefully.

重 点 小 结

一、亲核取代反应

1. 反应类型

(1) 与 H_2O 反应

$$R{-}X + H_2O \xrightarrow{\triangle} ROH + HX$$

$$R{-}X + NaOH \xrightarrow{H_2O} ROH + NaX$$

$$CH_3CH_2CH_2{-}Br + NaOH \xrightarrow{H_2O} CH_3CH_2CH_2OH + NaX$$

(2) 与 ROH 反应

$$R{-}X + NaOR' \longrightarrow ROR' + NaX$$

$$CH_3CH_2Cl + (CH_3)_3CONa \longrightarrow CH_3CH_2OC(CH_3)_3 + NaCl$$

(3) 与 NaCN 反应

$$R{-}X + NaCN \longrightarrow RCN + NaX$$

$$\text{Ph}{-}CH_2Cl + NaCN \longrightarrow \text{Ph}{-}CH_2CN \xrightarrow[H^+]{H_2O} \text{Ph}{-}CH_2COOH$$

(4) 与 $NaC \equiv CR$ 反应

$$R{-}X + R'C{\equiv}CNa \longrightarrow RC{\equiv}CR' + NaX$$

$$CH_3CH_2Cl + CH_3C{\equiv}CNa \longrightarrow CH_3C{\equiv}CCH_2CH_3 + NaCl$$

(5) 与 NH_3 反应

$$R{-}X + NH_3 \longrightarrow R\overset{+}{N}H_3 X^-$$

$$R\overset{+}{N}H_3X^- + NaOH \longrightarrow RNH_2 + H_2O + NaX$$
$$\quad\quad\quad\quad\quad \xrightarrow{RX} R_2NH + HX$$
$$\quad\quad\quad\quad\quad\quad\quad\quad \xrightarrow{RX} R_3N + HX$$

(6) 与 $AgNO_3$ 反应

$$R{-}X + AgNO_3 \xrightarrow{EtOH} RONO_2 + AgX$$

反应活性：RI>RBr>RCl

$$\begin{array}{c} CH_2{=}CHCH_2X \\ PhCH_2X \end{array} > R_3CX > R_2CHX > RCH_2X > \begin{array}{c} CH_2{=}CHX \\ PhX \end{array}$$

（7）与 NaI 反应

$$\begin{matrix} RCl \\ RBr \end{matrix} + NaI \xrightarrow{\text{acetone}} RI + \begin{matrix} NaCl \\ NaBr \end{matrix} \downarrow$$

2. 反应机制

（1）S_N1 反应　反应特点：两步反应、反应速率只与卤烃浓度成正比，出现碳正离子中间体、具有重排现象，产物是外消旋体。

（2）S_N2 反应　反应特点：一步反应、反应速率与卤烃浓度和碱的浓度成正比，无中间体，产物是构型翻转。

（3）S_N1 与 S_N2 反应的竞争

1）底物的影响　S_N1 反应：$R_3CX > R_2CHX > RCH_2X$

　　　　　　　　S_N2 反应：$RCH_2X > R_2CHX > R_3CX$

2）离去基团的影响　$RI > RBr > RCl$

3）试剂的影响　S_N1 反应不受试剂影响；增强试剂亲核性，有利于 S_N2 反应。

4）溶剂的影响　增大溶剂极性，有利于 S_N1 反应，不利于 S_N2 反应。

二、消除反应

1. 反应类型

$$CH_3CH\overset{\beta}{-}\underset{\underset{Br}{|}}{\overset{\overset{CH_3}{|}}{C}}-\overset{\beta}{CH_2} \xrightarrow[\text{alcohol, }\triangle]{\text{sodium ethoxide}} CH_3CH=\overset{CH_3}{\overset{|}{C}}CH_3 + CH_3CH_2\overset{CH_3}{\overset{|}{C}}=CH_2$$

2-methylbut-2-ene(71%)　2-methyl-but-1-ene(29%)

$$\text{（环己烯）} \xrightarrow[\text{alcohol, }\triangle]{\text{sodium ethoxide}} \text{（环己二烯）} + \text{（环己烯）}$$

major product

2. 反应机制

（1）E1 反应　反应特点：两步反应、反应速率只与卤烃浓度成有关，出现碳正离子中间体、具有重排现象，以重排产物为主。

（2）E2 反应　反应特点：一步反应、反应速率与卤烃浓度和碱的浓度成有关，无中间体，卤原子与氢原子必须处于共面反位。

（3）E1 与 E2 反应的竞争

1）底物的影响　α-C 上支链越多，有利于消除反应，E1 和 E2 反应活性次序一致，$R_3CX > R_2CHX > RCH_2X$。

2）离去基团的影响　$RI > RBr > RCl$

3）试剂的影响　E1 反应不受试剂影响；增强试剂亲核性或碱性，有利于 S_N2 反应。

4）溶剂的影响　增大溶剂极性，有利于 E1，不利于 E2。

三、取代与消除反应的竞争

消除反应趋势增强 →

1. 底物的影响　CH_3X　　CH_3CH_2X　　$(CH_3)_2CHX$　　$(CH_3)_3CX$

← 取代反应趋势增强

2. 试剂的影响　强碱性试剂有利于消除反应,弱碱性试剂有利于取代反应。

3. 溶剂的影响　弱极性溶剂有利于消除反应,强极性溶剂有利于取代反应。

4. 温度的影响　高温有利于消除反应,低温有利于取代反应。

四、有机金属卤化物的反应

1. 有机镁化物

$$RX + Mg \xrightarrow{C_2H_5OC_2H_5} RMgX \quad \text{organomagnesium halide}$$

$$\text{C}_6\text{H}_5\text{Br} + Mg \xrightarrow{THF} \text{C}_6\text{H}_5\text{MgBr}$$

2. 有机锂化物

$$RX \xrightarrow[\text{benzene}]{Li} RLi \xrightarrow{CuI} R_2CuLi \xrightarrow{R'X} RR'$$

$$CH_2{=}CHBr \xrightarrow[\text{benzene}]{Li} CH_2{=}CHLiBr \xrightarrow{CuI} (CH_2{=}CH)_2CuI \xrightarrow{CH_3CH_2Cl} CH_2{=}CHCH_2CH_3$$

3. 与Na反应

$$2RX + 2Na \longrightarrow R{-}R + 2NaX$$

题库

Problems
目标检测

1. Give systematic (IUPAC) names for the following compounds.
用系统命名法命名下列化合物。

(1) $\underset{Br}{\overset{CH_3CH_2}{>}}C{=}C\underset{CH_3}{\overset{CH(CH_3)_2}{<}}$

(2) 带Cl与CH$_3$的螺环结构

(3) 带C(CH$_3$)$_3$和Br的环己烷

(4) $\underset{CH=CHCH_2CH_3}{\overset{CH_3}{Cl{-}{-}H}}$

(5) $\overset{CH_3}{\underset{CH_2CH_3}{\overset{Cl{-}{-}H}{Br{-}{-}H}}}$

(6) 桥环结构带Br

2. Draw the structures of the following compounds.
写出下列化合物的结构式。

(1) trans-1-chloro-4-methylcyclohexane　　(2) isobutyl bromide

(3) 1,2-dibromo-3-methylpentane　　　　(4) 1-bromo-3-methylbut-2-ene

3. List the following carbocations in decreasing sequence of their stability.

排列下列碳正离子稳定性从大到小的次序。

(1)　A. 　B.　C. $CH_3 \atop CH_3$$C = \overset{+}{C}H$　D. $\overset{+}{C}H_2$

(2)　A.　B.　C. $CH_3CH_2\overset{+}{C}CH_3 \atop CH_3$　D. $CH_3\overset{+}{C}HCH_3$

(3)　A. $CH_3\overset{+}{C}CH_3 \atop CH_3$　B. $CH_3\overset{+}{C}H_2$　C. $CH_3CH_2\overset{+}{C}CH_3$　D. $CH_3\overset{+}{C}HCH_3$

(4)　A.　B.　C.　D.

4. Propose mechanisms to account for these products.

请写出下列反应的机理。

(1) $CH_3CH_2\underset{CH_3}{\overset{CH_3}{C}}-CH_2Br \xrightarrow[C_2H_5OH]{C_2H_5ONa} CH_3CH_2\underset{CH_3}{C}=CHCH_3$

(2) 　$\xrightarrow[C_2H_5OH, \triangle]{C_2H_5ONa}$　 + 　 +

5. Predict the products.

预测反应产物。

(1) $CH_3CH_2CH_2Cl + CH_3CH_2ONa \xrightarrow{EtOH}$

(2) $(CH_3CH_2)_3CBr + CH_3CH_2ONa \xrightarrow{EtOH}$

(3) $CH_3CH_2CH_2\underset{Cl}{C}HCH_3 + NaOH \xrightarrow{EtOH}$

(4) 　$\xrightarrow[EtOH, \triangle]{EtONa}$　$\xrightarrow{KMnO_4}$

(5) $CH_3CH_2CH_2Cl + CH_3C{\equiv}CNa \longrightarrow$

6. Which compound will be immediately precipitated with silver nitrate solutions?

下列化合物与硝酸银醇溶液立即出现沉淀的是（　　　）。

A.　B.　C. $CH_3 \atop CH_3$$C = CHCl$　D. CH_2Cl

7. Which is the fastest in S_N1 reaction?

下列化合物发生S_N1反应，反应速度最快的是（　　　）。

A. 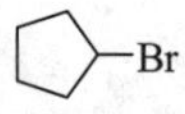 ⬠—Br　　B. 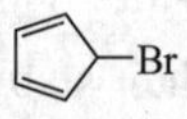 ⬠—Br　　C. $CH_3CH_2\overset{Br}{\underset{CH_3}{C}}CH_3$　　D. $CH_3\overset{}{\underset{Br}{C}HCH_3}$

8. Which is the most stable carbocation?

下列最稳定的碳正离子是（　　　）。

A. $CH_3\overset{+}{\underset{CH_3}{C}}CH_3$　　B. $CH_3\overset{+}{C}H_2$　　C. $CH_3CH_2\overset{+}{\underset{CH_3}{C}}CH_3$　　D. $CH_3\overset{+}{C}HCH_3$

9. Zaitsev's rule is applicable to (　　　).

扎依采夫规则适用于（　　　）。

    A. reaction of alkene with HBr

    B. substitution reaction of alkyl halides

    C. elimination reaction of alkyl halides

    D. substitution reaction of aromatic hydrocarbon

Discussion Topic

Discuss about the reactions involving alkyl halides and configuration of the product in the synthesis of Chloroamphenicol.

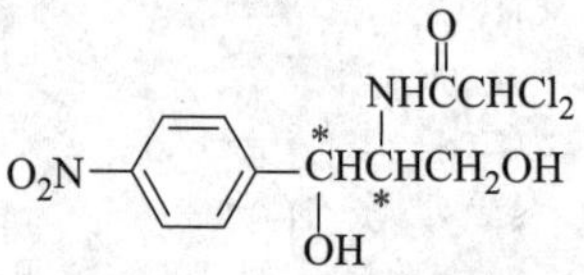

$$O_2N-\!\!\!\langle\ \rangle\!\!\!-\overset{*}{\underset{OH}{C}}H\overset{\overset{O}{\overset{|}{NHCCHCl_2}}}{\overset{|}{\underset{}{C}}}H\overset{*}{C}H_2OH$$

Chloroamphenicol, also called chloramphenicol, is an amidol. It is an antibiotic with optical activity.

（林玉萍）

第八章 醇、酚和醚
Chapter 8 Alcohols, Phenols and Ethers

学习目标

1. **掌握** 醇、酚、醚的主要化学性质。
2. **熟悉** 醇、酚、醚的命名、结构及异构。
3. **了解** 重要的醇、酚、醚类化合物及其在医药领域的应用。

Alcohols, phenols, ethers are a class of organic compounds that have functional groups containing oxygen. **Alcohols** (醇) contain a **hydroxyl group (OH, 羟基)** bonded to an sp^3-hybridized carbon atom. Compounds with a hydroxyl group bonded directly to an aromatic ring are called **phenols** (酚). Phenols have a hydroxyl group bonded to an sp^2-hybridized carbon atom of an aromatic ring. The functional group of a phenol is a hydroxyl group bonded to a benzene ring. **Ethers** (醚) contain two groups, which may be alkyl or aryl groups, bonded to an oxygen atom.

$$H-O-H \qquad R-O-H \qquad Ar-OH \qquad R-O-R'$$
$$\text{water} \qquad \text{alcohol} \qquad \text{phenol} \qquad \text{ether}$$

Alcohols, phenols, ethers are widely distributed in nature. Some of them are the most common and useful compounds in nature or in industry. Alcohol is found in alcoholic beverages, cosmetics, and drug preparations, it is also an important industrial and laboratory solvent. Phenol can **denature** (变性) the protein of bacterial cells and kills bacteria. 2% phenol **ointment** (软膏) was used for skin **antisepsis** (防腐) and **antipruritic** (止痒). **Diethyl ether** (乙醚) was the first inhalation **anesthetic** (麻醉剂) used in general surgery. It is also an important laboratory and industrial solvent.

PPT

8.1 Alcohols

8.1.1 Structure of Alcohols

The structure of an alcohol resembles the structure of water, with an alkyl group replacing one of the hydrogen atoms of water. The functional group of an alcohol is an -OH bonded to an sp^3-hybridized carbon. The oxygen atom of an alcohol is also sp^3-hybridized. Two sp^3-hybrid orbitals of oxygen form σ-bonds to atoms of carbon and hydrogen, and the remaining two sp^3-hybrid orbitals each contain an **unshared pair of electrons** (未共用电子对). Figure 8.1 shows a Lewis structure of **methanol (CH₃OH, 甲醇)**, the simplest alcohol. The measured C-O-H bond angle in methanol is 108.9°, very close to the perfectly **tetrahedral (四面体的)** angle of 109.5°.

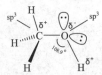

Figure 8.1   Structure of methanol
图8-1   甲醇的结构

We classify alcohols as primary (1°), secondary (2°), or tertiary (3°), depending on the number of carbon groups bonded to the carbon bearing the -OH group. If this carbon atom is primary (bonded to one other carbon atom), the compound is a **primary alcohol (伯醇)**. A **secondary alcohol (仲醇)** has the group attached to a secondary carbon atom, and a **tertiary alcohol (叔醇)** has it bonded to a tertiary carbon atom.

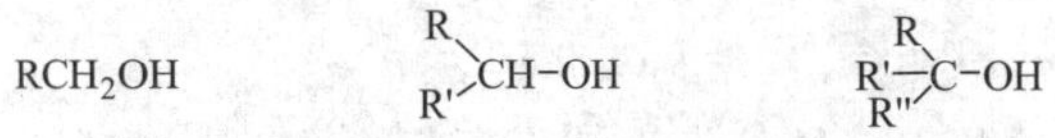

$$RCH_2OH \qquad \underset{R'}{\overset{R}{\diagdown}}CH-OH \qquad \underset{R''}{\overset{R}{\diagup}}R'-\!\!C-OH$$

primary alcohols   secondary alcohols   tertiary alcohols

8.1.2   Nomenclature of Alcohols

1. Common Names of Alcohols   Common nomenclature is suitable for alcohols with simple structures. The common name of an alcohol is derived from the common name of the alkyl group and the word *alcohol*. This system pictures an alcohol as a molecule of water with an alkyl group replacing one of the hydrogen atoms.

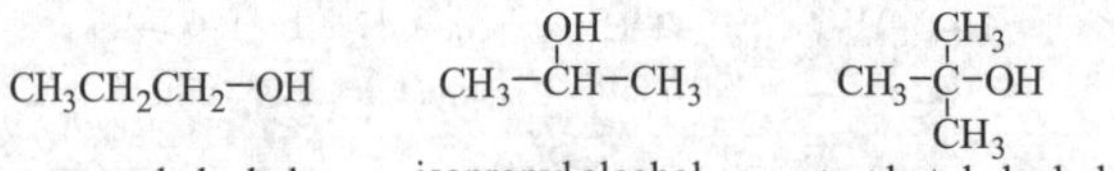

$$CH_3CH_2CH_2\!-\!OH \qquad CH_3\!-\!\overset{OH}{CH}\!-\!CH_3 \qquad CH_3\!-\!\overset{CH_3}{\underset{CH_3}{C}}\!-\!OH$$

n-propyl alcohol   isopropyl alcohol   *tert*-butyl alcohol

$$H_2C\!=\!CH\!-\!CH_2\!-\!OH \qquad\qquad C_6H_5CH_2\!-\!OH$$

allyl alcohol   benzyl alcohol

2. Systematic Nomenclature of Alcohols IUPAC   The IUPAC system provides names for alcohols, based on rules that are similar to those for other classes of compounds. In general, the name carries the **suffix (后缀)** *-ol*, together with a number to give the location of the hydroxyl group. The formal rules are summarized in the following three steps.

(1) Select the longest carbon chain that contains the carbon atom bearing the -OH group, and number it from the end that gives -OH the lower number. Change the ending of the parent alkane from *-e* to *-ol* and use a number to show the location of the -OH group.

(2) Number the longest carbon chain starting at the end nearest the hydroxyl group, and use the appropriate number to indicate the position of the group. The hydroxyl group takes precedence over double and triple bonds, so the chain is numbered in order to give the lowest possible number to the carbon atom bonded to the hydroxyl group.

(3) Name and number **substituents (取代基)** and list them in **alphabetical order (字母顺序)**.

CH₃-C(CH₃)(CH₃)-CH(OH)-CH₂-Br

1-bromo-3,3-dimethylbutan-2-ol

trans-pent-2-en-1-ol

R-2-chloropentan-2-ol

Cyclic alcohols (环醇) are named using the **prefix** (前缀) *cyclo*. Numbering begins at the carbon bearing the -OH group. This carbon is automatically numbered as carbon 1.

1-ethylcyclopropanol

cyclopent-2-en-1-ol

trans-4-chlorocyclohexanol

In the IUPAC system, a compound containing two hydroxyl groups is named as a **diol** (二醇), the one containing three hydroxyl groups as a **triol** (三醇), and so on. They are named like other alcohols except that the **suffix** *-diol* is used and two numbers are needed to tell where the two hydroxyl groups are located. As with many other organic compounds, common names for certain diols and triols have persisted.

CH₂—CH₂
OH   OH

ethane-1,2-diol
(glycol)

CH₂—CH—CH₃
OH   OH

propane-1,2-diol

CH₂—CH—CH₂
OH   OH   OH

propane-1,2,3-triol
(glycerol)

8.1.3   Physical Properties of Alcohols

Most of the common alcohols, up to eleven carbon atoms, are liquids at room temperature. Methanol and ethanol are free-flowing **volatile** (挥发性的) liquids with characteristic fruity odors. The higher alcohols (from butanols to decanols) are somewhat viscous, and some of the highly branched **isomers** (同分异构体) are solids at room temperature.

Alcohols are much more soluble in water than hydrocarbons with similar molecular weight, because alcohol molecules interact by hydrogen bonding with water molecules. Alcohols form **hydrogen bonds** with water, and several of the lower-molecular-weight alcohols are miscible with water. Alcohols with one-, two-, or three-carbon alkyl groups are miscible with water. The water solubility decreases as the alkyl group becomes larger.

Alcohols have higher **boiling points** than alkanes, alkenes, and alkynes with similar molecular weight, because alcohol molecules associate with one another in the liquid state by hydrogen bonding. For example, hydrogen bonding is the major intermolecular attraction responsible for ethanol's high boiling point (78℃). The hydroxyl hydrogen of ethanol is strongly **polarized** by its bond to oxygen, and it forms a hydrogen bond with a **pair of nonbonding electrons** (未成键电子对) from the oxygen atom of another alcohol molecule. With the increase of molecular weight, the solubility of alcohol is more and more close to that of hydrocarbons with similar molecular weight. The physical properties of some common alcohols are shown in Table 8.1.

Table 8.1　The physical properties of some common alcohols

表8-1　常见醇的物理性质

Structural formula	Name	Melting point/ °C	Boiling point / °C	Density/ (g · ml⁻¹)	Solubility water/ (g · 100ml⁻¹)
CH₃OH	methanol	−97	65	0.7914	∞
C₂H₅OH	ethanol	−114	78	0.7893	∞
CH₃(CH₂)₂OH	*n*-propyl alcohol	−126	97	0.8035	∞
CH₃CHOHCH₃	*iso-* propyl alcohol	−89	82	0.7855	∞
CH₃(CH₂)₃OH	n-butyl alcohol	−90	117	0.8098	8.0
C₂H₅CHOHCH₃	*sec-* butyl alcohol	−115	100	0.8063	12.5
(CH₃)₂CHCH₂OH	*iso-* butyl alcohol	–	108	0.8021	11.1
(CH₃)₃C-OH	*tert-* butyl alcohol	25	82	0.7887	∞
CH₃(CH₂)₄OH	*n*-pentyl alcohol	−79	138	0.8144	2.2
C₂H₅(CH₃)₂COH	*tert-* pentyl alcohol	−8.4	102	0.8059	∞
C₃H₇ CHOHCH₃	2-pentanol	–	119	0.8090	4.9
C₂H₅CHOHC₂H₅	3- pentanol	–	115	0.8150	5.6
(CH₃)₃CCH₂OH	*neo-* pentanol	53	114	0.8120	∞
CH₃(CH₂)₅OH	*n*-hexyl alcohol	−47	158	0.1360	0.7
⬡—OH	cyclohexyl alcohol	2.3	161	0.9624	3.6
CH₂= CHCH₂OH	allyl alcohol	−129	97	0.8555	∞
(C₆H₅)₃COH	triphenyl carbinol	164	380	1.1994	—
CH₂OHCH₂OH	glycol	−11	198	1.1088	∞
(CH₂OH)₂CHOH	glycerol	20	290	1.2613	∞

8.1.4　Reactions of Alcohols

1. Acidity of Alcohols　Alkoxide ions are strong **nucleophiles** and strong bases. Alcohols react with Na, K, and other active metals to liberate hydrogen and form **metal alkoxides** (金属醇盐). This is an **oxidation-reduction** (氧化还原反应), with the metal being oxidized and the hydrogen ion being reduced to form hydrogen gas. Hydrogen bubbles out of the solution, leaving the sodium or potassium salt of the alkoxide ion.

The acidic alcohols, like methanol and ethanol, react rapidly with sodium to form **sodium methoxide** (甲醇钠) and **sodium ethoxide** (乙醇钠). Secondary alcohols, such as **propan-2-ol** (异丙醇), react more slowly. Tertiary alcohols, such as *tert*-**butyl alcohol** (叔丁醇), react very slowly with sodium.

$$ROH \ + \ Na \longrightarrow RONa \ + \ 1/2\,H_2\uparrow$$

2. Reactions of C-O Bonds   In acidic solution, an alcohol is in equilibrium with its protonated form. **Protonation (质子化)** converts the hydroxyl group from a poor leaving group to a good leaving group. Once the alcohol is protonated, all the usual **substitution (取代)** and **elimination** reactions of are feasible, regardless of the structure (1°, 2°, 3°) of the alcohol.

$$R-\overset{\cdot\cdot}{\underset{\cdot\cdot}{O}}-H \;+\; H^{+} \;\rightleftharpoons\; R-\overset{\overset{H}{|}}{\underset{\cdot\cdot}{O}^{+}}-H \xrightarrow[\text{S}_\text{N}1 \text{ or } \text{S}_\text{N}2]{X^{-}} R-X$$

poor leaving group   good leaving group

(1) Reactions of alcohols with hydrohalic acids   General equation for the chemical reaction is:

$$ROH \;+\; HX \;\rightleftharpoons\; RX \;+\; H_2O$$

The order of HX reaction rate is HI > HBr > HCl.

Mechanism: A primary alcohol reacts with HBr by the S_N2.

Step 1: Protonation converts the hydroxyl group to a good leaving group.

butan-1-ol

Step 2: Bromide displaces water to give the alkylbromide.

1-bromobutane

Secondary alcohols also react with HBr to form alkyl bromides, usually by the S_N1 mechanism.

cyclohexanol   bromocyclohexane
                          (80%)

A tertiary alcohol reacts with HBr by the S_N1 mechanism.

Step 1: Protonation converts the hydroxyl group to a good leaving group.

tert-butyl alcohol

Step 2: Water leaves to form a carbocation.

193

Step 3: Bromide ion attacks the carbocation to form alkylbromide.

$$CH_3-\overset{\overset{CH_3}{|}}{\underset{\underset{CH_3}{|}}{C}}{}^+ \quad :\overset{..}{\underset{..}{Br}}:^- \longrightarrow H_3C-\overset{\overset{CH_3}{|}}{\underset{\underset{CH_3}{|}}{C}}-\overset{..}{\underset{..}{Br}}:$$

tert-butyl alcohol

Primary alcohols with extensive β-branching give a large amounts of products derived from rearrangement. There is a rearrangement of the primary alcohols with side chains on the carbon atoms, with the formation of a carbocation **intermediate** (中间体). For example, treatment of 2,2-dimethyl-1-propanol (neopentyl alcohol) with HBr gives a rearranged product almost exclusively.

$$CH_3-\overset{\overset{CH_3}{|}}{\underset{\underset{CH_3}{|}}{C}}-CH_2OH \;+\; HBr \longrightarrow CH_3-\overset{\overset{Br}{|}}{\underset{\underset{CH_3}{|}}{C}}-CH_2CH_3 \;+\; H_2O$$

2,2-dimethylpropan-1-ol   2-bromo-2-methybutane

Many secondary alcohols also form some rearranged product, giving evidence for the formation of carbocation intermediates during their reaction.

$$H_3C-\overset{\overset{CH_3}{|}}{\underset{\underset{CH_3}{|}}{C}}-\underset{\underset{OH}{|}}{CHCH_3} \underset{}{\overset{+HBr}{\rightleftharpoons}} H_3C-\overset{\overset{CH_3}{|}}{\underset{\underset{CH_3}{|}}{C}}-\underset{\underset{\overset{+}{O}H_2}{|}}{CHCH_3} \underset{}{\overset{-H_2O}{\rightleftharpoons}} H_3C-\overset{\overset{CH_3}{|}}{\underset{\underset{CH_3}{|}}{C}}-\overset{+}{C}HCH_3 \overset{rearrange}{\longrightarrow}$$

$$H_3C-\overset{\overset{CH_3}{|}}{\underset{\underset{CH_3}{|}}{\overset{+}{C}}}-CHCH_3 \quad Br^- \longrightarrow H_3C-\overset{\overset{CH_3}{|}}{\underset{\underset{Br}{|}}{C}}-\underset{\underset{CH_3}{|}}{CHCH_3} \;+\; H_3C-\overset{\overset{CH_3}{|}}{\underset{\underset{CH_3}{|}}{C}}-\underset{\underset{Br}{|}}{CHCH_3}$$

major product   minor product

A reagent composed of HCl and $ZnCl_2$ is called **Lucas reagent** (卢卡斯试剂), which reacts with primary, secondary, and tertiary alcohols at predictable rates, and these rates can distinguish among the three types of alcohols. Because the halohydrocarbon produced by the reaction is insoluble in hydrochloric acid and **turbid** (浑浊), the primary, secondary and tertiary alcohols with less than 6 carbon atoms can be distinguished according to the rate of turbid time. The relative ease of reaction of alcohols with HX is: primary (10°) alcohol < secondary (2°) alcohol < tertiary (30°) alcohol.

(2) Reactions of alcohols with phosphorus halides   Several **phosphorus halides** (卤化磷) are useful for converting alcohols to alkyl halides. **Phosphorus tribromide** (三溴化磷), **phosphorus trichloride** (三氯化磷), and **phosphorus pentachloride** (五氯化磷) work well.

$$3R-OH \;+\; PCl_3 \longrightarrow 3R-Cl \;+\; P(OH)_3$$
$$3R-OH \;+\; PBr_3 \longrightarrow 3R-Br \;+\; P(OH)_3$$
$$R-OH \;+\; PCl_5 \longrightarrow R-Cl \;+\; POCl_3 \;+\; HCl$$

Phosphorus tribromide is often the best reagent for converting a primary or secondary alcohol to alkyl bromide, especially if the alcohol might rearrange in strong acid. Phosphorus halides produce good **yields** (产率) of most primary and secondary alkyl halides, but not works well with tertiary alcohols.

(3) Reactions of alcohols with thionyl chloride   Thionyl chloride (亚硫酰氯, 氯化亚砜) is often the most widely used reagent for converting an alcohol to an alkyl chloride. For the synthesis of alkyl chlorides, thionyl chloride generally gives better yields, and rearrangements are seldom observed than PCl_3 or PCl_5 or especially with tertiary alcohols. The by-products (SO_2 and HCl) leave the reaction mixture and ensure there can be no reverse reaction.

$$R\!-\!OH + Cl\!-\!\underset{\underset{O}{\|}}{S}\!-\!Cl \xrightarrow{\text{heat}} R\!-\!Cl + SO_2 + HCl$$

A particular value of thionyl halides is that their reaction with alcohols is stereoselective. Reaction of thionyl chloride with (R)-pentane-2-ol, for example, in the presence of an diethyl ether occurs with **invariant of configuration** (构型保持) and gives (R)-2-chlorooctane. Because the chlorine atom which is attacked by nucleophiles is on the same side as the departing group SO_2, the configuration of $\alpha\text{-}C$ of alcohol remains unchanged during the reaction. However, in the presence of a **pyridine** (吡啶) occurs with **inversion of configuration** (构型翻转) and gives (S)-2-chlorooctane.

$$
\begin{array}{c}
CH_3 \\
HO\!-\!\!\!\!-\!\!H \\
CH_2CH_2CH_3
\end{array}
\quad \text{(R)-pentane-2-ol}
$$

$\xrightarrow[\text{Diethyl ether}]{SOCl_2}$
$\begin{array}{c} CH_3 \\ Cl\!-\!\!\!\!-\!\!H \\ CH_2CH_2CH_3 \end{array}$ (R)-2-chloropentane

$\xrightarrow[\text{Pyridine}]{SOCl_2}$
$\begin{array}{c} CH_3 \\ H\!-\!\!\!\!-\!\!Cl \\ CH_2CH_2CH_3 \end{array}$ (S)-2-chloropentane

3. Esterification of Alcohols   In addition to forming esters with **carboxylic acids**, alcohols form **inorganic esters** (无机酸酯) with inorganic acids such as sulfuric acid, nitric acid, and phosphoric acid.

(1) Sulfate ester (硫酸酯)   In an alkyl sulfate esters, alkoxy groups are bonded to sulfur through oxygen atoms.

$$C_2H_5\!-\!OH + H\!-\!OSO_3H \rightleftharpoons CH_3CH_2OSO_3H + H_2O$$
$$\text{etherosulfuric acid}$$

$$2CH_3CH_2OSO_3H \xrightarrow[\text{distillation}]{\text{reduced pressure}} C_2H_5\!-\!OSO_2OC_2H_5 + H_2SO_4$$
$$\text{diethyl sulfate}$$

(2) Nitrate Esters (硝酸酯)   Nitrate esters are formed from alcohols and nitric acid. The best-known nitrate ester is "nitroglycerine", whose systematic name is **glyceryl trinitrate** (甘油三硝酸酯). Glyceryl trinitrate results from the reaction of glycerol with three molecules of nitric acid.

$$
\begin{array}{c}
CH_2OH \\
CHOH \\
CH_2OH
\end{array}
+ 3HNO_3 \longrightarrow
\begin{array}{c}
CH_2ONO_2 \\
CHONO_2 \\
CH_2ONO_2
\end{array}
+ 3H_2O
$$
$$\text{glycerol} \qquad\qquad\qquad \text{glyceryl trinitrate}$$
$$\text{(nitroglycerine)}$$

(3) Phosphate esters (磷酸酯)   Alkyl phosphates are composed of 1 mole of phosphoric acid combined with 1, 2, or 3 moles of an alcohol. For example, methanol forms three phosphate esters.

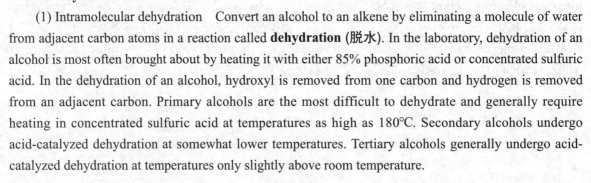

Phosphate esters play a central role in biochemistry. Phosphate ester linkages compose the backbone of the **nucleic acids** (核酸), RNA (**ribonucleic acid,** 核糖核酸) and DNA (**deoxyribonucleic acid,** 脱氧核糖核酸), which carry the **genetic information** (遗传信息).

4. Dehydration of Alcohols

(1) Intramolecular dehydration　Convert an alcohol to an alkene by eliminating a molecule of water from adjacent carbon atoms in a reaction called **dehydration** (脱水). In the laboratory, dehydration of an alcohol is most often brought about by heating it with either 85% phosphoric acid or concentrated sulfuric acid. In the dehydration of an alcohol, hydroxyl is removed from one carbon and hydrogen is removed from an adjacent carbon. Primary alcohols are the most difficult to dehydrate and generally require heating in concentrated sulfuric acid at temperatures as high as 180℃. Secondary alcohols undergo acid-catalyzed dehydration at somewhat lower temperatures. Tertiary alcohols generally undergo acid-catalyzed dehydration at temperatures only slightly above room temperature.

$$-\underset{H}{\overset{|}{C}}-\underset{OH}{\overset{|}{C}}-\quad\xrightarrow{H^+}\quad -C=C- \;+\; H_2O$$

Thus, the ease of acid-catalyzed dehydration of alcohols follows this order:

tertiary (3°) alcohol > secondary (2°) alcohol > primary (1°) alcohol

When isomeric alkenes are obtained in acid-catalyzed dehydration of an alcohol, the alkene having the greater number of substituents on the double bond (the more stable alkene) generally predominates (**Zaitsev's rule** 扎伊采夫规则). In the acid-catalyzed dehydration of butan-2-ol, for example, the major product is but-2-ene, which has two alkyl groups (two methyl groups) on its double bond. The minor product is but-1-ene, which has only one alkyl group (an ethyl group) on its double bond.

$$\underset{\text{butan-2-ol}}{CH_3\underset{OH}{\overset{OH}{CH}}CH_2CH_3}\xrightarrow[180℃]{H_2SO_4}\underset{\substack{\text{but-2-ene}\\(80\%)}}{CH_3CH{=}CHCH_3}\;+\;\underset{\substack{\text{but-1-ene}\\(20\%)}}{CH_3CH_2CH{=}CH_2}+H_2O$$

(2) Intermolecular dehydration　Diethyl ether and several other commercially available ethers are synthesized by the acid-catalyzed intermolecular dehydration of primary alcohols. Primary alcohols generally require heating in concentrated sulfuric acid at temperatures as high as 140℃. For example, intermolecular dehydration of ethanol gives diethyl ether with S_N2 mechanism.

$$\underset{\text{ethanol}}{2CH_3CH_2OH}\xrightarrow[140℃]{H_2SO_4}\underset{\text{diethyl ether}}{CH_3CH_2OCH_2CH_3}+H_2O$$

Mechanism (S_N2):

Step 1: Add a proton. Proton transfer from the acid catalyst to the hydroxyl group gives an **oxonium ion** (离子), which converts -OH, a poor leaving group, into $-OH_2^+$, a better leaving group.

$$CH_3CH_2-\ddot{O}-H + H-\ddot{O}-\underset{\underset{O}{\parallel}}{\overset{\overset{O}{\parallel}}{S}}-O-H \rightleftharpoons CH_3CH_2\overset{+}{\ddot{O}}H + \ddot{:}\ddot{O}-\underset{\underset{O}{\parallel}}{\overset{\overset{O}{\parallel}}{S}}-O-H$$

oxonium ion

Step 2: Make a new bond between a nucleophile and an electrophile and simultaneously break a bond to give stable molecules or ions.

$$CH_3CH_2-\ddot{O}-H + CH_3CH_2\overset{+}{\overset{|}{\ddot{O}}}-H \xrightarrow{S_N2} CH_3CH_2-\ddot{O}-CHCH_2 + \ddot{:}\overset{|}{\ddot{O}}-H$$

a new oxonium ion

Step 3: Take a proton away. Proton transfer from the new oxonium ion to H_2O to complete the reaction.

$$CH_3CH_2-\overset{|}{\ddot{O}}-CH_2CH_2 + \ddot{:}\overset{|}{\ddot{O}}-H \underset{\text{transfer}}{\overset{\text{proton}}{\rightleftharpoons}} CH_3CH_2-\ddot{O}-CH_2CH_3 + H-\overset{+}{\ddot{O}}H$$

5. Oxidation of Alcohols   The reagent most commonly used in the laboratory for the oxidation of alcohol is **potassium dichromate** (重铬酸钾) or **potassium permanganate** (高锰酸钾), dissolved in aqueous sulfuric acid.

Oxidation of a primary alcohol initially forms an **aldehyde** (醛). Then, an aldehyde is easily oxidized further to give a carboxylic acid.

$$R-\underset{\underset{\text{primary alcohol}}{}}{\overset{\overset{OH}{|}}{C}H-H} \xrightarrow{[O]} R-\underset{\text{aldehyde}}{\overset{\overset{O}{\parallel}}{C}-H} \xrightarrow{[O]} R-\underset{\text{carboxylic acid}}{\overset{\overset{O}{\parallel}}{C}-OH}$$

Secondary alcohols are easily oxidized to give excellent yields of **ketones**.

$$\underset{\text{cyclohexanol}}{\overset{\overset{H}{\overset{|}{\text{OH}}}}{\bigcirc}} \xrightarrow[H_2SO_4]{Na_2Cr_2O_7} \underset{\substack{\text{cyclohexanone} \\ (90\%)}}{\bigcirc}$$

Tertiary alcohols cannot be oxidized because the carbon bearing the -OH is bonded to three carbon atoms and, therefore, cannot form a carbon–oxygen double bond.

Oppenauer oxidation (Oppenauer氧化), refers to the reaction between secondary alcohol and acetone in the presence of $Al[OCH(CH_3)_2]_3$ or $Al[OC(CH_3)_3]_3$, in which alcohol is oxidized to ketone and acetone is reduced to isopropanol.

$$\underset{\underset{OH}{|}}{R\overset{|}{C}HR'} + CH_2CCH_3 \xrightarrow[\text{or } Al[OC(CH_3)_3]_3]{Al[OCH(CH_3)_2]_3} R-\overset{\overset{O}{\parallel}}{O}-R' + CH_3\overset{\overset{OH}{|}}{C}HCH_3$$

The C=C or C≡C bonds of alcohol are not affected, so it can be used for the preparation of unsaturated ketone.

$$H_3C-CH-C=CHCH_3 + CH_2CCH_3 \xrightarrow{Al[OCH(CH_3)_2]_3} H_3C-C-C=CHCH_3 + CH_3CHCH_3$$

3-methyl pent-3-en-2-ol　　acetone　　　　　　　3-methylpent-3-en-2-one　　isopropanol

6. Unique Reactions of Diols Compounds containing hydroxyl groups on two adjacent carbon atoms are called **vicinal diols** (邻二醇).

(1) Periodic acid oxidation of diols The major use of **periodic acid** (**HIO₄,** 高碘酸) in organic chemistry is for the cleavage of a diol to two carbonyl groups. In the process, periodic acid is reduced to **iodic acid** (碘酸).

$$\underset{\text{glycol}}{-C-C-} \xrightarrow{HIO_4} \underset{\text{ketones or aldehydes}}{>C=O + O=C<}$$

Whenever one C-C bond of the diol is broken, one molecule of periodic acid is consumed. According to the amount of periodic acid, the molecular structure formula of diols can be inferred.

$$\underset{\substack{OH\,OH\\ \text{propane-1,2-diol}}}{H_2C-CHCH_3} + \underset{\text{periodic acid}}{HIO_4} \longrightarrow \underset{\text{formaldehyde}}{HCHO} + \underset{\text{acetaldehyde}}{CH_3CHO} + \underset{\text{iodic acid}}{HIO_3} + H_2O$$

Mechanism

Step 1: Reaction of the *diol* with periodic acid gives a five-membered cyclic **periodate** (高碘酸盐).

Step 2: Break bonds to give stable molecules or ions. **Redistribution** (再分配) of valence electrons within the cyclic periodate gives HIO₃ and two carbonyl groups. A result of this electron redistribution is an oxidation of the organic component and a reduction of the iodine-containing component.

(2) The pinacol rearrangement The products of acid-catalyzed dehydration of vicinal diols are quite different from those of acid-catalyzed dehydration of alcohols. The following dehydration is an example of the **pinacol rearrangement** (频哪醇重排).

$$\underset{\substack{OH\,OH\\ \text{pinacol}\\ \text{(2,3-dimethylbutane-2,3-diol)}}}{H_3C-C-C-CH_3} \xrightarrow[100\,°C]{H_2SO_4} \underset{\substack{\text{pinacolone}\\ \text{(3,3-dimethylbutan-2-one)}}}{H_3C-C-C-CH_3} + H_2O$$

Mechanism

Step 1: Add a proton. Proton transfer from the acid catalyst to one of the -OH groups gives an oxonium ion, which converts -OH, a poor leaving group, into $-OH_2^+$, a better leaving group.

Step 2: Break a bond to give stable molecules or ions. Loss of H_2O from the oxonium ion gives a 3° carbocation intermediate.

Step 3: 1, 2 Shift. **Migration (迁移)** of a methyl group from the adjacent carbon with its bonding electrons gives a new, more stable resonance-stabilized cation intermediate. Of the two contributing structures we can draw for it, the one on the right makes the greater contribution because, in it, both carbon and oxygen have complete octets of valence electrons.

Step 4: Take a proton away. Proton transfer to solvent gives **pinacolone (频哪酮)**.

When the hydrocarbon groups connected by the carbon atoms on the pinacol are different, which hydroxyl group leaves first and which hydrocarbon group migrates, the general migration order is as follow: aryl > alkyl > hydrogen

8.1.5   Application of Alcohols in the Pharmacy Study

Glycerol, is a colorless and viscous liquid. It can be miscible with water or ethanol. In medicine, it can be used as a solvent, such as phenol glycerin, iodine glycerin, etc. It can also be used in constipation patients.

$$CH_2{-}CH{-}CH_2$$
$$OH \quad OH \quad OH$$
propane-1,2,3-triol
(glycerol)

(1) Mannitol (甘露醇) is a white crystalline powder with sweet taste. It is found in vegetables, fruits and many plants. In medical clinic, 20% mannitol solution is used to produce **hyperosmotic (高渗透的)** effect of blood, dehydrate surrounding tissue and brain **parenchyma (脑实体)** and excrete with urine, so as to reduce **intracranial pressure (颅内压)** and eliminate **edema (水肿)**.

199

$$HOH_2C-\overset{\overset{\displaystyle OH}{|}}{\underset{\underset{\displaystyle H}{|}}{C}}-\overset{\overset{\displaystyle OH}{|}}{\underset{\underset{\displaystyle H}{|}}{C}}-\overset{\overset{\displaystyle H}{|}}{\underset{\underset{\displaystyle OH}{|}}{C}}-\overset{\overset{\displaystyle H}{|}}{\underset{\underset{\displaystyle OH}{|}}{C}}-CH_2OH$$

mannitol

(2) Benzyl alcohol (苯甲醇, 苄醇) is a colorless liquid with aromatic odor, which can be dissolved in water and easily dissolved in organic solvents such as ethanol. Because of its weak paralytic **effect** (麻痹作用), it is often used as a **painkiller** (止痛药) in injection, such as **penicillin** (青霉素) diluent, which is 20% benzyl alcohol aqueous solution. Benzyl alcohol is also used as **preservative** (防腐剂).

⬡—CH₂OH

benzyl alcohol

(3) Misoprostol (米索前列醇) has a strong inhibitory effect on **gastric acid secretion** (胃酸分泌). As a second-line drug for the treatment of **gastric ulcer** (胃溃疡), it is mainly used for refractory ulcer or recurrent ulcer.

misoprostol

(4) Erythritol (赤藓糖醇), is a newly developed 4-carbohydrates, which is not involved in glucose metabolism and blood glucose changes, so it is suitable for diabetic patients.

(2R, 3S)-butane 1,2,3,4- tetraol
(erythritol)

There are also some more complex alcohols in nature, which have important physiological functions, such as vitamin A, menthol and cholesterol, etc.

8.2 Phenols

PPT

8.2.1 Structure of Phenols

The functional group of a phenol is a hydroxyl group bonded to a benzene ring. The carbon atom is

sp^2-hybridized and the C-O bond in phenols is shorter and stronger than the C-O bond of alcohols, where the carbon atom is sp^3-hybridized. Figure 8.2 shows p-π conjugate structure of phenol.

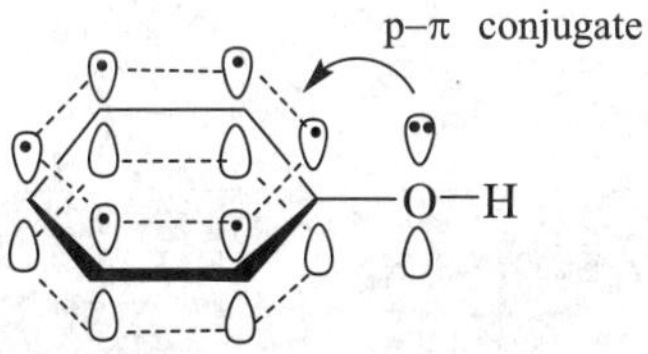

Figure 8.2   Structure of phenol
图8-2   苯酚的结构

8.2.2   Nomenclature of Phenols

Because the phenol structure involves a benzene ring, we usually take phenol as the parent and other groups as the substituents.

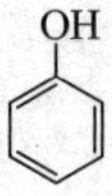

phenol 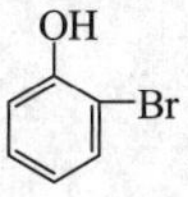   2-bromophenol 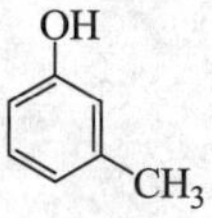   3-methylphenol 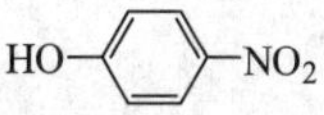   4-nitrophenol

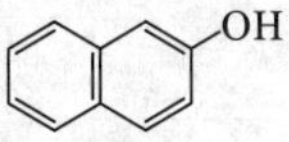

β-naphthol   α-naphthol

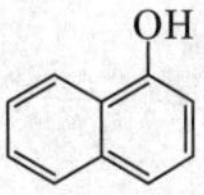

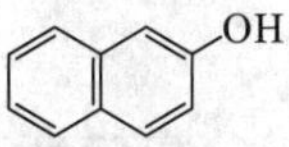

α-anthracene   β-anthracene   γ-anthracene

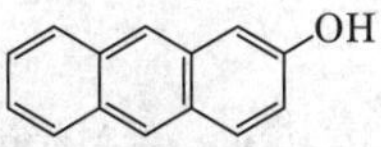

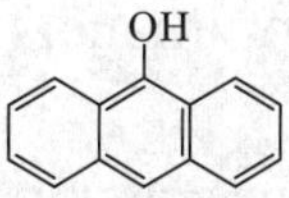

Dihydroxyphenol (二元酚):

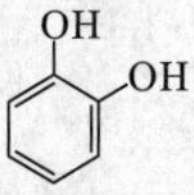

benzene-1,2-diol 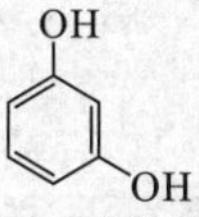   benzene-1,3-diol 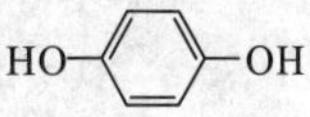   benzene-1,4-diol

Trihydric phenol (三元酚):

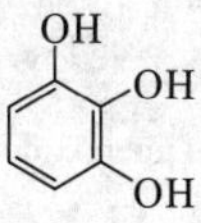

benzene-1,2,3-triol 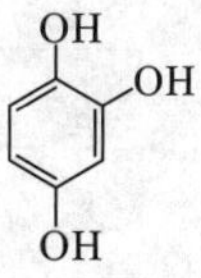   benzene-1,2,4-triol 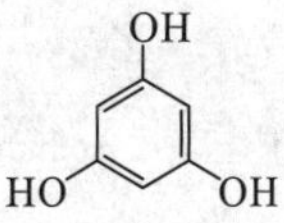   benzene-1,3,5-triol

8.2.3   Physical Properties of Phenols

Most phenols are crystalline solids at room temperature, which have special smell and toxicity. Phenol can form intermolecular hydrogen bond through -OH, so that it has higher boiling point. Phenol and water can also form hydrogen bond, so phenol has certain solubility in water. The physial properties of some common phenols are shown in Table 8.2.

Table 8.2　The physical properties of some common phenols

表8-2　常见酚的物理性质

Name	Melting point/ °C	Boiling point/ °C	pK_a/ 25°C	Solubility water/ (g/100ml)
phenol	41	182	9.96	9
o-cresol	31	191	9.92	2.5
m-cresol	11	201	9.90	2.6
p-cresol	35	202	9.92	2.3
o-nitrophenol	45	217	7.21	0.2
m-nitrophenol	96	–	8.30	1.4
p-nitrophenol	114	–	7.16	1.7
2,4-dinitrophenol	113	–	4.00	0.6
2,4,6-dinitrophenol	122	–	0.71	1.4

8.2.4　Reactions of Phenols

Phenols have many properties similar to those of alcohols, however due to the p-π conjugation of oxygen on phenol hydroxyl with benzene ring, the carbon oxygen bond of phenol is not easy to break as in the alcohols. While other properties derive from their **aromatic character** (芳香性).

1. Acidity of Phenols　Phenols are weak acids, with pK_a values of approximately 10.

	H$_2$CO$_3$	C$_6$H$_5$OH	H$_2$O	ROH
pK_a	~6.35	10	15.7	16~19

Most phenols are insoluble in water, but they are soluble in basic solutions. They react with strong bases such as **sodium hydroxide** (氢氧化钠) to form water-soluble salts. But, they are not sufficiently acidic to react with aqueous **sodium bicarbonate** (碳酸氢钠).

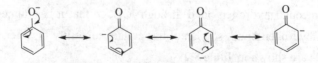

phenol　　　sodium hydroxide　　　　　　　sodium phenoxide

Formation of phenoxide ions is particularly easy because phenols are more acidic than water; aqueous sodium hydroxide deprotonated phenols to give phenoxide ions. The phenoxide ion is more stable than a typical alkoxide ion because a large part of the negative charge in the resonance hybrid still resides on the oxygen atom. In a phenoxide ion, the negative charge is **delocalized** (离域的) over the benzene ring which means it is resonance-stabilized.

Because of this special stability phenol is much more acidic than alcohol. For example, the acidity of cyclohexanol (pK_a=18) is much weaker than that of phenol (pK_a=10).

If carbon dioxide is introduced into the aqueous solution of phenol, which will make sodium phenol liberate. This property can be used to separate and purify phenol compounds.

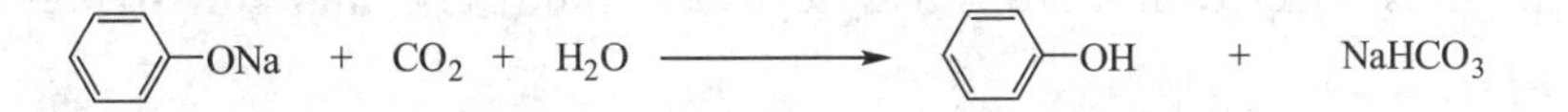

$$\text{C}_6\text{H}_5\text{—ONa} + \text{CO}_2 + \text{H}_2\text{O} \longrightarrow \text{C}_6\text{H}_5\text{—OH} + \text{NaHCO}_3$$

sodium phenoxide   phenol   sodium bicarbonate

2. Synthesis of Phenol Ether   Under acidic condition, alcohol can dehydrate to form ether, but the dehydration of phenol is more difficult than that of alcohol. P-π conjugation results in positive charge of oxygen and negative charge of benzene ring in phenol, the higher bond energy makes it difficult for phenols to dehydrate into ether.

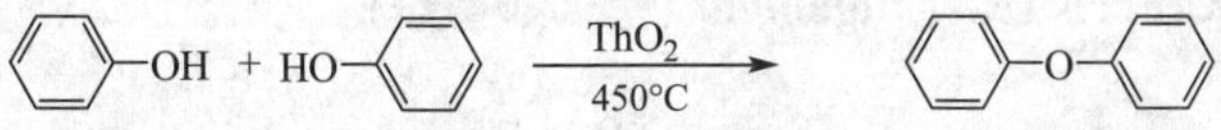

$$\text{C}_6\text{H}_5\text{—OH} + \text{HO—C}_6\text{H}_5 \xrightarrow[450°C]{\text{ThO}_2} \text{C}_6\text{H}_5\text{—O—C}_6\text{H}_5$$

diphenyl ether

Phenol can react with halogenated hydrocarbon to form phenol ether with alkaline condition.

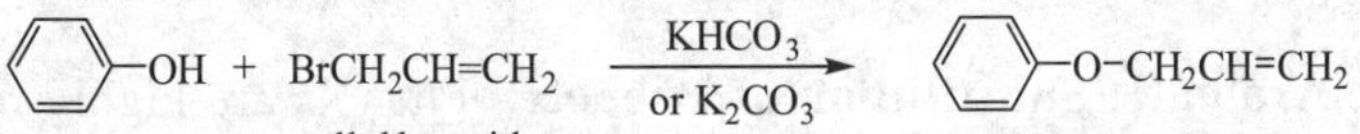

$$\text{C}_6\text{H}_5\text{—OH} + \text{BrCH}_2\text{CH=CH}_2 \xrightarrow[\text{or K}_2\text{CO}_3]{\text{KHCO}_3} \text{C}_6\text{H}_5\text{—O–CH}_2\text{CH=CH}_2$$

allyl bromide   allyl phenyl ether

Heating allyl phenyl ether, the **Claisen rearrangement** (克莱森重排) transforms allyl phenyl ethers to *o*-allylphenols. For example, the simplest member of this class of compounds, at 200~250°C results in a Claisen rearrangement to form *o*-allylphenol.

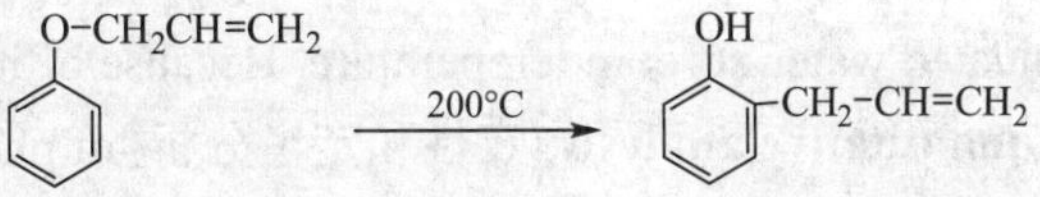

allyl phenyl ether   *o*-allylphenol

The mechanism of a Claisen rearrangement involves a concerted redistribution of six electrons in a cyclic transition state as described above. The product of this rearrangement is a substituted **cyclohexadienone** (环己二烯酮), which undergoes keto-enol **tautomerism** (互变异构) to reform the aromatic ring. A new carbon-carbon bond is formed in the process.

Mechanism:

Step 1: Sigmatropic shift. **Redistribution** (重新分布) of six electrons in a cyclic transition state gives a cyclohexadienone intermediate. Dashed red lines indicate bonds being formed in the transition state, and dashed blue lines indicate bonds being broken.

Step 2: Keto-enol tautomerism. Keto-enol tautomerism restores the aromatic character of the ring.

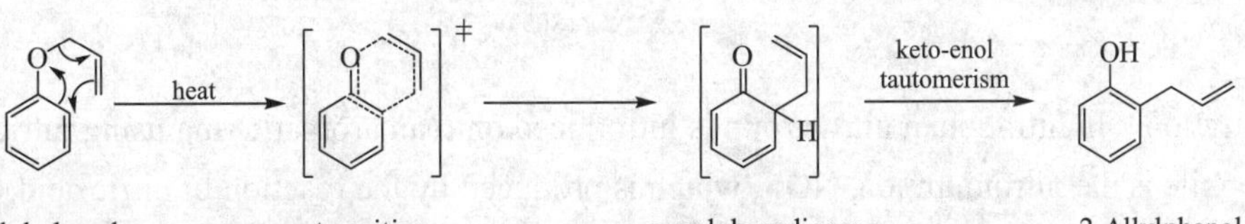

allyl phenyl   transition   a cyclohexadienone   2-Allylphenol
ether   state   intermediate

3. Synthesis of Phenol Ester Breaking of the phenolic O-H bond is a common way for phenols to react. All the alcohol-like reactions involve breaking of the phenolic O-H bond, such as, esterification. Therefore, phenol must react with more active **acyl chloride** (酰氯) and **anhydrides** (酸酐) to form esters.

phenyl acetate   acetic acid

When heating phenol ester in the presence of **aluminum trichloride** (三氯化铝), the acyl group can be rearranged to the *ortho* position or *para* position of the hydroxyl group to obtain phenol ketone. This rearrangement is called **K. Fries rearrangement** (傅瑞斯重排).

phenyl acetate   *o*-hydroxyacetophenone   *p*-hydroxyacetophenone

4. Electrophilic Aromatic Substitution of Phenols Phenols are highly reactive substrates for electrophilic aromatic substitution because the nonbonding electrons of the hydroxyl group stabilize the sigma complex formed by attack at the *ortho* or *para* position. Therefore, the hydroxyl group is strongly activating and *ortho*, *para*-directing. Phenols are excellent substrates for halogenation, nitration, sulfonation, and some Friedel Crafts reactions.

(1) Halogenation The white precipitate of **2,4,6-tribromophenol** (三溴苯酚) can be formed by the reaction of phenol and brominated water at room temperature. Because of its sensitivity, this reaction can be used for **qualitative and quantitative analysis** (定性和定量分析) of phenols.

2,4,6-tribromophenol

When phenol and bromine water are at low temperature and nonpolar solution, a mono-substitution product is obtained.

4-bromophenol (80%)   2-bromophenol (20%)

(2) Nitration In nitration, a nitro group is introduced onto an aromatic ring using nitric acid, HNO_3. The electrophile is the nitronium ion, NO_2^+, which is produced by the reaction of nitric acid.

OH → OH—NO₂ + OH—NO₂

20%HNO₃
25°C

2-nitrophenol　4-nitrophenol

O-nitrophenol can form intramolecular hydrogen bond and no longer form hydrogen bond with water, so it has low water solubility, low boiling point and high volatility, it can be evaporated with water vapor; while *p*-nitrophenol can form a complex through intermolecular hydrogen bond, so it has high boiling point, low volatility, and does not volatilize with water vapor. These two nitrophenols can be separated by **steam distillation** (水蒸气蒸馏).

hydrogen bond

o-nitrophenol　　　　*p*-nitrophenol

(3) Sulfonation　A sulfonic acid group, -SO₃H, can be introduced onto a phenol ring by electrophilic aromatic substitution. The process, called sulfonation, requires sulfuric acid, to form -SO₃H. The product of phenol sulfonation is closely related to the reaction temperature. Generally, *ortho* products are mainly obtained at lower temperature (15~25°C), and *para* products are mainly obtained at higher temperature (80~100°C).

OH
concentrated H₂SO₄

25°C → OH—SO₃H
2-hydroxybenzenesulfonic acid

100°C → OH—SO₃H
4-hydroxybenzenesulfonic acid

concentrated H₂SO₄
100°C

OH—SO₃H with SO₃H
4-hydroxybenzene-1,3-sulfonic acid

(4) Friedel-Crafts alkylation　Because of the highly reactive, phenols are usually alkylated or acylated using relatively weak Friedel Crafts catalysts (such as BF₃, H₃PO₄, HF) to avoid over alkylation or over acylation.

OH + H₃C-C(=O)-OH → OH + OH

BF₃

acetic acid

O=C-CH₃
(95%)
4-hydroxyacetophenone

C(=O)CH₃
2-hydroxyacetophenone

5. Oxidation of Phenols to Quinines   Phenols are easy to be oxidized, but compared with aliphatic alcohols, the oxidation products are different. **Chromic acid** (铬酸) oxidation of a phenol gives a conjugated 1,4-diketone called a **quinone** (醌). In the presence of air, many phenols will be slowly autoxidized to dark mixtures containing quinines.

phenol   1,4-benzoquinone

Hydroquinone (benzene-1,4-diol, 对苯二酚或苯-1,4-二酚) is easily oxidized because it already has two hydroxyl bonded to the ring. Even very weak oxidants like **silver bromide** (AgBr, 溴化银) can oxidize hydroquinone. Silver bromide is reduced to black metallic silver in a light sensitive reaction. Any grains of silver bromide that have been exposed to light react faster than unexposed grains.

benzene-1,4-diol   1,4-benzoquinone

8.2.5   Application of Phenols in the Pharmacy Study

A number of phenolic compounds have medicinal properties and have long been used as drugs.

(1) Salicylic acid (水杨酸) is white crystalline powder, which exists in willow bark, white pearl leaves and sweet birch in nature.

2-hydroxybenzoic acid
(salicylic acid)

It is an important chemical raw material and can be used for the preparation of **aspirin** (阿司匹林) and other drugs.

salicylic acid   acetic acid   acetylsalicylic acid
(aspirin)

(2) Phenolphthalein (酚酞), a common **nonprescription laxative** (非处方泻药), is also an acid base indicator that is colorless in acid and red in base.

phenolphthalein   red dianion

(3) Thymol (百里香酚或麝香草酚) is the aroma component of thyme. It is a colorless crystal, slightly soluble in water, which is used as antiseptic, **disinfectant** (消毒剂) and **insect repellent** (杀虫剂) in medicine.

2-isopropyl-5-methylphenol
(thymol)

(4) Propofol (丙泊酚) is a milky white liquid. As a powerful **sedative** (镇静剂), it can be used as **general anesthetics** (全身麻醉剂), as well as **analgesics** (止痛药), **muscle relaxants** (肌松剂) and **inhaled anesthetics** (吸入麻醉剂).

propofol

(5) Quercetin (槲皮素) is yellow needle-like crystalline powder. It is slightly soluble in water, soluble in an alkaline aqueous solution. Quercetin has various kinds of pharmacological functions such as having a good **expectorant** (祛痰剂), cough effect, also having certain **anti-asthma** (平喘) effect, and having further effects of lowering blood pressure, enhancing capillary resistance, reducing capillary fragility, reducing blood fat, expansion of **coronary artery** (冠状动脉), increasing coronary blood flow.

quercetin

8.3 Ethers

PPT

Ethers are compounds of formula R-O-R', where R and R' may be alkyl groups or aryl (benzene ring) groups. In an ether molecule, both hydrogens are replaced by alkyl groups. The two alkyl groups are the same in a **symmetrical** (对称的) ether and different in an unsymmetrical ether.

8.3.1　Structure of Ethers

The oxygen atom of ether is sp^3 hybridized, and the C-O-C bond angle is approximately the tetrahedral bond angle. In dimethyl ether, two sp^3 hybrid orbitals of oxygen form σ-bonds to the two carbon atoms. The other two sp^3 hybrid orbitals of oxygen each contain an unshared pair of electrons. The two O-C bonds are directed to two of the corners of a **tetrahedron (四面体)**, the **lone pair electrons (孤对电子)** in the remaining two sp^3 hybrid orbitals are directed to the remaining corners of the tetrahedron. The C-O-C bond angle in **methyl ether (甲醚)** is 110°, a value close to the tetrahedral angle of 109.5°. Figure 8.3 shows the structure of dimethyl ether.

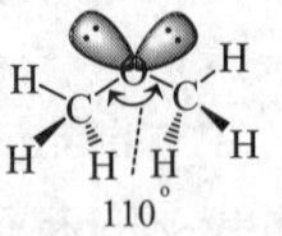

Figure 8.3　Structure of methyl ether
图8-3　甲醚的结构

8.3.2　Nomenclature of Ethers

1. Simple Ethers　Simple ethers are common named as alkyl alkyl ethers. Symmetrical ethers are named by using the prefix *di-* with the name of the alkyl group to indicate that the alkyl groups are the same. For example, the name of an ether with two methyl groups bonded to an oxygen atom is dimethyl ether. We list the alkyl (or aryl) groups in alphabetical order to name unsymmetrical ether.

$H_3C-O-CH_3$
dimethyl ether

$C_2H_5-O-C_2H_5$
diethyl ether

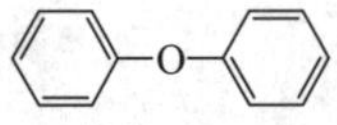

diphenyl ether

$CH_3-O-C_2H_5$
ethyl methyl ether

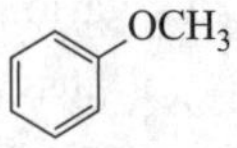

methyl phenyl ether

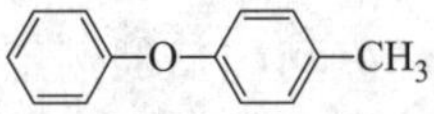

phenyl *p*-methyl phenyl ether

2. Complex Ethers　IUPAC names use the more complex alkyl group as the root name, selecting the longest carbon chain as the parent alkane and naming the -OR group bonded to it as an **alkoxy group (烷氧基)**. This systematic nomenclature is often the only clear way to name complex ethers.

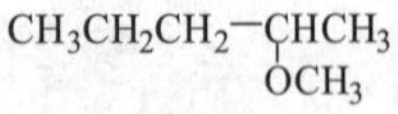

$CH_3CH_2CH_2-\overset{|}{C}HCH_3$
$\overset{|}{O}CH_3$
2-methoxypentane

$CH_3O-CH_2CH_2-OC_2H_5$
1-ethoxy-2-methoxy ethane

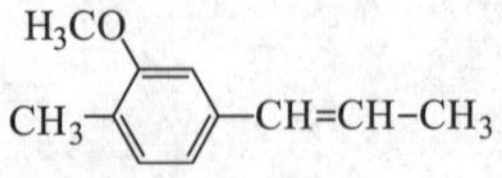

2-methoxy-4-propenyl toluene

3. Cyclic Ethers　Cyclic ethers (环醚) are given special names. The presence of an oxygen atom in a saturated ring is indicated by the prefix *ox-*, and ring sizes from three to six are indicated by the endings -irane, -etane, -olane, and -ane, respectively. Several of these smaller ring cyclic ethers are more often referred to by their common names. Numbering of the atoms of the ring begins with the oxygen atom.

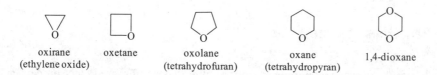

oxirane
(ethylene oxide)　　oxetane　　oxolane
(tetrahydrofuran)　　oxane
(tetrahydropyran)　　1,4-dioxane

4. Crown Ethers   Crown ethers (冠醚) are macrocyclic ethers with —(OCH$_2$CH$_2$)$_n$— repeat units in molecules, which named because one of their most stable conformations resembles the shape of a crown. The parent name crown is preceded by a number describing the size of the ring and followed by a number describing the number of oxygen atoms in the ring, as, X-crown-Y.

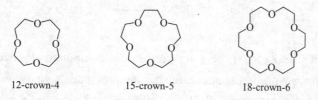

12-crown-4　　　　15-crown-5　　　　18-crown-6

8.3.3   Physical Properties of Ethers

Ethers are polar molecules in which oxygen bears a partial negative charge and each attached carbon bears a partial positive charge. Ethers have two alkyl groups bonded to their oxygen atoms, so they have no hydrogen atoms to form O-H bonds. Only weak dipole-dipole interactions exist between ether molecules in the liquid state. Consequently, boiling points of ethers are much lower than those of alcohols of comparable molecular weight, and are close to those of hydrocarbons of comparable molecular weight. Because ethers cannot act as **hydrogen bond donors** (氢键供体), they are much less soluble in water than alcohols. However, they can act as **hydrogen bond acceptors** (氢键受体), which makes them more water-soluble than hydrocarbons of comparable molecular weight and shape. The physical properties of some common ethers are shown in Table 8.3.

Table 8.3   The physical properties of some common ethers
表8-3   常见醚的物理性质

Structural formula	Name	Melting point (°C)	Boiling point (°C)	density (g · ml^{-1})
CH$_3$OCH$_3$	dimethyl ether	−138.5	−23	−
(C$_2$H$_5$)$_2$O	diethyl ether	−116.6	34.5	0.7137
(CH$_3$CH$_2$CH$_2$)$_2$O	di-*n*-propyl ether	−12.2	90.1	0.7360
[(CH$_3$)$_2$CH]$_2$O	di-*iso*-propyl ether	−85.9	68	0.7241
[CH$_3$(CH$_2$)$_3$]$_2$O	di-*n*-butyl ether	−95.3	142	0.7689
C$_6$H$_5$—O—CH$_3$	methyl pheny ether	−37.5	155	0.9961
C$_6$H$_5$-O—C$_6$H$_5$	dipheny ether	26.8	257.9	1.0748
(tetrahydrofuran structure)	tetrahydrofuran	−65	67	0.8892

8.3.4 Reactions of Ethers

1. Oxonium Salt Because the oxygen atom of ether has an unshared electron pair, as a base, it can react with strong acid or Lewis acid (such as BF_3, $AlCl_3$) to form oxonium salt. The oxonium salt is very unstable. It can be decomposed into ether and acid immediately when encountering water.

$$C_2H_5-O-C_2H_5 \xrightleftharpoons[H_2O]{浓H_2SO_4} \left[C_2H_5-\overset{+}{\underset{H}{O}}-C_2H_5 \right] HSO_4^-$$

$$C_2H_5-O-C_2H_5 \ + \ BF_3 \longrightarrow C_2H_5-\overset{+}{\underset{-BF_3}{O}}-C_2H_5$$

2. Acid-Catalyzed Cleavage Ethers are very stable compounds that react with few common reagents. They do not react with bases, but do react with strong acids whose conjugate bases are good nucleophiles. For example, ethers react with HI (or with HBr) with cleavage of the carbon-oxygen bond to produce alkyl iodides (or bromides).

$$R-O-R' \ + \ HX \xrightarrow{\triangle} RX \ + \ R'-OH \xrightarrow{HX} R'X \ + \ H_2O$$

In general, the less substituted halide is formed by an S_N2 reaction. The halide ion attacks the less hindered carbon atom, and the oxygen atom of the displaced alkoxy group remains bonded to the more substituted carbon atom.

In the case of phenol ether, the phenol oxygen bond is not easy to break, so it can only break from the alkyl end to form phenol and alkyl halide.

$$\text{C}_6\text{H}_5-O-CH_3 \ + \ HCl \xrightarrow{\triangle} \text{C}_6\text{H}_5-OH \ + \ CH_3Cl$$

3. Formation of Hydroperoxides **Hydroperoxidation** (氢过氧化反应) proceeds by a radical chain mechanism. Rates of **hydroperoxide** (氢过氧化物) formation increase dramatically if the C-H bond adjacent to oxygen is secondary (e.g., in diisopropyl ether) because of favored generation of a relatively stable 3° **radical** (自由基) intermediate next to oxygen.

$$\underset{H_3C}{\overset{H_3C}{>}}CH-O-CH\underset{CH_3}{\overset{CH_3}{<}} \xrightarrow{O_2} \underset{H_3C}{\overset{H_3C}{>}}CH-O-\underset{\underset{O-O-H}{|}}{C}\underset{CH_3}{\overset{CH_3}{<}}$$

Hydroperoxides in ethers can be detected by shaking a small amount of the ether with an acidified 10% aqueous solution of **potassium iodide** (KI, 碘化钾), Peroxides oxidize iodide ion to iodine (I_2), which gives a yellow color to the solution. Or by using **starch** (淀粉) potassium iodine paper, which shows a blue color in the paper. Hydroperoxides can be removed by treating them with a reducing agent. One effective procedure is to shake the hydroperoxide-contaminated ether with a solution of iron (Ⅱ) sulfate in dilute aqueous sulfuric acid.

4. Epoxides　Although **epoxies** (环氧化合物) are technically classified as ethers, we discuss them separately because of their exceptional chemical reactivity compared with other ethers which can be attributed to the reactivity of the three-membered ring.

Ethylene oxide is the simplest cyclic ether. It is a colorless gas or liquid and has a sweet, etheric odor. It is a flammable, very reactive and explosive chemical substance. Because of the strain associated with the three-membered ring, epoxides undergo a variety of ring-opening reactions, the characteristic feature of which is nucleophilic substitution at one of the carbons of the epoxide ring with the oxygen atom as the leaving group.

Unsymmetrical epoxides give different products under acid- and base-catalyzed conditions. Under acid-catalyzed conditions, the epoxide oxygen is protonated, and the ring opens in the opposite direction when the epoxide reacts with methanol. In the case of the ring opening by a nucleophile under basic conditions, the reaction is controlled by features of the S_N2 displacement reactions. The nucleophile attacks the less hindered primary carbon atom instead of the tertiary carbon atom. The resulting alkoxide ion then exchanges a proton with the solvent, and the methoxide base is regenerated.

8.3.5　Application of Ethers in the Pharmacy Study

Large cyclic polyethers, that specifically solvate metal cations by complexing the metal in the center of the ring. Complexation by crown ethers often helps polar inorganic salts to dissolve in nonpolar organic solvents. This enhanced solubility allows polar salts to be used under aprotic conditions, where the uncomplexed anions may show greatly enhanced reactivity. When a potassium ion is inserted into the cavity of 18-crown-6, the unshared electron pairs on the six oxygens of the crown ether are close enough to the potassium ion to provide very effective **solvation** (溶剂化) for K^+. Inserted Hg^{2+} (or Cr^{3+}) ion into the cavity of crown, it can be used as a **heavy metals antidote** (重金属解毒剂).

A complex of K^+ and 18-crown-6

Triclosan (三氯生) is a broad-spectrum antibacterial agent, which is insoluble in water, soluble in alkali solution and organic solvent.

2,4,4'-trichloro-2'-hydroxydiphenyl ether
(triclosan)

Penehyclidine hydrochloride (盐酸戊乙奎醚) is a selective **anticholinergic drug** (抗胆碱能药物), which is used for emergency treatment of **organophosphorus pesticide poisoning** (有机磷农药中毒).

penehyclidine hydrochloride

重 点 小 结

1. 醇、酚、醚的结构特征　价键理论与分子轨道理论。

(1) 醇的结构特征　R–O–H 中氧原子为 sp^3 杂化;C–O 和 O–H 键都有较强的极性。

(2) 酚的结构特征　酚羟基的氧为 sp^2 杂化;p–π 共轭体系。

(3) 醚的结构特征　R–O–R (R') 中氧原子为 sp^3 杂化;C–O 键具有较强的极性。

2. 醇、酚、醚的命名　普通命名法和 IUPAC 系统命名法。

3. 醇的化学性质

(1) 酸性　醇的酸性比水弱,醇钠的碱性比氢氧化钠强。

(2) 羟基的反应　①取代反应:与氢卤酸、卤化磷和亚硫酰氯的反应。②酯化反应:与含氧无机酸(硫酸、硝酸、磷酸)成酯。③脱水反应:分子内脱水成烯烃;分子间脱水成醚。

(3) 氧化反应　①强氧化剂氧化:$KMnO_4$ 或 K_2CrO_7 将醇氧化成羧酸或酮。②Oppenauer 选择性氧化:C=C 和 C ≡ C 不被氧化。

(4) 邻二醇的特殊反应　①高碘酸氧化:可推测邻二醇的结构。②频哪醇重排:在酸性催化剂作用下脱水形成频哪酮。

4. 酚的化学性质

(1) 酸性　$H_2CO_3 > C_6H_5OH > H_2O > ROH$

(2) 酚羟基的反应　①酚醚的形成：Claisen 重排（机理）。②酚酯的形成：K. Fries 重排。

(3) 芳环上的反应　①卤代：与溴水的鉴别反应。②硝化、磺化。③傅－克反应。

(4) 氧化反应　酚被氧化成醌。

5. 醚的化学性质

(1) 烊盐的生成　与强酸成盐反应。

(2) 醚键的断裂　与氢卤酸反应。

(3) 过氧化物的生成（酚的贮存）。

(4) 环醚的性质　不对称环氧化物的开环反应（开环机理）。

Problems
目 标 检 测

1. Write IUPAC names for the following compounds.

写出下列化合物的IUPAC名称。

(1) C_6H_5–CH–CH–CH$_2$CH$_3$
　　　　　│　│
　　　　　OH　CH$_3$

(2)
　　　CH$_3$　　OH
　　　│　　　│
CH$_3$–C–CH$_2$CH–◁
　　　│
　　　CH$_3$

(3)
OH
HO——Cl （苯环，3,5位）

(4) CH_2=CH–O–C≡CH

2. Predict the products of the following reactions.

预测下列反应产物。

(1) ⌁OH $\xrightarrow{SOCl_2}$

(2)
　　　C_6H_5　C_6H_5
　　　│　　　│
H_3C–C———C–CH$_3$ $\xrightarrow[\triangle]{H_2SO_4}$
　　　│　　　│
　　　OH　　OH

(3)
OH
　CH$_2$OH $\xrightarrow[H_2O]{NaOH}$

(4) HO——Br $\xrightarrow{FeCl_3 / Cl_2}$

(5) ⌁O⌁ $\xrightarrow[\triangle]{HBr}$

(6) （环己烷氧化物）O $\xrightarrow[H^+]{CH_3CH_2OH}$

3. What are the characteristics of the structure of alcohol, phenol and ether?

醇、酚、醚的结构有什么特点？

4. Explain the acidity of alcohol and phenol.

请解释醇和酚的酸性特点。

Discussion Topic

Both of alcohols and halocarbons have elimination reactions. Please describe their similarities and differences.

（余宇燕）

第九章　醛和酮
Chapter 9　Aldehydes and Ketones

学习目标

1. **掌握**　醛和酮的结构特点，主要类型的化学反应。
2. **熟悉**　醛、酮的分类与物理性质，IUPAC 命名方法以及常用的命名方法。
3. **了解**　醛和酮的特点、研究方法以及在医药领域的重要性。

9.1　Structure of Aldehydes and Ketones

PPT

The functional group of an aldehyde is a carbonyl group bonded to a hydrogen atom and a carbon atom. In **methanal** (甲醛) (always called formaldehyde), the simplest aldehyde, the carbonyl group is bonded to two hydrogen atoms. In other aldehydes, it is bonded to one hydrogen atom and one carbon atom. The functional group of a ketone is one carbonyl group bonded to two carbon atoms. The simplest ketone is **propanone** (丙酮), which is always called acetone. The following are Lewis structures for methanal, **ethanal** (乙醛) (always called acetaldehyde) and propanone.

$$\overset{\ddot{O}}{\underset{HCH}{\|}} \qquad \overset{\ddot{O}}{\underset{CH_3CH}{\|}} \qquad \overset{\ddot{O}}{\underset{CH_3CCH_3}{\|}}$$

Methanal　　　Ethanal　　　Propanone
(Formaldehyde)　(Acetaldehyde)　(Acetone)

According to valence bond theory, the carbon-oxygen double bond consists of one σ bond formed by the overlap of sp^2 hybrid orbitals of carbon and oxygen and one π bond formed by the overlap of parallel $2p$ orbitals. The two pairs of non-bonding electrons on oxygen are located on the remaining sp^2 hybrid orbitals.

An aldehyde can be written with the condensed formula RCHO or ArCHO, where the symbol CHO indicates that both the hydrogen and oxygen atoms are bonded to the carbonyl carbon atom. A ketone has the condensed formula RCOR′. In this condensed formula the symbol CO represents the carbonyl group, and the two R groups flanking the CO group are bonded to the carbonyl carbon atom.

$$R-\overset{\displaystyle \overset{..}{\underset{\|}{O}}}{C}H \qquad R-\overset{\displaystyle \overset{..}{\underset{\|}{O}}}{C}-R'$$

general formulas for an aldehyde or a ketone

PPT

9.2 Nomenclature of Aldehydes and Ketones

9.2.1 IUPAC Names of Aldehydes

1. Aldehydes are named by IUPAC rules similar to those outlined for alcohols. The final *-e* of the parent hydrocarbon corresponding to the aldehyde is replaced by the ending *-al*.

2. The functional group of an aldehyde has a higher priority than **alkyl (烷基)**, **halogen (卤素)**, **hydroxyl (羟基)**, and **alkoxy groups (烷氧基)**. The names and positions of these groups are indicated as prefixes to the name of the parent aldehyde.

3. The functional group of an aldehyde has a higher priority than double or triple bonds. When the parent chain contains a double or triple bond, replace the final -e of the name of the parent **alkene (烯烃)** or **alkyne (炔基)** with the suffix -al. Indicate the position of the multiple bond with a prefix. And the relative priorities of common functional groups are shown in the following Table 9.1.

Table 9.1   Relative priorities of common functional groups
表 9-1   常见官能团的相对优先次序

Functional group priority	Structure	Class of compound
1	—COOH	carboxylic acid
2	—SO_3H	sulfoacid
3	—COOR	ester
4	—COX	acyl halide
5	—CONHR	amide
6	—CN	cyanide
7	—CHO	aldehyde
8	$R-\overset{\displaystyle \overset{O}{\|}}{C}-R'$	ketone
9	—OH	alcohol or phenol
10	—NH_2	amine
11	≡	alkyne
	═	alkene
12	—OR	ether
13	—X	halide
14	—NO_2	nitrocompound

4. If an aldehyde group is attached to a ring, use the suffix *-carbaldehyde*.

5. If an aldehyde or ketone contains other groups with a higher priority, such as **carboxylic acids (羧**

酸), give the carbonyl group the prefix *-oxo*. Use a number to indicate the position of the *-oxo* group. The priority order is carboxylic acid > aldehyde > ketone.

2, 3-dimethybutanal

3-hydroxy-2-methylbutanal

4-methylpent-2-ynal

2-methy-3-oxobutanal

cyclohexanecarbaldehyde

cis-3-bromocyclopentanecarbaldehyde

3-oxobutanoic acid

3-oxopropanoic acid

9.2.2   IUPAC Names of Ketones

The IUPAC rules for naming ketones are similar to those used for aldehydes. The final *-e* of the parent hydrocarbon is replaced with the ending *-one*. However, if the carbonyl group in a ketone is not on a terminal carbon atom, we must indicate its position.

1. Except for the situation described in number 5 for naming aldehyde, number the carbon chain so that the carbonyl carbon atom has the lower number. This number appears as a prefix to the parent name. The identity and location of substituents are indicated with a prefix to the parent name.

2. Name cyclic ketones as **cycloalkanones** (环烷酮). The carbonyl carbon is C-1. Number the ring in the direction that gives the lower number to the first substituent encountered.

3. Halogen, hydroxyl, alkoxy groups, and multiple bonds have lower priorities than the ketone group. These substituted ketones are named using the same method described for aldehydes.

4-methylpentan-2-one

3-methylcyclohexanone

2-bromocyclopentanone

9.2.3   Common Names

The common name for an aldehyde is derived from the common name of the corresponding carboxylic acid by dropping the word *acid* and changing the suffix *-ic* or *-oic* to *-aldehyde*. Because we have not yet studied common names for carboxylic acids, we are not in a position to discuss common names for aldehydes. We can illustrate how they are derived, however, by reference to a few common names with which you are familiar. The name *formaldehyde* is derived from **formic acid** (甲酸); the name *acetaldehyde* is derived from **acetic acid** (乙酸).

Common names for ketones are derived by naming the two alkyl or **aryl** (芳基) groups bonded to the carbonyl group as separate words followed by the word ketone.

$$
\begin{array}{cccc}
\overset{O}{\overset{\|}{HCH}} & \overset{O}{\overset{\|}{HCOH}} & \overset{O}{\overset{\|}{CH_3CH}} & \overset{O}{\overset{\|}{CH_3COH}} \\
\text{Formaldehyde} & \text{Formic acid} & \text{Acetaldehyde} & \text{Acetic acid}
\end{array}
$$

Ethyl isoproyl ketone   Diethyl ketone   Dicyclohexyl ketone

9.3 Physical Properties of Aldehydes and Ketones

The carbonyl oxygen atom is a Lewis base, which can be readily **protonated** (质子化) in the presence of an acid. The polar nature of the C=O group is due to the different electronegativity of the carbon and oxygen atoms. The C=O group cannot form **intermolecular hydrogen bonding** (分子间氢键), but it can accept hydrogen from hydrogen bond donors, e.g. water, alcohols and amines. Therefore, aldehydes and ketones have higher melting and boiling points compared with analogous alkanes, and much lower boiling points than analogous alcohols. They are much more soluble than alkanes but less soluble than analogous alcohols in aqueous media; e.g., acetone and acetaldehyde are miscible with water.

9.4 Reactions of Aldehydes and Ketones

9.4.1 Nucleophilic Addition

1. Addition of Carbon Nucleophiles

(1) Reactions of aldehydes and ketones with **grignard reagents** (格氏试剂)　　The carbon atom of the Grignard reagent resembles a carbanion. It reacts as a **nucleophile** and adds to the electrophilic carbon atom of a carbonyl group in an aldehyde or ketone. The **magnesium ion** (镁离子) forms a salt with the negatively charged oxygen atom. The resulting product is a magnesium alkoxide, which is hydrolyzed to obtain an alcohol.

$$
R-MgX \quad \overset{\delta^-}{\underset{\delta^+}{C=\ddot{O}}} \longrightarrow R-\overset{}{\underset{}{C}}-\ddot{O}-MgX \longrightarrow R-\overset{}{\underset{}{C}}-\ddot{O}-H \ + \ HOMgX
$$

A Grignard reagent adds to various types of carbonyl compounds to give primary, secondary, and tertiary alcohols. Primary alcohols are synthesized by reacting the Grignard reagent, R-MgX, with formaldehyde.

Secondary alcohols are obtained by reacting the Grignard reagent with an aldehyde, RCHO. Note that the carbon atom bearing the hydroxyl group is bonded to the alkyl groups from both the Grignard reagent and the aldehyde.

a secondary alcohol

Tertiary alcohols are formed by reacting the Grignard reagent with a ketone. Two of the alkyl groups bonded to the carbon atom bearing the hydroxyl group come from the ketone; one alkyl group comes from the Grignard reagent.

a tertiary alcohol

(2) Addition of anions of terminal alkynes   The anion of a terminal alkyne is a nucleophile, which is added to the carbonyl group of an aldehyde or a ketone to form a tetrahedral carbonyl addition compound. In the following example, addition of **sodium acetylide** (乙炔钠) to cyclohexanone followed by **hydrolysis** (水解作用) in aqueous acid gives 1-ethynylcyclohexanol.

Sodium acetylide   Cyclohexanone   a sodium alkoxide   1-ethynylcyclohexanol

(3) Addition of hydrogen cyanide   Hydrogen cyanide, HCN, adds to the carbonyl group of an aldehyde or a ketone to form a tetrahedral carbonyl addition compound called **cyanohydrin** (氰醇).

2-hydroxypropanenitrile

(4) Addition of organolithium compounds   Organolithium compounds have greater negative charge character on carbon, so they are generally more reactive in nucleophilic acyl addition reactions than organomagnesium compounds, and typically give higher yields of products. They are more troublesome to use, however, because they must be prepared and used under an atmosphere of **nitrogen** (氮气) or another **inert gas** (惰性气体). The following synthesis illustrates the use of an organolithium compound

to form a sterically hindered tertiary alcohol.

Phenyllithium　　　3, 3-dimethyl-　　　A lithium alkoxide　　　3, 3-dimethyl-
　　　　　　　　　2-butanone　　　　　　　　　　　　　　2-phenyl-2-butanol

　　(5) The Wittig reaction　In 1954, Georg Wittig reported a method for the synthesis of alkenes from aldehydes and ketones in the presence of a compound called **phosphonium** (磷) ylide. For his pioneering study and development of this reaction into a major synthetic tool, Professor Wittig shared the 1979 Nobel Prize in Chemistry. (The other recipient was Herbert C. Brown for his studies of hydroboration and the chemistry of organoboron compounds.) The synthesis of wittig reagent is illustrated by the conversion of cyclohexanone to methylenecyclohexane. In this reaction, C=O double bond is converted to C=C double bond. One noteworthy aspect of this reaction is that a strong **thermodynamic** (热力学的) driving force is provided by formation of the very strong P=O bonding interaction in the phosphine oxide product.

Cyclohexanone　　　A phosphonium　　　Methylenecyclohexane　　Triphenylphosphine
　　　　　　　　　　　ylide　　　　　　　　　　　　　　　　　　　oxide

　　We study wittig reaction in two stages: first, the formation and structure of phosphonium ylides ①, and second, the reaction of a phosphonium ylide with the carbonyl group of an aldehyde or a ketone to give an alkene ②. Wittig reaction is especially valuable as a synthetic tool because it takes place under mild conditions and the position of the carbon-carbon bond is unambiguously determined. The only disadvantage of Wittig reaction is that it is subject to steric hindrance. Yields are generally highest with aldehydes due to the lowest carbonyl hindrance, and are relatively low with ketones in which the carbonyl group is more hindered.

　　①

　　②

2. Addition of Oxygen Nucleophiles

(1) Addition of water: Formation of Carbonyl Hydrates　Nucleophilic acyl addition of water (hydration) to a carbonyl group of an aldehyde or a ketone forms a **geminal diol** (偕二醇).

$$\text{C=O} \quad + \quad H_2O \quad \underset{}{\overset{\text{acid or base}}{\rightleftarrows}} \quad \text{C}\begin{smallmatrix}OH\\OH\end{smallmatrix}$$

Carbonyl group　　　　　　　　　　　A hydrate
of an aldehyde or ketone　　　　　　　(a gem-diol)

A gem-diol is commonly referred to as the hydrate of the corresponding aldehyde or ketone. These compounds are unstable and rarely isolated. This reaction is catalyzed by acid or base. The **mechanism** (机制) is identical to that for the addition of alcohols, which will be discussed next. Hydration of an aldehyde or a ketone is readily reversible, and the diol can eliminate water to regenerate the aldehyde or ketone. In most cases, equilibrium strongly favors the carbonyl group. For a few simple aldehydes, however, the hydrate is favored. For example, when formaldehyde is dissolved in water at 20℃, the position of equilibrium is such that it is more than 99% hydrated.

$$\text{H}_2\text{C=O} \quad + \quad H_2O \quad \rightleftarrows \quad \text{HC}\begin{smallmatrix}OH\\OH\end{smallmatrix}$$

Formaldehyde　　　　　　　　　Formaldehyde
　　　　　　　　　　　　　　　hydrate (>99%)

A 37% solution of formaldehyde in water, called **formalin** (福尔马林), which is commonly used to preserve biological specimens. In contrast, an aqueous solution of acetone consists of less than 0.1% of the hydrate at **equilibrium**.

$$\text{C=O} \quad + \quad H_2O \quad \rightleftarrows \quad \text{C}\begin{smallmatrix}OH\\OH\end{smallmatrix}$$

Acetone　　　　　　　　　　　2, 2-propanediol
(99.9%)　　　　　　　　　　　(0.1%)

(2) Addition of alcohols: Formation of Acetals　Alcohols add to aldehydes and ketones in the same manner as described for water. Nucleophilic acyl addition of one molecule of alcohol to the carbonyl group of an aldehyde or a ketone forms a **hemiacetal** (半缩醛) (a half-acetal).

$$\text{C=O} \quad + \quad H-OEt \quad \underset{}{\overset{\text{acid or base}}{\rightleftarrows}} \quad \text{C}\begin{smallmatrix}OH\\OEt\end{smallmatrix}$$

A hemiacetal

The functional group of a hemiacetal is a carbon bonded to an -OH group and an -OR group.

$$\text{R}-\overset{OH}{\underset{H}{\text{C}}}-OR' \qquad \text{R}-\overset{OH}{\underset{R''}{\text{C}}}-OR'$$

From an aldehyde　　From a ketone

Hemiacetals are generally unstable and are only minor components of an equilibrium mixture, except in one very important type of compound. When a hydroxyl group is part of the same molecule that contains the carbonyl group and can form a five- or six-membered ring, the compound exists almost entirely in the cyclic hemiacetal form. Recall that five- and six-membered rings have relatively little ring strain. In the following example, (S)-4-hydroxypentanal already has a chiral center and a new chiral center is created upon formation of the hemiacetal.

(S)-4-hydroxypentanal

Cyclic hemiacetals

Formation of hemiacetals can be catalyzed by acid or base. For acid, the main function of catalyst is to enhance the positive charge of carbonyl carbon, so that it is easier to be attacked by oxygen atoms. In the case of base, the function of the catalyst is to remove a proton from the alcohol, making it a better nucleophile.

acid catalyst

base catalyst

Hemiacetals are often not stable relative to starting materials, but they can further react with alcohols to form acetals and a molecule of water. Acetals are considerably more stable than hemiacetals and can be isolated in good yield under the proper conditions.

A hemiacetal + H—OEt $\rightleftharpoons$ A diethyl acetal + H_2O

The formation of acetals and its reverse is catalyzed by acids, not by bases, because the OH group cannot be substituted directly by nucleophiles.

From an aldehyde   From a ketone

Acetals are unreactive to bases, hydride- reducing agents such as $LiAlH_4$ and $NaBH_4$, Grignard and other **organometallic** (有机金属) reagents, oxidizing agents (except, of course, those involving the use of aqueous acid), and catalytic reduction. This lack of reactivity is because acetals have no sp^2 hybridized electrophilic carbon atom to react with nucleophiles. Because of their lack of reactivity toward these reagents and easily hydrolyszed in aqueous acid, acetals are often used to reversibly "protect" the carbonyl groups of aldehydes and ketones while reactions are carried out on other functional groups in the molecule.

3. Addition of Nitrogen Nucleophiles

(1) Ammonia and its derivatives　Ammonia, primary **aliphatic** (脂肪族) amines (RNH_2), and primary aromatic amines ($ArNH_2$) react with the carbonyl group of aldehydes and ketones to give an imine, often referred to as a **Schiff base** (希夫碱). Imines are usually unstable unless the C=N group is part of an extended system of conjugation (e.g., **rhodopsin** 视紫红质) and are generally not isolated.

Secondary amines react with aldehydes and ketones to form **enamines** (烯胺). The name enamine is derived from *-en-* to indicate the presence of a carbon-carbon double bond and *-amine* to indicate the presence of an amino group. An example is enamine formation between cyclohexanone and **piperidine** (哌啶), a cyclic secondary amine. Water is removed by a **Dean-Stark trap** (分水器), which forces the equilibrium to the right.

Briefly, the mechanism for formation of an enamine is very similar to that for the formation of an imine. In the first step, nucleophilic addition of the secondary amine to the carbonyl carbon of the aldehyde or ketone followed by proton transfer from nitrogen to oxygen gives a **tetrahedral** (四面体的) carbonyl addition compound. Acid-catalyzed **dehydration** (脱水) gives the enamine. At this stage, enamine formation differs from imine formation. The nitrogen has no proton to lose. Instead, a proton is lost from the α-carbon of the ketone or aldehyde portion of the molecule in an elimination reaction.

有机化学 | **Organic Chemistry**

(reaction mechanism scheme)

$$Ph-CH_2-CHO + H-N\diagdown \longrightarrow Ph-CH_2-\underset{OH}{\underset{|}{C}}H-N\diagdown \xrightarrow[-H_2O]{H^+} Ph-CH=CH-N\diagdown$$

phenylacetaldehyde   piperidine   a tetrahedral carbonyl   enamine
                                            addition compound

(2) Hydrazine (肼) and related compounds   Aldehydes and ketones react with hydrazine to form compounds called **hydrazones** (腙) as illustrated by treating cyclopentanone with hydrazine.

$$\text{(cyclopentanone)} =O + H_2NNH_2 \longrightarrow \text{(cyclopentane)}=NNH_2 + H_2O$$

hydrazine   hydrazone

9.4.2   Oxidation

Aldehydes have hydrogen atoms on their carbonyl carbon atoms, making them easier to oxidize than ketones. Aldehydes are oxidized to carboxylic acids under strong or weak oxidants, while ketones can react only under strong oxidants. So weak oxidants can be used to distinguish aldehydes from ketones.

1. Oxidation of Aldehydes   Aldehydes react with strong oxidants [such as **potassium permanganate (高锰酸钾) and potassium dichromate (重铬酸钾)**].

$$R-\underset{\overset{\parallel}{O}}{C}-H \xrightarrow{[O]} R-\underset{\overset{\parallel}{O}}{C}-OH$$

Aldehydes react with weak oxidants [such as **Tollens reagent (多伦试剂) and Fehling's solution (斐林试剂)**].

Tollens reagent is a basic solution of a silver ammonia complex ion. When Tollens reagent is added to a test tube that contains an aldehyde, the aldehyde is oxidized and metallic silver is deposited as a mirror on the wall of the test tube.

$$R-\underset{\overset{\parallel}{O}}{C}-H + Ag(NH_3)_2^+ + OH^- \longrightarrow R-\underset{\overset{\parallel}{O}}{C}-O^- + Ag + NH_3 + H_2O$$

Fehling's solution contains cupric ion, Cu^{2+}, as a complex ion in a basic solution. It oxidizes aldehydes to carboxylic acids as the Cu is reduced to Cu^+, which forms a brick-red precipitate, Cu_2O. Fehling's solution has the characteristic blue color of Cu^{2+}, which fades as the red precipitate of Cu_2O forms. Because Fehling's solution is basic, the carboxylic acid product is formed as its conjugate base.

$$R-\overset{\overset{O}{\|}}{C}-H \ + \ Cu^{2+} \ + \ OH^- \ \longrightarrow \ R-\overset{\overset{O}{\|}}{C}-O^- \ + \ Cu_2O \ + \ H_2O$$

2. Oxidation of Ketone   **Cyclohexanone** (环己酮) can oxidize adipic acid.

9.4.3   Reduction

1. Reduction of Aldehydes and Ketones to Alcohols   Aldehydes and ketones can be catalytically reduced to alcohols by hydrogen gas and catalysts such as palladium and Raney nickel. However, the reduction of aldehydes and ketones with hydrogen gas requires more severe conditions than reducing alkenes. We saw that lithium aluminum hydride, LiAlH$_4$, and sodium borohydride, NaBH$_4$, reduce carbonyl groups, but neither reagent reduces carbon-carbon double or triple bonds. Thus, these reagents may be used to reduce a carbonyl group selectively in compounds with carbon-carbon multiple bonds.

2. Reduction of a Carbonyl Group to a Methylene Group   A carbonyl group can be reduced directly to a methylene group by using either the **Clemmensen reduction** (克莱门森还原反应) or the **Wolff-Kishner reduction** (沃尔夫–凯惜纳尔还原反应). The former uses a zinc amalgam (Zn/Hg) and HCl, and the latter uses hydrazine (NH$_2$NH$_2$) and base.

3. Cannizzaro Reaction   Cannizzaro reaction refers to the intermolecular redox reaction of an aldehyde without α-hydrogen under the action of a strong base and generates a molecule of carboxylic acid and a molecule of alcohol.

9.4.4　Keto-Enol Tautomerism

1. Acidity of Hydrogens　A carbon atom adjacent to a carbonyl group is called α-carbon, and hydrogen atoms bonded to it are called α-hydrogens.

$$\alpha\text{-hydrogens} \qquad H_3C-\overset{\overset{O}{\|}}{C}-CH_2-CH_3 \qquad \alpha\text{-carbons}$$

Because carbon and hydrogen have comparable electronegativities, a C-H bond normally has little polarity. In addition, carbon does not have a high electronegativity (compare it, for example, with oxygen, which has an electronegativity of 3.5), so that an anion based on carbon is relatively unstable. As a result, a hydrogen bonded to carbon usually shows very low acidity. The situation is different, however, for α-hydrogens of a carbonyl group. α-Hydrogens are more acidic than acetylenic, vinylic, and alkane hydrogens but less acidic than -OH hydrogens of alcohols.

The greater acidity of a-hydrogens arises because the negative charge on the resulting enolate anion is delocalized by resonance, thus stabilizing it relative to an alkane, alkene, or alkyne anion.

$$H_3C-\overset{\overset{:O:}{\|}}{C}-CH_2-H \quad :B^- \rightleftharpoons H_3C-\overset{\overset{:O:}{\|}}{C}-CH_2^-$$

resonance-stabilized enolate anion

$$H-B \quad + \quad H_3C-\overset{\overset{:\ddot{O}:^-}{\|}}{C}=CH_2$$

When an enolate anion reacts with a proton donor, it may do so either on oxygen or on the α-carbon. Protonation of the enolate anion on the α-carbon gives the original molecule in what is called the keto form. Protonation on oxygen gives an enol form. In this way, the keto form of an aldehyde or a ketone can be converted into the enol catalyzed by base.

Enol formation can also be catalyzed by acid. The only difference between the base catalyzed and acid-catalyzed reactions is the order of proton addition and elimination. In acid-catalyzed reactions, a proton is added first; in base-catalyzed reactions, a proton is removed first.

2. The Position of Equilibrium in Keto-Enol Tautomerism　Aldehydes and ketones with at least one α-hydrogen are in equilibrium with their enol forms. We first encountered this type of equilibrium in our study of the **hydroborationoxidation** (硼氢化氧化作用) and acid-catalyzed hydration of alkynes. As we see, the position of keto-enol equilibrium for simple aldehydes and ketones lies far on the side of the keto form, primarily because carbon-hydrogen single bonds are about as strong as oxygen-hydrogen single bonds but a carbon-oxygen double bond is stronger than a carbon-carbon double bond.

For certain types of molecules, the enol form may be the major form and, in some cases, the only form present at equilibrium. For **β-diketones** (二酮) such as 1,3-cyclohexanedione and **2,4-pentanedione**

(乙酰丙酮), where an α-carbon lies between two carbonyl groups, the position of equilibrium shifts in favor of the enol form. These enols are stabilized by conjugation of the systems of the carbon-carbon double bond and the carbonyl group. The enol of 2,4-pentanedione, an open-chain β-diketone, is further stabilized by intramolecular hydrogen bonding.

cyclohexane-1,3-dione

pentane-2,4-dione
(Acetylacetone)

9.4.5   Reaction at the *α*-Carbon

1. Racemization   When enantiomerically pure (either *R* or *S*) 3-phenyl-2-butanone is dissolved in ethanol, no change occurs in the optical activity of the solution over time. If, however, a trace of either acid (e.g., aqueous or gaseous HCl) or base (e.g., sodium ethoxide) is added, the optical activity of the solution begins to decrease gradually and eventually drops to zero. When 3-phenyl-2-butanone is isolated from this solution, it is found to be a racemic mixture. Furthermore, the rate of **racemization (外消旋化)** is proportional to the concentration of acid or base. These observations can be explained by a rate-determining acid- or base-catalyzed formation of an achiral enol intermediate. Tautomerism of the achiral enol to the chiral keto form generates the *R* and *S* enantiomers with equal probability.

(*R*)-3-phenyl-2-butanone   an enol   (*S*)-3-phenyl-2-butanone

Racemization by this mechanism occurs only at α-carbon chiral centers with at least one α-hydrogen.

2. Aldol Condensation   Under the catalysis of dilute acid or alkali, one aldehyde containing α-hydrogen adds to another aldehyde to form a **β - hydroxyaldehydes (β-羟基醛)**.

In the most common case, base promotes the loss of *α* - hydrogen in one aldehyde, then it attacks the carbonyl carbon of another aldehyde, and the resulting oxygen anion takes the hydrogen ion from water to obtain the corresponding product.

Of course, this reaction can also occur under the catalysis of acid. The role of acid is to promote the enolization of aldehydes and enhance the activity of the other aldehyde group which is to be attacked.

$$H_3C-\overset{O}{\overset{\|}{C}}-H \xrightarrow{H^+} H_3C-\overset{+OH}{\overset{\|}{C}}-H \qquad H_3C-\overset{+O}{\overset{\|}{C}}-H \rightleftharpoons H_2C=\overset{OH}{\overset{\|}{C}}-H$$

$$H_3C-\overset{+OH}{\overset{\|}{C}}-H + H_2C=\overset{O-H}{\overset{\|}{C}}-H \xrightleftharpoons{-H^+} H_3C-\overset{HO}{\underset{H}{C}}-\overset{H_2}{C}-\overset{O}{\overset{\|}{C}}-H$$

If the product can form a better conjugated system, or under the condition of heating or acid, it may dehydrate to form α, β - unsaturated aldehydes.

$$H_3C-\overset{HO}{\underset{H}{C}}-\overset{H_2}{C}-\overset{O}{\overset{\|}{C}}-H \xrightarrow[\text{or } H^+]{\text{heat}} H_3C-\overset{\beta}{\underset{H}{C}}=\overset{\alpha}{\underset{H}{C}}-\overset{O}{\overset{\|}{C}}-H$$

Ketones can also undergo similar condensation reactions under catalysis, but are usually more difficult than aldehydes. In addition, the intramolecular aldol condensation can take place in the binary aldehydes and ketones.

$$2 \overset{H_3C}{\underset{CH_3}{\diagdown}}\!C{=}O \xrightarrow{OH^-} \text{...}$$

The condensation between two different aldehydes and ketones is called cross aldol condensation. Because both of them contain α - hydrogen, there may be four combinations, which makes the reaction less ideal. The solution is that only one of the two molecules with alpha hydrogen, normally formaldehyde or benzaldehyde.

$$\text{PhCHO} + H_3C-\overset{H_2}{C}-\overset{O}{\overset{\|}{C}}-H \xrightarrow[\text{heat}]{H^+} \text{PhCH=C(CH_3)CHO}$$

$$\text{PhCHO} + \text{PhCOCH}_3 \xrightarrow{OH^-} \text{PhCH=CHCOPh}$$

$$H_3C-\overset{O}{\overset{\|}{C}}-H + 3HCHO \xrightarrow{OH^-} (HOH_2C)_3C-\overset{O}{\overset{\|}{C}}-H$$

$$H_3C-\overset{O}{\overset{\|}{C}}-H + 1HCHO \xrightarrow[\text{heat}]{OH^-} H_2C=\overset{O}{\underset{H}{C}}-\overset{\|}{C}-H$$

3. α-Halogenation Aldehydes and ketones with at least one α-hydrogen react at the α-carbon with **bromine** (溴) and **chlorine** (氯) to form α-haloaldehydes and α-haloketones as illustrated by bromination of acetophenone.

$$\text{PhCOCH}_3 + Br_2 \xrightarrow{CH_3COOH} \text{PhCOCH}_2Br + HBr$$

Acetophenone   α-bromoacetophenone

Bromination or chlorination at a α-carbon is catalyzed by both acid and base. Under basic condition, all of the three hydrogen atoms can be substituted and haloform may be formed. The mechanism is shown as follows. First, the Hydroxide ion removes an acidic alpha hydrogen. This results in the formation of an enolate ion. The enolate anion then goes on to displace an iodide ion from the iodine molecule. This process repeats twice to give R-CO-CI$_3$. Now, a hydroxide ion forms a bond with the carbonyl carbon. This leads to the reformation of the carbonyl group and the elimination of the CI$_3$- anion. An R-COOH group is also formed. The carboxylic acid group and the basic CI$_3$- ion neutralize each other. Thus the iodoform is precipitated. Iodoform test is used to check the presence of carbonyl compounds with the structure R-CO-CH$_3$ or alcohols with the structure R-CH(OH)-CH$_3$, which can be oxidized to R-CO-CH$_3$, in a given unknown substance. For example in Chinese Pharmacopoeia, iodoform test is used to distinguish methanol and ethanol.

9.4.6   Reaction of *α, β*-unsaturated aldehydes and ketones

Aldehydes and ketones whose π bond and unsaturated bond are separated by a single bond in the carbonyl group and can generate a conjugated system are called α, β-unsaturated aldehydes or ketones.

1. Nucleophilic Addition   Reaction with Grignard Reagent

α, β-unsaturated aldehydes and ketones can undergo addition reactions with Grignard reagents, There are two types of 1,2-addition and 1,4-addition.

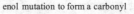

enol mutation to form a carbonyl

Reaction with Hydrogen Cyanide

The addition reaction of carbonyl with HCN is mainly based on 1,4-addition.

$$H_3C-C=C-CH \xrightarrow{HCN} H_3C-C-C=CH \longleftrightarrow H_3C-C-C-CH$$

2. Electrophilic Addition

$$CH_3-C=CHCCH_3 \xrightarrow{HCl} CH_3C-CH_2CCH_3$$

$$CH_3-C=CHCCH_3 \xrightarrow[CCl_4]{Br_2} CH_3C-CHCCH_3$$

3. Michael Reaction

$$\underset{}{} + H_2C=CHCCH_3 \xrightarrow{KOH} \underset{}{}$$

$$\xrightarrow[C_6H_6]{NH} + H_2O$$

4. Reduction Reaction

$$\xrightarrow[(2)H_2O]{(1)LiAlH_4, Et_2O}$$

$$\xrightarrow{H_2/Pd-C}$$

9.5 Application of Aldehydes in the Pharmacy Study

Oral Contraceptives

Alkynide ions, the conjugate bases of alkynes, react with carbonyl groups in much the same way as the female sex hormones that are collectively called **estrogens** (雌激素). Estrogens, such as estradiol, are released during pregnancy and inhibit further **ovulation** (排卵). Oral contraceptives are designed to mimic this effect of pregnancy by inhibiting ovulation.

Estradiol (雌二醇) itself is not an effective oral contraceptive because its C-17 hydroxyl group is rapidly oxidized to estrone in metabolic reactions. This oxidation product has greatly reduced estrogenic activity.

The hormonal action of estradiol is related to the C-17 hydroxyl group, which is located above the plane of the five-membered ring. Thus, it was decided to synthesize a compound structurally related to estradiol that is a tertiary alcohol with the correct stereochemistry. Such compound would survive metabolic oxidation because tertiary alcohols cannot be oxidized. Based on the reactivity of Grignard reagents, one might propose to add methyl Grignard reagent to estrone to produce a tertiary alcohol. However, **estrone** (雌激素酮) has a phenolic hydroxyl group that would react with the Grignard reagent. This difficulty has been cleverly bypassed. Instead of making a tertiary alcohol by adding an alkyl group derived from a Grignard reagent, a tertiary alcohol is made by adding acetylide anion-the conjugate base of **acetylene** (乙炔). The reaction of the acetylide ion with a carbonyl compound produces an acetylenic alcohol. Sodium acetylide reacts with estrone to give ethynyl estradiol. Although the acetylide ion could potentially attack from either side of the C-17 carbonyl group of estrone, a methyl group extends above the plane of the ring in the vicinity of the carbonyl group. The acetylide anion thus approaches the "bottom" of the ring, so the product has its -OH group "up", which is the stereochemistry required for hormonal activity. This tertiary alcohol cannot be oxidized and is an effective oral contraceptive.

estradiol

17-ethynylestradiol

estrone

重 点 小 结

一、醛、酮的结构

醛的官能团是一个羰基，连着一个氢原子和一个碳原子。在最简单的醛甲醛中，羰基连着两个氢原子。在其他醛中，它连着一个氢原子和一个碳原子。酮的官能团是一个连着两个碳原子的羰基。最简单的酮是丙酮，它通常被称为丙酮。下面是甲醛、乙醛（通常称为乙醛）和丙酮的路易斯结构。

醛可以用缩合式 RCHO 或 ArCHO 来表示，其中 CHO 表示氢原子和氧原子都与羰基碳原子相连。酮的分子式是 RCOR。在这个缩合式中，符号 C=O 表示羰基，与 C=O 基团相邻的两个 R 基团与羰基碳原子成键。

二、醛、酮的命名

1. IUPAC规则命名法

（1）醛的命名　　①醛类物质的命名规则与醇类物质的命名规则相似。②醛官能团优先于烷基、卤素、羟基和烷氧基。这些基团的名称和位置表示为母醛名称的前缀。③醛官能团优先于双键或三键。当母链含有双键或三键时，用后缀代替母链烯烃或炔名的最后一个。用前缀表示多键的位置。④如果一个醛基连在一个环上,就用后缀-醛。⑤如果醛或酮含有优先级更高的基团,优先顺序为羧酸>醛>酮。

（2）酮的命名 命名酮的IUPAC规则类似于醛类的命名规则。酮中的羰基不在末端碳原子上,我们必须指出它的位置。①碳链上的数字使羰基碳原子的数字更低。这个数字作为父名称的前缀出现。取代基的标识和位置用父名的前缀表示。②将环酮命名为环烷酮。羰基碳是C-1。在遇到第一个取代基的方向上给环编号。③卤素、羟基、烷氧基和多键比酮基的优先级低。这些取代酮的命名方法与醛类的命名方法相同。

2.习惯命名法 醛的俗名是由相应的羧酸的俗名衍生而来的,我们可以通过引用一些熟悉的常见名称来说明它们是如何派生出来的。甲醛的名称来源于甲酸;乙醛的名字来源于乙酸。酮的常见名称是通过将两个与羰基相连的烷基或芳基分别命名的。

三、醛的物理性质

羰基氧原子是路易斯碱,可以在酸的存在下容易地质子化。C=O基团的极性性质是由于碳和氧原子的电负性差异。

C=O基团不能形成分子间氢键,但它可以接受氢键供体例如氢的氢,如水,酒精和胺。因此,与类似的烷烃相比,醛和酮具有更高的熔点和沸点,并且比类似的醇具有更低的沸点。

它们在水性介质中的溶解度比烷烃高得多,但比类似的醇低。例如丙酮和乙醛可与水混溶。

四、醛、酮的反应

1. 亲核加成反应 ①碳的亲核试剂（格氏试剂、炔基钠、氢氰酸、烷基锂、Wittig 试剂）;②氧的亲核试剂（水、醇）;③氮的亲核试剂（氨／胺、肼）。

2. 氧化反应 ①高锰酸钾氧化;②吐伦试剂和斐林试剂氧化。

3. 还原反应 ①还原为醇的反应;②还原为亚甲基;③歧化反应。

4. α-碳上反应 ①烯醇互变与 α-外消旋化;②羟醛缩合;③卤代反应。

5. α,β-不饱和醛和酮反应 ① 1,2- 与 1,4- 加成;② Michael 加成;③还原反应。

Problems
目标检测

题库

1. What is the IUPAC name for the following compound, which is an alarm pheromone in some species of ants?

其是某些蚂蚁体内的一种报警信息素，请问该化合物的 IUPAC 名称是什么？

2. The European bark beetle produces a pheromone that causes beetles to aggregate. Describe two ways that the compound could be synthesized in the laboratory by a Grignard reagent.

欧洲树皮甲虫产生一种信息素，使甲虫聚集在一起。描述两种在实验室中用格氏试剂合成该化合物的方法。

3. Draw a structural formula for the enol form of each keto.

为下列每个醛、酮化合物绘制其烯醇式。

4. Why is addition of aldehyde or ketone nudeophilic while for alkene it is electrophilic?

为什么醛和酮发生加成反应时，反应类型是亲核加成，而烯烃的加成则为亲电加成？

5. Complete the equations.

完成下面反应式。

6. Complete the equations.

完成下面反应式。

7. Complete the equations.

完成下面反应式。

8. Complete the equations.

完成下面反应式。

9. Complete the equations.

完成下面反应式。

10. Complete the equations.

完成下面反应式。

11. Complete the equations.

完成下面反应式。

12. Complete the equations.

完成下面反应式。

13. Complete the equations.

完成下面反应式。

14. Complete the equations.
完成下面反应式。

$$\text{CH}=\text{CHO} \xrightarrow[\text{② } H_2O]{\text{① } LiAlH_4}$$

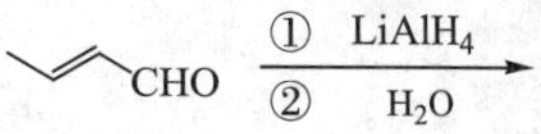

15. Complete the equations.
完成下面反应式。

$$\xrightarrow[\text{② } H_2O/H^+]{\text{① 异丙醇铝/异丙醇}}$$

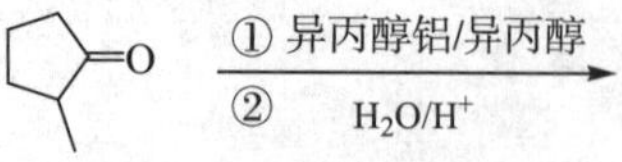

16. Complete the equations.
完成下面反应式。

$$\xrightarrow[\text{② } H_2O/H^+]{\text{① } Cl_2/OH^-}$$

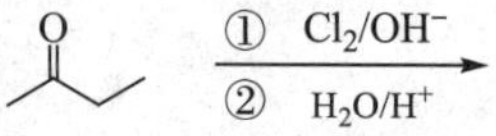

17. Complete the equations.
完成下面反应式。

$$\xrightarrow{Cl_2/H_2O/H^+}$$

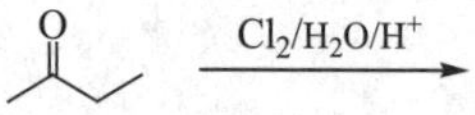

18. Complete the equations.
完成下面反应式。

$$\xrightarrow{H_2O/OH^-}$$

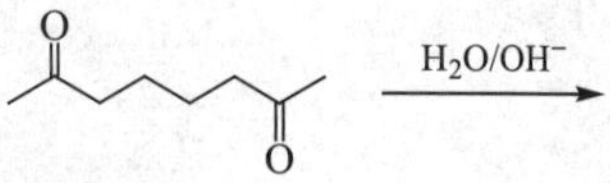

19. Complete the equations.
完成下面反应式。

$$+ \text{ } CH_2O \text{ } + \text{ } NH_3 \xrightarrow{NH_4Cl/H_2O}$$

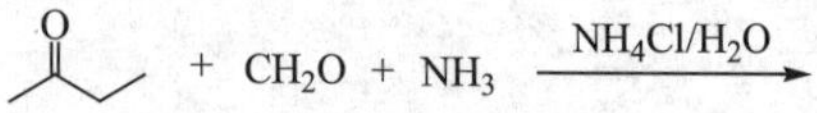

Discussion Topic

From the results, the following equation is a reduction reaction, but based on the reaction mechanism, what is the type of it? Can you list a few other reactions of this mechanism?

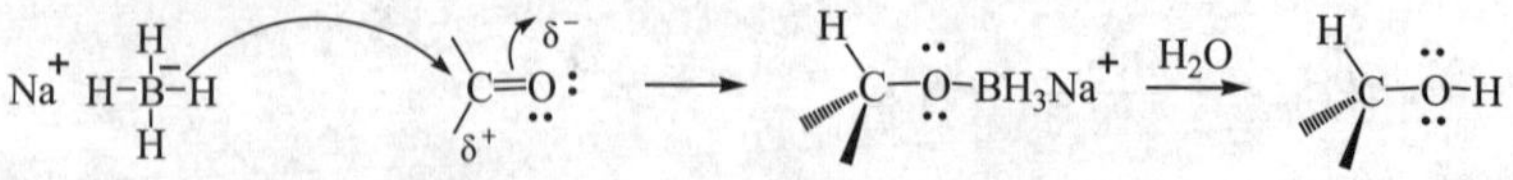

$$Na^+ \text{ } H-\overset{\overset{H}{|}}{\underset{\underset{H}{|}}{B}}-H \quad \overset{\delta^-}{C=\overset{..}{\underset{..}{O}}:} \longrightarrow \overset{H}{C}-\overset{..}{\underset{..}{O}}-BH_3Na^+ \xrightarrow{H_2O} \overset{H}{C}-\overset{..}{\underset{..}{O}}-H$$

（韩　波　谢达春）

第十章 羧酸及衍生物
Chapter 10 Carboxylic Acids and Derivatives

 学习目标

1. **掌握** 羧酸类化合物及衍生物的结构与命名;诱导效应、共轭效应对羧酸类化合物酸性的影响;羧酸衍生物亲核取代反应机理。
2. **熟悉** 羧酸类化合物及衍生物的主要化学性质以及应用。
3. **了解** 羧酸类化合物及衍生物在医药领域的应用。

A compound produced by replacing a hydrogen atom in a hydrocarbon molecule with a **carboxylic group** (羧基) is called a carboxylic acid. The general formula for a carboxylic acid is RCOOH(R in formic acid is H). If the hydroxy group in carboxylic acid molecule are substituted by other atoms or groups such as halogen, alkoxyl or amino, the compounds are called substituted carboxylic acids.

Carboxylic acids and derivatives are abundant in nature and are closely related to human life. For example, the vinegar used in daily life contains 5% acetic acid. The flavoring agent in beverages is citric acid, and the fruit acids in fruits contain malic acid. Some carboxylic acids are intermediate products of plant and animal metabolism and are involved in the life processes of plants and animals. For instance, the metabolic **tricarboxylic acid cycle** (三羧酸循环) is common in the mitochondria of aerobic organisms, and the main intermediate metabolites in this cycle are citric acid containing three carboxyl groups. Many carboxylic acids are used as raw materials or intermediates for the synthesis of drugs and other organic compounds. For example:

aspirin

阿司匹林(解热镇痛药)

(R)-naproxen

萘普生(解热镇痛抗炎药)

malic acid

苹果酸

10.1 Structure of Acids

The functional group of a carboxylic acid is the carboxyl group which is made up of a carbonyl group and a hydroxyl group. The carbon atom of a carboxyl group is sp^2-hybridized, so the carboxyl group has a planar structure and the bond angles are all approximately 120°. The unhybridized p orbit on the carbon atom of the carboxyl forms a π bond with the p orbit on the oxygen atom of the carbonyl group. In addition, the oxygen atom of the hydroxyl in the carboxyl group has a lone pair and forms a p-π conjugated system with the π bond of the carbonyl group.

Due to p-π conjugation, the bond lengths of C-O and C=O in carboxylic acid molecules tend to be equalized, which is different from the independent carbon-oxygen single bond and carbon-oxygen double bond. By X-ray diffraction and electron diffraction, the C=O bond length in formic acid is 0.123 nm (C=O bond length in aldehyde and ketone is 0.120 nm), and the C-O bond length is 0.136 nm (the C-O bond length in alcohol is 0.143 nm).

The chemical reactions of carboxylic acids mainly occur on the carboxyl groups. Due to the interaction between the two groups of hydroxy and carbonyl, the carboxyl groups of the carboxylic acid have unique chemical properties. For example, carboxylic acids are more acidic than alcohols and phenols, and they are not prone to similar nucleophilic addition reactions like aldehyde and ketone compounds.

PPT

10.2 Nomenclature of Carboxylic Acids

10.2.1 Common Name

Several aliphatic carboxylic acids have been known for centuries, and their common names reflect their historical sources. For example, **formic acid** (甲酸) is extracted from ants: formica in Latin. **Acetic acid** (醋酸) is isolated from vinegar, called acetum ("sour") in Latin. **Cinnamic acid** (肉桂酸) is obtained from cinnamon.

HCOOH	CH₃COOH	⬡—CH=CHCOOH
formic acid	acetic acid	cinnamic acid
甲酸	醋酸	肉桂酸

10.2.2 System Nomenclature

1. Nomenclature of Aliphatic Acid   To name simple open-chain carboxylic acids, select the longest continuous carbon chain that includes the acid group as the main chain and replace the final "-e"

ending of alkane (烷烃) with the suffix "-oic acid" . The -COOH carbon is always numbered C_1.

hexanoic acid

己酸（羊油酸）

3-iodopropanoic acid

3-碘丙酸

Aliphatic acids are given common names using Greek letters. Greek letters are assigned beginning with the carbon atom next to the carboxyl group.

$$CH_3 \text{------} CH_2 — CH_2 — CH_2 — COOH$$
$$\omega \qquad \gamma \qquad \beta \qquad \alpha$$

α-amino-β-phenylpropionicacid

α-氨基-β-苯丙酸

2. Nomenclature of Aliphatic Dicarboxylic Acid   If there are two carboxyl groups, the suffix "-dioic acid" is used.

2-ethyl-3-methylbutanedioic acid

2-乙基-3-甲基丁二酸

2,3-dihydroxybutanedioic acid

2,3-二羟基丁二酸（酒石酸）

3. Nomenclature of Cyclic Aliphatic and Aromatic Carboxylic Acids   Names of cyclic aliphatic and aromatic carboxylic acids both use fatty acids as the parent. When there are several carboxyl groups on the ring, the relative positions of the carboxyl groups should be indicated. Cycloalkanes with -COOH substituents are generally named as cycloalkanecarboxylic acids.

benzene-1,4-dicarboxylic acid

terephthalic acid

苯-1,4-二甲酸

trans-cyclohexane-1,2-dicarboxylic acid

反-环己烷-1,2-二甲酸

4. Nomenclature of Unsaturated Acids   Unsaturated acids are named using the name of the corresponding alkene, with the final "-e" replaced by "-oic acid". The carbon chain is numbered starting with the carboxyl carbon, and a number gives the location of the double bond. The stereochemical terms *cis* and *trans*(Z or E)are used as they are with other alkenes.

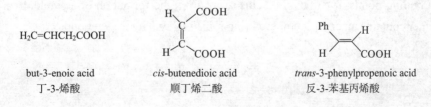

but-3-enoic acid

丁-3-烯酸

cis-butenedioic acid

顺丁烯二酸

trans-3-phenylpropenoic acid

反-3-苯基丙烯酸

5. Nomenclature of Substituted Carboxylic Acid   In the IUPAC system, a carboxyl group takes precedence over most other functional groups, including hydroxyl groups, amino groups, and the carbonyl groups of aldehydes and ketones. As illustrated in the following examples, an -OH group is indicated by the prefix hydroxy-; an -NH$_2$ group, by amino-; and the C=O group of an aldehyde or ketone, by oxo-.

4-bromobenzoic acid
4-溴苯甲酸

4-aminobenzoic acid
4-氨基苯甲酸

H$_3$C—CHCOOH
     |
     OH

2-hydroxypropanoic acid
2-羟基丙酸
乳酸

3-oxobutanoic acid
3-氧亚基丁酸
3-丁酮酸

PPT

10.3   Physical Properties of Carboxylic Acids

10.3.1   State

Lower molecular weight acids are liquids at room temperature. Methanoic acid, acetic acid and propanoic acid have sharp, irritating odors. From *n*-butyric acid to *n*-nonanoic acid they are oily liquids with putrid odor. Aliphatic carboxylic acids containing more than ten carbon atoms are odorless waxy solids. Both **polybasic (多元酸)** and **aromatic acids (芳香酸)** are solids at room temperature.

10.3.2   Boiling Point

Boiling points of carboxylic acids are considerably higher than alcohols, ketones, or aldehydes of similar molecular weights. For example, acetic acid boils at 118℃, propan-1-ol boils at 97℃, and propanal (MW 58) boils at 49℃.

The high boiling points of carboxylic acids result from the formation of a stable, hydrogen-bonded dimer. This dimer contains an eight-membered ring joined by two hydrogen bonds.

10.3.3　Melting Point

The melting points of some common carboxylic acids are given in Table 10.1. Acids containing more than eight carbon atoms are generally solids, unless they contain double bonds. The presence of double bonds (especially *cis* double bonds) in a long chain impedes formation of a stable crystal lattice, resulting in a lower melting point. For example, both **stearic acid (硬脂酸，octadecanoic acid)** and **linoleic acid (亚油酸，*cis,cis*-octadeca-9,12-dienoic acid)** have 18 carbon atoms, but stearic acid melts at 70℃ and linoleic acid melts at -5℃.

stearic acid, mp 70℃　　　　linoliec acid, mp −5℃

The melting points of dicarboxylic acids (Table 10.1) are relatively high. With two carboxyl groups per molecule, the forces of hydrogen bonding are particularly strong in diacids.

Table 10.1　Physical properties of some carboxylic acids
表 10-1　一些羧酸的物理常数

Names	Structure	Melting point/℃	Boiling point/℃	Solubility (g/100gH$_2$O)
methanoic acid (formic acid)	HCOOH	8.4	100.5	∞ (miscible)
ethanoic acid (acetic acid)	CH$_3$COOH	16.6	118	∞
propanoic acid (propionic acid)	CH$_3$CH$_2$COOH	−22	141	∞
butanoic acid (butyric acid)	CH$_3$CH$_2$CH$_2$COOH	−4.7	162.5	∞
pentanoic acid (valeric acid)	CH$_3$(CH$_2$)$_3$COOH	−35	187	3.7
hexanoic acid (caproic acid)	CH$_3$(CH$_2$)$_4$COOH	−1.5	205	0.4
heptanoic acid	CH$_3$(CH$_2$)$_5$COOH	−11	223.5	0.24
octanoic acid (caprylic acid)	CH$_3$(CH$_2$)$_6$COOH	16.5	237	0.25
nonanoic acid	CH$_3$(CH$_2$)$_7$COOH	12.5	254	–
decanoic acid	CH$_3$(CH$_2$)$_8$COOH	31.5	268	–
hexadecanoic acid	CH$_3$(CH$_2$)$_{14}$COOH	62.9	269	–
octadecanoic acid	CH$_3$(CH$_2$)$_{16}$COOH	69.9	287	–
propenoic acid	CH$_2$=CHCOOH	13.0	141	–
oxalic acid	HOOC-COOH	189	–	8.6
hexanedioic acid	HOOC(CH$_2$)$_4$COOH	151	276	1.5

10.3.4　Solubility

Carboxylic acids form hydrogen bonds with water, and the lower-molecular-weight acids (up through four carbon atoms) are miscible with water. As the length of the hydrocarbon chain increases, water solubility decreases until acids with more than 10 carbon atoms are nearly insoluble in water. The water solubilities of some simple carboxylic acids and diacids are given in Tables 10.1.

Carboxylic acids are very soluble in alcohols because the acids form hydrogen bonds with alcohols. Also, alcohols are not as polar as water, so the longer-chain acids are more soluble in alcohols than they are in water. Most carboxylic acids are quite soluble in relatively nonpolar solvents such as chloroform because the acid continues to exist in its dimeric form in the nonpolar solvent. Thus, the hydrogen bonds of the cyclic dimer are not disrupted when the acid dissolves in a nonpolar solvent.

10.3.5　Relative Density

The relative density of formic acid and acetic acid is greater than 1, while the relative density of other carboxylic acids is less than 1. The relative density of binary carboxylic acid and aromatic carboxylic acid is greater than 1.

PPT

10.4　Chemical Properties of Carboxylic Acid

In the **carboxylic acid molecule** (羧酸分子), the **unshared electron pair** (未共用电子对) of the hydroxyl oxygen atom and the carbonyl carbon-oxygen double bond form a **p-π conjugate(p-π共轭)**, which reduces the positive charge of the carbonyl carbon atom and the activity of the reaction between the carboxylic acid and the nucleophile. Carboxylic acids react with nucleophiles with lower activity than **aldehydes** (醛) and ketones. Also due to the p-π conjugation, the electron cloud on the oxygen atom in the **hydroxyl** (羟基) group is transferred to the **carbonyl** (羰基) group, which reduces the density of the electron cloud on the hydroxyl oxygen atom and makes the carboxylic acid more acidic than the alcohol.

10.4.1　Acidity of Carboxylic Acids

1. Salt Formation Reaction and Application　Carboxylic acid has obvious acidity, it can form salt and water with **alkali** (碱).

$$RCOOH + NaOH \rightarrow RCOONa + H_2O$$
$$RCOOH + NaHCO_3 \rightarrow RCOONa + CO_2 \uparrow + H_2O$$

Carboxylate (羧酸盐) is an ionic compound. **Potassium** (钾), **sodium** (钠) and **ammonium salts** (铵盐) of monocarboxylic acids with less than 10 carbon atoms are soluble in water. The carboxylic acid salt will release the carboxylic acid when it meets the strong inorganic acid. Therefore, the salt formation reaction and the acidification reaction of the salts are an effective method for separating and purifying the water-insoluble carboxylic acid. For example, it has been used to extract the active ingredient containing carboxyl group from the Chinese medicinal materials.

Carboxylic acid is strongly acidic because **electron-withdrawing conjugation effect(-C，吸电子共轭效应)** of carbonyl group not only makes the –OH more polar, but also stabilizes the dissociated negative ion $RCOO^-$. Because the negative charge is delocalized and dispersed on the two oxygen atoms of the carboxylate, the carboxyl negative ion is far more stable than the alkoxyl negative RCH_2O^- and is easier to generate.

The acidic strength of a carboxylic acid can be expressed by the K_a or pK_a. The larger K_a (or smaller pK_a), the stronger is the acidity. The acidity of some compounds is shown in Table 10.2.

Table 10.2   Acidity

表10-2   酸性

	RCOOH	H_2CO_3	ArOH	HOH	ROH	$HC\equiv CH$	RH
pK_a	4~5	6.38	9~10	~15.74	16~19	~25	~50

pK_a of most unsubstituted carboxylic acids is 4~5. Their acidity is stronger than carbonic acid (pK_a=6.38). Therefore, carboxylic acids can decompose carbonates, while **phenol (苯酚)** (pK_a=10) cannot decompose carbonates. Therefore, this property can be used to distinguish carboxylic acid from phenol.

2. Effect of Substituents on Acidity   Any factor that can strengthen the stability of carboxylic anions will make the acidity stronger; conversely, any factor that can weaken the stability of carboxylic anions will weaken the acidity. These factors include **induced effects** (诱导效应), **conjugate effects** (共轭效应) and **stereo effects** (立体效应).

(1) Induced effect   In a saturated carboxylic acid, if the hydrogen atom is replaced with a highly electronegative group such as **halogen (卤素)** the acidity will get stronger. Substituents with electron-withdrawing induction effect (-I) will make the carboxylic anions more stable by dispersing the charge.

The stronger the electron-withdrawing ability (-I effect) of a substituent, the more the number of substituents, the greater the influence.

1) The effect of different electronegativity on acidity   For example, the electronegativity of halogen atom is F> Cl> Br> I, so the acidity of monohaloacetic acid is：

$$FCH_2COOH > ClCH_2COOH > BrCH_2COOH > ICH_2COOH$$

pK_a 2.66   2.86   2.90   3.18

2) The effect of different numbers of electron withdrawing groups on acidity   The acidity of acetic acid and three chloroacetic acids is:

$$Cl_3CCOOH > Cl_2CHCOOH > ClCH_2COOH > CH_3COOH$$

pK_a 0.65   1.29   2.86   4.76

3) The effect of different positions of electron withdrawing group from carboxyl group on acidity

$CH_3CH_2CHClCOOH$ $CH_3CHClCH_2COOH$ $CH_2ClCH_2CH_2COOH$

pK_a 2.86   4.41   4.70

The transfer of the inductive effect of the substituents on the saturated chain decreases rapidly with the increase of distance, and usually has a little effect after three bonds.

4) The effect of different hybrid atoms on acidity　The more s components, the stronger the electron-absorbing ability of the unsaturated substituent has, the stronger the acid is. For example:

$$HC\equiv C\text{-}CH_2COOH > CH_2=CHCH_2COOH > CH_3CH_2CH_2COOH$$

pK_a　　　　3.32　　　　　　　4.35　　　　　　　4.82

5) The effect of electron-repelling group on acidity　The acidity is reduced due to the +I effect of the alkyl group. For example:

$$HCOOH \quad CH_3COOH \quad CH_3CH_2COOH \quad (CH_3)_2CHCOOH \quad (CH_3)_2CCOOH$$

pK_a　　3.77　　　4.76　　　4.88　　　　4.86　　　　　5.05

There are two carboxyl groups in the dicarboxylic acid molecule, and the hydrogen atom of the dicarboxylic acid can be dissociated in two steps:

$$\begin{array}{c} COOH \\ | \\ (CH_2)n \\ | \\ COOH \end{array} \Longleftrightarrow \begin{array}{c} COO^- \\ | \\ (CH_2)n \\ | \\ COOH \end{array} + H^+ \Longleftrightarrow \begin{array}{c} COO^- \\ | \\ (CH_2)n \\ | \\ COO^- \end{array} + H^+$$

The first step of dissociation is affected by the -I effect of the other carboxyl group. The closer the two carboxyl groups are, the greater the -I effect is. For example, oxalic acid (pKa_1=1.46) is more acidic than malonic acid (pKa_1=2.80). When one carboxyl group dissociates to form a negative ion, the +I effect is generated, making the second carboxyl group difficult to dissociate, so the pKa_2 of the dibasic acid is always greater than pKa_1. The pK_a of some dibasic acids are shown in Table 10.3.

Table 10.3　pK_a of some dibasic acids
表10-3　一些二元酸的pK_a

Name	Structure	pKa_1	pKa_2
ethanedioic acid	HOOC-COOH	1.46	4.46
propanedioic acid	$HOOCCH_2COOH$	2.80	5.85
butanedioic acid	$HOOCCH_2CH_2COOH$	4.17	5.64
pentanedioic acid	$HOOC(CH_2)_3COOH$	4.33	5.57
hexanedioic acid	$HOOC(CH_2)_4COOH$	4.43	5.52
phthalic acid	COOH / COOH (ortho)	3.00	5.25
isophthalic acid	COOH / COOH (meta)	3.28	4.46
terephthalic acid	COOH / COOH (para)	3.28	4.40

(2) Conjugation effect　Among the **substituted aromatic acids** (取代芳香酸), the induced effect and the conjugation effect often coexist, and the strength of the acidity depends on the combined effect of the two, sometimes steric effects and hydrogen bonding are also considered.

The acidity of substituted aromatic carboxylic acids is affected by the structure of substituents and the relative position of substituents on aromatic rings.

Generally speaking, **electron withdrawing groups** (吸电子基团) in the *para* positions (对位) of a carboxyl group, such as $-NO_2$、$-CN$、$-COOH$、$-CHO$、$-COR$, have both electron-withdrawing conjugation effect (-C effect) and induction effect (-I), which increases the acidity of aromatic acids. However, the group like $-NH_2$、$-NHR$、$-OH$、$-OR$、$-OCOR$、$-Cl$、$-Br$, has the opposite **electron-releasing** (供电子的) conjugation effect (+C) and electron-withdrawing induction effect (-I). However, due to the distance from the carboxyl group, the induction effect is very weak. The main effect is +C conjugation effect, which reduces the acidity of aromatic acid.

For example, when the nitro group is attached to the *para* position, it has both -I and -C effects. When the methoxy group is attached to the benzene ring, it has electron-withdrawing -I and electron-releasing conjugation effects +C. Due to the distance from the carboxyl group, the induction effect is very weak, and the main influence is from +C conjugation effect. As a result the -C effect of nitro group make the acidity increase, while the +C effect of methoxy group make the acidity decrease.

$$
\begin{array}{ccc}
\text{C}_6\text{H}_5\text{—COOH} & \text{O}_2\text{N—C}_6\text{H}_4\text{—COOH} & \text{H}_3\text{CO—C}_6\text{H}_4\text{—COOH} \\
pK_a \quad 4.17 & 3.40 & 4.47
\end{array}
$$

According to the **vinylogy rule** (插烯规则),when the substituent is in the *meta* position (间位) , the electrical effect on carboxyl group is mainly induced effect. However because of the separation of three carbon atoms, the -I effect is weakened. The -I effect of nitro group increases the acidity of *m*-**nitrobenzoic acid** (间硝基苯甲酸). The methoxy group in the meta position also shows the -I effect, but its strength is weaker than that of the nitro group, so the acidity of *m*-**methoxybenzoic acid** (间甲氧基苯甲酸) is weaker than *m*-nitrobenzoic acid, and slightly higher than benzoic acid.

$$
\begin{array}{cc}
\text{O}_2\text{N—C}_6\text{H}_4\text{—COOH} & \text{H}_3\text{CO—C}_6\text{H}_4\text{—COOH} \\
pK_a \quad 3.49 & 4.09
\end{array}
$$

When the substituents are in the *ortho* position (邻位), except the amino group, whether they attract electrons or repel electrons, the acidity of aromatic acids is enhanced, and both of them are stronger than the *meta* or *para* substituted benzoic acid (Table 10.4). The special effect of this *ortho* group on the active center is called *ortho* effect, and its mechanism is complex, which can be regarded as the sum of the effects of electron effect, steric effect and hydrogen bonding effect.

In the benzoic acid molecule, the carboxyl group and the benzene ring are **coplanar** (共平面的), forming a conjugated system. The benzene ring has a +C effect on the carboxyl group. Therefore, the acidity of benzoic acid is weaker than that of **formic acid** (甲酸). When there is a substituent in the *ortho* position, because the substituent occupy a certain position, to a certain extent, exclude the carboxyl, the carboxy group deviates from the benzene ring plane, this weakens the +C conjugation effect of the benzene ring to the carboxyl group, the migration of benzene ring electron cloud to carboxyl group is reduced, making carboxylic hydrogen atoms are more easily dissociated, so the acidity increased. The more space the *ortho* substituents occupy, the greater the effect. However, the electrical effect is still shown to be effective. The stronger the electrical absorption of substituents, the more acidic they become. For example:

pK_a 3.21   3.46   3.89   2.89

Table 10.4   Acidity (pK_a) of substituted benzoic acids
表10-4   取代苯甲酸的pK_a

Substituent group	o-	m-	p-
-H	4.17	4.17	4.17
-CH$_3$	3.89	4.28	4.35
-Cl	2.89	3.82	4.03
-Br	2.82	3.85	4.18
-NO$_2$	2.21	3.46	3.40
-OH	2.98	4.21	4.54
-OCH$_3$	4.09	4.09	4.47
-NH$_2$	5.00	4.82	4.92

(3) Field effect (场效应)   The relatively small **secondary ionization constant** (二级电离常数) of mono anion of dibasic acid is caused by both the +I effect of carboxyl anion and the field effect. The carboxyl anion in a molecule inhibits the dissociation of hydrogen atoms in another carboxyl group through spatial electrical attraction. This special electrical effect transmitted through space is called the field effect (F effect). The size of the field effect is inversely proportional to the square of the distance between atoms or atomic groups.

Induction effect and field effect often exist at the same time, the direction of action may be the same (such as dibasic acid mono-anion), or may be opposite. In many cases, field effects depend on the geometry of the molecule, while induction effects depend on the relative length of the σ bond. For example, in the following two isomers, Ⅰ and Ⅱ, the chlorine atoms should have the same induction effect on the carboxyl group. However, from the spatial arrangement, the chlorine atoms in compound Ⅰ are closer to the carboxyl group than those in Ⅱ, so they have different field effects. The -Ⅰ effect of chlorine increases acidity, but the field effect of the C-Cl **dipole** (偶级) reduces acidity by inhibiting the dissociation of hydrogen in the carboxyl group. Because the field effect in Ⅰ is greater than that in Ⅱ, Ⅰ is less acidic than Ⅱ.

pKa   6.07
compound Ⅰ

pKa   5.69
compound Ⅱ

10.4.2   Reaction of The Carbonyl Group in the Carboxyl Group

Although carbonyl group of carboxylic acid molecule is not as active as the carbonyl group of aldehyde and ketone, it can still be attacked by nucleophile under certain conditions, resulting in addition-elimination reaction. As a result, carbon-oxygen bond is broken, other groups replace hydroxy group of carboxylic acids to form carboxylic acid derivatives. It can also be reduced to produce corresponding alcohols.

1. Substitution of Hydroxyl Groups on Carboxylic Acids   Hydroxy group of carboxylic acid can be substituted by halogens, **acyloxy groups** (酰氧基), **alkoxy groups** (烷氧基) and **amino groups** (氨基) to form compounds such as acyl halides, anhydrides, esters and amides, which are referred to as carboxylic acid derivatives.

(1) Conversion to **Acid Chlorides (酰卤的生成)**   When carboxylic acids react with PX_3, PX_5 or $SOCl_2$, halogen atoms replace hydroxy group of carboxylic acid to form acyl halides.

$$CH_3COOH + PCl_3 \longrightarrow CH_3COCl + H_3PO_3$$

$$RCOOH + SOCl_2 \longrightarrow RCOCl + SO_2 + HCl$$

(2) **Esterification (酯化反应)**   An **ester (酯)** can be prepared by treating a **carboxylic acid** with an **alcohol (醇)** in the presence of an acid **catalyst (催化剂)**. Carboxylic acids can also be obtained by esters and water under the same condition, which is called the ester **hydrolysis (水解作用)** reaction. So esterification is a typical **reversible reaction (可逆反应)**.

$$RCOOH + R'OH \underset{}{\overset{H^+}{\rightleftharpoons}} RCOOR' + H_2O$$

Esterification is very slow without catalyst. It will take a long time (several days) to get **equilibrium (平衡)**. Catalysts like **sulphuric acid (硫酸)**, **chloric acid (盐酸)** and **phenylsulfonic acid (苯磺酸)**, etc. can accelerate the reaction. However, while accelerating the forward reaction (esterification), these conditions also accelerate the **reverse reaction (逆反应)** (hydrolysis). In order to increase the yield of esters, one of the cheap ingredients can be added, so that it can force the equilibrium to the product. In addition, a continuous removal of a product (such as water) from the reaction system can be used to force the equilibrium to the product. In practice, both ways are usually used together.

In the esterification of carboxylic acids with alcohols, the **dehydration** (脱水) between carboxylic acids and alcohols occur in two different ways.

$$\text{R-C(=O)-}\boxed{\text{O-H}}\quad\boxed{\text{H-O}}\text{-R'}\qquad\qquad\text{R-C(=O)-}\boxed{\text{O-H}}\quad\boxed{\text{H-O}}\text{-R'}$$

I II

Method (I) is called **acyl-oxygen bond cleavage** (酰氧键断裂), (II) is called **alkyl-oxygen bond cleavage** (烷氧键断裂). In most cases, esterification proceeds via acyl-oxygen bond cleavage (I), which means the **hydroxyl group** (羟基) in the carboxylic acid dehydrates by combining with the hydrogen of the hydroxyl group in the alcohol.

The mechanism of esterification reaction is complicated and often changes due to various reaction conditions and structures of reactant. The process of esterification in the presence of an acid catalyst is usually a **nucleophilic addition-elimination reaction** (亲核加成-消除反应). The first step is the protonation of the oxygen-atom of carbonyl group in the carboxylic acid, which enhances the **electropositivity** (正电性) of carbon atoms in the **carbonyl group** (羰基). Then nucleophilic addition occurs with alcohol as the nucleophile and produce a **tetrahedron** (四面体) intermediate. Then proton translocation happens on the intermediate to make the hydroxy group into a better leaving group, which is followed by elimination of water. At last the product is obtained with deprotonation of the positively charged carbonyl group.

$$\text{RCOOH} + \text{H}^+ \rightleftharpoons \underset{1}{\text{R-C(}^+\text{OH)-OH}} \xrightarrow{\text{R'OH}} \underset{2}{\text{R-C(OH)(}^+\text{OR'H)-OH}}$$

$$\rightleftharpoons \underset{3}{\text{R-C(OH)(OR')(}^+\text{OH}_2)} \xrightarrow{-\text{H}_2\text{O}} \underset{4}{\text{R-C(}^+\text{OH)-OR'}} \xrightarrow{-\text{H}^+} \underset{5}{\text{RCOOR}}$$

According to this mechanism, the steric effect of acids and alcohols will affect the reaction rate of esterification, and the activity that the esterification reaction of different carboxylic acids with alcohols is as follows:

The reactivity of the acid: $HCO_2H > CH_3CO_2H > RCH_2CO_2H > R_2CHCO_2H > R_3CCO_2H$

The reactivity of the alcohol: $CH_3OH > 1°ROH > 2°ROH > 3°ROH$

Due to the stable tertiary **carbocation** (碳正离子) formed by **tertiary alcohols** (叔醇), the esterification reaction of tertiary alcohols proceeds generally via alkyl-oxygen cleavage (II).

$$(CH_3)_3C-O^{18}H \xrightarrow{H^+} (CH_3)_3C-O^{18}H_2^+ \rightleftharpoons (CH_3)_3C^+ + H_2O^{18}$$

$$H_3C-C(=O)-OH + (CH_3)_3C^+ \rightleftharpoons H_3C-C(^+OH)-O-C(CH_3)_3 \xrightarrow{-H^+} H_3C-C(=O)-O-C(CH_3)_3$$

(3) Decarboxylation　**Decarboxylation** (脱羧反应) is the loss of CO_2 from the **carboxyl** (羧基) group of a molecule. Simple **aliphatic** (脂肪族) and **aromatic** (芳香族) carboxylic acids are usually stable up to high temperatures. The most common decarboxylation method is to heat the **carboxylate** (羧酸盐) and solid **sodium hydroxide** (氢氧化钠) to high temperature. Decarboxylation however occur very readily when there are **electron-withdrawing groups** (吸电子基团) in the a-position of aliphatic acids.

$$CH_3COONa + NaOH \longrightarrow CH_4 + Na_2CO_3$$

$$Cl_3C\text{-}COOH \xrightarrow{\Delta} CHCl_3 + CO_2$$

$$\begin{matrix} COOH \\ | \\ COOH \end{matrix} \xrightarrow{160\text{-}180^oC} HCOOH + CO_2$$

When the electron-withdrawing groups, such as carbonyl group, is in the β-position of carboxyl group , decarboxylation occurs easily. This is because carbonyl and carboxyl groups are easy to chelate with hydrogen bonds, electron transfer occurs with heat. Carbon dioxide is lost, **enol** (烯醇) is formed, and then the **ketone** (酮) is formed.

The decarboxylation of ketone acids is widely used in **organic synthesis** (有机合成). The decarboxylation of **malonic acid** (丙二酸) compounds and α、β-supersaturated carboxylic acids generally falls into this category. When **butanone diacid** (丁酮二酸) is heated, **pyruvate** (丙酮酸) is easily obtained by losing its carboxyl group.

$$\underset{\displaystyle \text{HOOC}\overset{\textstyle O}{\overset{\|}{C}}CH_2COOH}{} \xrightarrow{heat} CH_3\overset{O}{\overset{\|}{C}}COOH + CO_2$$

An important example of decarboxylation of a β-ketoacid in the biological world occurs during the oxidation of food in the tricarboxylic acid(TCA)cycle. One of the intermediates in this cycle is **oxalosuccinic acid** (草酰琥珀酸), which undergoes spontaneous decarboxylation to produce a-ketoglutaric acid. Only one of the three carboxyl groups of oxalosuccinic acid has a carbonyl group in the β-position to it, and this carboxyl group is lost as CO_2.

The decarboxylation of aromatic acids is easier than that of aliphatic acids, especially 2,4,6-trinitrobenzoic acid (三硝基苯甲酸), which makes the bond between carboxyl group and benzene more easily broken due to the strong electron withdrawing effect of the three nitro groups.

Decarboxylation can also take place with the help of **enzymes** (酶), which are encountered in **biochemistry** (生物化学).

10.5 Functional Derivatives of Carboxylic Acids

Compounds formed by replacing hydroxy groups in carboxylic acid molecules are **carboxylic acid derivatives** (羧酸衍生物), including **acyl halides** (酰卤), **anhydrides** (酸酐), **esters** (酯), and **amides** (酰胺). Since nitriles are chemically similar to the above-mentioned compounds, they are also included in such compounds. The structural formula is as follows.

acyl halide 酰卤   anhydride 酸酐   ester 酯   nitrile 腈

amide 酰胺

Carboxylic acid derivatives are an important class of compounds. Acyl halides and acid anhydrides have high reactivity and are widely used in the synthesis of pharmaceutical intermediates and drugs; while esters and amides have low reactivity and the compounds are relatively stable. They can stably exist in the flora and fauna and chemical drugs in nature, they have shown good physiological activity, eg. :

isoamyl acetate 乙酸异戊酯   camptothecin 喜树碱   amoxicillin 阿莫西林

10.5.1 Structure of Carboxylic Acid Derivatives

Except for nitrile, carboxylic acid derivatives are similar, and they all contain acyl groups，which can be expressed as the following general formula.

$$\underset{\text{R-C-L}}{\overset{\overset{\displaystyle O}{\parallel}}{}} \quad L = -X, \quad -O\overset{\overset{\displaystyle O}{\parallel}}{C}R', \quad -NR'R''$$

Due to the formation of p-π conjugates, the structures of acid halides, anhydrides, esters, and amides can be expressed by the resonance structural formula.

$$\left[\underset{\text{R}}{\overset{\overset{\displaystyle O}{\parallel}}{C}}\overset{..}{L} \quad \longleftrightarrow \quad \underset{\text{R}}{\overset{\overset{\displaystyle O^-}{|}}{C}}\overset{+}{L} \right]$$

The contribution of the resonance structure formulas Ⅰ and Ⅱ of the carboxylic acid derivative to the resonance hybrid is determined by the electronegativity of L (the atom directly connected to the carbonyl carbon). The greater the electronegativity of L, the more unstable the resonance structure Ⅱ, and the greater the contribution of resonance structure Ⅰ to the resonance hybrid. Therefore, the structure of the acyl halide is mainly based on the resonance structural formula Ⅰ, and the structure of the amide is mainly based on the resonance structural formula Ⅱ.

The amide molecule mainly exists in the resonance structure Ⅱ, indicating that the C-N bond has the properties of the double bond in this part. The spectral data shows carbon-nitrogen bond length in formamide is about 138pm, which is much shorter than the normal carbon-nitrogen bond (bond length 147pm).

The carbon and nitrogen atoms of the cyano group in the nitrile molecule are *sp* hybrids, and their structure is similar to that of an alkyne. The carbon-nitrogen triple bond consists of a 6 bond and two π bonds. The resonance structure is as follows.

$$\left[R-C\equiv N \quad \longleftrightarrow \quad R-C^+\equiv N^- \right]$$

10.5.2   Nomenclature of Acid Derivatives

1. Acid Halides   Acid halides (acyl halides) are named by changing the suffix *-ic acid* to *-yl halide*. The following are some common acid halides.

acetyl chloride

乙酰氯

p-methylbenzoyl chloride

对甲基苯甲酰氯

butanedioyl chloride or succinyl chloride

丁二酰氯

cyclopentanecarbonyl chloride

环戊基甲酰氯

3-bromobutanoyl bromide

3-溴丁酰溴

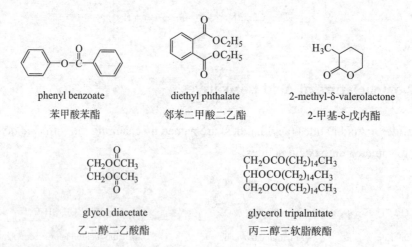

terephthaloyl bromide
对苯二甲酰溴

cyclopropyl formyl bromide
环丙基甲酰溴

2. Acid Anhydrides   Acid anhydrides are named by changing the suffix *-oic acid* to *-oic anhydride*. They consist of two carboxylic acids connected by losing water. The following are common acid anhydrides containing two identical acyl groups.

acetic anhydride
乙酸酐

benzoic anhydride
苯甲酸酐

Mixed anhydrides composed of two different acids are named by following alphabetical order of the parent acids and then adding the word "anhydride".

acetic butyric anhydride
乙酸丁酸酐

benzoic formic anhydride
苯甲酸甲酸酐

3. Esters   An **ester** (酯) is named by a combination of a carboxylic acid and an alcohol. Cyclic esters are called **lactones** (内酯) and named by changing the *-ic acid* ending of the hydroxy acid to *-olactone*. A Greek letter designates the carbon atom. The following are some common esters .

phenyl benzoate
苯甲酸苯酯

diethyl phthalate
邻苯二甲酸二乙酯

2-methyl-δ-valerolactone
2-甲基-δ-戊内酯

glycol diacetate
乙二醇二乙酸酯

glycerol tripalmitate
丙三醇三软脂酸酯

4. Amides   Amides (酰胺) are named by changing the suffix *-ic acid* from the name of the parent acid to *-amide*. If the nitrogen atom of an amide is bonded to an alkyl or aryl group, the position of the substituent should be indicate by the prefix *N-*. Two alkyl or aryl groups on nitrogen are indicated by *N,N-di-*. Following are some common amides.

$$H_3C-\overset{\overset{O}{\|}}{C}-NH_2 \qquad\qquad H-\overset{\overset{O}{\|}}{C}-N(CH_3)_2$$

acetamide   N,N-dimethyl formamide (DMF)

乙酰胺   N,N-二甲基甲酰胺

Lactams (内酰胺) are named by changing the *-ic acid* ending of the amino acid to *-olactam* ,which is similar to cyclic esters.

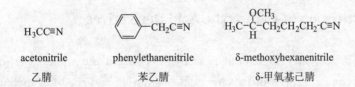

3-methyl-δ-valerolactam   ε-caprolactam

3-甲基-δ-戊内酰胺   ε-己内酰胺

An imide contains two acyl groups bonded to nitrogen.

phthalic imidine or phthalimide   N-bromosuccinimide

邻苯二甲酰亚胺   N-溴代丁二酰亚胺

5. Nitriles   Nitriles (腈) are named by changing the suffix *-ic acid* to the suffix *-nitrile* and following the pattern alkanenitrile. Its functional group is a cyano group. Following are some common nitriles.

$$H_3CC{\equiv}N \qquad\qquad \text{—}CH_2C{\equiv}N \qquad\qquad H_3C-\overset{\overset{OCH_3}{|}}{\underset{H}{C}}-CH_2CH_2CH_2\text{-}C{\equiv}N$$

acetonitrile   phenylethanenitrile   δ-methoxyhexanenitrile

乙腈   苯乙腈   δ-甲氧基己腈

10.6   Reactions of Acid Derivative

PPT

10.6.1   Acidity of Amides, Imides, and Sulfonamides

The following are structural formulas of a primary amide, a sulfonamide, and two cyclic imides, along with pK_a values for each.

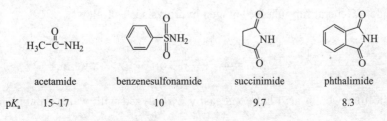

	acetamide	benzenesulfonamide	succinimide	phthalimide
pK_a	15~17	10	9.7	8.3

Values of pK_a of amides are in the range of 15~17, which means that they are comparable in acidity to alcohols. Amides show no evidence of acidity in aqueous solution; that is, water-insoluble amides do not react with aqueous solutions of NaOH or other alkali metal hydroxides to form water-soluble salts.

Imides (pK_a 8~10) are considerably more acidic than amides and readily dissolve in 5% aqueous NaOH by forming water-soluble salts. We account for the acidity of imides in the same manner as for the acidity of carboxylic acids; namely, the imide anion is stabilized by delocalization of its negative charge. The more important contributing structures for the anion formed by ionization of an imide delocalize the negative charge on nitrogen and the two carbonyl oxygen atoms.

a resonance-stabilized anion

Sulfonamides derived from ammonia and primary amines are also sufficiently acidic to dissolve in aqueous solutions of NaOH or other alkali metal hydroxides by forming water-soluble salts. The pK_a of **benzenesulfonamide** (苯磺酰胺) is approximately 10. We account for the acidity of sulfonamides in the same manner as for imides, namely the resonance stabilization of the resulting anion.

benzenesulfonamide   A resonance-stabilized anion

10.6.2　Substitution of Carboxylic Acid Derivatives

Carboxylic acid derivatives (except nitrile) all contain the same functional acyl group in their structures and thus exhibit similar chemical properties and reaction mechanisms. Due to the electronegativity of the halogen, oxygen, or nitrogen atoms directly attached to the carbonyl group, the degree of p-π conjugation is different, making them chemically different. Some carboxylic acid derivatives also exhibit special chemical properties.

Carboxylic acid derivatives (羧酸衍生物) have a polar acyl group, and the partially positively charged carbon atoms are vulnerable to the attack by nucleophiles (H_2O, ROH, RNH_2), resulting in nucleophilic substitution reactions. General formula is as follows.

$$R-\overset{O}{\overset{\|}{C}}-L + Nu^- \longrightarrow R-\overset{O}{\overset{\|}{C}}-Nu + L^-$$

1. Reaction with Water: Hydrolysis（水解反应）

(1) Acyl halide　General formula for halogen hydrolysis is as follows.

$$R-\overset{O}{\overset{\|}{C}}-X + H_2O \longrightarrow R-\overset{O}{\overset{\|}{C}}-OH + HX \qquad X=Cl, Br, I$$

The low molecular weight acyl halide is easily hydrolyzed with water without catalyst. However, with the increase of molecular weight of acyl halide, the hydrolysis rate will gradually slow down and the reaction needs to be catalyzed by base. For example.

$$H_3C-\overset{O}{\underset{}{C}}-Cl \ + \ H_2O \longrightarrow H_3C-\overset{O}{\underset{}{C}}-OH + HCl$$

$$(Ph)_2CHCH_2\overset{O}{\underset{}{C}}-Cl \ + \ H_2O \xrightarrow[0\,℃]{Na_2CO_3,\ H_2O} (Ph)_2CHCH_2\overset{O}{\underset{}{C}}-OH + \ HCl$$

(2) Acid anhydride　General formula for acid anhydride hydrolysis is as follows.

$$R-\overset{O}{\underset{}{C}}-O-\overset{O}{\underset{}{C}}-R \ + \ H_2O \longrightarrow R-\overset{O}{\underset{}{C}}-OH$$

The hydrolysis of anhydride can be carried out in acidic, neutral or basic solution, and the reaction is less active than that of acyl halide. Such as.

$$H_3C-\overset{O}{\underset{}{C}}-O-\overset{O}{\underset{}{C}}-C_2H_5 \ + \ H_2O \longrightarrow H_3C-\overset{O}{\underset{}{C}}-OH \ + \ C_2H_5-\overset{O}{\underset{}{C}}-OH$$

(3) Ester　General formula for ester hydrolysis is as follows.

$$R-\overset{O}{\underset{}{C}}-OR' \ + \ H_2O \xrightarrow{OH^-} R-\overset{O}{\underset{}{C}}-OH \ + \ R'OH$$

The hydrolysis of ester catalyzed by acid is reversible. In contrast, in the basic-catalyzed hydrolysis, the acid and base formed salt, so the reaction is relatively complete. Therefore, enzymatic hydrolysis is usually carried out under basic condition. For example:

Natural animal and vegetable oils are glycerides of **higher aliphatic acids** (高级脂肪酸).The salts of higher aliphatic acids are obtained by hydrolysis of oils under alkaline conditions, and higher aliphatic acids with different carbon chains are obtained after acidification.

$$\begin{array}{l} H_2C-OOCR \\ HC-OOCR' \\ H_2C-OOCR'' \end{array} + \ H_2O \ \rightleftharpoons \ \begin{array}{l} CH_2OH \\ CHOH \\ CH_2OH \end{array} + \ \begin{array}{l} RCOOH \\ R'COOH \\ R''COOH \end{array}$$

$$\begin{array}{l} H_2C-OOCR \\ HC-OOCR' \\ H_2C-OOCR'' \end{array} + \ 3NaOH \xrightarrow{\triangle} \ \begin{array}{l} CH_2OH \\ CHOH \\ CH_2OH \end{array} + \ \begin{array}{l} RCOONa \\ R'COONa \\ R''COONa \end{array}$$

$$CH_3(CH_2)_{10}CO_2H$$

The higher aliphatic acids obtained by the hydrolysis of fat are mainly saturated or unsaturated acids with 12~18 carbons, and all contain even number of carbon atoms, such as **lauric acid** (月桂酸), **nutmyl acid** (肉豆蔻酸), **palmitic acid** (软脂酸), **stearic acid** (硬脂酸) etc.

253

2. Reaction with Alcohols

(1) Acyl halide   The general formula of acyl halide **alcoholysis (醇解)** is as follows.

$$R-\overset{O}{\overset{\|}{C}}-X + R'OH \longrightarrow R-\overset{O}{\overset{\|}{C}}-OR' + HX \quad X=Cl, Br, I$$

Reaction of acyl halides with alcohols or phenols gives an ester. Acyl chlorides react readily with alcohols without **catalysis (催化剂)**. For **tertiary alcohols (叔醇)** with higher **steric hindrance (空间位阻)** or phenols with weaker **nucleophilicity (亲核性)**, base catalysis can promote the reactions. For example:

$$H_3C-\overset{O}{\overset{\|}{C}}-Cl + CH_3CH_2OH \longrightarrow H_3C-\overset{O}{\overset{\|}{C}}-OCH_2CH_3$$

$$CH_3-\overset{O}{\overset{\|}{C}}-Cl + (H_3C)_3COH \xrightarrow{C_6H_5N(CH_3)_2} H_3C-\overset{O}{\overset{\|}{C}}-OC(CH_3)_3$$

(2) Anhydride   The general formula of anhydride alcoholysis is as follows.

$$R-\overset{O}{\overset{\|}{C}}-O-\overset{O}{\overset{\|}{C}}-R + R'OH \longrightarrow R-\overset{O}{\overset{\|}{C}}-OR' + R-\overset{O}{\overset{\|}{C}}-OH$$

The reactivity of acid anhydrides is moderate. Acid or base catalysis can promote the reactions, the reaction of an alcohol with an anhydride is a useful method for the **synthesis (合成)** of esters. Anhydride and acyl halide are active **acylating agents (酰化剂)**. And compared with acyl halides, acid anhydride is more convenient to use. For example, using anhydride as acylating agent in the synthesis of ethyl acetate has the advantages of convenient operation and no **corrosive (腐蚀性的) hydrogen halide (卤化氢)** is produced in the reaction.

$$CH_3-\overset{O}{\overset{\|}{C}}-O-\overset{O}{\overset{\|}{C}}-CH_3 + C_2H_5OH \longrightarrow CH_3-\overset{O}{\overset{\|}{C}}-OC_2H_5 + CH_3-\overset{O}{\overset{\|}{C}}-OH$$

(3) Ester   The general formula of ester alcoholysis is as follows.

$$R-\overset{O}{\overset{\|}{C}}-OR' + R''OH \xrightarrow{OH^-} R-\overset{O}{\overset{\|}{C}}-OR'' + R'OH$$

Alcoholysis of esters needs to be catalyzed by acid or base. Esters can undergo alcoholysis to obtain another ester and alcohol, so the alcoholysis of esters is usually called as ester exchange reaction. **Transesterification (酯基转移作用)** is an **equilibrium reaction (平衡反应)** that can be driven in either direction by control of experimental conditions. For example, in the reaction of methyl acrylate with butanol, transesterification is carried out at a temperature slightly above the boiling point of methanol (the lowest boiling component in the mixture). Methanol distills from the reaction mixture, thus shifting the position of equilibrium in favor of butyl acrylate. Conversely, reaction of butyl acrylate with a large excess of methanol shifts the equilibrium to favor formation of methyl acrylate.

$$H_2C=CHCO_2CH_3 + CH_3(CH_2)_3OH \xrightarrow{H^+} H_2C=CHCO_2(CH_2)_3CH_3 + CH_3OH$$

Transesterification is often used in the synthesis of drugs and their **intermediates** (中间体). When the ester structure is too complex to be synthesized by direct esterification, methyl or ethyl ester can be prepared first, and then the ester with complex structure can be obtained by transesterification. For example, the synthesis of the **local anesthetic procaine** (局部麻醉剂普鲁卡因).

$$\text{(structure with } NH_2, CO_2C_2H_5) + HOCH_2CH_2N(C_2H_5)_2 \xrightarrow{H^+} \text{(structure with } NH_2, CO_2CH_2CH_2N(C_2H_5)_2)$$

3. Reactions with Ammonia and Amines

(1) Acyl halides

$$R-\overset{O}{\underset{}{C}}-X + R'NH_2 \longrightarrow R-\overset{O}{\underset{}{C}}-NHR' + HX \qquad X=Cl, Br, I$$

Acyl halides can easily react with ammonia or amines to obtain amides. The reaction is usually carried out under basic conditions. Cheap base is used to neutralize the hydrogen halide generated in the reaction and avoid consumption of the reagent ammonia or amine. Such as:

$$H_3C-\overset{O}{\underset{}{C}}-Cl + H-N \text{(ring)} \xrightarrow{NaOH} H_3C-\overset{O}{\underset{}{C}}-N \text{(ring)}$$

(2) Acid anhydrides

$$R-\overset{O}{\underset{}{C}}-O-\overset{O}{\underset{}{C}}-R + R'NH_2 \longrightarrow R-\overset{O}{\underset{}{C}}-NHR' + R-\overset{O}{\underset{}{C}}-OH$$

Aminolysis of acid anhydrides produces amides with lower reaction rate than aminolysis of acyl halides. Such as:

$$H_3C-\overset{O}{\underset{}{C}}-O-\overset{O}{\underset{}{C}}-CH_3 + H_2N-\text{(ring)} \longrightarrow H_3C-\overset{O}{\underset{}{C}}-\overset{H}{\underset{}{N}}-\text{(ring)} + CH_3COOH$$

$$HO-\text{(ring)}-NH_2 + (CH_3CO)_2O \xrightarrow{AcOH} HO-\text{(ring)}-NHCOCH_3 + CH_3COOH$$

(3) Esters

$$R-\overset{O}{\underset{}{C}}-OR' + R''NH_2 \longrightarrow R-\overset{O}{\underset{}{C}}-NHR'' + R'OH$$

Esters can also undergo **ammonolysis** (氨解反应) with ammonia to form amides or amide derivatives. Such as:

$$\underset{CH_3\overset{OH}{\underset{}{C}}HCO_2C_2H_5}{} + NH_3 \xrightarrow{25℃/24h} \underset{CH_3\overset{OH}{\underset{}{C}}HCONH_2}{} + C_2H_5OH$$

10.6.3　Interconversion of Functional Derivatives

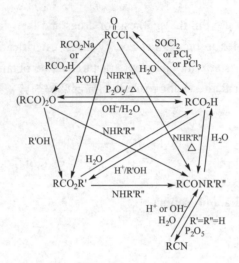

Interconversion (相互转化) of functional **derivatives** (衍生物) can be achieved by the above substitution reactions and their relationships are shown here.

10.6.4　Reactivity of Acid Derivatives

The four carboxylic acid derivatives that are the focuses of this chapter have the relative reactivities toward nucleophilic acyl substitution as follows. The differences in this trend are dramatic. For example, at common ambient temperatures and neutral pH, acid halides will react with water within seconds to minutes, while anhydrides will do so over minutes to hours. However, esters do not react with water at appreciable rates under these conditions, taking many years to hydrolyze; amides take centuries to react. Hence, acid halides and acid anhydrides are so reactive that they are not found in nature, whereas esters and amides are universally present.

$$\underset{}{R-\overset{\overset{O}{\|}}{C}-Cl} \;>\; \underset{}{R-\overset{\overset{O}{\|}}{C}-O-\overset{\overset{O}{\|}}{C}-R'} \;>\; \underset{}{R-\overset{\overset{O}{\|}}{C}-OR'} \;>\; \underset{}{R-\overset{\overset{O}{\|}}{C}-NH_2}$$

Two effects lead to this trend. One is relative leaving group ability. We show below the leaving groups as anions in order to illustrate an important point: the weaker the base (i.e. the more stable the anion), the better the leaving group. The weakest base in the series and the best leaving group is the halide ion; acid halides are most reactive toward nucleophilic acyl substitution. The strongest base and the poorest leaving group is the amide ion; amides are least reactive toward nucleophilic acyl substitution.

$$R_2N^- < RO^- < \overset{\overset{O}{\|}}{R C O^-} < X^- \qquad\qquad R_2N^- > RO^- > \overset{\overset{O}{\|}}{R C O^-} > X^-$$

leaving ability　　　　　　　　　　　　basicity

The second effect derives from the relative resonance stabilization of the carboxylic acid derivatives. As follows, each derivative can be written with contributing structures that will be stabilizing to some extent. The second contributing structure that we show for each carboxylic acid derivative has a positive charge on the carbonyl carbon. This structure reflects the electrophilicity of these carbons. However, for each derivative, it is the other contributing structures that reflect the relative resonance stabilization of the derivatives.

Let's start with an analysis of the acid chloride. The third contributing structure for an acid chloride has a carbon-to-chlorine double bond whose π bond is weak due to poor orbital overlap between the differently sized p orbitals on these two atoms. Further, there is a positive charge on the electronegative chlorine atom. Both of these factors make this a poor contributing structure for the acid chloride. An acid anhydride has five contributing structures; the last two shown place positive charges on the central oxygen. However, these positive charges are adjacent to an electron-withdrawing carbonyl group. Hence, these two contributing structures are not very reasonable depictions of an acid anhydride. But the analogous contributing structure for an ester places the positively charged oxygen near an electron-donating alkyl group, which stabilizes this charge. Accordingly, this contributing structure is a reasonable depiction of an ester; it is stabilizing, and it lowers the susceptibility of the carbonyl carbon to nucleophilic attack. Finally, the third contributing structure for an amide has a positive charge on the less electronegative nitrogen (relative to oxygen as with an ester), making this an even more reasonable structure and thereby increasingly stabilizing. In fact, the C-N double bond character of an amide is significant. This increased stability makes the amide least susceptible to nucleophilic attack.

Acid chloride contributing structures

Acid anhydride contributing structures

Ester contributing structures

Amide contributing structures

Taken together, the combined effects of leaving group ability and susceptibility to nucleophilic attack reinforce each other, thereby resulting in the order of reactivity given below.

$$\underset{\text{Amide}}{R-\overset{\displaystyle O}{\overset{\|}{C}}-NH_2} \; < \; \underset{\text{Ester}}{R-\overset{\displaystyle O}{\overset{\|}{C}}-OR'} \; < \; \underset{\text{Anhydride}}{R-\overset{\displaystyle O}{\overset{\|}{C}}-O-\overset{\displaystyle O}{\overset{\|}{C}}-R'} \; < \; \underset{\text{Acid hylide}}{R-\overset{\displaystyle O}{\overset{\|}{C}}-X}$$

10.6.5 Reduction of Carboxylic Acid Derivatives

The unsaturated bond (C=O or C≡N) in carboxylic acid derivatives can be reduced. There are many reduction methods for carboxylic acid derivatives, and different reduction products can be obtained by different reduction methods.

(1) Acyl halide   Acyl halides can be reduced to primary alcohols by lithium aluminum hydride (LiAlH₄). For example:

$$R-\overset{\displaystyle O}{\overset{\|}{C}}-Cl \xrightarrow[\text{②}H_2O]{\text{①}LiAlH_4,Et_2O} RCH_2OH \; + \; HX$$

It can selectively reduce acyl chloride to aldehyde without further reduction by Rosenmund reduction reaction. For example:

$$CH_3COCH_2CH_2COCl \xrightarrow[\text{quinoline-S}]{H_2,Pd/BaSO_4} CH_3COCH_2CH_2CHO$$

What's more, acyl chloride can be reduced to aldehyde by lithium tri-tert-butoxyaluminum hydride or lithium triethoxyamminium hydride. For example:

$$\text{C}_6\text{H}_5\text{-COCl} \xrightarrow[\text{or } LiAlH(OC_2H_5)_3]{LiAlH[OC(CH_3)_3]_3} \xrightarrow{H_2O} \text{C}_6\text{H}_5\text{-CHO}$$

(2) Anhydride   Anhydrides can be reduced to primary alcohols by lithium aluminum hydride(LiAlH₄). For example:

$$R-\overset{\displaystyle O}{\overset{\|}{C}}-O-\overset{\displaystyle O}{\overset{\|}{C}}-R' \xrightarrow[\text{②}H_2O]{\text{①}LiAlH_4,Et_2O} RCH_2OH \; + \; R'CH_2OH$$

(3) Ester   Esters can be reduced to primary alcohols by lithium aluminum hydride (LiAlH₄). For example:

$$R-\overset{\displaystyle O}{\overset{\|}{C}}-OR' \xrightarrow[\text{②}H_2O]{\text{①}LiAlH_4,Et_2O} RCH_2OH \; + \; R'OH$$

Esters can also be reduced to primary alcohols by heating and refluxing in sodium alcohol solution. This reaction is called Bouveault-Blanc reduction reaction.

$$CH_3CH{=}CHCH_2CO_2C_2H_5 \xrightarrow{Na,C_2H_5OH} CH_3CH{=}CHCH_2CH_2OH$$

(4) Amide   Amides can be reduced to Amines by lithium aluminum hydride (LiAlH₄). According to the number of substituents on nitrogen, primary amines, secondary amines and tertiary amines can be reduced respectively.

$$R-\overset{\overset{\displaystyle O}{\|}}{C}-NR'R'' \xrightarrow[\text{②H}_2\text{O}]{\text{①LiAlH}_4,\text{Et}_2\text{O}} RCH_2NR'R''$$

$$Ph-\overset{\overset{\displaystyle O}{\|}}{C}-NHCH_3 \xrightarrow[\text{②H}_2\text{O}]{\text{①LiAlH}_4,\text{Et}_2\text{O}} PhCH_2NHCH_3$$

(5) Nitrile　Nitriles can be reduced to primary amines by lithium aluminum hydride (LiAlH₄)or catalytic hydrogenation.

$$RCN \xrightarrow[\text{(2)H}_2\text{O}]{\text{(1)LiAlH}_4,\text{Et}_2\text{O}} RCH_2NH_2$$

$$\text{CH}_2\text{CN} + \text{H}_2 \xrightarrow[120℃]{\text{Ni,pressure}} \text{CH}_2\text{CH}_2\text{NH}_2$$

The reduction products of various carbonyl compounds are compared as follows in Table 10.5.

Table 10.5　Reduction products of various carbonyl compounds

表 10-5　典型羰基化合物的还原

Name	Molecular structure	NaBH₄/Et₂O	LiAlH₄/Et₂O	H₂/catalytic hydrogenation
Carboxylic acid	RCOOH	(−)	RCH₂OH	(−)
Acyl chloride	RCOCl	RCH₂OH	RCH₂OH	RCH₂OH
Ester	RCOOR'	(−)	RCH₂OH,R'OH	RCH₂OH, R'OH
Amide	RCONH₂	(−)	RCH₂NH₂	RCH₂NH₂(hard)
Substituted amide	RCONHR	(−)	RCH₂NHR	RCH₂NHR
Ketone	R₂CO	R₂CHOH	R₂CHOH	R₂CHOH
Aldehyde	RCHO	RCH₂OH	RCH₂OH	RCH₂OH

10.7　Application of Carboxylic Acids and Derivatives in the Pharmacy Study

PPT

10.7.1　Application of Carboxylic Acids in the Pharmacy Study

Carboxylic acids and substituted carboxylic acids exist widely in nature and are closely related to human life. Carboxylic acids are also widely used in medicine. Low fatty acids such as formic acid can be used as disinfectants directly. Senior fatty acids are basic to Oils and Fats Industry. Oils are applied widely in the production of medicines, and are important excipients, either to improve the quality of the form of a drug, or to improve the bioavailability of drugs.

There are also many drugs with carboxylic acid structure. Some natural carboxylic acid compounds in plants have good pharmacological activities. For example, caffeic **acid** (咖啡酸), which has hemostatic (止血的) effect, exists in many Traditional Chinese Medicines, such as Daucus Carota (Caroule Fructus), Tridentata Merremia, Buckwheat, Cherry Elaeagnus, etc. **Salicylic acid** (水杨酸) is extracted from willow bark, which is used in bactericide and preservatives, and has antipyretic, analgesic and anti-rheumatic effects.

Protocatechuic acid (原儿茶酸) is called 3,4-dihydroxybenzoic acid, which is a major active ingredient in Chinese traditional medicine Purple flower Holly Leaf. Purple flower Holly Leaf has a good effect in treating burns. In addition, it can also be used to treat bacillary dysentery, pyelonephritis and some canker. Chemical synthesis drugs also contain many compounds containing carboxyl groups, typically penicillin, thiomycin, cephalosporins, ibuprofen.

In the field of medicine, binary carboxylic acid such as **succinic acid** (琥珀酸), also known as succinic acid, succinic acid in medicine has antispasmodic, expectation and diuretic effects. Traditional Chinese medicines such as Pheretima and aster contain succinic acid and are reported to have antiasthmatic effects.

10.7.2　Application of Carboxylic Acid Derivatives in the Pharmacy Study

Carboxylic Acid Derivatives are an important organic compound, widely used in medicine, pesticides and other fields, which play an important role in social development and human health.

Acyl chloride and anhydride are important acylating agents, which play an important role in pharmaceutical synthesis. The active groups in acyl chloride and anhydride can be used to synthesize a series of pharmaceutical chemical products and drug intermediates. For example, acyl chloride is used in the synthesis of enoxacin and jimifloxacin ; Aspirin is made from acetic anhydride.

$$\text{(COOH, OH)} + (CH_3CO)_2O \xrightarrow{\text{浓}H_2SO_4} \text{(COOH, OCOCH}_3) + CH_3COOH$$

Esters and amides are found in many drugs. Ester of **aspirin** (阿司匹林)(antipyretic analgesics), **procaine hydrochloride** (盐酸普鲁卡因)(narcotic), **camptothecin** (喜树碱) (anticancer drugs) and **clofibrate** (安妥明) (heart medicine). And the synthesis of this class of drugs is mostly involved in the formation of ester bond. Following is how to synthesize the procaine hydrochloride.

$$NO_2-\text{C}_6H_4-CH_3 \xrightarrow[\triangle]{[O]} NO_2-\text{C}_6H_4-CO_2H \xrightarrow[\triangle]{HOCH_2CH_2N(C_2H_5)_2}$$

$$NO_2-\text{C}_6H_4-CO_2CH_2CH_2N(C_2H_5)_2 \xrightarrow[Ph=4]{Fe,HCl} NH_2-\text{C}_6H_4-CO_2CH_2CH_2N(C_2H_5)_2$$

$$\xrightarrow[pH=4]{HCl,Na_2SO_4,NaCl} [NH_2-\text{C}_6H_4-CO_2CH_2CH_2N(C_2H_5)_2]\cdot HCl$$

重点小结

一、羧酸

1. 羧酸结构(–COOH) 羧酸官能团是羧基,羧基由羰基($-\overset{O}{\underset{||}{C}}-$)和羟基(–OH)。组成。羧基中碳为 sp^2 杂化。羧基中羟基氧原子有一对未共用电子,它和羰基的 π 键形成 p-π 共轭体系。

2. 羧酸命名 羧基级别优先于绝大多数官能团。命名编号时羧基碳原子编为 1 号碳。羧酸 IUPAC 名称来源于烷烃,命名时将烷烃后缀 e 去掉,改为 oic acid。二元羧酸名称后缀是 dioic acid。

3. 羧酸物理性质 羧酸室温下通常以液体、固体形式存在。羧酸分子间形成氢键作用,形成双分子缔合二聚体,因此羧酸沸点比分子量相近醇、醛、酮和醚高。羧酸由极性亲水羧基和非极性疏水烃链组成。其中极性亲水的羧基,它能增加水的溶解度,低分子量羧酸在水中是可无限溶解的。随着碳链增长,疏水烃链比例增大,在水中溶解度减小。

4. 羧酸化学性质

(1) 羧酸的酸性 大多数脂肪族羧酸的 pK_a 为 4.0~5.0。羧酸酸性强于醇,是因为羧基中羰基的吸电子作用使 O–H 极性增大,也使解离后的负离子 $RCOO^-$ 稳定,由于负电荷离域而分散于羧酸根两个氧原子上,因此羧酸根负离子更稳定。羧基附近连有吸电子基,会增大羧酸酸性。

(2) 羧酸酯化反应 羧酸和醇在酸催化下作用生成羧酸酯和水,这是制备酯的一种方法。在同种条件下,酯和水反应也可生成羧酸和醇,酯化反应是一个可逆反应。

(3) 酰卤的生成 羧酸与 PX_3、PX_5、$SOCl_2$ 作用,羧酸中羟基被卤素取代,生成酰卤。

(4) 脱羧反应 羧酸分子中脱去羧基放出二氧化碳的反应称为脱羧反应。β-酮酸的脱羧反应和丙二酸型化合物的脱羧反应在有机合成上应用广泛。

二、羧酸衍生物

1. 羧酸衍生物结构与命名 羧酸衍生物除腈外,酰卤($\underset{R-\overset{||}{C}-X}{}$)、酸酐($\underset{R-\overset{||}{C}-O-\overset{||}{C}-R'}{}$)、酯($\underset{R-\overset{||}{C}-OR'}{}$)、酰胺($\underset{R-\overset{||}{C}-NH_2}{}$)中都含有酰基($\underset{R-\overset{||}{C}-}{}$)。酰卤名称将羧酸后缀 –ic acid 改为 –yl halide。酸酐名称将羧酸后缀 acid 改为 anhydride。酯名称将羧酸后缀 –ic acie 改为 –ate。酰胺名称将羧酸后缀 –ic acid 改为 –amide。

2. 羧酸衍生物的化学性质

(1) 酰胺、亚胺、磺胺的酸性 亚胺分子中,氮原子与两个羰基相连,氮原子上电子云密度大大降低,导致亚胺有一定的酸性。

(2) 羧酸衍生物水解反应 酰卤和酸酐与水自发反应生成羧酸和 HX 或两个羧酸分子,反应是由酸催化的,但没有酸也会发生反应,因为反应中产生的酸催化了这个过程。酯和酰胺的水解需要酸或碱催化,同样会生成相应羧酸或羧酸盐。

(3) 羧酸衍生物醇解反应 酰卤与醇反应生成酯和 HX,酸酐与醇反应生成一个酯分子和一个分子羧酸。酯与醇发生酸催化反应,称为酯交换反应,其中一个酯或基团被交换给另一个。酰胺的反应性不足以与醇发生反应。

（4）羧酸衍生物氨解反应　酰卤与氨（胺）很容易反应生成酰胺。酸酐氨（胺）解生成酰胺，其反应活性比酰卤低，反应速率更慢。酯发生氨（胺）解生成酰胺或酰胺衍生物。

（5）羧酸衍生物反应活性

$$\underset{R-C-NH_2}{\overset{O}{\|}} < \underset{R-C-OR'}{\overset{O}{\|}} < \underset{R-C-O-C-R'}{\overset{O\quad\ O}{\|\quad\ \|}} < \underset{R-C-X}{\overset{O}{\|}}$$

（6）羧酸衍生物的还原反应　酰卤、酸酐和酯都能被四氢铝锂（$LiAlH_4$）还原为伯醇。酰胺被四氢铝锂还原为胺类。

Problems
目 标 检 测

1. Which of the following structural formulae is correct for (1R,4R)-3,3-dichloro-4-hydroxycyclohexane-1-carboxylic acid?

化合物 (1R, 4R)-3,3-二氯-4-羟基环己基甲酸可以用下列（　　　）结构式表示。

A. ![structure A]　B. ![structure B]

C. ![structure C]　D. ![structure D]

2. Which of the following compounds is the strongest acid?

下列化合物中酸性最强的是（　　　）。

A. ![structure A]　B. ![structure B]

C. ![structure C]　D. ![structure D]

3. Which of the following names is correct for 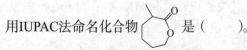 ?

用IUPAC法命名化合物 ![structure] 是（　　　）。

    A. α-甲基-ε-己内酯 (α-methyl-ε-caprolactone)

    B. β-甲基-δ-戊内酯 (β-methyl-δ-valerolactone)

C. α-甲基-δ-己内酯 (α-methyl-δ-caprolactone)

D. β-甲基-ε-戊内酯 (β-methyl-ε-valerolactone)

4. Complete the equation.

$$ \begin{array}{c} \diagdown\!\!\!= \end{array} + \begin{array}{c} =\!\!\!\diagup^{COOC_2H_5} \end{array} \xrightarrow{\triangle} \xrightarrow[②H_2O]{①LiAlH_4} $$

方程式 $\diagdown\!\!\!=$ + $=\!\!\!\diagup^{COOC_2H_5}$ $\xrightarrow{\triangle}$ $\xrightarrow[②H_2O]{①LiAlH_4}$ 产物为（ ）。

A. 〔cyclohexene with CH_2OH〕

B. 〔cyclohexene with CH_2OH〕

C. 〔cyclohexadiene with CH_2OH〕

D. 〔cyclohexene with CH_2OH〕

5. Complete the equation.

〔phthalic anhydride〕 $\xrightarrow{NH_3}$ $\xrightarrow[②H_3O^+]{①Br_2, OH^-}$

方程式 〔phthalic anhydride〕 $\xrightarrow{NH_3}$ $\xrightarrow[②H_3O^+]{①Br_2, OH^-}$ 产物为（ ）。

A. 〔benzene with $CONH_2$ and Br〕

B. 〔benzene with $COBr$ and NH_2〕

C. 〔benzene with $COBr$ and $NHBr$〕

D. 〔benzene with $COOH$ and NH_2〕

6. Which of the following compounds can discolor bromine water and acidic $KMnO_4$ solution and hydrolyze to form alcohols?

下列能使溴水和酸性$KMnO_4$溶液褪色，又能发生水解反应生成醇的是（ ）。

A. $CH_2=CH-COOC_2H_5$

B. $\underset{\underset{Br}{|}}{CH_3CHCOOH}$

C. $CH_2=CH-CH_2Br$

D. $CH_3COOCH_2CH_3$

7. Which of the following compounds is most readily decarboxylated?

下列化合物最易脱羧的是（ ）。

A. 〔cyclohexane with OH and COOH〕

B. 〔cyclohexane with O= and COOH〕

C. 〔cyclohexane with =O and COOH〕

D. 〔cyclohexane with COOH and COOH〕

8. Complete the equation.

〔acetic anhydride〕 $+ H_2N-$〔phenyl〕 $\longrightarrow$

方程式 〔acetic anhydride〕 $+ H_2N-$〔phenyl〕 $\longrightarrow$ 产物为（ ）。

A. H_2N-〔benzene〕$-\overset{\overset{O}{\|}}{C}CH_3 + CH_3COOH$

B. CH₃CONH—⬡ + CH₃COOH

C. H₂N—⬡—OCCH₃ + CH₃COOH

D. CH₃CNH—⬡ + CH₃COOH

9. Complete the equation. ⬡—C—¹⁸OH + HOCH₃ ⇌ (H⁺)

方程式 ⬡—C—¹⁸OH + HOCH₃ ⇌ (H⁺) 产物为( )。

A. ⬡—C(O¹⁸)—OCH₃ + H₂O

B. ⬡—C—OCH₃ + H₂¹⁸O

C. ⬡—C(O¹⁸)—OCH₃ + H₂¹⁸O

D. ⬡—C—¹⁸OCH₃ + H₂O

10. Complete the equation. ⬡—CCl + CH₃OH ⟶

方程式 ⬡—CCl + CH₃OH ⟶ 产物为( )。

A. H₃CO—⬡—CCl

B. H₃CO—⬡—CCl

C. ⬡—COCH₃ + HCl

D. ⬡—CCH₂OH + HCl

Discussion Topic

By heating α-hydroxypropanoic acid, we get two diastereoisomeric lactide. What are their structures? Can they be resolved?

（李贺敏　赵珊珊）

第十一章　取代羧酸
Chapter 11　Substituted Carboxylic Acids

 学习目标

　　1. 掌握　取代羧酸的类型及酸性变化规律；乙酰乙酸乙酯在合成中的应用。掌握氨基酸的两性、等电点等性质。

　　2. 熟悉　α、β、γ-卤代酸、羟基酸和氨基酸受热后的反应。

　　3. 了解　β-酮酸酯的酸式分解和酮式分解。

When the hydrogen atoms on the hydrocarbon in carboxylic acid molecules are replaced by other atoms or groups, the products are called **substituted carboxylic acids** (取代羧酸). According to the type of the substituent group, substituted carboxylic acid can be divided into **Halogenated acid** (卤代酸), **Hydroxy acid** (羟基酸), **Carbonyl acid** (羰基酸) and **Amino acid** (氨基酸). Hydroxy acid can be fatherly divided into **Alcohol acid** (醇酸) and **Phenolic acid** (酚酸), and carbonyl acid can be divided into **Aldehyde acid** (醛酸) and **Ketoacid** (酮酸).

$\overset{Cl}{\underset{}{CH_2COOH}}$	$\overset{OH}{\underset{}{CH_3CHCOOH}}$	$HO\!-\!\!\!\bigcirc\!\!\!-COOH$
chloroacetic acid	α-hydroxypropanoic acid	*p*-hydroxybenzoic acid
氯乙酸	α-羟基丙酸	对羟基苯甲酸
$\overset{O}{\underset{}{HCCOOH}}$	$\overset{O}{\underset{}{CH_3CCOOH}}$	$\overset{NH_2}{\underset{}{CH_3CHCOOH}}$
glyoxylic acid	pyruvic acid	α-aminoxypropanoic acid
乙醛酸	丙酮酸	α-氨基丙酸

Substituted carboxylic acids are compounds with two or more functional groups, generally called complex functional group compounds. They have not only typical properties of halogen or the functional groups, but also some special properties resulting from the interaction between functional groups.

PPT

11.1 Halogenated acid

11.1.1 Structure and Nomenclature

When the hydrogen atom in the carboxylic acid molecule is replaced by halogen, the resulting product is called halogenated acid. The nomenclature of halogenated acids are based on the name of relevant carboxylic acids. Remember that when giving systematic name of an organic molecule the halogen is always taken as substituent group. The position of the substituent groups can be expressed in **Arabic numerals** (阿拉伯数字) or **Greek letters** (希腊字母), for example, ω is the last one of the Greek alphabet, therefore ω- is commonly used to represent the position at the end of a long main chain.

ICH_2CH_2COOH
3-iodopropanoic acid
3-碘丙酸（β-碘丙酸或ω-碘丙酸）

$CH_3CBr_2CH_2COOH$
3,3-dibromobutanoic acid
3,3-二溴丁酸（β,β-二溴丁酸）

$ClCH_2CH_2CH_2COOH$
4-chlorobutanoic acid
4-氯丁酸（γ-氯丁酸或ω-氯丁酸）

3-bromobenzoic acid
3-溴苯甲酸（间溴苯甲酸或 m-溴苯甲酸）

11.1.2 Physical Properties of Halogenated Acids

Halogenated acids contain both halogen (-X) and carboxyl (-COOH) . Due to the -I induction effect, the halogen can enhance the acidity of carboxylic acid. For example: $CH_2ClCOOH$, $CHCl_2COOH$ and CCl_3COOH, the more halogen atoms contain in the molecule, the greater the acidity is.

11.1.3 Reactions of Halogenated Acids

Halogenated carboxylic acid molecules contain both carboxyl and halogen functional groups, so they have general reactions of carboxylic acid and alkyl halides (e.g. Carboxylic acid can form salt, ester, amide and so on; halogen atoms can be replaced by hydroxyl, amino and so on). However, due to the interaction between the carboxyl group and halogen, halogenated acid also shows some special properties. For example, due to the influence of halogen, the acidity of halogenated carboxylic acid is stronger than that of the corresponding carboxylic acid and under the influence of the carboxyl, there are also some changes in the nature of halogen atoms, such as the halogen atom in α -halogenated carboxylic acid is easy to be replaced and so on.

1. Acidity When the hydrogen atom is replaced by an halogen atom, since the halogen atom causes-I effect, the acidity of halogenated carboxylic acid becomes stronger than original carboxylic acid. The strength of acidity depends on the position, the type and the number of halogen atoms.

(1) **The position of the halogen atom**　The effect of halogen atom varies significantly depending on its location. For example if the α-hydrogen is replaced, the acidity is much stronger, while the hydrogen atoms on the β-or γ-carbon atoms is replaced, the acidity is enhanced but the difference with non-replaced carboxylic acid is not very large.

(2) **The number of halogens**　When the hydrogen atoms in a carboxylic acid are replaced by halogen atoms, the more halogens there are, the more acidic it becomes.

(3) **The type of halogen**　Different kinds of halogen atoms will cause different acidity. Fluoroacid is the most acidic, chloroacid and bromoacid are weaker, iodoacid is the weakest, which is consistent with the electronegativity of halogen.

微课

2. Reaction with Alkali　Based on the relative position of halogen and the carboxyl group, different product can be obtained.

(1) **α-Halogenated carboxylic acid**　The reaction of α-halogenated carboxylic acid in boiled water or alkali solution can produce hydroxyl carboxylic acid. The halogen atoms become more active under the influence of the carboxyl group, so the hydrolysis of α-halogenated carboxylic acid is easier than that of alkyl halide.

$$H_3C-\underset{\underset{Cl}{|}}{CH}-COOH + H_2O \xrightarrow{\triangle} H_3C-\underset{\underset{OH}{|}}{CH}-COOH + HCl$$

(2) **β-Halogenated carboxylic acid**　When β-halogenated carboxylic acid reacts with NaOH solution, it loses a molecule of HX and produces α, β-unsaturated carboxylic acid. This is because the α-hydrogen atom in β-halogenated carboxylic acid is more active and prone to elimination reactions due to the influence of the two electron-withdrawing groups.

$$\underset{\underset{Cl\ H}{|\ |}}{H_2C-CH}-COOH + NaOH \longrightarrow CH_2=CHCOOH + NaCl + H_2O$$

(3) **γ-and δ-Halogenated carboxylic acid**　When mixed with boiling water or solution of NaOH, γ- or δ-halogenated carboxylic acid first produces unstable γ- or δ-hydroxy carboxylic acid, and then the carboxyl and hydroxyl in the γ- or δ-hydroxy carboxylic acid immediately occur intramolecular esterification reaction, resulting in stable five- or six-membered cyclic ester.

$$\underset{\underset{Cl}{|}}{CH_2CH_2CH_2COOH} \xrightarrow{Na_2CO_3} \underset{\underset{Cl}{|}}{CH_2CH_2CH_2COOH} \xrightarrow{-H_2O}$$

γ-hydroxybutyric acid　　γ-butyrolactone(1,4-butyrolactone)
γ-羟基丁酸　　　　γ-丁内酯(1,4-丁内酯)

$$\underset{\underset{Cl}{|}}{CH_2CH_2CH_2CH_2COOH} \xrightarrow{Na_2CO_3} \underset{\underset{OH}{|}}{CH_2CH_2CH_2CH_2COOH} \xrightarrow{-H_2O}$$

δ-hydroxypyric acid　　δ-valerolacton(1,5-valerolacton)
δ-羟基戊酸　　　　δ-戊内酯(1,5-戊内酯)

3. Darzens Reaction　α-Halogen ester containing α-hydrogen atoms, can react with aldehydes or ketones in a similar way as **Claisen ester condensation reaction** (克莱森酯缩合). The oxygen anions quickly replace the adjacent halogen atoms with S_N2 reactions and produce **epoxyester** (环氧酸酯). This phenomenon, which is prompted by the direct involvement of the adjacent group, is called the **neighboring participation** (邻基参与), and the reaction is called the **Darzens reaction** (达森反应).

$$ClCH_2COOC_2H_5 + C_6H_5COCH_3 \xrightarrow[\text{or NaNH}_2]{C_2H_5ONa} C_6H_5-\underset{O}{\overset{CH_3}{C}}-CHCOOC_2H_5$$

The reaction process is:

$$ClCH_2COOC_2H_5 + C_2H_5ONa \rightleftharpoons {}^-CHClCOOC_2H_5 + C_2H_5OH$$

$$^-CHClCOOC_2H_5 + C_2H_5OH \rightleftharpoons \left[C_6H_5-\underset{O^-}{\overset{CH_3}{C}}-\overset{Cl}{CHCOOC_2H_5} \right]$$

$$\longrightarrow C_6H_5-\underset{O}{\overset{CH_3}{C}}-CHCOOC_2H_5 + Cl^-$$

At last, aldehyde or ketone can be obtained by **saponification** (皂化反应), acidizing and heating of the α, β-epoxyester.

$$C_6H_5-\underset{O}{\overset{CH_3}{C}}-CHCOOC_2H_5 \xrightarrow[H_2O]{OH^-} C_6H_5-\underset{O}{\overset{CH_3}{C}}-CHCOO^- + C_2H_5OH$$

$$C_6H_5-\underset{O}{\overset{CH_3}{C}}-CHCOO^- + H^+ \rightleftharpoons C_6H_5-\underset{\overset{O^+}{H}}{\overset{CH_3}{C}}-CHCOO^- \rightleftharpoons$$

$$C_6H_5-\underset{\overset{+}{OH}}{\overset{CH_3}{C}}-CHCOO^- \xrightarrow{-CO_2} C_6H_5-\underset{OH}{\overset{CH_3}{C}}=C\overset{H}{} \xrightarrow{重排} C_6H_5-\overset{CH_3}{C}-CHO$$

4. Reformatsky Reaction α-Halogenated acid esters may react with with zinc powder and subsequently with carbonyl compounds (aldehydes, ketones, esters), and β-hydroxyacid can be obtained after hydrolysis. The reaction is called **Reformatsky reaction** (雷福尔马斯基反应).

$$BrCH_2COOC_2H_5 + Zn \xrightarrow{醚} BrZnCH_2COOC_2H_5$$

Organic zinc compounds are similar to Grignard reagents and can react similarly with aldehyde and ketone. However they are not as active as Grignard reagents and will not react with esters. Grignard reagents react quickly with esters, so in Reformatsky reaction the metal of Zn cannot be replaced by Mg.

$$BrZnCH_2COOC_2H_5 \ + \ C_6H_5CHO \longrightarrow \underset{\underset{OZnBr}{|}}{C_6H_5CHCOOC_2H_5}$$

$$\underset{\underset{OZnBr}{|}}{C_6H_5CHCOOC_2H_5} \ + \ H_2O \longrightarrow \underset{\underset{OH}{|}}{C_6H_5CHCOOC_2H_5}$$

$$C_6H_5COCH_3 \ + \ BrCH_2COOC_2H_5 \ \xrightarrow{Zn} \ \underset{\underset{OH}{|}}{\overset{\overset{CH_3}{|}}{C_6H_5CCH_2COOC_2H_5}}$$

11.1.4　Preparation of Halogenated Acids

1. Preparation of α–Halogenated Carboxylic Acids　Direct bromination to monocarboxylic acid can produce α-bromine substituted carboxylic acid, direct chlorination however will often lead to mixture. This is due to the low activity and thus high selectivity of bromine. For example:

$$CH_3CH_2COOH \ \xrightarrow{Cl_2} \ \underset{\underset{Cl}{|}}{CH_3CHCOOH} \ + \ \underset{\underset{Cl}{|}}{CH_2CH_2COOH}$$

α-chloropropionic acid　　　　　β-chloropropionic acid
α - 氯丙酸　　　　　　　　　β - 氯丙酸

$$CH_3CH_2COOH \ \xrightarrow{Br_2} \ \underset{\underset{Br}{|}}{CH_3CHCOOH}$$

α-bromopropionic acid
α - 溴丙酸

The above reaction is slow, but if a little red phosphorus (or halogenphosphate) is added as a catalyst, the reaction can go smoothly, the reaction is called **Hell-Volhard-Zelinski reaction** (赫尔–乌尔哈–泽林斯基反应).

$$RCH_2COOH \ + \ Br_2 \ \xrightarrow{PBr_3} \ \underset{\underset{Br}{|}}{RCHCOOH} \ + \ HBr$$

α - brominated carboxylic acid
α - 溴代酸

Generally α-iodized carboxylic acid cannot be prepared by direct iodization, but it can be produced using KI to substitute halogen in α-chlorocarboxylic acid or α-bromocarboxylic acid.

$$\underset{\underset{Cl}{|}}{RCHCOOH} \ + \ KI \longrightarrow \underset{\underset{I}{|}}{RCHCOOH} \ + \ KCl$$

2. Preparation of β–Halogenated Carboxylic Acids　α, β-Unsaturated carboxylic acid and HX can be combined to produce β-halogenated carboxylic acids. Because of the absorption electron withdrawing induction effect (-I) and the conjugate effect (-C) of the carboxylic group (-COOH), the addition reaction is always in the direction of anti-Malkovnikov rule.

$$\text{CH}_2\text{=CHCOOH} + \text{HBr} \longrightarrow \underset{\overset{|}{\text{Br}}}{\text{CH}_2\text{CH}_2\text{COOH}}$$

acrylic acid   β - bromopropylene acid
丙烯酸   β - 溴丙酸

β-halogenated carboxylic acids can also be produced by β-hydroxy carboxylic acid reacting with HX or phosphorus halide.

$$\underset{\overset{|}{\text{OH}}}{\text{RCHCH}_2\text{COOH}} + \text{HBr} \longrightarrow \underset{\overset{|}{\text{Br}}}{\text{RCHCH}_2\text{COOH}} + \text{H}_2\text{O}$$

β-hydroxy carboxylic acid   β-halogenated carboxylic acid
β-羟基酸   β-卤代酸

3. Preparation of γ-、 δ- or even farther Halogenated Carboxylic Acid   γ-、δ- or even farther halogenated carboxylic acid can be prepared by corresponding diacid monoester by Hunsdiecker reaction.

$$\text{CH}_3\text{OOC(CH}_2)_4\text{COOH} \xrightarrow[\text{KOH}]{\text{AgNO}_3} \text{CH}_3\text{OOC(CH}_2)_4\text{COOAg}$$

$$\xrightarrow[\text{CCl}_4]{\text{Br}_2} \text{CH}_3\text{OOC(CH}_2)_3\text{CH}_2\text{Br} \xrightarrow[\text{H}_2\text{O}]{\text{H}^+} \text{HOOC(CH}_2)_3\text{CH}_2\text{Br}$$

PPT

11.2   Hydroxy Acid

Hydroxy acid can be divided into alcoholic acid and phenolic acid.

11.2.1   Structure and Nomenclature

In the aliphatic dicarboxylic acids whose carbon chain is numbered in Greek letters, there may be two alpha position carbon atoms, which should be expressed in α and α′, correspondingly by α, α′, β, β′, γ, γ′, δ, δ′…, for example:

$$\underset{\overset{|}{\text{COOH}}}{\overset{\text{COOH}}{\underset{|}{\overset{|}{\text{CHOH}}}}} \qquad \underset{\overset{|}{\text{COOH}}}{\overset{\text{COOH}}{\text{CHOH}}} \qquad \underset{\overset{|}{\text{CH}_2\text{COOH}}}{\overset{\text{CH}_2\text{COOH}}{\text{HO–C–COOH}}}$$

2-hydroxysuccinic acid   2,3-dihydroxysuccinic acid   2-hydroxypropane-1,2,3-tricarboxylic acid
α-羟基丁二酸（苹果酸）   α,α′-二羟基丁二酸（酒石酸）   2-羟基丙烷-1,2,3-三甲酸（柠檬酸）

Many hydroxy acids are natural products, therefore they are generally named according to their sources.

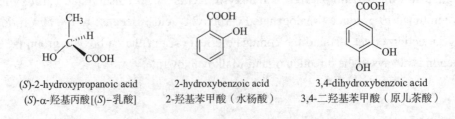

(S)-2-hydroxypropanoic acid   2-hydroxybenzoic acid   3,4-dihydroxybenzoic acid
(S)-α-羟基丙酸[(S)-乳酸]   2-羟基苯甲酸（水杨酸）   3,4-二羟基苯甲酸（原儿茶酸）

11.2.2 Physical Properties of Hydroxy Acids

Hydroxyl acid generally has higher boiling point and better water solubility than corresponding carboxylic acid. The solubility of hydroxy acid is also higher than corresponding carboxylic acid, which is due to the formation of hydrogen bonds between hydroxyl group and water. Phenolic acid is found in nature in the form of salts, esters or **glycosides** (苷). Examples of important phenolic acids are **Salicylic acid** (水杨酸) and **Gallic acids** (五倍子酸).

11.2.3 Reactions of Hydroxy Acids

Hydroxy acids have not only typical chemical properties of alcohols and acids, but also have some special properties due to the interaction between the two groups.

1. Alcoholic Acids

(1) Acidity   Alcoholic acid has a general reaction of alcohol and carboxylic acid (e.g. to form salt, esterification, amide, etc.). Hydroxyl is an inductive electron withdrawing group, which can increase the acidity of the carboxyl group. The closer the hydroxyl is, the stronger the acidity is. The pK_a values of several hydroxy acids are shown in Table 11.1.

Table 11.1   The pK_a values of several hydroxy acids
表 11-1   几种羟基酸的 pK_a

Hydroxy Acids	pK_a	Hydroxy Acids	pK_a
CH_3COOH	4.76	$\begin{matrix}CH_2COOH\\ \vert\\ OH\end{matrix}$	3.85
CH_3CH_2COOH	4.87	$\begin{matrix}CH_3CHCOOH\\ \vert\\ OH\end{matrix}$	3.86
$\begin{matrix}CH_2CH_2COOH\\ \vert\\ OH\end{matrix}$	4.51		

(2) Oxidation reaction   Hydroxy groups in alcoholic acid can be oxidized to produce aldehyde or ketoacids. The hydroxy group in α-alcoholic acid is more susceptible to oxidation than that in alcohol.

$$HOCH_2COOH \xrightarrow{[O]} \underset{\underset{O}{\|}}{H}CCOOH \xrightarrow{[O]} HOOC\text{-}COOH$$

hydroxyacetic acid   acetaldehyde   acetate
羟基乙酸   乙醛酸   乙二酸

$$\underset{\underset{OH}{\vert}}{CH_3CH}COOH \xrightarrow{[O]} \underset{\underset{O}{\|}}{CH_3C}COOH$$

α-hydroxypropionic acid   acetone acid
α-羟基丙酸   丙酮酸

$$CH_3CHCH_2COOH \quad \xrightarrow{[O]} \quad CH_3CCH_2COOH$$
$$| \qquad\qquad\qquad\qquad\qquad ||$$
$$OH \qquad\qquad\qquad\qquad\qquad O$$

β - hydroxybutyric acid   β - buttone acid
β-羟基丁酸   β -丁酮酸

(3) Dehydration reaction after heating   Dehydration reaction of an alcoholic acid may lead to different products depending on the relative position of the carboxyl and the hydroxyl.

Dehydration of α-alcoholic acid produces cyclic diester namely **lactide** (交酯).

$$R\text{-}CH\text{-}OH \quad HO{-}C{=}O \qquad \xrightarrow{\triangle} \qquad R\text{-}CH \quad C{=}O \qquad + \quad 2H_2O$$

Most lactides are crystalline substance. Similar to other esters, they are prone to hydrolysis and the original alcoholic acid will be formed when heated with acid or alkali:

$$R{-}HC \quad C{=}O \qquad \xrightarrow[H^+ \text{ 或 } OH^-]{H_2O} \qquad 2\ R{-}CH{-}COOH$$
$$O{=}C \quad CH{-}R \qquad\qquad\qquad\qquad\qquad\qquad |$$
$$\qquad\qquad\qquad\qquad\qquad\qquad\qquad OH$$

In β-alcohol acid the α-hydrogen is more active because it is affected by both carboxyl and hydroxyl, so with heating β-alcohol acid produces α, β-unsaturated acid by intramolecular dehydration.

$$R{-}\overset{H}{\underset{OH}{C}}{-}\overset{H}{\underset{H}{C}}{-}COOH \qquad \xrightarrow{\triangle} \qquad R{-}\overset{H}{C}{=}\overset{}{\underset{H}{C}}{-}COOH$$

α , β - unsaturated aci
α ,β - 不饱和酸

γ-alcoholic acid is very easy to lose water, at room temperature it can automatically dehydrate to produce a five-membered **lactone** (内酯).

$$\begin{array}{c} H_2C{-}CH_2 \\ | \qquad | \\ CH_2 \quad C{=}O \\ | \qquad | \\ OH \quad OH \end{array} \qquad \longrightarrow \qquad \begin{array}{c} H_2C{-}CH_2 \\ | \qquad | \\ H_2C \qquad C{=}O \\ \diagdown \ O \diagup \end{array} \quad + \quad H_2O$$

Therefore, γ-alcohol acid is stable only after it becomes salt. Some γ-alcohol acids are not available because they lose water and produce lactone. The lactone, like normal ester, reacts with alkali solution and hydrolyze to produce the original hydroxy carboxylate.

$$\begin{array}{c} H_2C{-}CH_2 \\ | \qquad | \\ H_2C \qquad C{=}O \\ \diagdown \ O \diagup \end{array} \quad + \quad NaOH \quad \longrightarrow \quad \begin{array}{c} CH_2CH_2CH_2COONa \\ | \\ OH \end{array} \quad + \quad H_2O$$

Sodium γ-hydroxybutyrate has an anaesthetic effect and can be used as an anesthetic. Dehydration of acid produces the δ-lactone of the six-membered ring.

Some Chinese medicines contain the structure of lactones as the active ingredients. For example, **Pasqueflower** (白头翁), **Anemonin** (白头翁脑) and the **Protoanemonin** (原白头翁脑) are compounds of unsaturated lactones.

protoanemonin
原白头翁脑

anemonin
白头翁脑

Another example is antibacterial drug **Andrographolide** (穿心莲内酯), one of the main active component of **Andrographis paniculata** (穿心莲).

2. Phenolic Acids　Phenolic acids have the typical reaction of phenol and acid. For example, when reacting with the solution of $FeCl_3$, it can occur **chromogeni reaction** (显色反应) (the properties of phenols). And they can form esters by the reaction with compounds of hydroxy groups (the properties of carboxylic acid).

(1) Decarboxylation may happen with heating especially when hydoxyl and carboxyl groups are in *ortho* or *para* position. For example when Gallic acid is heated to more than 200℃, a molecule of carbon dioxide is lost and it becomes **Gallic phenol** (没食子酚):

Salicylic acid molecule contains hydroxyl and carboxyl, so it has the general properties of phenol and acid. Acetyl salicylic acid is commonly known as **Aspirin** (阿司匹林), which is prepared by acylation of salicylic acid with acetic anhydride in acetic acid at 80℃.

$$\text{(salicylic acid)} + (CH_3CO)_2O \longrightarrow \text{(acetylsalicylic acid)}$$

11.2.4　Preparation of Hydroxy Acid

1. Alcohol Acid

(1) Halogenated acids hydrolysis　This is a common method to prepare α-hydroxy acids.

$$\underset{Cl}{\overset{CH_2COOH}{|}} + H_2O \xrightarrow{\Delta} \underset{OH}{\overset{CH_2COOH}{|}} + HCl$$

α- hydroxyacetic acid
α-羟基乙酸

(2) **Hydroxytril** (羟基腈) hydrolysis　Reaction of aldehyde or ketone with HCN produces hydroxytril, and hydolysis of hydroxytril will produce α-hydroxy acid. This is a common method to prepare α-hydroxy acids.

$$RCHO + HCN \longrightarrow R-\underset{H}{\overset{OH}{\underset{|}{C}}}-CN \xrightarrow[H_2O]{H^+} R-\underset{H}{\overset{OH}{\underset{|}{C}}}-COOH$$

$$R-\overset{O}{\overset{\|}{C}}-R + HCN \longrightarrow R-\underset{R}{\overset{OH}{\underset{|}{C}}}-CN \xrightarrow[H_2O]{H^+} R-\underset{R}{\overset{OH}{\underset{|}{C}}}-COOH$$

Reaction of alkene with **hypochloric acid** (次氯酸) produces β-hydroxytril, and the β-hydroxytril can be hydrolyzed to β-hydroxyacid. For example:

$$RCH=CH_2 \xrightarrow{HOCl} R-\underset{OH}{\underset{|}{C}}H-\underset{Cl}{\underset{|}{C}}H_2 \xrightarrow{KCN} R-\underset{OH}{\underset{|}{C}}H-CH_2CN \xrightarrow[H_2O]{H^+} RCHCH_2COOH$$

2. Phenolic Acid　Many phenolic acids are extracted from natural products. The general method for the synthesis of phenolic acid is to use the **Kolbe-Schmidt reaction** (科尔贝-许密特反应), by reacting the dry sodium phenoxide and carbon dioxide at 405~709kPa and 120~140℃, and finally acidifying of the products.

$$\text{(sodium phenoxide)} + CO_2 \xrightarrow[405\sim709\ KPa]{120\sim140℃} \text{(sodium salicylate)} \xrightarrow{H^+} \text{(salicylic acid)}$$

11.3　Carbonyl Acids

PPT

The carbonyl acid with carbonyl at the end of the carbon chain is an aldehyde acid, while carbonyl group in the middle of the carbon chain is a ketoacid. Among them the **β-ketoacid esters** (**β-酮酸酯**) is the most important and has many kinds of applications in the synthesis of medicinal compounds.

11.3.1   Structure and Nomenclature

When naming a carbonyl acid you should chose the longest carbon chain containing both carbonyl group and carboxyl group as the main chain, a carbonyl acid should be called a aldehyde acid or a ketone acid, or an oxo subsituted carboxyl acid.

$$HCCH_2COOH$$

3-oxopropanoic acid

β-氧亚基丙酸或ω-氧亚基丙酸（丙醛酸）

$$HCCH_2CH_2CH_2CH_2COOH$$

6-oxohexanoic acid

ε-氧亚基己酸或ω-氧亚基己酸（己醛酸）

$$CH_3CCH_2COOH$$

3-oxobutanoic acid

β-氧亚基丁酸（丁-3-酮酸或乙酰乙酸）

$$CH_3CCH_2CH_2CH_2COOH$$

5-oxohexanoic acid

δ-氧亚基己酸（己-5-酮酸或乙酰丁酸）

11.3.2   Physical Properties of Carbonyl Acids

1. α–Carbonyl Acid   The simplest α-carbonyl acid is acetone acid. It is an intermediate product of carbohydrate and protein metabolism in plants and animals. Oxidation of lactic acid can produce acetone acid:

$$CH_3CHOHCOOH \xrightarrow{[O]} CH_3CCOOH + H_2O$$

Acetone acid is a colorless, irritating odor liquid with a boiling point of 165℃ (decomposition), soluble in water, ethanol and ether.

2. β-Carbonyl Acid (β-丁酮酸)   Acetylacetic acid (乙酰乙酸) (CH_3CCH_2COOH), namely β-butyl acid is the simplest β-ketone acid. Acetyl acid is an intermediate product of fat metabolism in organisms. Acetylacetic acid is a viscous liquid.

3. Ethyl Acetoacetate (乙酰乙酸乙酯)   Ethyl acetoaceta ($CH_3CCH_2COOC_2H_5$), namely ethyl-3-ketobutyric acid, is a colorless transparent liquid with pleasant smell, with melting point of 45℃ and boiling point of 181℃. It is slightly soluble in water, easy to dissolve in ethanol, ether, chloroform and other organic solvents.

4. Diethyl Malonate (丙二酸二乙酯)   Diethyl malonate ($H_5C_2OOC\text{-}CH_2\text{-}COOC_2H_5$) is a colorless, sweet smell liquid with a boiling point of 199℃. It is slightly soluble in water, soluble in organic solvents such as ethanol, ether, chloroform and benzene.

11.3.3　Reactions of Carbonyl Acids

1. Acidity and Tautomerization of Ethyl Acetylacetate

(1) The acidity of α-hydrogen on the methylene. In ethyl acetylacetate and β-diketone compounds, due to the influence of two carbonyl groups, the acidity of α-hydrogen atom is stronger than that of general aldehydes, ketones or esters.

Ethyl acetylacetate anion that is obtained by losing α-hydrogen can be stablized by the resonance structures:

$$\left[CH_3-\overset{O}{\overset{\|}{C}}-\overset{-}{C}H-\overset{O}{\overset{\|}{C}}-OC_2H_5 \longleftrightarrow CH_3-\overset{O^-}{\overset{|}{C}}=CH-\overset{O}{\overset{\|}{C}}-OC_2H_5 \longleftrightarrow CH_3-\overset{O}{\overset{\|}{C}}-CH=\overset{O^-}{\overset{|}{C}}-OC_2H_5 \right]$$

$$\left[\overset{-}{C}H_2-\overset{O}{\overset{\|}{C}}-OC_2H_5 \longleftrightarrow CH_2=\overset{O^-}{\overset{|}{C}}-OC_2H_5 \right]$$

$$\left[\overset{-}{C}H_2-\overset{O}{\overset{\|}{C}}-CH_3 \longleftrightarrow CH_2=\overset{O^-}{\overset{|}{C}}-CH_3 \right]$$

(2)The tautomerism of ethyl acetylacetate. Ethyl acetylacetate has similar properties as ketones, such as, reaction with hydrocyanic acid, sodium sulphate etc. But there are some special reactions. For example, adding Br_2/CCl_4 to ethyl acetylacetate can make the color of bromine disappear, indicating that double bonds exist in the molecule. Ethly acetylacetate can also react with solution $FeCl_3$, indicating that the molecule has an enol structure. According to the above experimental facts, it can be considered that ethyl acetylacetate is a **tautomer** (互变异构体) with keto and enol structures in a dynamic equilibrium.

Both chemical and physical methods have proved that ethyl acetylacetate is a mixture of ketone and **enol** (烯醇式), and that they can transform from each other. In acetone solution at room temperature, ketone type accounts for 93%, and enol type accounts for 7%.

$$H_3C-\overset{O}{\overset{\|}{C}}-CH_2-\overset{O}{\overset{\|}{C}}-OC_2H_5 \rightleftharpoons H_3C-\overset{OH}{\overset{|}{C}}=\overset{}{\underset{H}{C}}-\overset{O}{\overset{\|}{C}}-OC_2H_5$$

ketone (93%)   enol (7%)

酮式 (93%)   烯醇式 (7%)

The interchanging of ketone and enol isomers of ethyl acetylacetate is caused by the reversible rearrangement of acidic hydrogen atoms on methylene.

Theoretically, there always should be two forms of tautomerism of the compounds with $-\overset{H}{\overset{|}{\underset{|}{C}}}-\overset{O}{\overset{\|}{C}}-$ structure. But there are big differences in the proportion of enol form for different compounds.

The higher enol content of β-dicarbonyl compounds is due to the formation of a stable six membered ring through intramolecular hydrogen bond, on the other hand, the C=O double bond and C=C double bond in the enol form a larger conjugate system and thus reduce the energy of the molecule. These factors make the stability of the enol form increase, and the content of the enol form increase at equilibrium.

ketone
酮式

enol
烯醇式

In addition, solvents, concentrations, temperature, etc. can also affect the content of enol. In the mixture of ethyl acetylacetate that achieves equilibrium, the isomer content varies with the difference of solvent: in water or other proton-containing polar solvents, the enol content is less, while in nonpolar solvents, the content of enol is more.

2. Acid Decomposition (酸式分解) and Ketone Decomposition (酮式分解) of Ethyl Acetylacetate

(1) Acid decomposition. When ethyl acetylacetate is heated in a concentrated basic solution, both the α-and β-carbon bond break and produce two molecules of carboxylate which can be acidified in the next step to produce acetic acid.

$$CH_3 - \overset{O}{\overset{||}{C}} \vdots CH_2 - \overset{O}{\overset{||}{C}} \vdots OC_2H_5 \xrightarrow{40\%NaOH} 2CH_3 - \overset{O}{\overset{||}{C}} - OH \ + \ C_2H_5OH$$

(2) Ketone decomposition. When ethyl acetylacetate is heated in dilute alkali solution, the ester hydrolyze and sodium acetylacetate acid is obtained, and with acidification acetylacetic acid is obtained. Acetylacetic acid is unstable and will decarboxylate to ketone immediately under heating condition. So it is called ketone decomposition. In fact the susceptibility to decarboxylation is another important characteristic of β-ketoacid.

$$CH_3 - \overset{O}{\overset{||}{C}} - CH_2 - \overset{O}{\overset{||}{C}} - OC_2H_5 \xrightarrow{5\%NaOH} CH_3\overset{O}{\overset{||}{C}}CH_2COONa \xrightarrow[\text{②}CO_2, \triangle]{\text{①}H^+} CH_3\overset{O}{\overset{||}{C}}CH_3$$

3. Alkylation and Acylation of α−Hydrogen   The hydrogen atoms on the methylene in the ethyl acetoacetate molecule are acidic due to the electron withdrawing effect from the two adjacent carbonyl groups. Therefore, strong base such as alkoxide or sodium metal, can remove the hydrogen atom on methylene and in the next step an in-situ substitution reaction of carbanion with alkyl halide or acyl halide will lead to alkylation or acylation of α-hydrogen :

$$CH_3 - \overset{O}{\overset{||}{C}} - CH_2 - \overset{O}{\overset{||}{C}} - OC_2H_5 \xrightarrow{C_2H_5ONa} \left[CH_3 - \overset{O}{\overset{||}{C}} - CH - COOC_2H_5 \right]^- Na^+$$

$$\left[CH_3 - \overset{O}{\overset{||}{C}} - CH - COOC_2H_5 \right]^- Na^+ \begin{cases} \xrightarrow{RX} CH_3\overset{O}{\overset{||}{C}}CH\overset{O}{\overset{||}{C}}OC_2H_5 \underset{R}{|} \xrightarrow[\text{②}R'X]{\text{①}C_2H_5ONa} CH_3\overset{O}{\overset{||}{C}}\overset{RO}{\overset{|}{C}}\overset{O}{\overset{||}{C}}OC_2H_5 \underset{R'}{|} \\ \\ \xrightarrow[RCX]{\overset{O}{\overset{||}{}}} CH_3\overset{O}{\overset{||}{C}}CH\overset{O}{\overset{||}{C}}OC_2H_5 \underset{\underset{R}{|}}{\overset{C=O}{|}} \end{cases}$$

The alkyl halides used in the alkyl substitution reaction are prefered to be primary halides.

The above-mentioned substituted ethyl acetoacetate can be fatherly keto or acid decomposed as follows:

$$CH_3 - \overset{O}{\overset{\|}{C}} - \overset{R}{\overset{|}{\underset{|}{C}}} - COC_2H_5 \quad \begin{cases} \xrightarrow{\text{ketone decomposition}} CH_3 \overset{O}{\overset{\|}{C}} - \overset{}{\underset{R'}{\overset{|}{CHR}}} \\ \\ \xrightarrow{\text{acid decomposition}} \overset{R'}{\underset{}{RCH}} - COOH \end{cases}$$

$$CH_3 - \overset{O}{\overset{\|}{C}} - \overset{}{\underset{\underset{R}{\overset{|}{C=O}}}{CH}} - COC_2H_5 \quad \begin{cases} \xrightarrow{\text{ketone decomposition}} CH_3\overset{O}{\overset{\|}{C}} - CH_2 - \overset{O}{\overset{\|}{C}}R \\ \\ \xrightarrow{\text{acid decomposition}} R - \overset{O}{\overset{\|}{C}}CH_2 - COOH \end{cases}$$

4. Reactions of Diethyl Malonate The hydrogen on the methylene of diethyl malonate (丙二酸二乙酯) is also acidic due to the influence of two ester groups, and has similar properties as ethyl acetylacetate.

Alkylation

$$CH_2 \overset{COOC_2H_5}{\underset{COOC_2H_5}{}} \xrightarrow[C_2H_5OH]{C_2H_5ONa} \left[CH(COOC_2H_5)_2 \right]^- Na^+ \xrightarrow{RX} R - CH \overset{COOC_2H_5}{\underset{COOC_2H_5}{}}$$

The two acidic hydrogen atoms in diethyl malonate may be both substituted by alkyl group.

$$R - CH \overset{COOC_2H_5}{\underset{COOC_2H_5}{}} \xrightarrow[C_2H_5OH]{C_2H_5ONa} \left[RC(COOC_2H_5)_2 \right]^- Na^+ \xrightarrow{RX} \overset{R}{\underset{R'}{}} C \overset{COOC_2H_5}{\underset{COOC_2H_5}{}}$$

In the alkylation reaction, primary haloalkanes are the best, while tertiary haloalkanes mostly have side reactions (elimination reaction) and generate olefins.

Acylation

$$H_2C \overset{COOC_2H_5}{\underset{COOC_2H_5}{}} \xrightarrow[C_2H_5OH]{C_2H_5ONa} \left[CH(COOC_2H_5)_2 \right]^- Na^+ \xrightarrow{\overset{O}{\overset{\|}{RCCl}}} RC \overset{O}{\overset{\|}{}} - CH \overset{COOC_2H_5}{\underset{COOC_2H_5}{}}$$

Hydrolysis

Diethyl malonate is very easy to be hydrolyzed and decarboxylated to produce monocarboxylic acid. Therefore, diethyl malonate is often used to synthesize substituted carboxylic acid.

11.3.4 Synthesis Using β-dicarbonyl Compounds

1. Synthesis Using Ethyl Acetoacetate

(1) Synthesis of methylketone. Under the action of strong base, ethyl acetoacetate can be mono or

dialkylated, and then with ketone decomposition, methyl ketone is obtained.

$$\text{CH}_3\overset{O}{\underset{\|}{\text{C}}}\text{CH}_2\overset{O}{\underset{\|}{\text{C}}}\text{OC}_2\text{H}_5 \xrightarrow[\text{②CH}_3\text{Br}]{\text{①C}_2\text{H}_5\text{ONa}} \text{CH}_3\overset{O}{\underset{\|}{\text{C}}}\underset{\underset{\text{CH}_3}{|}}{\text{CH}}\overset{O}{\underset{\|}{\text{C}}}\text{OC}_2\text{H}_5 \xrightarrow[\text{②CH}_2=\text{CHCH}_2\text{Br}]{\text{①C}_2\text{H}_5\text{ONa}} \text{CH}_3\overset{O}{\underset{\|}{\text{C}}}\underset{\underset{\text{CH}_3}{|}}{\overset{\overset{\text{CH}_2\text{CH}=\text{CH}_2}{|}}{\text{C}}}\text{COOC}_2\text{H}_5$$

$$\xrightarrow[\text{②H}^+, \triangle]{\text{①5\%NaOH}} \text{H}_3\text{C}-\overset{O}{\underset{\|}{\text{C}}}-\underset{\underset{}{}}{\overset{\overset{\text{CH}_3}{|}}{\text{CH}}}\text{CH}_2\text{CH}=\text{CH}_2$$

(2) Synthesis of monocarboxylic acid. Monocarboxylic acid can be obtained by acid decomposition of ethyl acetoacetate after alkyl substitution.

$$\text{CH}_3\overset{O}{\underset{\|}{\text{C}}}\text{CH}_2\overset{O}{\underset{\|}{\text{C}}}\text{OC}_2\text{H}_5 \xrightarrow[\text{②CH}_3\text{CH}_2\text{CH}_2\text{Br}]{\text{①C}_2\text{H}_5\text{ONa}} \text{CH}_3\overset{O}{\underset{\|}{\text{C}}}\underset{\underset{\text{CH}_2\text{CH}_2\text{CH}_3}{|}}{\text{CH}}\overset{O}{\underset{\|}{\text{C}}}\text{OC}_2\text{H}_5 \xrightarrow[\text{②CH}_3\text{Br}]{\text{①C}_2\text{H}_5\text{ONa}} \text{CH}_3\overset{O}{\underset{\|}{\text{C}}}\underset{\underset{\text{CH}_3}{|}}{\overset{\overset{\text{CH}_2\text{CH}_2\text{CH}_3}{|}}{\text{C}}}\text{COOC}_2\text{H}_5$$

$$\xrightarrow[\text{②H}^+]{\text{①40\%NaOH}} \text{CH}_3\text{CH}_2\text{CH}_2\underset{\underset{\text{CH}_3}{|}}{\text{CH}}\text{COOH}$$

(3) Synthesis of ketone acid.

微课

$$\text{CH}_3\overset{O}{\underset{\|}{\text{C}}}\text{CH}_2\overset{O}{\underset{\|}{\text{C}}}\text{OC}_2\text{H}_5 \xrightarrow[\text{②CH}_3\text{Cl}]{\text{①C}_2\text{H}_5\text{ONa}} \text{CH}_3\overset{O}{\underset{\|}{\text{C}}}\underset{\underset{\text{CH}_3}{|}}{\text{CH}}\overset{O}{\underset{\|}{\text{C}}}\text{OC}_2\text{H}_5 \xrightarrow[\text{②ClCH}_2\text{COOC}_2\text{H}_5]{\text{①C}_2\text{H}_5\text{ONa}} \text{CH}_3\overset{O}{\underset{\|}{\text{C}}}\underset{\underset{\text{CH}_3}{|}}{\overset{\overset{\text{CH}_2\text{COOC}_2\text{H}_5}{|}}{\text{C}}}\text{COOC}_2\text{H}_5$$

$$\xrightarrow[\text{②H}^+, \triangle]{\text{①5\%NaOH}} \text{CH}_3\overset{O}{\underset{\|}{\text{C}}}\underset{\underset{\text{CH}_3}{|}}{\text{CH}}\text{CH}_2\text{COOH}$$

$$\text{CH}_3\overset{O}{\underset{\|}{\text{C}}}\text{CH}_2\overset{O}{\underset{\|}{\text{C}}}\text{OC}_2\text{H}_5 \xrightarrow[\text{②CH}_2=\text{CHCOOC}_2\text{H}_5]{\text{①C}_2\text{H}_5\text{ONa}} \text{CH}_3\overset{O}{\underset{\|}{\text{C}}}\underset{\underset{}{}}{\overset{\overset{\text{CH}_2\text{CH}_2\text{COOC}_2\text{H}_5}{|}}{\text{CH}}}\text{COOC}_2\text{H}_5$$

$$\xrightarrow[\text{②H}^+, \triangle]{\text{①5\%NaOH}} \text{CH}_3\overset{O}{\underset{\|}{\text{C}}}\text{CH}_2\text{CH}_2\text{CH}_2\text{COOH}$$

(4) Synthesis of diketone.

$$\text{CH}_3\overset{O}{\underset{\|}{\text{C}}}\text{CH}_2\overset{O}{\underset{\|}{\text{C}}}\text{OC}_2\text{H}_5 \xrightarrow[\text{② ClCH}_2\text{COCH}_3]{\text{①C}_2\text{H}_5\text{ONa}} \text{CH}_3\overset{O}{\underset{\|}{\text{C}}}\underset{\underset{}{}}{\overset{\overset{\text{CH}_2\text{COCH}_3}{|}}{\text{CH}}}\text{COOC}_2\text{H}_5 \xrightarrow[\text{② H}^+, \triangle]{\text{①5\%NaOH}} \text{CH}_3\overset{O}{\underset{\|}{\text{C}}}\text{CH}_2\text{CH}_2\overset{O}{\underset{\|}{\text{C}}}\text{CH}_3$$

(5) Synthesis of dicarboxylic acids.

$$\text{CH}_3\overset{O}{\underset{\|}{\text{C}}}\text{CH}_2\overset{O}{\underset{\|}{\text{C}}}\text{OC}_2\text{H}_5 \xrightarrow[\text{② CH}_3\text{CHClCOOC}_2\text{H}_5]{\text{①C}_2\text{H}_5\text{ONa}} \text{CH}_3\overset{O}{\underset{\|}{\text{C}}}\underset{\underset{}{}}{\overset{\overset{\text{CH}_3\text{CHCOOC}_2\text{H}_5}{|}}{\text{CH}}}\text{COOC}_2\text{H}_5 \xrightarrow[\text{② H}^+]{\text{①40\%NaOH}} \text{CH}_3\overset{O}{\underset{\|}{\text{C}}}\text{CH}_2\text{CH}_2\overset{O}{\underset{\|}{\text{C}}}\text{CH}_3$$

2. Synthesis Using Diethyl Malonate

(1) Synthesis of monocarboxylic acids.

$$\underset{\text{COOC}_2\text{H}_5}{\overset{\text{COOC}_2\text{H}_5}{\text{CH}_2}} \xrightarrow[\text{C}_2\text{H}_5\text{OH}]{\text{C}_2\text{H}_5\text{ONa}} \left[\text{CH(COOC}_2\text{H}_5)_2\right]^- \text{Na}^+ \xrightarrow{\text{CH}_3\text{CH}_2\text{CH}_2\text{Br}} \text{CH}_3\text{CH}_2\text{CH}_2-\underset{\text{COOC}_2\text{H}_5}{\overset{\text{COOC}_2\text{H}_5}{\text{CH}}}$$

$$\xrightarrow[\text{C}_2\text{H}_5\text{OH}]{\text{C}_2\text{H}_5\text{ONa}} \left[\text{CH}_3\text{CH}_2\text{CH}_2-\underset{\text{COOC}_2\text{H}_5}{\overset{\text{CHOOC}_2\text{H}_5}{\text{C}}}\right]^- \text{Na}^+ \xrightarrow{\text{CH}_3\text{CH}_2\text{Br}} \underset{\text{CH}_3\text{CH}_2\text{CH}_2}{\overset{\text{CH}_3\text{CH}_2}{\text{C}}} \underset{\text{COOC}_2\text{H}_5}{\overset{\text{COOC}_2\text{H}_5}{}}$$

$$\xrightarrow[\text{H}_2\text{O}]{\text{H}^+} \underset{\text{CH}_3\text{CH}_2\text{CH}_2}{\overset{\text{CH}_3\text{CH}_2}{\text{C}}}\underset{\text{COOH}}{\overset{\text{COOH}}{}} \xrightarrow{\triangle} \underset{\text{CH}_3\text{CH}_2\text{CH}_2}{\overset{\text{CH}_3\text{CH}_2}{\text{CH}}}-\text{COOH} + \text{CO}_2$$

(2) Synthesis of dicarboxylic acids.

Cyclic diacid

$$\underset{\text{COOC}_2\text{H}_5}{\overset{\text{COOC}_2\text{H}_5}{\text{CH}_2}} \xrightarrow[\text{C}_2\text{H}_5\text{OH}]{\text{C}_2\text{H}_5\text{ONa}} 2\left[\text{CH}_2(\text{COOC}_2\text{H}_5)_2\right]^- \text{Na}^+ \xrightarrow{\text{BrCH}_2\text{CH}_2\text{CH}_2\text{Br}}$$

$$(\text{H}_5\text{C}_2\text{OOC})_2\text{CHCH}_2\text{CH}_2\text{CH}_2\text{CH}(\text{COOC}_2\text{H}_5)_2 \xrightarrow[\text{②} \quad \text{I}_2]{\text{①} \text{C}_2\text{H}_5\text{ONa}} \text{(cyclopentane)}\underset{(\text{COOC}_2\text{H}_5)_2}{\overset{(\text{COOC}_2\text{H}_5)_2}{}}$$

$$\xrightarrow[\triangle]{\text{H}^+} \text{(cyclopentane)}\underset{\text{COOH}}{\overset{\text{COOH}}{}}$$

Glutaric acid

$$\underset{\text{COOC}_2\text{H}_5}{\overset{\text{COOC}_2\text{H}_5}{\text{CH}_2}} \xrightarrow[\text{C}_2\text{H}_5\text{OH}]{\text{C}_2\text{H}_5\text{ONa}} \left[\text{CH(COOC}_2\text{H}_5)_2\right]^- \text{Na}^+ \xrightarrow{\text{CH}_2=\text{CHCOOC}_2\text{H}_5}$$

$$\text{H}_5\text{C}_2\text{OOCCH}_2\text{CH}_2-\underset{\text{COOC}_2\text{H}_5}{\overset{\text{COOC}_2\text{H}_5}{\text{CH}}} \xrightarrow[\text{②} \text{H}^+, \triangle]{\text{①} \text{OH}^-} \underset{\text{CH}_2\text{COOH}}{\overset{\text{CH}_2\text{COOH}}{\text{CH}_2}}$$

Dicarboxylic acids. Synthesis of dicarboxylic acids from malonate is generally realized by the reaction of malonate anion with one molecule of dihalogenated hydrocarbon.

$$\begin{array}{c}\left[\text{CH(COOC}_2\text{H}_5)_2\right]^- \text{Na}^+ \\ \left[\text{CH(COOC}_2\text{H}_5)_2\right]^- \text{Na}^+\end{array} \quad \underset{\text{CH}_2-\text{Br}}{\overset{\text{CH}_2-\text{Br}}{|}} \longrightarrow \underset{\text{CH}_2\text{CH(COOC}_2\text{H}_5)_2}{\overset{\text{CH}_2\text{CH(COOC}_2\text{H}_5)_2}{|}}$$

$$\xrightarrow[\text{H}_2\text{O}]{\text{H}^+} \underset{\text{CH}_2\text{CH(COOH)}_2}{\overset{\text{CH}_2\text{CH(COOH)}_2}{|}} \xrightarrow[\triangle]{-\text{CO}_2} \underset{\text{CH}_2-\text{CH}_2-\text{COOH}}{\overset{\text{CH}_2-\text{CH}_2-\text{COOH}}{|}}$$

(3) Synthesis of ketones.

For example: synthesis of o-nitroacetophenone from two ethyl malonate, toluene and other necessary inorganic reagents.

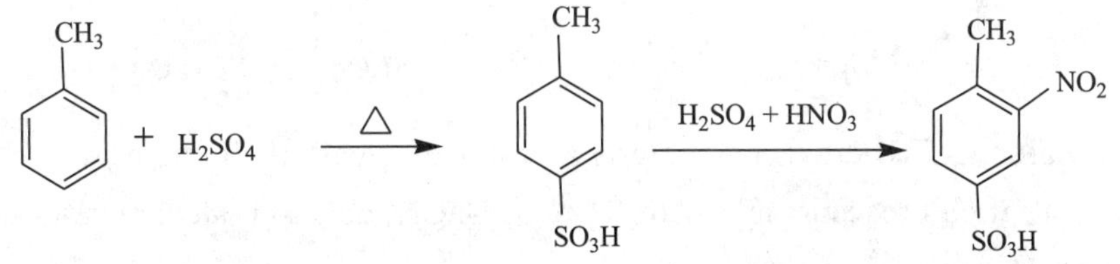

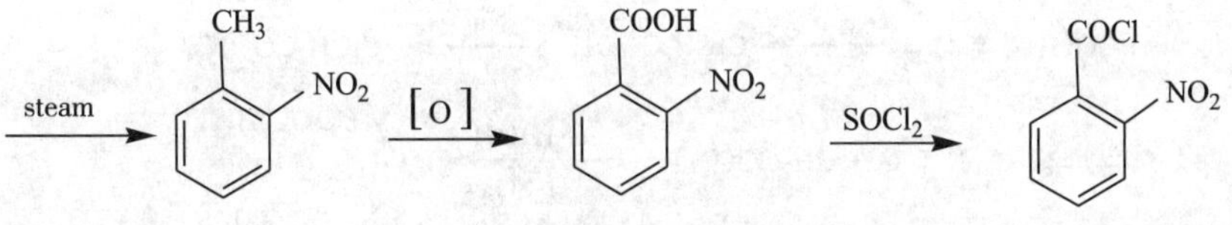

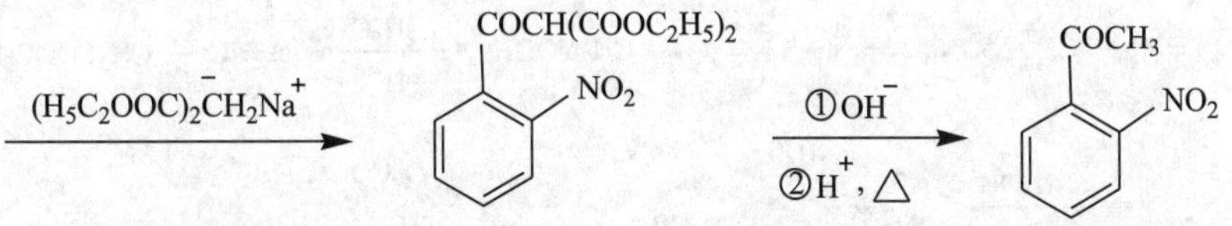

(4) Synthesis lipid ring compounds. Under the action of strong base, malonate reacts with one molecule of dihaloalkane to form cyclic derivatives.

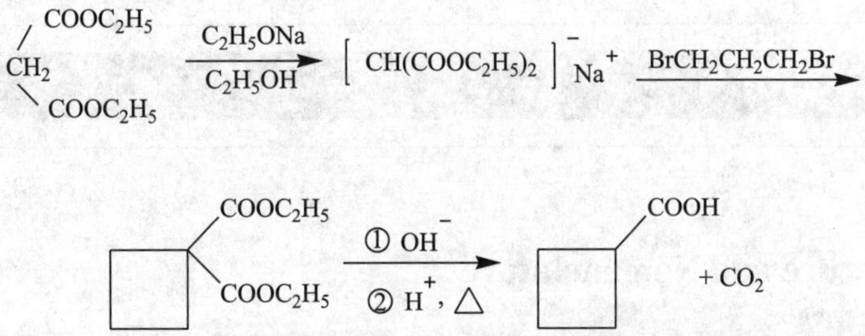

$$CH_2(COOC_2H_5)_2 \xrightarrow[C_2H_5OH]{C_2H_5ONa} \left[CH(COOC_2H_5)_2\right]^- Na^+ \xrightarrow{BrCH_2CH_2CH_2Br}$$

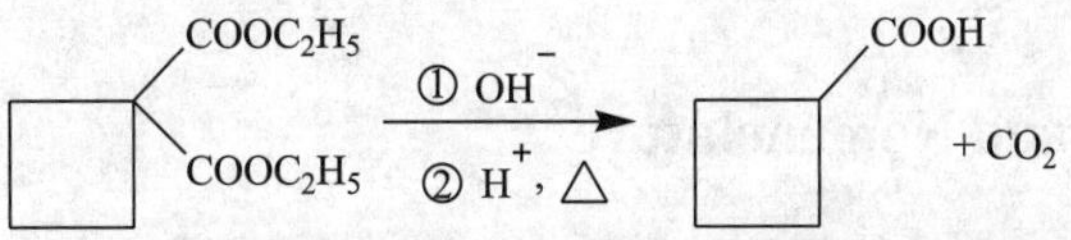

(5) Synthesis of α, β - unsaturated acids.

$$CH_2(COOC_2H_5)_2 \xrightarrow[C_2H_5OH]{C_2H_5ONa} \left[CH(COOC_2H_5)_2\right]^- Na^+ \xrightarrow{CH_3CHO}$$

$$CH_3CH(OH)CH(COOC_2H_5)_2 \xrightarrow[\text{②}\ H^+,\ \triangle]{\text{①}\ OH^-} CH_3CH{=}CHCOOH + CO_2 + H_2O$$

11.3.5 Preparation of Carbonyl Acids

1. Reaction of Diethylene Ketones with Ethanol Reaction of diethylene ketones with ethanol can produce ethyl acetylacetate.

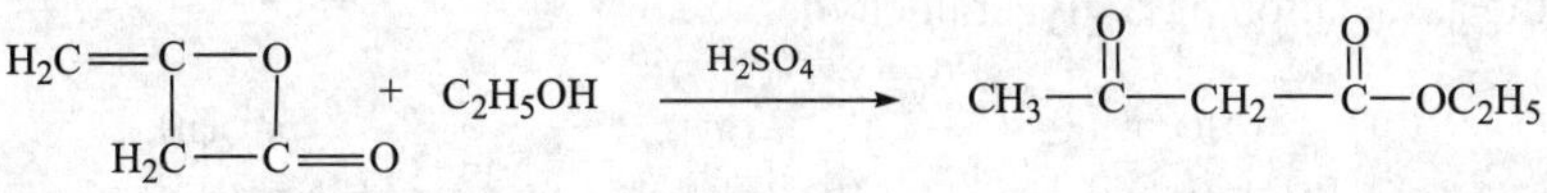

2. Claisen Reaction Under basic condition, ethyl acetate undergoes Claisen ester condensation reaction to produce ethyl acetoacetate.

$$H_3C-\overset{O}{\overset{\|}{C}}-OC_2H_5 + H_3C-\overset{O}{\overset{\|}{C}}-OC_2H_5 \xrightarrow{C_2H_5OH} H_3C-\overset{O}{\overset{\|}{C}}-CH_2-\overset{O}{\overset{\|}{C}}-COOC_2H_5$$

3. Diethyl Malonate Diethyl malonate is an ester of dicarboxylic acid, which is prepared by reacting sodium salt of chloroacetic acid with KCN or NaCN, and then alcoholization with ethanol in sulfuric acid (or dry HCl) .

$$CH_3COOH + Cl_2 \xrightarrow{P} ClCH_2COOH \xrightarrow{NaOH} ClCH_2COONa$$

$$\xrightarrow{NaCN} N\equiv C-CH_2COONa \xrightarrow[H_2SO_4]{C_2H_5OH} \underset{CH_2}{\overset{\diagup COOC_2H_5}{\underset{\diagdown COOC_2H_5}{|}}}$$

$$N\equiv C-CH_2COONa \xrightarrow{H^+} N\equiv C-CH_2COOH \xrightarrow[H^+]{C_2H_5OH} N\equiv C-CH_2COOC_2H_5$$

$$\xrightarrow[H^+]{C_2H_5OH} HN=\overset{OC_2H_5}{\overset{|}{C}}-CH_2COOC_2H_5 \xrightarrow[H^+]{H_2O} \underset{CH_2}{\overset{\diagup COOC_2H_5}{\underset{\diagdown COOC_2H_5}{|}}}$$

PPT

11.4   Amino Acids

11.4.1   Structure and Nomenclature

$$\overset{NH_2}{\overset{|}{HOOCCH_2CH_2CHCOOH}}$$

2-aminopentanedioic acid (glutamic acid)
α-氨基戊二酸（谷氨酸）

$$\overset{NH_2}{\overset{|}{H_2NCH_2CH_2CH_2CH_2CHCOOH}}$$

2,6-diaminohexanoic acid (lysine)
α,ω-二氨基己酸（赖氨酸）

The amino acids produced by hydrolyzation of natural proteins are almost all α-amino acids. They all have the same characteristics in chemical structure with an amino group on the α-carbon atom of the carboxyl group, so they are called α-amino acids.

$$R-\overset{H}{\underset{NH_2}{\overset{|}{\underset{|}{C}}}}-COOH$$

Amino acids, with the exception of a few (e.g., glycine), are chiral molecular and therefore are optically active. The natural α-amino acids all have L-configuration. The configuration of α-amino acids is determined by the association of L-glyceraldehyde.

CHO	COOH	COOH
HO——H	HO——H	H₂N——H
CH₂OH	CH₃	CH₃
L-glyceraldehyde	L-lactic acid	L-alanine

Depending on the relative position of amino and carboxyl, amino acids can be divided into α-amino acids, β-amino acids, γ -amino acids, etc. α-Amino acids are the most important amino acids because they are the basic unit of protein. In addition, according to the number of amino and carboxyl, Amino acids can be divided into three types: acidic, basic and neutral type. Amino acids with the same number of amino and carboxyl groups are called neutral amino acids; those with more carboxyl groups are acidic amino acids; those with more amino groups are basic amino acids. According to the type of the hydrocarbon group (R-), they can be divided into aliphatic, aromatic and heterocyclic amino acids.

There are more than 100 kinds of amino acids found in nature, but there are only more than 20 α-amino acids that make up proteins in organisms.

Amino acids obtained from protein hydrolysis are list in Table 11.2.

Table 11.2　Common α-amino acids
表 11-2　常见 α- 氨基酸

classification		Name	Structure	Isoelectric point
Aliphatic amino acid	Neutral amino-acid	Glycine	CH₂COOH / NH₂	5.79
		Alanine	CH₃CHCOOH / NH₂	6.00
		* Valine	(CH₃)₂HC—CHCOOH / NH₂	5.96
		* Leucine	(CH₃)₂CHCH₂—CHCOOH / NH₂	6.02
		* Lsoleucine	CH₃CH₂CH—CHCOOH / CH₃ NH₂	5.98
		Serine	HO—CH₂—CHCOOH / NH₂	5.68
		Threonine	HO—CH—CHCOOH / CH₃ NH₂	6.16
		Cysteine	HS—CH₂—CHCOOH / NH₂	5.05
		Methionine	CH₃S—CH₂CH₂—CHCOOH / NH₂	5.74
		Asparagine	H₂NCCH₂—CHCOOH / O NH₂	5.41
		Glutamine	H₂NCCH₂CH₂—CHCOOH / O NH₂	5.65

(Continued)

classification		Name	Structure	Isoelectric point
Aliphatic amino acid	Basic ammoniacyl acid	* Lysine	H₂NCCH₂(CH₂)₃—CHCOOH (with ‖ O and ⎸ NH₂)	9.74
		* Arginine	H₂NCNH(CH₂)₃CHCH₂COOH (with ‖ NH and ⎸ NH₂)	10.76
	Acid amino acid	Aspartic acid	HOOCCH₂CHCOOH (with ⎸ NH₂)	2.77
		Glutamic acid	HOOCCH₂CH₂CHCOOH (with ⎸ NH₂)	8.22
Aromatic amino acid		* Phenylalanine	CH₂CHCOOH (with ⎸ NH₂)	5.48
		Tyrosine	HO—CH₂CHCOOH (with ⎸ NH)	5.68
Heterocyclic ammoni-acylic acid		* Tryptophan	CH₂CHCOOH (with ⎸ NH₂)	5.89
		Histidine	CH₂CHCOOH (with ⎸ NH₂)	7.59
		Proline	COOH	6.30

Note: * refers to the amino acid that cannot be synthesized by human body and must be supplied by food, which is called "essential amino acid".

11.4.2　Physical Properties of Amino Acid

α-Amino acids are less volatile colorless crystals, with high melting point generally between 200℃-300℃. Generally amino acids are soluble in water, insoluble in ethanol, ether, benzene and other organic solvents. The high melting point and dissolution behavior of amino acids show the characteristics of salt compounds.

11.4.3　Reactions of Amino Acids

Amino acid molecules contain both amino and carboxyl groups, so amino acids not only have the typical properties of amino and carboxyl groups, but also have some special properties due to the interaction between the two groups in the molecule.

1. Acidity and Isoelectric Point (等电点)

(1) Acidity　Amino acids have both basic amino group (-NH$_2$) and acid carboxyl group (-COOH), which can react with strong acid or strong base to form salt, so amino acids are amphoteric compounds.

Since the amino acid molecule contains both carboxyl group and amino group, the two groups in one amino acid molecule can also interact with each other to form salt, which is called **internal salt (内盐)**.

$$R-\underset{\underset{NH_2}{|}}{CH}-COOH \rightleftharpoons R-\underset{\underset{NH_3^{\oplus}}{|}}{CH}-COO^{\ominus}$$

The internal salt molecule has both the positive ion part and the negative ion part, so it is also called **zwitterion (两性离子)**. The reason why amino acids have high melting point and are insoluble in organic solvents is that amino acids are internal salts and have the properties of salt.

In aqueous solution, the carboxyl and amino groups in amino acid molecules can be ionized like acids and bases, respectively.

$$R-\underset{\underset{NH_2}{|}}{CH}-COOH + H_2O \longrightarrow R-\underset{\underset{NH_2}{|}}{CH}-COO^{\ominus} + H_3O^{\oplus}$$

$$R-\underset{\underset{NH_2}{|}}{CH}-COOH + H_2O \longrightarrow R-\underset{\underset{NH_3^{\oplus}}{|}}{CH}-COOH + OH^{\ominus}$$

In fact, in the aqueous solution, amino acid molecules have the following ionization equilibrium.

$$R-\underset{\underset{NH_2}{|}}{CH}-COO^{\ominus}$$

(2) Isoelectric point　The degree of ionization of amino and carboxyl groups in amino acid molecules is not the same. Even for neutral amino acids, the degree of ionization of the two groups is not the same. In fact the degree of ionization of carboxyl group is slightly higher than that of amino group. If an acid is added to the aqueous solution of an amino acid, the ionization of the carboxyl group in the molecule can be inhibited. In strongly acidic (pH<1) solutions, amino acids exist mainly in the cationic state. Conversely, when alkali is added to an aqueous solution, the ionization of the amino group is inhibited, and amino acids will mainly exist in the anion state in the strongly alkaline (pH > 11) solution.

$$R-\underset{\underset{NH_2}{|}}{CH}-COO^{\ominus}Na^{\oplus} \xleftarrow{NaOH} R-\underset{\underset{NH_3^{\oplus}}{|}}{CH}-COO^{\ominus} \xrightarrow{HCl} R-\underset{\underset{NH_3^{\oplus}\ Cl^{\ominus}}{|}}{CH}-COOH$$

Because of the different electric charge of amino acids in different pH value of aqueous solution, the behavior of amino acids in electric field is also different. Generally, amino acids exist in the cationic state in acidic solution, and the cationic amino acid moves towards the cathode under the action of electric field. Amino acids exist as anions in alkaline solutions and move towards the anode under the action of an electric field. When the solution is adjusted to a specific pH value, the number of anions moving toward the anode in the electric field is exactly equal to the number of cations moving toward the cathode. And at this point, the pH value of the solution is called the isoelectric point (pI, isoelectric point) of the amino acid. The isoelectric points of various amino acids are shown in Table 11.2. The isoelectric points of neutral amino acids were 5.0~6.3, those of acidic amino acids were 2.8~3.2, and those of basic amino acids were 7.6~10.8.

The isoelectric point is not neutral point. At the isoelectric point, if the difference in the mobility of positive and negative amino acid ions is not considered (the difference is very small), then the concentration of positive and negative amino acid ions in the solution is the same, and the concentration of amphoteric ions and neutral molecules in the amino acid solution is the highest, while the solubility of amino acid is the lowest. Thus, mixtures of amino acids can be separated by adjusting the pH of the solvent.

2. Reaction with Heating Similar to hydroxy acids, reactions of α-、β-、 γ - or δ - amino acids with heating are different due to the relative difference of amino location in the molecules.

(1) When the α-amino acid is heated, the dehydration of the amino group and the carboxyl group between the two molecules can occur, resulting in the formation of the six-membered cyclic **crossamide (交酰胺)**.

$$\begin{array}{c}
\underset{HNH}{\overset{R}{HC}}-\overset{O}{C}\text{--}OH \\
HO-\underset{O}{C}-\underset{R}{HC}-HNH
\end{array} \longrightarrow
\begin{array}{c}
\underset{HN}{\overset{R}{HC}}-\overset{O}{C}\text{--}NH \\
\underset{O}{C}-\underset{R}{HC}
\end{array}$$

(2) When β-amino acids are heated, they lose one molecule of NH_3 and form α, β-unsaturated acids.

$$RCH\underset{NH_2}{-}CH\underset{H}{COOH} \xrightarrow{\triangle} RCH{=}CHCOOH + NH_3$$

(3) When γ - or δ - amino acids are heated, they are easily dehydrated to form five or six membered rings of **lactam (内酰胺)**.

$$\begin{array}{c}
H_2C-\overset{O}{C}\text{--}OH \\
| \quad\quad H \\
H_2C-CH_2NH
\end{array} \xrightarrow{\triangle}
\begin{array}{c}
H_2C-\overset{O}{C} \\
| \quad\quad NH \\
H_2C-CH_2
\end{array} + H_2O$$

$$\begin{array}{c}
H_2C-\overset{O}{C}\text{--}OH \\
H_2C \quad\quad H \\
H_2C-CH_2NH
\end{array} \xrightarrow{\triangle}
\begin{array}{c}
H_2C-\overset{O}{C} \\
H_2C \quad\quad NH \\
H_2C-CH_2
\end{array} + H_2O$$

3. Reaction with Ninhydrin（茚三酮） When α-amino acid and ninhydrin hydrate are heated in aqueous solution, they can react to give a blue-purple colour. This color reaction is often used for colorimetry of α-amino acid and as a color reagent for thin-layer analysis. This is a rapid, sensitive and effective method for the identification of α-amino acids.

$$\text{(phthalic trione)} + \text{H}_2\text{N}-\overset{\displaystyle R}{\underset{\displaystyle R}{\text{CH}}}-\text{COOH} \longrightarrow \text{(Ruhemann's purple)} + \text{RCHO} + \text{CO}_2$$

4. Reaction with Nitrous Acid   Most of amino acids containing free amino groups, so they may react with nitrous acid, giving off nitrogen. The amount of nitrogen released by the reaction can be used to calculate the amount of amino acid in a protein molecule.

$$\underset{\displaystyle \text{NH}_2}{\text{RCHCOOH}} + \text{HONO} \longrightarrow \underset{\displaystyle \text{OH}}{\text{RCHCOOH}} + \text{N}_2 + \text{H}_2\text{O}$$

5. Decarboxylation Reaction   Dearboxylation may occur with the help of some microbial enzymes, or in boiling solvent, producing corresponding amine. This reaction is also involved human metabolism. As with the help of intestinal bacteria, histidine may be decarboxylated to form histamine.

$$\underset{\displaystyle \text{NH}_2}{\text{(imidazole)}-\text{CH}_2\text{CHCOOH}} \xrightarrow{-\text{CO}_2} \text{(imidazole)}-\text{CH}_2\text{CH}_2\text{NH}_2$$

11.5   Application of Substituted Carboxylic Acid in the Pharmacy Study

1. Citric Acid (柠檬酸) ($\underset{\displaystyle \text{CH}_2\text{COOH}}{\overset{\displaystyle \text{CH}_2\text{COOH}}{\text{HO}-\text{C}-\text{COOH}}}$)   Citric acid exists in citrus, hawthorn and other fruits. Especially in lemon its content may reach to about 6%~10%, so it is commonly known as citric acid. It is commonly used as a flavoring agent, cool agent. Its salt ($\text{C}_6\text{H}_5\text{O}_7\text{K}_3\cdot6\text{H}_2\text{O}$) is white crystalline and soluble in water and is used as a phlegm agent and diuretic.

2. Acetyl Salicylic Acid (乙酰水杨酸) ([structure])   Acetyl salicylic acid is commonly known as **Aspirin** (阿司匹林). Aspirin is often used to treat fever, headache, joint pain, active rheumatism and so on. It can be used in combination with **phenacetin** (非那西丁), **caffeine** (咖啡因).

3. Gallic Acid (没食子酸) ([structure])   Gallic acid is a widely distributed organic acid in nature. It exists in free state in plants such as tea, or as the composition of tannins in plant such as **Chinese gall** (五倍子). Hydrolysis of tannins can produce Gallic acid.

Alkaline bismuth salt of gallic acid has the function of **convergence** (收敛) and **corrosion prevention** (防

腐). It is a protective agent for **gastrointestinal mucosa** (胃肠黏膜) for oral administration and an antiseptic astringent for external use. The structure is:

$$HO-C_6H_2(OH)_2-COOBi(OH)_2$$

4. Protocatechuic Acid (原儿茶酸) ( ) Protocatechuic acid is one of the effective ingredients in the **Chinese holly leaf** (四季青). Chinese holly leaf has therapeutic effect on scalding, in addition, it can also be used to treat bacterial dysentery, renal nephritis and some ulcer diseases.

5. Caffeic Acid (咖啡酸) ( ) Caffeic acid is found in many Traditional Chinese Medicine, such as Wild Carrot, **Marshy Betony Herb** (光叶水苏), **Buckwheat** (荞麦), **Elaeagnus longipes** (木半夏) and so on. Some Traditional Chinese Medicine do not contain caffeic acid, but contains caffeic acid ester, namely **Chlorogenic acid** (绿原酸) the antibacterial effective component of **Honeysuckle** (金银花). In addition, **Ramie** (苎麻), **Mulberry leaves** (桑叶), **Valerian** (缬草), etc, also contain chlorogenic acid. The structure of chlorogenic acid is as follows:

重 点 小 结

1. 典型的卤代酸、羟基酸 (醇酸、酚酸)、羰基酸及氨基酸的结构、命名和制备。

2. 不同取代位置的卤代酸、羟基酸及氨基酸受热后的反应性质。

3. 重要的羟基酸及其衍生物的用途。

4. 乙酰乙酸乙酯的酮式分解和酸式分解。

5. 乙酰乙酸乙酯在有机合成中的应用。

6. 氨基酸的两性和等电点。

本章学习取代羧酸，学习中应注意梳理、比较和归纳卤代酸、羟基酸和氨基酸等取代羧酸的共性特征和反应规律，举一反三，触类旁通，从根本上理解取代羧酸的理化性质。在羰基酸学习过程中，注意体会负离子的结构和反应特性。注意掌握乙酰乙酸乙酯酸式分解、酮式分解的断键部位和反应规律。

Problems

目 标 检 测

1. Which is the lactone formed after dehydration of γ -hydroxy acid?

γ - 羟基酸脱水后形成的内酯为（　　　）。

    A. Ternary ring　　　　　　　　　　B. Quaternary ring

    C. Five-membered ring　　　　　　　D. Six-membered ring

2. Which can generate lactide by heating among the following hydroxy acids and keto acids?

下列羟基酸或羰基酸能形成丙交酯的是（　　　）。

A. CH_3CHCH_2COOH with OH below

B. $CH_3CH-COOH$ with OH below

C. CH_3CCH_2COOH with O below

D. $HOCH_2CH_2CH_2COOH$

3. Which of the following compound solution can lead to colour changing adding $FeCl_3$ reagent?

下列化合物中加入 $FeCl_3$ 后能发生颜色变化的是（　　　）。

A. (benzene ring with COOH and OH)

B. (benzene ring with CH_2OH)

C. (cyclohexane ring with COOH and OH)

D. (cyclohexane ring with OH)

4. To identify ethyl acetoacetate and ethyl levulinate, use (　　　).

下列试剂能区分乙酰乙酸乙酯和乙酰丙酸乙酯的是（　　　）。

    A. Fehling reagent　　　　　　　　　B. Lucas reagent

    C. Carbonyl reagent　　　　　　　　D. $FeCl_3$ solution

5. Treatment of ethyl acetoacetate with (1) CH_3CH_2MgBr (2) H_2O gives (　　　).

乙酰乙酸乙酯经过 (1) CH_3CH_2MgBr (2) H_2O 的反应能制备的化合物是（　　　）。

    A. $CH_3COCH(C_2H_5)C(OH)(C_2H_5)_2$　　　　B. $CH_3COCH_2C(OH)(C_2H_5)_2$

    C. $CH_3CH_2C(OH)(CH_3)CH_2COOEt$　　　D. CH_3COCH_2COOEt

6. Among the following keto acids, the most susceptible to decarboxylation one is (　　　).

下列酮酸最易脱羧的是（　　　）。

A. $H_3C-C-CH_2CH_2CH_2COOH$ with O below

B. $H_3C-CH_2CH_2CH_2C-CH_2COOH$ with O above

C. $H_2CC-CHCH_2CH_2COOH$ with H_3C above and O below

D. $H_3C-CHCH_2COOH$ with OH below

7. Among the following compounds the one most likely to happen decarboxylation is (　　　).

下列化合物最易脱羧的是（　　　）。

A. (环己基COOH) B. $(CH_3)_3C—COOH$

C. $CH_3\text{-}\overset{O}{\underset{\|}{C}}\text{-}CH_2\text{-}COOH$ D. $H_3C\text{-}\underset{OH}{CH}CH_2COOH$

8. The reagent to be used to distinguish formalin from formic acid is (   ).

下列化合物可用于区分福尔马林和甲酸的是（   ）。

A. Tollens reagent B. NaHCO$_3$ solution
C. Fehling reagent D. I$_2$/NaOH

9. Which of the following compounds cannot be decarboxylated when heated (   ).

下列化合物在加热时不发生脱羧反应的是（   ）。

A. $(CH_3)_2C(COOH)_2$ B. (4-hydroxypentanoic acid)

C. (biphenyl-2,2'-dicarboxylic acid) D. $H_3C\text{-}\overset{O}{\underset{\|}{C}}\text{-}\overset{CH_3}{\underset{CH_3}{C}}\text{-}COOH$

10. Which can generate *cis* and *trans* isomers after heating the one of the following hydroxy acid?

下列羟基酸在加热时产生的化合物具有顺反异构体的是（   ）。

A. $(H_3C)_2\text{-}\underset{OH}{C}CH_2COOH$ B. $HOOCH_2CH_2C\text{-}\underset{OH}{CH_2}$

C. $H_3C\text{-}\underset{OH}{CH}CH_2COOH$ D. $HOOCH_2C\text{-}\underset{OH}{CH_2}$

11. What is the main product when adipic acid is heated?

下列化合物是加热己二酸时得到的主要产品的是（   ）。

A. Mononic acid B. Cyclic ketone
C. Lactone D. Anhydride

12. Which of the following is trans-1,2-cyclobutane dicarboxylic acid?

下列结构是反 -1，2- 环丁二酸的是（   ）。

A. B.
C. D.

13. Benzoic acid react with CH$_3$OH, CH$_3$CHOHCH$_2$CH$_3$, CH$_3$CH$_2$CH$_2$OH respectively under acid catalysis, please arrange the esterification reaction speed sorted by fast to slow.

苯甲酸在酸性条件下分别与 CH$_3$OH, CH$_3$CHOHCH$_2$CH$_3$, CH$_3$CH$_2$CH$_2$OH 反应，请将酯化反应

的速率按从大到小排列。

　　A. $CH_3OH>CH_3CHOHCH_2CH_3>CH_3CH_2CH_2OH$

　　B. $CH_3CHOHCH_2CH_3>CH_3CH_2CH_2OH>CH_3OH$

　　C. $CH_3OH>CH_3CH_2CH_2OH>CH_3CHOHCH_2CH_3$

　　D. $CH_3CH_2CH_2OH>CH_3OH>CH_3CHOHCH_2CH_3$

14. Arrange the acidities of the following compound（Ⅰ）butyric acid,（Ⅱ）maleic acid（Ⅲ）succinic acid（Ⅳ）butyl acetylene in order of high to low.

下列化合物（Ⅰ）butyric acid,（Ⅱ）maleic acid（Ⅲ）succinic acid（Ⅳ）butyl acetylene 的酸性由大到小的排序为（　　）。

　　A. Ⅰ>Ⅱ>Ⅲ>Ⅳ　　　　　　　　　　B. Ⅳ>Ⅰ>Ⅲ>Ⅱ

　　C. Ⅱ>Ⅲ>Ⅰ>Ⅳ　　　　　　　　　　D. Ⅲ>Ⅳ>Ⅱ>Ⅰ

15. In acid catalyzed esterification, 1-propanol reacts with benzoic acid，2,4,6-trimethylbenzoic acid, and 2, 4-dimethylbenzoic acid respectively. Which is the correct order?

1- 丙醇与下列化合物进行酸性条件下的酯化反应时，速率由快到慢的顺序为（　　）。

　　A. 2,4, 6-trimethylbenzoic acid> benzoic acid>2, 4-dimethylbenzoic acid

　　B. 2, 4-dimethylbenzoic acid> benzoic acid>2,4, 6-trimethylbenzoic acid

　　C. 2,4,6-trimethylbenzoic acid>2,4-dimethylbenzoic acid> benzoic acid

　　D. benzoic acid>2, 4-dimethylbenzoic acid>2,4, 6-trimethylbenzoic acid

16. Which can be heated to produce unsaturated acid of the following organic acids?

下列有机酸在加热时能产生不饱和酸的是（　　）。

　　A. acetoacetic acid　　　　　　　　　B. 2-hydroxypropionic acid

　　C. Phthalic acid　　　　　　　　　　D. 3-hydroxybutyric acid

17. Which can form intramolecular hydrogen bonds Among the following compounds.

下列化合物能形成分子内氢键的是（　　）。

　　A. p-hydroxybenzoic acid　　　　　　B. m-hydroxybenzoic acid

　　C. o-hydroxybenzoic acid　　　　　　D. p-nitrobenzoic acid

18. When esterified under acid catalysis, the reaction rates of ethanol and $(CH_3)_3CCOOH$, $(CH_3)_2CHCOOH$ and CH_3CH_2COOH are ranked from fast to slow（　　）.

在酸性条件下乙醇和下列化合物进行的酯化反应速率由快到慢的排序为（　　）。

　　A. $(CH_3)_3CCOOH>(CH_3)_2CHCOOH>CH_3CH_2COOH$

　　B. $(CH_3)_2CHCOOH>(CH_3)_3CCOOH>CH_3CH_2COOH$

　　C. $(CH_3)_3CCOOH>CH_3CH_2COOH>(CH_3)_2CHCOOH$

　　D. $CH_3CH_2COOH>(CH_3)_2CHCOOH>(CH_3)_3CCOOH$

19. Among the solutions of the following compounds, which product can be precipitated out by bubbling CO_2 gas into the solution（　　）.

在下列化合物的溶液中通入 CO_2 气体时能产生沉淀的是（　　）。

　　A. Phenol sodium　　　　　　　　　B. Sodium benzoate

　　C. Sodium salicylate　　　　　　　　D. Disodium salicylate

20. Which is the most acidic of the following compounds?

下列化合物酸性最强的是（　　）。

A.

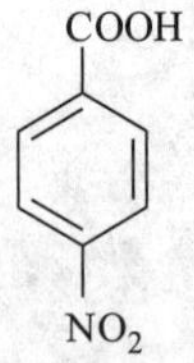

COOH

NO$_2$

B.

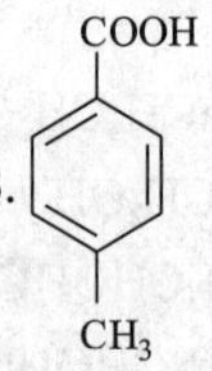

COOH

CH$_3$

C.

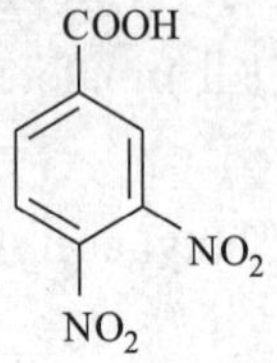

COOH

NO$_2$

NO$_2$

D.

COOH

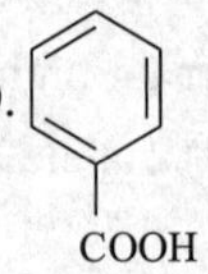

Discussion Topic

Using the dehydration reaction of hydroxy acid as an example to discuss how the functional groups in one molecule will interact with each other to influence the formation of different products.

（胡冬华）

第十二章 胺
Chapter 12　Amine

 学习目标

1. **掌握** 胺类化合物的结构,分类和命名,胺的碱性。
2. **熟悉** 与亚硝酸的反应,Hofmann消除反应以及Cope消除反应。
3. **了解** 胺的物理性质和药学应用。

Nitrogen-Containing Compounds are essential to our life activities, for the reason that nitrogen is essential in amino acids and proteins, in nucleotides and nucleic acids, and in scores of other cellular molecules. In addition, many nitrogen-containing compounds are important industrial products, including polymers such as nylon, many dyes, explosives, and pharmaceutical agents. The structures of some representative alkaloids are shown in Figure 12.1.

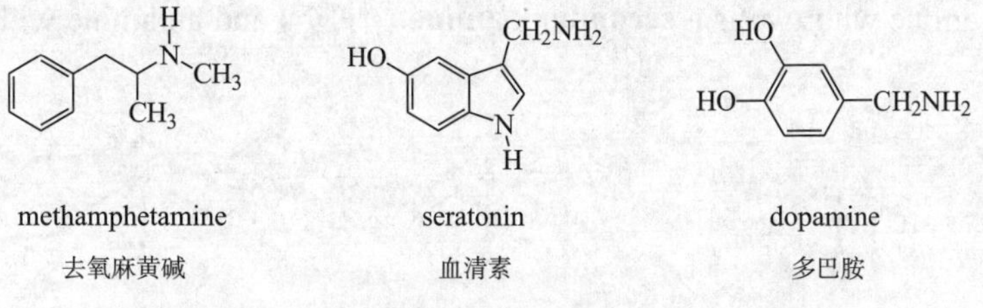

methamphetamine　　　　seratonin　　　　dopamine
去氧麻黄碱　　　　　　　血清素　　　　　　多巴胺

Figure 12.1　Examples of some biologically active amines
图12-1　一些具有生物活性的胺类化合物示例

Their ultimate source is atmospheric nitrogen that, by a process known as **nitrogen fixation** (固氮作用), is reduced to ammonia, then converted to organic nitrogen compounds. This chapter describes the chemistry of amines, organic derivatives of ammonia. Alkylamines have their nitrogen attached to sp^3 hybridized carbon; arylamines have their nitrogen attached to an sp^2 hybridized carbon of a benzene or benzene-like ring. Amines, like ammonia, are weak bases. They are, however, the strongest uncharged bases found in significant quantities under physiological conditions. Amines are usually the bases involved in biological acid–base reactions; they are often the nucleophiles in biological nucleophilic substitutions.

293

12.1　Structure of Amine

PPT

The structure of **amine** (胺) is similar to that of **ammonia** (氨). The nitrogen atom in the molecule also demonstrates heterogeneous sp^3 hybrided, thus the whole molecular structure of amine is **tetrahedral** (四面体的). One pair of unshared electrons occupies one sp^3 hybrid orbital, while the other three sp^3 hybrid orbitals each enjoys one electron. They can overlap with the s orbital of hydrogen atom or the hybrid orbital of carbon atom to form three σ bonds.

微课

12.2　Nomenclature of Amine

PPT

Unlike alcohols and alkyl halides, which are classified as primary, secondary or tertiary according to the degree of substitution at the carbon that bears the functional group, amines are classified according to their degree of substitution at nitrogen. An amine with one carbon attached to nitrogen is a **primary(1°) amine (伯胺)**, an amine with two is a **secondary amine (仲胺)** and an amine with three is a **tertiary amine** (叔胺).

primary amine	secondary amine	tertiary amine
伯胺	仲胺	叔胺

When primary amines can be named as alkylamines, the ending *–amine* is added to the name of the alkyl group that bears the nitrogen.

$CH_3CH_2NH_2$ 　　　　 ⬡—NH_2 　　　　 $CH_3CHCH_2CH_2CH_3$
　　　　　　　　　　　　　　　　　　　　　　　　　　 |
　　　　　　　　　　　　　　　　　　　　　　　　　　NH_2

ethylamine　　　　　　 cyclohexylamine　　　　　 1-methylbutylamine

(ethanamine)　　　　　 (cyclohexanamine)　　　 (2-pentanamine or pentan-2-amine)

乙胺　　　　　　　　　　 环己胺　　　　　　　　　 1–甲基丁胺

Aniline (苯胺) is the parent IUPAC name for amino-substituted derivatives of benzene. Substituted derivatives of aniline are numbered beginning at the carbon that bears the amino group. Substituents are listed in alphabetical order, and the direction of numbering is governed by the usual "first point of difference" rule.

医药大学堂
WWW.YIYAODXT.COM

F—⟨benzene⟩—NH₂

4-fluoroaniline
4-氟苯胺

Br—⟨benzene⟩(NH₂)(C₂H₅)

5-bromo-2-ethylaniline
5-溴-2-乙基苯胺

Compounds with two amino groups are named by adding the suffix -diamine to the name of the corresponding alkane or arene. The final -e of the parent hydrocarbon is retained.

NH₂CH₂CHCH₃
|
NH₂

propane-1,2-diamine
丙 -1,2-二胺

NH₂CH₂CH₂CH₂CH₂CH₂CH₂NH₂

hexane-1,6-diamine
己 - 1,6-二胺

H₂N—⟨benzene⟩—NH₂

benzene-1,4-diamine
1,4-苯二胺

Amino group ranks rather low in seniority when the parent compound is identified for naming purposes. The following compounds each contain a functional group of higher precedence than the amino group, and accordingly, the amino group is indicated by the prefix *amino*.

HOCH₂CH₂NH₂

2-aminoethanol

2- 氨基乙醇

$\overset{O}{\overset{\|}{HC}}$—⟨benzene⟩(1)(4)—NH₂

p-aminobenzaldehyde
(4-aminobenzenecarbaldehyde)

4-氨基苯甲醛

Secondary and tertiary amines are named as *N*-substituted derivatives of primary amines. The parent primary amine is taken to be the one with the longest carbon chain. Rings, however, take precedence over chains. The prefix *N*- is added as a locant to identify substituents on the amine nitrogen.

CH₃NHCH₂CH₃

N-Methylethylamine
(a secondary amine)
N-甲基乙基胺
仲胺

⟨benzene⟩(NHCH₂CH₃)(1)(4)(3)(NH₂)(Cl)

3-Amino-4-Chloro-N-ethylaniline
(a secondary amine)
3-氨基-4-氯-*N*-乙基苯胺
仲胺

⟨cycloheptane⟩—N(CH₃)₂

N,N-Dimethylcycloheptylamine
(a tertiary amine)
N,N-二甲基环庚胺
叔胺

A nitrogen that bears four substituents is positively charged and is named as an **ammonium ion** (铵 离子). The ending –*amine* is replaced with -*ammonium*，and the name of the anion is added.

CH₃N⁺H₃Cl⁻

Methylammoniumchloride

氯化甲铵

⟨cyclopentane⟩—$\overset{CH_3}{\underset{H}{N^+}}$CH₂CH₂ CF₃CO₃⁻

N-Ethyl-N-methylcyclopentyl-
ammoniumtrifluoroacetate
N-乙基-N-甲基 三氟乙酸环戊铵

C₆H₅CH₂N⁺(CH₃)₃I⁻

Benzyltrimethyl-
ammoniumiodide
碘化苯基三甲基铵

PPT

12.3   Chirality of Amines and Quaternary

The geometry of a nitrogen atom bonded to three other atoms or groups of atoms is trigonal pyramidal. The sp^3 hybridized nitrogen atom is at the apex of the pyramid, and the three groups bonded to it extend downward to form the triangular base of the pyramid. If we consider the unshared pair of electrons on nitrogen as a fourth group, then the arrangement of "groups" around nitrogen is approximately tetrahedral. Because of this geometry, an amine with three different groups bonded to nitrogen is **chiral** (手性的) and can exist as a pair of **enantiomers** (对映异构体), as illustrated by the nonsuperposable mirror images of ethylmethylamine. In assigning configuration to these enantiomers, the group of lowest priority on nitrogen is the unshared pair of electrons.

(S)-Ethylmethylamine   *(R)*-Ethylmethylamine

(S)-乙基甲基胺   *(R)*-乙基甲基胺

A most important concern is the rate of this inversion process. If it is slow at room temperature, then we should be able to isolate appropriately substituted optically active amines. If inversion is fast, however, racemization will also be fast as the inversion process interconverts enantiomers. In practice, barriers to inversion in simple amines are very low, the interconversion occurs on the order of a million times per second and isolation of one enantiomer, even at very low temperature, is difficult.

S enantiomer   planar transition state   *R* enantiomer

If the amine nitrogen is part of a rigid cyclic system, then pyramidal inversion is either impossible or extremely slow at room temperature. In such cases it is possible to isolate or observe the two enantiomeric forms of the amine. Troger's base is an example of a resolvable amine in which only the nitrogen atoms are chiral.

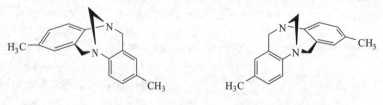

a pair of enantiomers of Troger's base

Quaternary ammonium ion (or asymmetric amine salt) with four different substituted groups on itsnitrogen atom is chiral molecule and can be resolved into two enantiomeric forms.

R Enantiomer　　　　*S* Enantiome

12.4　Physical Properties

Because nitrogen is less electronegative than oxygen, the N-H bond is less polar than the O-H bond and hydrogen bonding is weaker in amines than in alcohols. Amines are more polar than alkanes but less polar than alcohols. For similar molecular weights, alkylamines have boiling points higher than those of alkanes but lower than those of alcohols.

$CH_3CH_2CH_3$	$CH_3CH_2NH_2$	CH_3CH_2OH
Propane	Ethylamine	Ethanol
丙烷	乙胺	乙醇
bp-42℃	bp 17℃	bp 78℃

Primary amines have the highest boiling points, while tertiary amines the lowest. Primary and secondary amines form **intermolecular (分子间的)** hydrogen bonding and therefore have higher boiling points than those of tertiary amines (Table 12.1).

Table 12.1　Physical properties of amines
表 12-1　胺类化合物的物理性质

Name	Structure	Molecular Weight	mp/℃	bp/℃	H$_2$O Solubility
Primary amines					
methylamine	CH_3NH_2	31	−93	−7	very soluble
ethylamine	$CH_3CH_2NH_2$	45	−81	17	∞
isopropylamine	$(CH_3)_2CH_2NH_2$	59	−101	33	∞
Secondary amines					
dimethylamine	$(CH_3)_2NH$	45	−96	7	very soluble
di-*n*-propylamine	$(CH_3CH_2CH_2)_2NH$	101	−40	111	slightly soluble
N-methylaniline	$C_6H_5NHCH_3$	107	−57	196	slightly soluble
Tertiary amines					
trimethylamine	$(CH_3)_3N$	59	−117	3.5	very soluble
triethylamine	$(CH_3CH_2)_3N$	101	−115	90	14%
N,N-dimethylaniline	$C_6H_5N(CH_3)_2$	121	2	194	1.4%

PPT

12.5　Reactions of Amines

12.5.1　Basicity

There is a pair of unshared electron pairs on the nitrogen in the amine, which tends to be shared with other atoms, so the amine is alkaline and nucleophilic (Table 12.2).

Table 12.2　Basicity of amines
表12-2　胺类化合物的碱性

Compound	structure	pK_a of conjugated acid
Ammonia	NH_3	9.3
Primary amine		
Methylamine	CH_3NH_2	10.6
Ethylamine	$CH_3CH_3NH_2$	10.8
Isopropylamine	$(CH_3)_2CHNH_2$	10.6
terk-butylamine	$(CH_3)_3CNH_2$	10.4
Aniline	$C_6H_5NH_2$	4.6
Secondary amines		
Dimethylamine	$(CH_3)_2NH$	10.7
Diethylamine	$(CH_3CH_2)_2NH$	11.1
M-methylanlllne	$C_6H_5NHCH_3$	4.8
Tertaryamines		
Trimethylamine	$(CH_3)_3N$	9.7
Triethylamine	$(CH_3CH_2)_3N$	10.8
N, N-dimethylaniline	$C_6H_5N(CH_3)_2$	5.1

It is more useful to describe the basicity of amines in terms of the pK_a of their **conjugate acids** (共轭酸) than as basicity constants Kb. The more basic the amine, the weaker is its conjugate acid. The more basic the amine, the larger is the pK_a of its conjugate acid. Citing amine basicity according to the pK_a of the conjugate acid makes it possible to analyze acid–base reactions of amines according to the usual relationships.

$$CH_3\overset{..}{N}H_2 \ + \ H{-}\overset{..}{O}\overset{O}{\overset{\|}{C}}CH_3 \ \underset{}{\overset{K=10^6}{\rightleftharpoons}} \ CH_3\overset{+}{N}H_3 \ + \ {^-}\overset{..}{:}\overset{O}{\overset{\|}{O}}\overset{}{C}CH_3$$

Methylamine　　　Acetic acid　　　　　　　Methylammonium ion　　　Acetate ion

(stronger acid;pKa=4.7)　　　　　　(weaker acid;pKa=10.7)

Their basicity provides a means by which amines may be separated from neutral organic compounds. A mixture containing an amine is dissolved in **diethyl ether** (乙醚) and shaken with dilute **hydrochloric**

acid (盐酸) to convert the amine to an ammonium salt.

Most **aliphatic** (脂肪族的) amines are slightly more alkaline than ammonia, while aromatic amines are less alkaline than ammonia.Under the condition of gas phase, the basic order of amines is tertiary amines > secondary amines > primary amines.

For electron donating groups, such as-OH,-OR, -OCOR, -NH$_2$, -NHR, -NHCOR and so on, there are the induction effect of electron withdrawing and the conjugation effect of electron donating. The conjugation effect of electron donating is greater than the induction effect of electron absorption. The total result is electron donating, but it can only increase the density of ortho and para electron cloud of substituent, so these groups in para position enhance the basicity of **aniline** (苯胺).

If electron absorbing groups (such as -NO$_2$, -SO$_3$H, -COOH, etc.) are linked to aromatic ring of **aromatic** (芳香的) amine, there are **induction effect** (诱导效应) of electron withdrawing and the conjugation effect of electron withdrawing. The lone pair electrons of the amino group will be transferred to the withdrawing groups through the aromatic ring. Thus the basicity of the aromatic amine will be weakened. This phenomenon is more obvious if the electron withdrawing group is in the *ortho* or *para* position of amino group.

12.5.2　Alkylation of Ammonia

Alkylamines are, in principle, capable of being prepared by nucleophilic substitution reactions of alkyl halides with ammonia.

$$RX \ + \ 2NH_3 \ \longrightarrow \ RNH_2 \ + \ \overset{+}{N}H_4 \ X^-$$

alkyl　　Ammonia　　　　　Primary　　Ammonium
halide　　　　　　　　　　　amine　　halide salt

Although this reaction is useful for preparing amino acids，it is *not* a general method for the synthesis of amines. Its major limitation is that the expected primary amine product is itself a nucleophile and competes with ammonia for the alkyl halide.

$$RX \ + \ RNH_2 \ + \ NH_3 \ \longrightarrow \ RNHR \ + \ \overset{+}{N}H_4 \ X^-$$

alkyl　　Primary　　Ammonia　　　　Secondary　Ammonium
halide　　amine　　　　　　　　　　amine　　halide salt

When 1-bromooctane, for example, is allowed to react with ammonia, both the primary amine and the secondary amine are isolated in comparable amounts.

$$CH_3(CH_2)_6CH_2Br \xrightarrow{NH_3 \ (2 \ mol)} CH_3(CH_2)_6CH_2NH_2 \quad + \quad [CH_3(CH_2)_6CH_2]_2NH_2$$

1-Bromooctane	Octylamine	N,N-Dioctylamine
(1 mol)	(45%)	(43%)

Competitive alkylation may continue, resulting in the formation of tertiary amines and quaternary ammonium salts. Alkylation of ammonia is used to prepare primary amines only when the starting alkyl halide is not particularly expensive and the desired amine can be easily separated from the other components of the reaction mixture.

12.5.3　Acylation of Amines

Primary and secondary amines react with acid halides to form amides. This reaction is a *nucleophilic acyl substitution*: the replacement of a leaving group on a carbonyl carbon by a nucleophile. In this case, the amine replaces chloride ion.

$$R-NH_2 \quad + \quad R-\overset{\overset{O}{\|}}{C}-Cl \xrightarrow{Pridine} R-\overset{\overset{O}{\|}}{C}-NH-R \quad + \quad \text{(pyridine)} \overset{+}{N}-HCl^-$$

The amine attacks the carbonyl group of an acid chloride much like it attacks the carbonyl group of a ketone or aldehyde. The acid chloride is more reactive than a ketone or an aldehyde because the electronegative chlorine atom draws electron density away from the carbonyl carbon, making it more electrophilic. The chlorine atom in the tetrahedral intermediate is a good leaving group. The tetrahedral intermediate expels chloride to give the amide. A base such as pyridine or NaOH is often added to neutralize the HCl produced.

The amide produced in this reaction usually does not undergo further acylation. Amide is stabilized by a resonance structure that involves nitrogen's nonbonding electrons and places a positive charge on nitrogen. As a result, amides are much less basic and less nucleophilic than amines. The diminished basicity of amides can be used to advantage in electrophilic aromatic substitutions. For example, if the amino group of aniline is acetylated to give acetanilide, the resulting amide is still activating and *ortho*, *para*-directing. Unlike aniline, however, **acetanilide**（乙酰苯胺）may be treated with acidic (and mild oxidizing) reagents, as shown next. Aryl amino groups are frequently acylated before further substitutions are attempted on the ring, and the acyl group is removed later by acidic or basic hydrolysis.

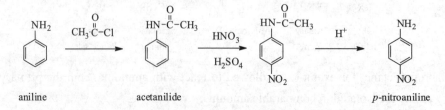

aniline	acetanilide		*p*-nitroaniline

12.5.4　Reaction with Nitrous Acid

Reactions of amines with nitrous acid (H-O-N=O) are particularly useful for synthesis. Because nitrous acid is unstable, it is generated in situ (in the reaction mixture) by mixing sodium nitrite (NaNO$_2$) with cold, dilute hydrochloric acid.

$$Na^+ \, {}^-O\text{-}N{=}O \; + \; H^+Cl^- \; \rightleftharpoons \; H\text{-}O\text{-}N{=}O \; + \; Na^+ \, Cl^-$$

$$\text{sodium nitrite} \hspace{4cm} \text{nitrous acid}$$

When primary amines are nitrosted, their *N*- nitroso compounds can not be isolated because they react further.

RNH$_2$ $\xrightarrow[\text{H}_3\text{O}^+]{\text{NaNO}_2}$ ...
Primary alkylamine　　(Not isolable)　　(Not isolable)

$-\text{H}_3\text{O}^+ \downarrow$

RN≡N: $\xleftarrow{-\text{H}_2\text{O}}$ ... $\xleftarrow{\text{H}_3\text{O}^+}$ ...
Alkyl diazonium ion　　(Not isolable)　　(Not isolable)

The product of this series of steps is an alkyl **diazonium ion** (重氮离子), and the amine is said to have been diazotized. Alkyl diazonium ions are not very stable, decomposing rapidly under the conditions of their formation. Molecular nitrogen is a leaving group par excellence, and the reaction products arise by **solvolysis** (溶剂分解) of the diazonium ion. Usually, a **carbocation** (碳正离子) intermediate is involved.

$$R\text{—}\overset{+}{N}{\equiv}N: \; \longrightarrow \; R^+ \; + \; N_2$$

Alkyl diazonium ion　　Carboncation　　Nitrogen

Reactions of an alkyl diazonium ion

CH$_3$CH$_2$CCH$_3$ (CH$_3$, NH$_2$) $\xrightarrow{\text{HONO}}$ CH$_3$CH$_2$CCH$_3$ (CH$_3$, N$_2^+$) $\longrightarrow$ CH$_3$CH$_2$CCH$_3$ (CH$_3$)$^+$ + :N≡N:

1,1-Eimethylpropylamine　　1,1-Dimethylpropyldiazoniumion　　1,1-Dimethylpropylcation　　Nitrogen

$-\text{H}^+ \downarrow$ 　　　　　　　　$\downarrow \text{H}_2\text{O}$

CH$_3$CH=C(CH$_3$)$_2$ + CH$_3$C=CCH$_2$ (CH$_3$) + CH$_3$CH$_2$CCH$_3$ (CH$_3$, OH)

2-Methylbut2-tene　　2-Methylbut-1-ene　　2-Methylbut-2-ene
(2%)　　　　　　(3%)　　　　　　(80%)

The diazonium ion generated by treatment of a primary alkylamine with nitrous acid loses nitrogen to give a carbocation. The isolated products are derived from the carbocation and include, in this example, alkenes (by loss of a proton) and an alcohol [(**nucleophilic** (亲核的) capture by water].

The secondary amine acts as a nucleophile toward the nitrogen of nitrosyl cation. The intermediate that is formed in the first step loses a proton to give an *N*- nitroso amine. Because there is no hydrogen on nitrogen, aliphatic tertiary amines react only with nitrous acid to form an unstable and hydrolyzable **nitrite** (亚硝酸盐).

$$R_2\ddot{N}: + :\overset{+}{N}-\ddot{O} \longrightarrow R_2\overset{+}{N}-\overset{\ddot{N}}{\overset{}{N}}=\ddot{O} \longrightarrow R_2\ddot{N}-\overset{\ddot{N}}{N}=\ddot{O}:$$

Secondary alkylamine	Nitrosyl cation	*N*-Nitroso amine

For example,

$$(CH_3)_2\ddot{N}H \xrightarrow[\text{H}_2\text{O}]{\text{NaNO}_2,\ \text{HCl}} (CH_3)_2\ddot{N}-\overset{\ddot{N}}{N}=\ddot{O}:$$

Dimethylamine   *N*-Nitrosodimethylamine

二甲胺   *N*-亚硝基二甲胺

Due to the strong activation of amino group, the electrophilic substitution reaction of aromatic tertiary amine and nitrous acid takes place on the benzene ring, which is called **nitrosylation reaction** (亚硝化反应). Nitroso will enter the *para* position of amino group, if the *para* position has been occupied, it will enter the neighborhood.

N,*N*-Diethylaniline   *N*,*N*-Diethyl-p-nitrosoaniline

N,*N*-二乙基苯胺   *N*,*N*-二乙基对亚硝基苯胺

12.5.5 Reaction of Arenediazonium Salts

The reaction of aromatic diazonium salt aromatic diazonium salt is widely used in synthesis. There are two kinds of reactions, one is the substitution reaction of releasing nitrogen, and the other is the reduction reaction and coupling reaction without releasing nitrogen.

Aromatic amines can be converted to phenols by first forming the arene diazonium salt in aqueous sulfuric acid and then heating the solution. In this manner, 2-bromo-4-methylaniline is converted to 2-bromo-4-methylphenol.

2-Bromo-4-methylaniline   2-Bromo-4-methylphenol

2-溴-4-甲基苯胺   2-溴-4-甲基苯酚

The intermediate in the decomposition of an arenediazonium ion in water is an aryl cation, which then undergoes reaction with water to form the phenol.

Benzenediazonium ion　　　Aryl cation　　　　Phenol
苯基重氮根离子　　　　　苯基正离子　　　　　苯酚

Diazonium salt chemistry provides the principal synthetic method for the preparation of aryl fluorides through a process known as the **Schiemann reaction (希曼反应)**. In this procedure the aryl diazonium ion is isolated as its fluoroborate salt, which then yields the desired aryl fluoride on being heated.

$$Ar\text{—}\overset{+}{N}\equiv N: \bar{B}F_4 \xrightarrow{heat} ArF \;+\; BF_3 \;+\; :N\equiv N:$$

Aryl diazonium fluoroborate　　　Arylfluoride　　　Borontrifluoride　　　Nitrogen

A standard way to form the aryl diazonium fluoroborate salt is to add fluoroboric acid (HBF_4) or a fluoroborate salt to the diazotization medium.

m-Aminophenyl ethyl ketone

1.NaNO₂,H₂O,HCl
2.HBF₄
3.heat

Ethyl m-fluorophenyl ketone(68%)

Although it is possible to prepare aryl chlorides and aryl bromides by electrophilic aromatic substitution, it is often necessary to prepare these compounds from an aromatic amine. The amine is converted to the corresponding diazonium salt and then treated with copper(I) chloride or copper(I) bromide as appropriate.

$$Ar\text{—}\overset{+}{N}\equiv N: \xrightarrow{CuX} ArX \;+\; :N\equiv N:$$

Aryl diazonium
ion

Aryl chloride
or bromide

Nitrogen

m-Nitroaniline

①NaNO₂, HCl, H₂O, 0-5℃
②CuCl,heat

m-Chloronitrobenzene
(68%-71%)

o-Chloroaniline

①NaNO₂, HCl, H₂O, 0-10℃
②CuBr, heat

o-Bromochlorobenzene
(89%-95%)

Reactions that use copper(I) salts to replace nitrogen in diazonium salts are called Sand-meyer reactions. The Sandmeyer reaction using copper(I) cyanide is a good method for the preparation of **aromatic nitriles** (芳香腈):

$$
\text{Ar—N}\equiv\text{N:} \xrightarrow{\text{CuCN}} \text{ArCN} + :\text{N}\equiv\text{N:}
$$

Aryl diazonium ion   Aryl nitrile   Nitrogen

$$
\text{o-Toluidine} \xrightarrow[\text{2.CuCN,heat}]{\text{1.NaNO}_2,\text{HCl,H}_2\text{O,0°C}} \text{o-Methylbenzonitile (64%~70%)}
$$

Because cyano groups may be hydrolyzed to carboxylic acids, the Sandmeyer preparation of aryl nitriles is a key step in the conversion of arylamines to substituted benzoic acids. In the example just cited, the o- methylbenzonitrile that was formed was subsequently subjected to acid-catalyzed hydrolysis to give o-methylbenzoic acid in 80%~89% yield. It is possible to replace amino groups on an aromatic ring by hydrogen by reducing a diazonium salt with hypophosphorous acid (H_3PO_2) or with ethanol.

$$
\text{Ar—N}\equiv\text{N:} \xrightarrow[\text{CH}_3\text{CH}_2\text{OH}]{\text{H}_3\text{PO}_2 \text{ or}} \text{ArH} + :\text{N}\equiv\text{N:}
$$

Aryl diazonium ion   Arene   Nitrogen

$$
\text{4-Isopropyl-2-nitroaniline} \xrightarrow[\text{CH}_3\text{CH}_2\text{OH}]{\text{NaNO}_2,\text{HCl,H}_2\text{O}} \text{m-Isoropylnitrobenzene (59%)}
$$

12.5.6 Azo Coupling

A reaction of aryl diazonium salts that does not involve loss of nitrogen takes place when they react with **phenols** (酚) and **arylamines** (芳香胺). Aryl **diazonium ions** (重氮离子) are relatively weak electrophiles but have sufficient reactivity to attack strongly activated aromatic rings. The reaction is known as **azo**(偶氮的) coupling; two aryl groups are joined together by an azo (-N=N-) function.

ERG is a powerful electron-releasing group such as-OH or-NR₂)   Aryl diazonium ion   Intermediate in electrophilic aromatic substitution   Azo compound

N,N-Dimethylaniline   Diazonium ion from   Methyl red (62%-66%)
o-aminobenzoic acid

The product of this reaction, as with many azo couplings, is highly colored. It is called methyl red and was a familiar acid–base indicator before the days of pH meters. It is red in solutions of pH 4 and below, yellow above pH 6.

Soon after azo coupling was discovered in the mid-nineteenth century, the reaction received major attention as a method for preparing dyes. Azo dyes fi rst became commercially available in the 1870s and remain widely used, with more than 50% of the synthetic dye market. **Chrysoidine** (碱性橙、橘红), an azo dye for silk, cotton, and wool, first came on the market in 1876 and remains in use today.

Chrysoidine

12.5.7 Hofmann Elimination

Amines can be converted to alkenes by elimination reactions, much like alcohols and alkyl halides undergo elimination to give alkenes. An amine cannot undergo elimination directly, however, because the leaving group would be an amide ion (or), which is a very strong base and a poor leaving group. When a quaternary ammonium halide is treated with moist silver(I) oxide (a slurry of Ag_2O in H_2O), silver halide precipitates, leaving a solution of a quaternary ammonium hydroxide.

(Cyclohexylmethyl)trimethyl-   Silver   (Cyclohexylmethyl) trimethyl-
ammoniumiodide   oxide   ammoniumhydroxide

In the mid-nineteenth century, Augustus Hofmann (奥格斯特·霍夫曼) discovered that when a quaternary ammonium hydroxide was heated, it decomposed to an alkene, a tertiaryamine, and water. Thermal decomposition of a **quaternary ammonium hydroxide** (季铵碱) to an alkene is known as the **Hofmann elimination** (霍夫曼消除).

(Cyclohexylmethyl)trimethyl-   Methylenecyclohexane   Trimethylamine
ammonium hydroxide

The Hofmann elimination has most of the characteristics of an E2 reaction. First, Hofmann eliminations are concerted, meaning that bond-breaking and bond-forming steps occur simultaneously or nearly so. Second, Hofmann eliminations are stereoselective for anti elimination, meaning that -H and the leaving group must be anti to each other. The following mechanism illustrates the concerted nature of bond forming and bond breaking, as well as the anti arrangement of -H and the **trialkylamino (三烷基氨基)** leaving group.

A novel aspect of the Hofmann elimination is its **regioselectivity (区域选择性)**. Elimination in alkyltrimethylammonium hydroxides proceeds in the direction that gives the less substituted alkene.

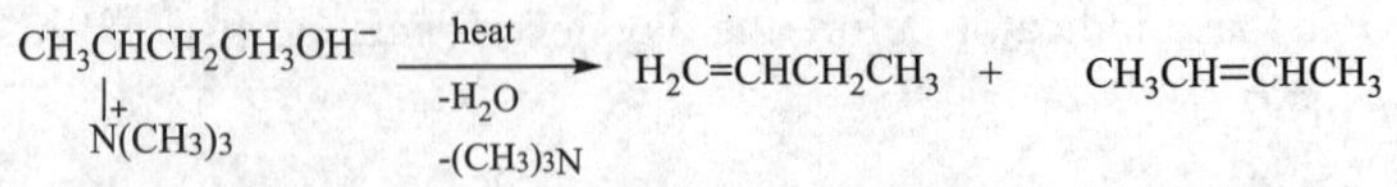
CH3CHCH2CH3OH−
N(CH3)3
heat
-H2O
-(CH3)3N
H2C=CHCH2CH3 +
CH3CH=CHCH3

sec-Butyltrimethylammonium hydroxide — But-1-ene(95%) — But-2-ene(5%) (cis and trans)

The least sterically hindered hydrogen is removed by the base in Hofmann elimination reactions. Methyl groups are deprotonated in preference to methylene groups, and methylene groups are deprotonated in preference to **methines (次甲基).** The regioselectivity of Hofmann elimination is opposite to that predicted by the **Zaitsev rule (查依采夫规则)**. Elimination reactions of alkyltrimethylammonium hydroxides are said to obey the Hofmann rule; they yield the less substituted alkene.

We can understand the regioselectivity of the Hofmann elimination by comparing steric effects in the E2 transition states for formation of but-1-ene and trans- but-2-ene from sec-butyltrimethyl ammonium hydroxide.

In terms of its size, $(CH_3)_3N$ (trimethylamino) is comparable to $(CH_3)_3C$ (tertbutyl). As Figure 12.2 illustrates, the E2 transition state requires an anti relationship between the proton that is removed and the trimethylamino group. No serious van der Waals repulsions are evident in the transition state geometry for formation of 1-butene. However, the conformation for the formation of trans-but-2-ene leads trimethylamino group and a methyl group gauche to each other. Thus, the activation energy for formation of trans- but-2-ene exceeds that of but-1-ene.

(a) Less crowded: Conformation leading to But-1-ene by anti elimination:

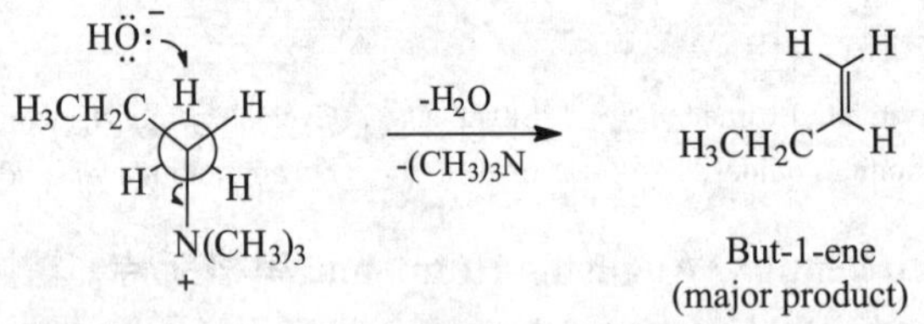
HÖ:−
H3CH2C
H
H
H
H
N(CH3)3
+
-H2O
-(CH3)3N
H
H
H3CH2C
H
But-1-ene
(major product)

(b) More crowded: Conformation leading to by But-2-ene anti elimination:

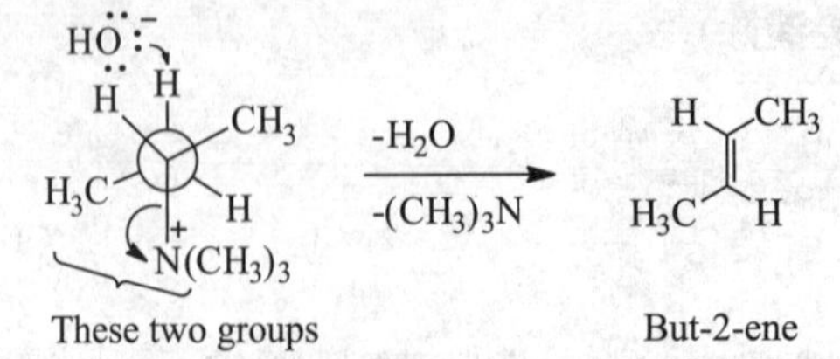
HÖ:−
H
H
CH3
H3C
H
+
N(CH3)3
These two groups
-H2O
-(CH3)3N
H
CH3
H3C
H
But-2-ene

Figure 12.2 Regioselectivity of the Hofmann elimination

图12-2 Hofmann消除反应的立体选择性

With a regioselectivity opposite to that of the Zaitsev rule, the Hofmann elimination is sometimes used in synthesis to prepare alkenes not accessible by dehydrohalogenation of alkyl halides.

12.5.8　Cope Elimination

Amines are notoriously easy to oxidize, and oxidation is often a side reaction in amine syntheses. Amines also oxidize during storage in contact with the air. Preventing air oxidation is one of the reasons for converting amines to their salts for storage or use as medicines. Treatment of a tertiary amine with hydrogen peroxide results in oxidation of the amine to an **amine oxide** (氧化胺).

A 3° amine   An amine oxide

When an amine oxide with at least one β-hydrogen is heated, it undergoes thermal decomposition to form an alkene and an N,N-dialkylhydroxylamine. Thermal decomposition of an amine oxide to an alkene is known as a Cope elimination after its discoverer **Arthur C. Cope** (亚瑟·科普).

Methylenecyclohexane　　　*N,N*-Dimethyl-
　　　　　　　　　　　　hydroxylamine

The Cope elimination is a one-step, concerted internal elimination using an amine oxide as both the base and the leaving group. Syn stereochemistry is required for the Cope elimination.

Cope elimination occurs under milder conditions than Hofmann elimination. It is particularly useful when a sensitive or reactive alkene must be synthesized by the elimination of an amine. Because the Cope elimination involves a cyclic transition state, it occurs with synstereochemistry and the regiochemistry of it is similar as Hoffmann eliminations.

minor

major

12.6 Applications of Amines in the Pharmacy Study

PPT

Nitrogen-containing natural products obtained from plants are called **alkaloids(生物碱).** The number of known alkaloids exceeds 5000. They are of special interest because most of them are characterized by a high level of biological activity.

(1) Atropine (阿托品) Atropine, molecular formula $C_{17}H_{23}NO_3$, is a tropane alkaloid,first isolated from the 'deadly nightshade' (Atropa belladonna), and also found in many other plants of the **Solanaceae(茄科植物)**. Atropine is a **racemic(外消旋的)**mixture of D-hyoscyamine and L-hyoscyamine. However, most of the pharmacological properties of atropine are due to its L-isomer, and due to its binding to **muscarinic (毒蕈碱)** acetylcholine receptors. Atropine is a competitive antagonist of the muscarinic acetylcholine receptors. The main medicinal use of atropine is as an opthalmic drug. Usually a salt of atropine, e.g. atropine sulphate, is used in pharmaceutical preparations. Atropine is used as an acycloplegic to paralyse accommodation temporarily, and as a **mydriatic (瞳孔放大剂)** to dilate the pupils. It is contraindicated in patients predisposed to narrow angle glaucoma. Injections of atropine are used in the treatment of **bradycardia(心动过缓)**(an extremely low heart-rate), asystole and pulseless electrical activity (PEA) in cardiac arrest. It is also used as an **antidote (解毒剂)** for poisoning by organophosphate insecticides and nerve gases.

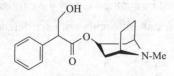

Atropine

(2) Cocaine (可卡因) Cocaine, molecular weight $C_{17}H_{21}NO_4$, is a white crystalline tropane alkaloid found mainly in coca plant. It is a potent central nervous system (CNS) stimulant and appetite suppressant. For its **euphoretic (快感的)** effect, cocaine is often used recreationally, and it is one of the most common drugs of abuse and addiction. Cocaine is also used as a topical **anaesthetic (麻醉**

308

剂) in eye, throat and nose surgery. Possession, cultivation and distribution of cocaine is illegal for non-medicinal and non-government sanctioned purposes virtually all over the world. The side-effects of cocaine include **twitching** (痉挛), paranoia and impotence, which usually increase with frequent usage. With excessive dosage it produces hallucinations, paranoid delusions, tachycardia, itching and formication. Cocaine overdose leads to tachyarrhythmias and elevated blood pressure, and can be fatal.

Cocaine

(3) Quinine (奎宁)　Quinine, molecular formula $C_{20}H_{24}N_2O_2$, is a white crystalline quinoline alkaloid, isolated from Cinchona bark (Cinchona succirubra), and is well known as an antimalarial drug. Despite being a wonder drug against malaria, quinine in therapeutic doses can cause various side-effects, e.g. **nausea**(恶心,呕吐), vomiting and cinchonism, and in some patients **pulmonary oedema**(肺水肿). It may also cause paralysis if accidentally injected into a nerve. An overdose of quinine may have fatal consequences. Non-medicinal uses of quinine include its uses as a flavouring agent in tonic water and bitter lemon.

(4) Quinidine (奎宁丁)　Quinidine, molecular formula $C_{20}H_{24}N_2O_2$, is a stereoisomer of quinine found in Cinchona bark. It is used as a Class 1 anti-arrhythmic agent. Intravenous injection of quinidine is also used in the treatment of P. falciparum malaria. Among the adverse effects, quinidine induces thrombocytopenia (low platelet counts) and may lead to thrombocytic purpurea.

Quinnine

Quinidine

重 点 小 结

一、胺的结构和分类

胺是氨上一个或多个氢被烷基和（或）芳基取代所形成的衍生物。伯胺是一个氨氢被烷基或芳基形式的碳取代。仲胺是两个氨氢被烷基和（或）芳基形式的碳取代。叔胺是氨上三个氢全部被烷基和（或）芳基形式的碳所取代。季铵离子是四个烷基和（或）芳基与氮结合产生正离子。脂肪胺结构中仅只有烷基与氮结合，而在芳香族胺结构中至少有一个芳香环与氮原子结合。杂环胺的氮原子是环的一部分，杂环芳胺的氮原子是芳香杂环的一部分。生物碱是源于植物的碱性含氮化合物，而其中许多在人体内具有生理活性。

二、胺的命名

简单的胺可以根据烃基的名称命名,即在烃基的名称后加上"胺"字。若氮原子上所连烃基 相同,用"二"或"三"表明烃基的数目:若氮原子上所连烃基不同,则按基团的次序规则由小到 大写出其名称,"基"字一般可省略脂肪胺英文名以-amine结尾,芳香胺通常视为苯胺的衍生 物来命名以-aniline结尾对于仲胺和叔胺,还可以选择最长的碳链作为母体,称为"某胺",氮上的其他烃基作为取 代基,以"N-某基"的形式写。

三、胺和季铵盐的手性

三种不同基团与氮结合的仲胺或叔胺是手性的,但它们通常不能被拆分,因为在室温下经历一个"金字塔反转"的过程,转换成其对映异构体。季铵盐不能进行金字塔形反转,因此它们可以被拆分。

四、胺的物理性质

胺是极性化合物,一级或二级胺能形成分子间氢键。 由于一级和二级胺可以与水分子通过氢键相互作用,因此比类似的碳氢化合物更易溶于水。因为H-N氢键比H-O氢键弱,因此,胺的沸点比类似的醇低。

芳香胺碱性没有脂肪胺强,因为氮原子上孤电子对与芳香环形成p-π共轭,降低了N原子上的电子云密度,使得其共轭酸电离质子变得容易。

五、胺的反应

1. **胺的碱性**　胺是弱碱,所以胺的水溶液是碱性的。脂肪胺共轭酸的pK_a值在10~11范围。烷基使胺碱性增强, 因为烷基对N原子的供电子作用使共轭酸电离质子变得困难。芳香胺的碱性比脂肪胺低得多, pK_a值它们的共轭酸在4~5范围内。

2. **与亚硝酸的反应**　叔芳胺可与亚硝酸进行亲电芳香取代。仲胺与亚硝酸反应生成N-亚硝基胺。伯胺与亚硝酸反应生成失去N_2的重氮离子中间体, 并能生成各种取代和消除产物。

3. **Hofmann 消除反应**　霍夫曼消除反应 (Hofmann elimination),指的是季铵碱与碘化钾、氢氧化银反应,从含氢较多的β-碳原子上消除氢,得到的主要产物是双键碳上含取代基比较少的烯烃,这一消除方式与卤代烃的消除方式 (Zaitsev's规则) 相反。

4. **Cope消除反应**　用过氧化氢处理叔胺得到一种胺氧化物,当加热时,它会产生烯烃和N,N-二烷基羟基胺。Cope消除是同面立体选择性的,如果能产生共轭双键,共轭产物占优势。

题库

医药大学堂
WWW.YIYAODXT.COM

Problems
目 标 检 测

1. Which one of the following compounds demonstrated the strongest basicity?
下列化合物中碱性最强的是 (　　　)。

A. 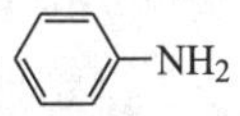 —NH₂

B.

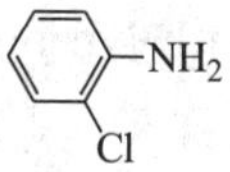

C. 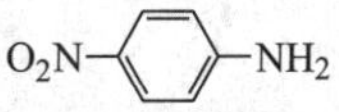 O₂N——NH₂

D. CH₃CH₂NH₂

2. Which one of the following compounds demonstrated the weakest basicity?

下列化合物碱性最弱的是（　　）。

A. 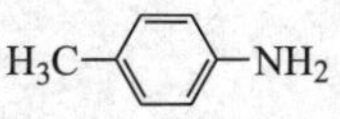 H₃C——NH₂　　B. 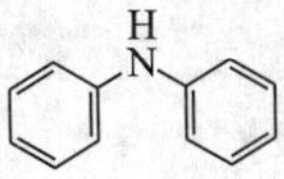 　　C. 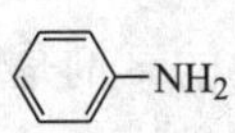 —NH₂　　D. CH₃CH₂NH₂

3. Which of the following four nitrogen-containing compounds can undergo Cope Elimination reaction?

下列四种含氮化合物中，能够发生Cope消除反应的是（　　）。

A.

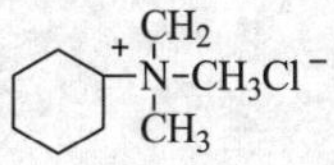

B.

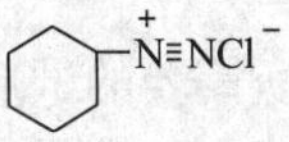

C.

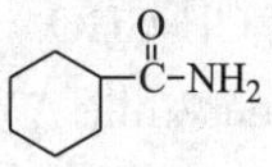

D.

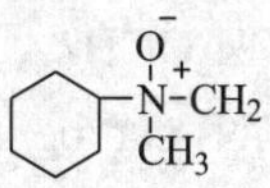

4. Which of the following compounds accord with Hofmann elimination rule in elimination?

下列物质发生消除反应，符合霍夫曼消除规则的是（　　）。

A.

B.

C.

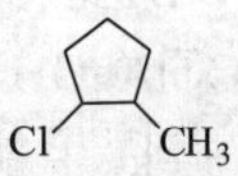

D.

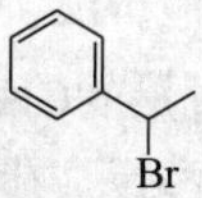

5. Among the following compounds, which belongs to quaternary ammonium?

下列化合物中属于季铵盐类的化合物是（　　）。

A.

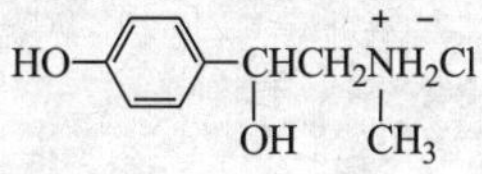

B.

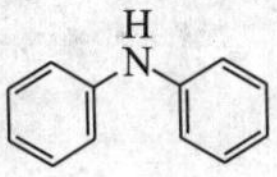

C.

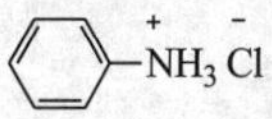

D.

6. The correct name for compound 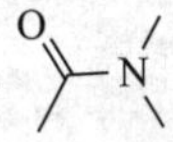 is (　　).

化合物 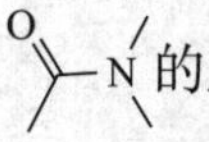 的正确名称是（　　）。

A. 二甲基乙酰胺　　　　　　　　B. N-甲基乙酰胺

C. N, N-二甲基乙酰胺　　　　　　D. 乙酰基二甲胺

7. Among the following reactions, which one can give rise to carbonium ion intermediate?

在下列反应过程中,生成碳正离子中间体的反应是（　　）。

A. Diels-Alder 反应　　　　　　　B. 卤代烃的双分子亲核取代反应

C. 芳环上的亲电取代反应　　　　D. 烯烃的催化加氢

8. Which compound can react with Nitrous acid to afford N-nitrosamines that are difficult to dissolve in water as yellow oil?

能与亚硝酸作用生成难溶于水的黄色油状物N-亚硝基胺化合物的是 (　　)。

 A. 二甲基卞基胺   B. N,N-二甲基甲酰胺

 C. 乙基胺   D. 六氢吡啶

9. To prepare ⬡—Br from ⬡—$\overset{+}{N}_2Br^-$, which reagent can be used ?

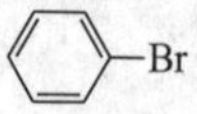

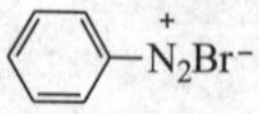

由 ⬡—$\overset{+}{N}_2Br^-$ 制备 ⬡—Br 所需要的试剂是 (　　)。

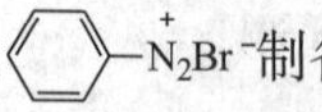

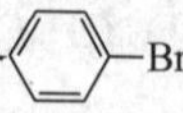

 A. CH_3CH_2Br B. HBr   C. CuBr   D. Br_2

10. To synthesize ⬡(OCH₃, NHCOCH₃) from ⬡(OCH₃, NH₂), which reagent can be used ?

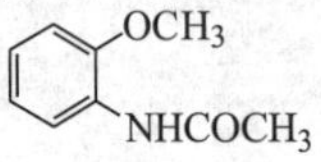

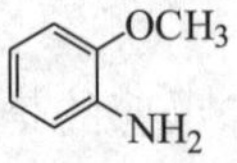

由 ⬡(OCH₃, NH₂) 合成 ⬡(OCH₃, NHCOCH₃) 所需要的试剂是 (　　)。

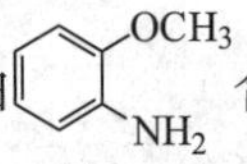

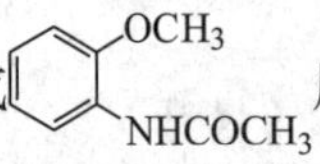

 A. CH_3COOH B. $CH_3CH_2OCH_3$ C. $(CH_3CO)_2O$ D. CH_3CH_2CHO

11. The basic substance A ($C_5H_{11}N$) can be deoxidized to afford formaldehyde. While A can change to B ($C_5H_{13}N$) after Catalytic hydrogenation. The compound B can also be obtained by the treatment of hexanamide and bromine in NaOH aqueous solution. A reacts with excess CH_3I and converts into salt C ($C_8H_{18}IN$). C react with AgOH to afford diene D (C_5H_8) after themolysis. D reacts with dimethyl but-2-ynedioate to form E ($C_{11}H_{14}O_4$). E is dehydrogenated by Pd to obtain dimethyl 3-methylphthalate. Write out the possible structures of A ~ E.

碱性物质A($C_5H_{11}N$)，臭氧化可生成甲醛，催化氢化得到化合物B($C_5H_{13}N$)，B也能由己酰胺与溴在NaOH水溶液中处理而得到。A与过量的CH_3I反应，转化为盐C($C_8H_{18}IN$)，C同AgOH进行热解，得到二烯D(C_5H_8)，D与丁炔二酸二甲酯反应生成E($C_{11}H_{14}O_4$)，E通过Pd脱氢得到3-甲基邻苯二甲酸二甲酯。写出A~E可能的结构式。

Discussion Topic

The difference between Hoffman elimination and Zaitsev elimination: the different regioselectivity and the reason behind it.

（左振宇）

第十三章　糖类化合物
Chapter 13　Carbohydrates

 学习目标

　　1. 掌握　单糖的结构和特殊的化学性质,低聚糖和多糖的组成及连接方式和性质结构、分类、命名及主要化学性质。

　　2. 熟悉　糖类化合物的性质和有关立体化学知识,糖的结构和立体异构问题,还原糖和非还原糖的含义和结构特点以及糖的变旋现象。

　　3. 了解　糖的定义、分类以及淀粉和纤维素的结构特点。

Carbohydrates (糖类) occur in every living organism. The sugar and **starch** (淀粉) in food, and the cellulose in wood, paper, and cotton are nearly pure carbohydrates. Modified carbohydrates form part of the coating around living cells, other carbohydrates are part of the nucleic acids that carry our **genetic information** (遗传信息), and still others are used as medicines.

The word carbohydrate derives historically from the fact that glucose, the first simple carbohydrate to be obtained pure, has the molecular formula $C_6H_{12}O_6$ and was originally thought to be a "hydrate of carbon, $C_6(H_2O)_6$". This view was soon abandoned, but the name persisted. Today, the term carbohydrate is used to refer loosely to the broad class of polyhydroxylated aldehydes and ketones commonly called sugars. Glucose, also known as dextrose in medical work, is the most familiar example.

Carbohydrates are synthesized by green plants during photosynthesis, a complex process in which sunlight provides the energy to convert CO_2 and H_2O into glucose plus oxygen. Many molecules of glucose are then chemically linked for storage by the plant in the form of either **cellulose** (纤维素) or **starch** (淀粉). It has been estimated that more than 50% of the dry weight of the earth biomass-all plants and animals-consists of glucose polymers. When eaten and metabolized, carbohydrates then provide animals with a source of readily available energy. Thus, carbohydrates act as the chemical intermediaries by which solar energy is stored and used to support life.

Carbohydrate chemistry is one of the most interesting areas of organic chemistry. Many chemists are employed by companies that use carbohydrates to make foods, building materials, and other consumer products. All biologists must understand carbohydrates, which play pivotal roles throughout the plant and animal kingdoms. At first glance, the structures and reactions of carbohydrates may seem complicated. We will learn how these structures and reactions are consistent and predictable, however, and we can study carbohydrates as easily as we study the simplest organic compounds.

It is common to have carbohydrates in Chinese herbal medicines, and some are even the main active ingredients. Some nutritional medicines, such as **ginseng** (人参), **ganoderma** (灵芝), **astragalus** (黄芪) and **shiitake mushrooms** (香菇), all contain a lot of saccharides, which are also the effective ingredients among them.

According to the composition or hydrolysis of saccharides, they are generally divided into three categories: monosaccharides, oligosaccharides and polysaccharides.

Monosaccharides (单糖) are the simplest saccharides. They are the basic structural units that make up oligosaccharides and polysaccharides. They cannot hydrolyze to smaller saccharides molecules. The above-mentioned polyhydroxy aldehydes and polyhydroxy ketones are monosaccharides.

Oligosaccharides (低聚糖) usually refer to carbohydrates formed by 2~10 monosaccharides, or releasing 2~10 monosaccharides when hydrolyzed, such as **maltose** (麦芽糖), **lactose** (乳糖) and **sucrose** (蔗糖).

Polysaccharides (多聚糖) refer to carbohydrates containing more than 10 or even tens or hundreds of monosaccharide units, often called high glycan, such as starch, cellulose and **chitosan** (甲壳素). In practical work, the molecular weight of oligosaccharide and high glycan is only relative, and there is no strict definition.

13.1 Monosaccharides

PPT

Monosaccharides are the basic structural units of polysaccharides. In order to study and understand carbohydrate compounds, we must first understand the structure and chemical properties of monosaccharides. Among all monosaccharides, **hexose** (己糖) is the most important. Next, we take aldohexose and ketohexose as examples to illustrate the chemical structure and chemical properties of monosaccharides.

13.1.1 Classification of Monosaccharides

Monosaccharides have the general formula $C_nH_{2n}O_n$ with one of the carbons being the carbonyl group of either an aldehyde or a ketone. The number of carbon atoms in the sugar generally ranges from three to seven, designated by the terms *triose* (three carbons), *tetrose* (four carbons), *pentose* (five carbons), *hexose* (six carbons), and *heptose* (seven carbons).

Terms describing sugars often reflect these first two criteria. For example, **glucose** (葡萄糖) has an aldehyde and contains six carbon atoms, so it is an **aldohexose** (己醛糖). **fructose** (果糖) also contains six carbon atoms, but it is a ketone, so it is called a **ketohexose** (己酮糖). Most ketoses have the ketone on C_2, the second carbon atom of the chain. The most common naturally occurring sugars are aldohexoses and **aldopentoses** (戊醛糖).

The aldehyde group of aldose is at C_1, while the carbonyl group of ketose is always at C_2. For example, the important monosaccharides found in nature-pentose and hexose, their chemical structural formulas are as follows:

CHO   CH₂OH   CHO   CH₂OH
| | | |
CHOH   C=O   CHOH   C=O
| | | |
CHOH   CHOH   CHOH   CHOH
| | | |
CHOH   CHOH   CHOH   CHOH
| | | |
CH₂OH   CH₂OH   CHOH   CHOH
                              | |
                              CH₂OH   CH₂OH

aldopentose   ketopentose   aldohexose   ketohexose

It can be seen from the above structures that monosaccharide molecules contain multiple chiral carbons, so stereoisomerism and optical rotation exist commonly.

There are only two trioses: the aldotriose **glyceraldehyde** (甘油醛) and the ketotriose dihydroxyacetone. The designations aldo- and keto- are often omitted, and these molecules are referred to simply as trioses, tetroses, and the like.

CHO   CH₂OH
| |
CHOH   C=O
| |
CH₂OH   CH₂OH

glyceraldehyde   dihydroxyacetone

13.1.2   Physical Properties of Monosaccharides

Monosaccharides have the following physical properties: They are all sweet tasting, but their relative sweetness varies a great deal. They are polar compounds with high melting points. The presence of so many polar functional groups capable of hydrogen bonding makes them water soluble. Unlike most other organic compounds, monosaccharides are so polar that they are insoluble in organic solvents like diethyl ether. Monosaccharides are colorless crystalline solids, although they often crystallize with difficulty.

When the sugar solution is concentrated, it is easy to obtain a thick syrup and it is not easy to crystallize, which indicates that the sugar has a high tendency for supersaturation and it is difficult to precipitate crystals. Solving the problem of sugar crystallization is a difficult problem. Sugar is usually crystallized by physical or chemical methods. The physical method is to change the solvent or freeze, rub the container wall or introduce seeds, etc., while leaving it for a few days or more, waiting for the crystals to grow. The chemical method is to convert the sugar into a suitable derivative, such as acylation of a hydroxyl group, or preparation into an acetal (ketone), etc., to change the molecular structure and increase the molecular weight in order to facilitate crystallization.

13.1.3   Open Chain Structure and Relative Configuration of Hexose

In nature, glucose is the most important representative of aldohexose, and fructose is an important representative of ketohexose. We will use glucose and fructose as examples to discuss the chemical structure of monosaccharides.

Glucose is the earliest sugar found in nature and has the most comprehensive research data. The research on the structure of glucose has laid the foundation for the research of sugar chemistry. After elemental analysis and molecular weight determination, the molecular formula of glucose was found to be $C_6H_{12}O_6$; through a multi-step chemical reaction, it was confirmed that glucose has a pentahydroxyaldehyde structure.

$$HOH_2C-\underset{\underset{OH}{|}}{CH}-\underset{\underset{OH}{|}}{CH}-\underset{\underset{OH}{|}}{CH}-\underset{\underset{OH}{|}}{CH}-CHO$$

glucose

The structural formula of glucose contains 4 chiral carbons, and theoretically there are 16 kinds of optical isomers. However, dextrose, which widely exists in nature and can be used by our bodies, is just one of them. Its structure is as follows.

$$\begin{array}{c} CHO \\ H-OH \\ HO-H \\ H-OH \\ H-OH \\ CH_2OH \end{array}$$

(+)-glucose

Fructose (果糖) is a kind of ketohexose, which contains three chiral carbon in its molecular structure, and there are eight optical isomers in theory, while the natural fructose is only (-)-fructose. Fructose has the following structure.

$$\begin{array}{c} CH_2OH \\ C=O \\ HO-H \\ H-OH \\ H-OH \\ CH_2OH \end{array}$$

(-)-fructose

Each monosaccharide has a pair of enantiomers and has the same name (usually the common name, which is named after each source). For example, the enantiomers of fructose and glucose are as follows.

(+)-glucose	(−)-glucose	(+)-fructose	(−)-fructose
CHO	CHO	CH₂OH	CH₂OH

In order to distinguish the two enantiomers of monosaccharide, it is necessary to add some symbol before the name of monosaccharide to determine its configuration isomerism. Because monosaccharide molecules contain more chiral carbon, it is inconvenient to express its configuration by the absolute configuration representation, i.e. *R/S* configuration representation, so the relative configuration representation, i.e. D/L configuration representation, is usually used to express its configuration. The details are as follows: Taking glyceraldehyde as the standard, the configuration of the chiral carbon with the largest number in the Fisher projection formula of monosaccharide is compared with that of the chiral carbon in the glyceraldehyde. If the configuration is the same as that of D-glyceraldehyde, it is specified as D-type; if the configuration is the same as that of L-type, it is specified as L-type.

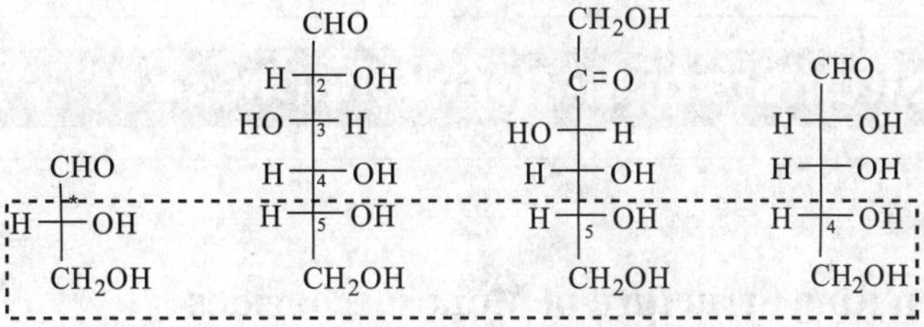

D-(+)-glyceraldehyde　　D-(+)-glucose　　D-(-)-fructose　　D-(-)-ribose

In the above structural formula, the C_5 of (+)-glucose and (-)-fructose and C_4 of (-)-ribose are similar to the chiral carbon C* of D-(+)-glyceraldehyde, respectively, and they are all D-series monosaccharides, while their enantiomers are all L-series.

Other monosaccharide configurations are also determined by this method. The structure of the eight D-monosaccharides of aldohexose is as follows.

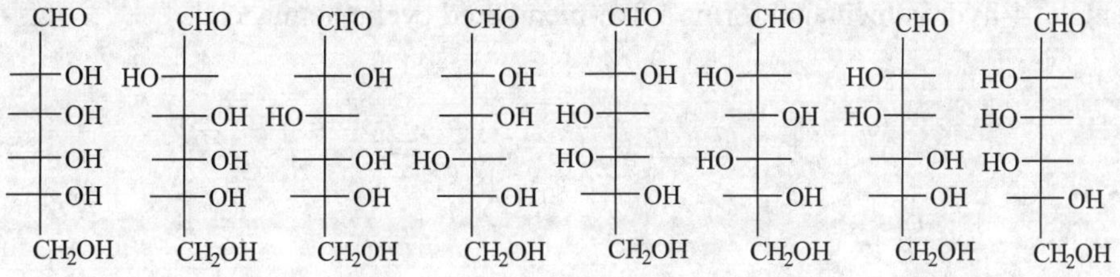

allose　　altrose　　glucose　　mannose　　gulose　　idose　　galactose　　talose

The L-Series of aldohexose and the above structures are enantiomers.

Of the 16 optical isomers of aldohexose, only (+)-glucose, (+)-**mannose** (甘露糖), and (+)-**galactose** (半乳糖) are naturally occurring, and the rest are artificially synthesized.

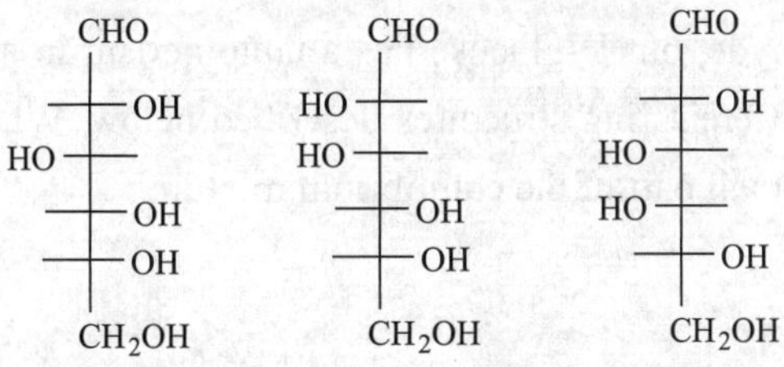

(+)-glucose　　(+)-gulose　　(+) galactose

So far, the natural monosaccharides found are all D-series, such as (+)-glucose, (+)-mannose, (+)-galactose, (-)-fructose and (-)-ribose. The chiral carbons that determine the configuration in the D-series monosaccharides are all *R* configurations. Therefore, the chiral carbon with the largest number in the monosaccharides is generally referred to as the chiral carbon that determines the configuration.

The configuration of glucose and fructose was determined by Fischer, a German chemist. Fischer also chemically synthesized several monosaccharides that are not found in nature. Due to his great contribution to the study of sugars, Fischer became the first organic chemist in history to win the Nobel Prize in chemistry.

13.2 Cyclic Structure of Monosaccharides

PPT

13.2.1   Oxygen Ring Structure of Monosaccharides

In Chapter 9, we saw that an aldehyde reacts with one molecule of an alcohol to give a **hemiacetal** (半缩醛), and with a second molecule of the alcohol to give an **acetal** (缩醛). The hemiacetal is not as stable as the acetal, and most hemiacetals decompose spontaneously to the aldehyde and the alcohol. Therefore, hemiacetals are rarely isolated. Cyclic hemiacetals however are particularly stable if they result in five or six-membered rings. In fact, five- and six-membered cyclic hemiacetals are often more stable than their open-chain forms.

For example, 4-hydroxypentanal forms a five-membered cyclic hemiacetal.

4-ydroxypentanal cyclic hemiacetal

As mentioned earlier, the glucose molecule has an open-chain structure of pentahydroxyaldehyde, but D-glucose has a **tautomerism** (互变异构) in an aqueous solution. In fact, it is mainly a tautomer of two different ring structures described below. When equilibrium is reached, the open-chain structure is only a small part of the equilibrium mixture as mentioned earlier, the glucose molecule has an open-chain structure of pentahydroxyaldehyde, but D-glucose has a tautomerism in an aqueous solution. In fact, it is mainly a tautomer of two different ring structures described below. When equilibrium is reached, the open-chain structure is only a small part of the equilibrium mixture.

β-D-(+)-glucose　　　　D-(+)-glucose　　　α-D-(+)-glucose
(cyclic hemiacetal)　　　　(chain)　　　(cyclic hemiacetal)

Because D-glucose has both a hydroxyl group and an aldehyde group, an intramolecular addition reaction can occur to generate a cyclic hemiacetal structure. Studies have shown that glucose reacts with a C_1 hydroxy group with a C_5 hydroxyl group to form a six-membered oxygen ring structure. When the C_5 hydroxyl group and the aldehyde group are added in the open-chain D-glucose molecule, C_1 becomes a chiral carbon atom. There are two configurations. One is the hydroxyl group of C_1 (that is, hemiacetal hydroxyl group, also known as a glycoside hydroxyl group). On the same side as the C_5 hydroxyl group that determines the configuration, it is called the α-former; another C_1 hydroxyl group and C_5 hydroxyl group occupy the two sides of the carbon chain, and it is called the β-former.

The α- and β-forms of glucose differ only in the configuration of C_1. They are C_1 epimers and are a pair of diastereomers. They are also known in the literature as end-group-isomerism or anomer.

Because **anomers** (异头物) are diastereomers, they generally have different properties. For example, α-D-glucopyranose has a melting point of 146℃ and a specific rotation of +112.2° while β-D-glucopyranose has a melting point of 150℃ and a specific rotation of +18.7°. When glucose is crystallized from water at room temperature, pure crystalline α-D-glucopyranose results. If glucose is crystallized from water by letting the water evaporate at a temperature above 98℃, crystals of pure β-D-glucopyranose are formed.

equilibrium mixture of α and β
$[\alpha] = +52.6°$

pure α anomer
mp 146℃, $[\alpha] = + 112.2$

pure β anomer
mp 150℃, $[\alpha] = +18.7°$

An aqueous solution of D-glucose contains an equilibrium mixture of α-D-glucopyranose, β-D-glucopyranose, and the intermediate open-chain form. Crystallization below 98℃ gives the α anomer, and crystallization above 98℃ gives the β anomer.

In each of these cases, all the glucose in the solution crystallizes as the favored anomer. In the solution, the two anomers are in equilibrium through a small amount of the open-chain form, and this equilibrium continues to supply more of the anomer that is crystallizing out of solution. When one of the pure glucose anomers dissolves in water, an interesting change in the specific rotation is observed. When the anomer dissolves, its specific rotation gradually decreases from an initial value of +112.2° to +52.6°. When the pure anomer dissolves, its specific rotation gradually increases from +18.7° to the same value of +52.6°. This change ("mutation") in the specific rotation is called **mutarotation** (变旋现象). Mutarotation occurs because the two anomers interconvert in solution. When either of the pure anomers dissolves in water, its rotation gradually changes to an intermediate rotation that results from equilibrium concentrations of the anomers. The specific rotation of glucose is usually listed as the value for the equilibrium mixture of anomers. The positive sign of this rotation is the source of the name **dextrose** (右旋糖), an old common name for glucose.

As a solid, glucose can be in the α- or β-formation. However, when dissolved in water, whether in the α- or β-structure, a part of the ring will be converted into an aldehyde structure, and the aldehyde structure and the hemiacetal can be converted into each other. When cyclic hemiacetals are formed by the aldehyde structure, since C_1 is converted into a chiral carbon from a planar structure, both an α-form and a β-form can be formed. When the α-, β- and aldehyde structures of glucose reach equilibrium, the α-former accounts for 36%, the β-former accounts for 64%, and the open-chain structure accounts for only 0.1%.

Like glucose, fructose has an oxygen ring structure. The difference is that the cyclic structure of fructose is **hemiketal** (半缩酮), and the free fructose is added by the C_6 hydroxyl and C_2 carbonyl group in the molecule. When fructose is in an aqueous solution, the oxygen ring and open chain structures are also in dynamic equilibrium:

α-D-(+)-fructose   D-(+)-fructose   β-D-(+)-fructose

(cyclic hemiketal)   (chain)   (cyclic hemiketal)

13.2.2　Haworth Projections of Carbohydrates

A common way of representing the cyclic structure of monosaccharides is the Haworth projection, named after the English chemist Sir Walter N. Haworth (1937 Nobel Prize in Chemistry). In a Haworth projection, a five- or six-membered cyclic hemiacetal is represented as a planar pentagon or hexagon, as the case may be, lying perpendicular to the plane of the paper. Groups bonded to the carbons of the ring then lie either above or below the plane of the ring. The new chiral center created in forming the cyclic structure is called an **anomeric carbon** (异头碳). Stereoisomers that differ in configuration only at the anomeric carbon are called anomers. The anomeric carbon of an aldose is carbon 1; that of D-fructose, the most common ketose, is carbon 2. Haworth projections are most commonly written with the anomeric carbon to the right and the hemiacetal oxygen to the back. In the terminology of carbohydrate chemistry, the designation β means that the -OH on the anomeric carbon of the cyclic hemiacetal is on the same side of the ring as the terminal -CH₂OH. Conversely, the designation α means that the -OH on the anomeric carbon of the cyclic hemiacetal is on the side of the ring opposite the terminal -CH₂OH.

D-glucose   β-D-glucopyranose   α-D-glucopyranose

A six-membered hemiacetal ring is indicated by the infix -pyran-, and a five membered hemiacetal ring is indicated by the infix -furan-. The terms furanose and pyranose are used because monosaccharide five- and six-membered rings correspond to the heterocyclic compounds furan and pyran, respectively.

furan　　　pyran

Because the α and β forms of glucose are six-membered cyclic hemiacetals, they are named α-D-**glucopyranose** (吡喃葡萄糖) and β-D-glucopyranose. These infixes are not always used in monosaccharide names, however. Thus, the glucopyranoses, for example, are often named simply α-D-glucose and β-D-glucose. You would do well to remember the configuration of groups on the Haworth projections of α-D-glucopyranose and β-D-glucopyranose as reference structures. Knowing how the open-chain configuration of any other aldohexose differs from that of D-glucose, you can then construct its Haworth projection by reference to the Haworth projection of D-glucose.

Aldopentoses (五碳醛糖) also form cyclic hemiacetals. The most prevalent forms of D-ribose and other pentoses in the biological world are furanoses. Following is Haworth projections for α-D-ribofuranose (α-D-ribose) and β-2-deoxy- D-ribofuranose (β-2-deoxy-D-ribose). The prefix 2-deoxy indicates the absence of oxygen at carbon 2. Units of D-ribose and 2-deoxy-D-ribose in nucleic acids and most other biological molecules are found almost exclusively in the β-configuration.

Other monosaccharides also form five-membered cyclic hemiacetals. Following are the five-membered cyclic hemiacetals of fructose. The β-D-fructofuranose form is found in the disaccharide sucrose.

α-D-fructofuranose　　　D-fructose　　　β-D-fructofuranose

In order to establish the connection between the open-chain structure of the monosaccharide and the platform structure, the following briefly describes how to rewrite the open-chain structure into the platform structure. Place the vertical open-chain structural model of the monosaccharide horizontally to the right, with the aldehyde group on the right; bend C_4 and C_1 backwards into a ring, with C_2-C_3 sides in front; rotate the C_4-C_5 bond to bring the hydroxyl group closer to the carbonyl group.

β-D-glucopyranose   α-D-glucopyranose

From the rewriting process of the monosaccharide, it can be seen that the group on the left in the Fisher projection will be on the platform ring; the group on the right will be under the ring; The hydroxyl groups that determine the configuration in the sugar are all on the right side, so the hydroxymethyl group is always upward after the ring formation, and the glycoside hydroxyl group of the β-former is also upward, but the glycoside hydroxyl group of the α-former is always downward. In the L-series monosaccharides the situation is exactly the opposite.

When writing a monosaccharide structure, the hydroxyl group can usually be represented by a short line, and the hydrogen atom can be omitted. When the C_1 configuration does not need to be emphasized or only the two isomer mixtures are indicated, the hydrogen atom and the hydroxyl group on C_1 can be written side by side, or C_1 and the hydroxyl group can be connected with a dashed line.

13.2.3 Conformation Representations and Anomeric Effect

Haworth project describes the ring structure of sugar as a plane, in fact, for pyranoses, however, the six-membered ring is more accurately represented as a chair conformation. There are two forms of chair conformation of pyranose: $^4C_1$ and $^1C_4$.

The $^4C_1$ structural formula means that C_4 is above the ring plane and C_1 is below the ring plane. Structural formula $^1C_4$ means that C_1 is above the ring plane and C_4 is below the ring plane. The chair conformation in which a monosaccharide exists is related to the substituent attached to each carbon atom. D-glucose adopts the $^4C_1$-type conformation:

β-D-glucopyranose   α-D-glucopyranose

In the $^4C_1$ conformation of D-glucose, all the larger groups of the β-isomer are in the flat bond position, and the steric hindrance is relatively small, which is a very stable dominant conformation; in the α-form except the glycoside hydroxyl, other large groups are also on the flat bond. Among D-aldohexoses, only D-glucose can maintain this most dominant conformation, and no other monosaccharide has this structural feature. But when D-glucose adopts the $^1C_4$ conformation, all the larger groups will occupy upright bond positions, which is an impossible conformation. So far, it is easy to understand why D-glucose exists in the largest amount in nature, and its β-isomers account for a larger proportion than α-isomers in an equilibrium aqueous solution.

Following are structural formulas for α-D-glucopyranose and β-D-glucopyranose drawn as chair conformations. Also shown is the open-chain or free aldehyde form with which the cyclic hemiacetal forms are in equilibrium in aqueous solution.

β-D-glucopyranose

rotate about
C-1 to C-2 bond

α-D-glucopyranose

Notice that each group, including the anomeric-OH, on the chair conformation of β-D-glucopyranose is equatorial. Notice also that the -OH group on the anomeric carbon is axial in α-D-glucopyranose. Because of the equatorial orientation of the -OH on its anomeric carbon, β-D-glucopyranose is more stable and predominates in aqueous solution.

At this point, you should compare the relative orientations of groups on the D-glucopyranose ring in the Haworth projection and the chair conformation. The orientations of groups on carbons 1 through 5 of β-D-glucopyranose, for example, are up, down, up, down, and up in both representations.

β-D-glucopyranose
(Haworth procetion)

β-D-glucopyranose
(chair conformation)

In addition to D-glucose, the stable conformation of D-pyranose is also the conformation of the largest group (hydroxymethyl) at the e-bond position, such as galactose, mannose, etc.

β-D-galactopyranose

β-D-glucopyranomannose

The three methods of representing the monosaccharide structure have been discussed above. Although the platform and conformation are closer to the true image of the molecule, when discussing some chemical properties of the monosaccharide, especially the aldehyde or ketocarbonyl properties, the Fischer projection is more convenient to write. It can be used arbitrarily according to the situation, but you must be familiar with the relationship between the three representations of the monosaccharide structure.

When the C_1 hydroxyl group becomes a methoxy group, an acyloxy group, or is substituted with a halogen atom, the conformation of this substituent at the a-bond is often the optimal conformation, that is, the α-isomer is more stable than the β-isomer. This anomaly is called an anomeric effect or end group effect.

The **anomeric effect** (异头效应) is caused by the mutual repulsion between the unshared electron pair on the oxygen atom in the sugar ring and the unshared electron pair on the oxygen atom or other heteroatoms on C_1. This repulsion is similar to 1,3-disturbance. Some people call this 1,3-interference the rabbit ear effect. When a methoxy, acetoxy or halogen atom is in the a-bond, the repulsion effect of such intra-external oxygen atoms without shared electrons is relatively small.

13.3 Reactions of Monosaccharides

PPT

Monosaccharides are polyhydroxy aldehydes or polyhydroxy ketones, which should have the inherent properties of hydroxyl and carbonyl groups; on the other hand, multiple functional groups are in the same molecule, and due to mutual influence, they will exhibit some special properties. In the following sections, we focus on the special chemical properties of monosaccharides and some important chemical reactions.

13.3.1 Epimerization

α-H in monosaccharides is similar to that in aldehyde and ketone molecules which shows a certain activity under the influence of carbonyl group. When monosaccharides are treated with dilute bases, α-H is taken away by the catalyst base to form carbanions, which is rearranged by the enol intermediates, partly converted into ketoses, and the other part becomes a pair of epimers This process is called epimerization. For example, in the presence of dilute alkali, glucose can be converted into a balanced mixture of mannose and fructose:

Whether fructose or mannose is treated with dilute alkali, the same equilibrium mixture is obtained. Therefore, ketose often exhibits the same properties as aldose under weakly alkaline conditions.

In organisms, glucose and fructose are also converted into each other under the catalysis of isomerase. In the modern food industry, starch is often used to produce fructose syrup through a biological and biochemical process.

13.3.2　Oxidation Reaction

Monosaccharides can be oxidized by various oxidants to produce various products. Only a few specific oxidation reactions are discussed in the following sections.

1. Oxidation in Alkaline Solutions　As we saw in Chapter 9, aldehydes (RCHO) are oxidized to carboxylic acids (RCOOH) by several oxidizing agents, including oxygen, O_2. Similarly, the aldehyde group of an aldose can be oxidized under basic conditions to a carboxylate group. Oxidizing agents for this purpose include bromine in aqueous calcium carbonate (Br_2, $CaCO_3$, H_2O) and Tollens' solution $[Ag(NH_3)_2^+]$. Under these conditions, the cyclic form of an aldose is in equilibrium with the open-chain form, which is then oxidized by the mild oxidizing agent. D-glucose, for example, is oxidized to D-gluconate (the anion of D-gluconic acid).

Any carbohydrate that reacts with an oxidizing agent to form an **aldonic acid** (醛糖酸) is classified as a **reducing sugar** (it reduces the oxidizing agent). Surprisingly, 2-ketoses are also reducing sugars. Carbon 1 (a CH_2OH group) of a 2-ketose is not oxidized directly. Rather, under the basic conditions of this oxidation, a 2-ketose is in equilibrium with an aldose by way of an enediol intermediate. The aldose

325

is then oxidized by the mild oxidizing agent.

$$
\underset{\text{a 2-ketose}}{\overset{\displaystyle \overset{H}{\underset{\overset{\displaystyle ||}{R}}{H-C-OH}}}{}} \; \underset{OH^-}{\rightleftharpoons} \; \underset{\text{an enediol}}{\overset{\displaystyle \overset{HO}{\underset{R}{C}}\overset{H}{\underset{OH}{C}}}{}} \; \underset{OH^-}{\rightleftharpoons} \; \underset{\text{an aldose}}{\overset{\displaystyle \overset{O}{\underset{R}{HO-C-H}}}{}} \; \xrightarrow{\text{oxidizing agent}} \; \underset{\text{an aldonate}}{\overset{\displaystyle \overset{O}{\underset{R}{HO-C-H}}}{}}
$$

Aldose or ketose can reduce Tollens reagent to produce silver mirror; it can also reduce Fehling reagent or Benedict reagent to produce brick red cuprous oxide precipitate. The reason that ketose can react positively with Tollens reagent or Benedict reagent is tautomerism under basic conditions.

In sugars chemistry, sugars that can react as described above are referred to as reducing sugars, and sugars that cannot undergo this reaction are called non-reducing sugars. This reaction is simple and sensitive, and is often used for qualitative testing of monosaccharides.

2. Oxidation in Acidic or Neutral Solutions Under acidic or neutral conditions, the aldehyde group in aldose can be selectively oxidized by bromine or other halogens to form a sugar acid with a carboxyl group, and then the sugar acid quickly forms a lactone. For example, the reaction of bromine and glucose.

The actual process of the reaction is relatively complicated and is related to the hemiacetal hydroxyl group. Keto sugars do not undergo epimerization under acidic or neutral conditions, so keto sugars cannot be oxidized by the weak oxidant bromine. This reaction can be used to distinguish between aldose and ketose.

Under the action of warm dilute nitric acid, the aldehyde group and primary alcohol hydroxyl group of aldose can be simultaneously oxidized to form **saccharic acid** (糖二酸). For example, D-**galactose** (半乳糖) is oxidized by nitric acid to form **galactaric acid** (半乳糖二酸), which is usually called **mucinic acid** (黏酸). Mucinic acid has low solubility and crystals are precipitated in water. This reaction is therefore often used to test for the presence of galactose.

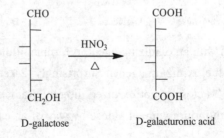

D-galactose   D-galacturonic acid

D-glucose is oxidized to **glucaric acid** (葡萄糖二酸) by dilute nitric acid, and then reduced by appropriate methods to obtain glucuronic acid. Under the above conditions, the C_2-C_3 chain of ketose breaks to form small molecule binary acid.

In organisms, glucose can also produce glucuronic acid under the action of enzymes. Glucuronic acid is easily combined with toxic substances such as alcohol or phenol to form glycosides, and the molecules after glycoside formation are relatively polar and are easily excreted from the body. For example, in the human liver, glucuronic acid can be excreted in combination with metabolites of foreign substances or drugs, which plays a detoxifying and detoxifying role.

3. Oxidation by Periodic Acid (高碘酸)   Oxidation by periodic acid, $HIO_4 \cdot 2H_2O$ or H_5IO_6, has been proven useful in structure determinations of carbohydrates, particularly in determining the size of glycoside rings. The periodic acid cleaves the carbon-carbon bond of a glycol in a reaction that proceeds through a cyclic periodic ester. In this reaction, iodine (Ⅶ) of periodic acid is reduced to iodine(V) of iodic acid.

Periodic acid also cleaves carbon-carbon bonds of α-hydroxyketones and da-hydroxyaldehydes by a similar mechanism. Following are abbreviated structural formulas for these functional groups and the products of their oxidative cleavage by periodic acid. As a way to help you understand how each set of products is formed, each carbonyl in a starting material is shown as a hydrated intermediate that is then oxidized. In this way, each oxidation can be viewed as being analogous to the oxidation of a glycol.

As an example of the usefulness of this reaction in carbohydrate chemistry, oxidation of methyl β-D-glucoside consumes two moles of periodic acid and produces one mole of formic acid. This stoichiometry and the formation of formic acid are possible only if -OH groups are on three adjacent carbon atoms.

This is evidence that methyl β-D-glucoside is indeed a pyranoside.

4. Oxidation to Uronic Acids（醛酸） Enzyme-catalyzed oxidation of the primary hydroxyl group at carbon 6 of a hexose yields a uronic acid. Enzyme-catalyzed oxidation of D-glucose, for example, yields D-glucuronic acid, shown here in both its open-chain and cyclic hemiacetal forms.

D-Glucuronic acid is widely distributed in both the plant and the animal world. In humans, it is an important component of the glucosaminoglycans of connective tissues. It is also used by the body to detoxify foreign hydroxylcontaining compounds such as phenols and alcohols. In the liver, these compounds are converted to glycosides of glucuronic acid (glucuronides) and excreted in the urine. The intravenous anesthetic propofol, for example, is converted to the following glucuronide and excreted in the urine.

propofol

a urine-soluble glucuronide

13.3.3　Formation of Glycosides (Acetals)

We saw in Chapter 9 that treatment of an aldehyde or a ketone with one molecule of alcohol gives a hemiacetal and that treatment of the hemiacetal with a molecule of alcohol gives an acetal. Treatment of monosaccharides, all of which exist almost exclusively in a cyclic hemiacetal form, also gives acetals, as illustrated by the reaction of β-D-glucopyranose with methanol.

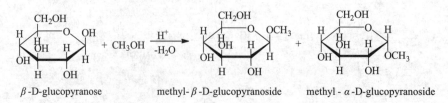

β-D-glucopyranose　　　methyl-β-D-glucopyranoside　　　methyl-α-D-glucopyranoside

A cyclic acetal derived from a monosaccharide is called a **glycoside** (糖苷), and the bond from the anomeric carbon to the -OR group is called a **glycosidic bond** (糖苷键). Glycosides are composed of two parts: sugar and non sugar. The sugar part is called glycosidic group, and the non sugar part is called aglycone or glycosidic ligand. For example, methyl in methyl glucoside is aglycone or glycosidic ligand, and the bond between sugar and non sugar part is called glycosidic bond.

Mutarotation is not possible in a glycoside because an acetal is no longer in equilibrium with the open-chain carbonyl-containing compound. Glycosides are stable in water and aqueous base, but like other acetals, they are hydrolyzed in aqueous acid to an alcohol and a monosaccharide. Glycosides are named by listing the alkyl or aryl group bonded to oxygen followed by the name of the carbohydrate in which the ending -*e* is replaced with -*ide*. For example, the glycosides derived from β-D-glucopyranose are named β-D-glucopyranosides; those derived from β-D-ribofuranose are named β-D-ribofuranosides.

Glycosides are widely distributed in nature and many are biologically active. In glycosides, the presence of sugar molecules can increase their solubility. Therefore, in modern drug research, sugar is often added to some water-insoluble drug molecules to increase its solubility.

According to the different atoms attached to the glycoside bonds, we classify the glycosides, such as oxoside, glucosinolate, nitrogen glycoside, and carbon glycoside, and the most common is oxoside.

nitroside (uridine)　　carbon glycoside (pseudouridine)　　thioside (myrosin)

Just as the anomeric carbon of a cyclic hemiacetal undergoes reaction with the -OH group of an alcohol to form a glycoside, it also undergoes reaction with the -NH- group of an amine to form an N-glycoside. Especially important in the biological world are the N-glycosides formed between D-ribose and 2-deoxy-D-ribose, each as a furanose, and the heterocyclic aromatic amines uracil, cytosine, thymine, adenine, and guanine. N-glycosides of these pyrimidine and purine bases are structural units of nucleic acids.

uracil　　cytosine　　thymine　　adenine　　guanine

13.3.4　Osazone Formation

Monosaccharides can undergo addition reactions with various carbonyl reagents. For example, under the mild conditions, an equimolar amount of phenylhydrazine can form sugar phenylhydrazone; however, when the phenylhydrazine is excessive (1 : 3), the α-position hydroxyl group can be oxidized by phenylhydrazine (phenylhydrazine does not show oxidative properties to other organic substances) to form a carbonyl group, which is then reacted with one mole of phenylhydrazine to form crystals of yellow **osazone** (糖脎). This reaction is unique to α-hydroxyaldehyde or ketone. Because of the simple reaction, it is often used as a qualitative test for monosaccharides. Here is the reaction of glucose with phenylhydrazine:

D-glucose   D-glucose osazone

Sugars do not form the simple phenylhydrazone derivatives we might expect, however. Two molecules of phenylhydrazine condense with each molecule of the sugar to give an osazone, in which both C_1 and C_2 have been converted to phenylhydrazones. The term osazone is derived from the -ose suffix of a sugar and the last half of the word hydrazone. Most osazones are easily crystallized, with sharp melting points. Melting points of osazone derivatives provide valuable clues for the identification and comparison of sugars.

In the formation of an osazone, both C_1 and C_2 are converted to phenylhydrazones. Therefore, a ketose gives the same osazone as its related aldose. Also notice that the stereochemistry at C_2 is lost in the phenylhydrazone. Thus, C_2 epimers give the same osazone.

The osazone formation only changes on C_1 and C_2, and does not involve other carbon atoms. Therefore, as long as the sugars have the same carbon atom configuration other than C_1 and C_2, the same osazone can be produced. For example, D-glucose, D-mannose, and D-fructose all produce the same osazone, which is valuable for determining sugar configuration.

13.3.5 Reduction to Alditols

The carbonyl group of a monosaccharide can be reduced to a hydroxyl group by a variety of reducing agents, including sodium borohydride and hydrogen in the presence of a transition metal catalyst. The reduction products are known as **alditols** (糖醇). Reduction of D-glucose gives **D-glucitol** (葡萄糖醇), more commonly known as D-**sorbitol** (山梨醇). Note that D-glucose is shown here in the open-chain form. Only a small amount of this form is present in solution, but as it is reduced, the rapid equilibrium between cyclic hemiacetal forms and the open-chain form replaces it.

D-glucopyranose   D-glucose   D-glucitol

Sorbitol is found in the plant world in many berries and in cherries, plums, pears, apples, seaweed, and algae. It is about 60% as sweet as sucrose (table sugar) and is used in the manufacture of candies and as a sugar substitute for diabetics. D-Sorbitol is an important food additive, usually added to prevent dehydration of foods and other materials upon exposure to air because it binds water strongly. Other alditols common in the biological world are erythritol, D-mannitol, and xylitol. Xylitol is used as a sweetening agent in "sugarless" gum, candy, and sweet cereals.

erythritol　　　D-mannitol　　　xylitol

13.3.6　Esterification

Sugar molecules are rich in hydroxyl groups, and are easily acidified by organic or inorganic acids like ordinary alcohols. Phosphates of alcohols have important biological significance. In the body, many sugar molecules exist as phosphates and participate in biochemical reactions, such as D-6-phosphate glucose, D-1,6-diphosphate fructose, and D-1-phosphate ribose. The phosphorylation reagent in the organism is adenosine triphosphate instead of phosphoric acid. Phosphorylation of alcohols with ATP is much faster than with phosphoric acid.

13.3.7　Formation of Cyclic Acetals or Ketals

The *cis-O*-dihydroxy group on the sugar ring can form cyclic acetals or ketals with aldehydes or ketones. This property is often used for the protection of hydroxyl groups on sugars in certain synthetic reactions. *Trans-O*-diols cannot undergo such reactions.

13.3.8 Dehydration and Chromogenic Reaction

Under the action of strong acid (sulfuric acid or hydrochloric acid), pentose or hexose is dehydrated in multiple steps to form furfural or furfural derivatives, respectively. This reaction can also occur when the polysaccharide is subjected to acid hydrolysis.

The furfural and its derivatives formed by the reaction can be condensed with phenols or aromatic amines to form colored compounds, so this property is often used to identify sugars. Frequently used are the Moritz reaction and the Silivanov reaction.

The Morrisch reaction uses concentrated sulfuric acid as a dehydrating agent, dehydrates monosaccharides or polysaccharides, and then reacts with α-naphthol to form purple condensates. The reaction is simple and sensitive, and it is often used for sugar testing.

The Silivanov reaction uses hydrochloric acid as a dehydrating agent, and the furfural derivative formed is reacted with resorcinol to form a bright red condensate. Because ketose reacts significantly faster than aldose, this reaction is often used to identify ketose and aldose.

13.3.9 Important Monosaccharides

Glucose, mannose, galactose and fructose are all important hexoses in nature. Here we introduce two important pentoses and glucosamine.

1. D-ribose and D-deoxyribose D-ribose (核糖) and D-**deoxyribose** (脱氧核糖) are both aldopentose. They also have α- and β-isomers. They also have reducibility and mutarotation phenomena. Their chemical structures are as follows.

α-D-ribofuranose   D-(-)-ribose   β-D-ribofuranose

α-D-2-deoxyribofurose   D-(-)-2-deoxyribose   β-D-2-deoxyfuranose

They do not exist in the free state in nature, and most of them are combined into glycosides. For example, croton contains crotonoside and releases ribose after hydrolysis. Ribose is a component of ribonucleic acid. Deoxyribose is an essential component of deoxyribonucleic acid. They play a very important role in life activities.

2. Amino Sugars　Amino sugars contain an -NH$_2$ group in place of an -OH group. Only three amino sugars are common in nature: D-glucosamine, D-mannosamine, and D-galactosamine. 2-Glucosamine and 2-acetylglucosamine, which have extremely large biological reserves, can be regarded as derivatives of 2-OH of glucose replaced by amino or acetylamino, respectively, and the structure is as follows.

β-D-2-glucosamine　　　β-D-2-acetylglucosamine

D-glucosamine　D-mannosamine　D-galactosamine　N-acetyl-D-glucosamine

N-acetyl-D-glucosamine, a derivative of D-glucosamine, is a component of many polysaccharides, including chitin, the hard shell-like exoskeleton of lobsters, crabs, shrimp, and other shellfish. Many other amino sugars are components of naturally occurring antibiotics.

3. Vitamin C　Vitamin C can be regarded as a derivative of monosaccharides. It is prepared by reducing D-glucose to obtain L-sorbitol, then oxidizing it to L-sorbose, and then oxidizing and lactonizing. Vitamin C is found in fresh vegetables and fruits, and is more abundant in citrus, lemon, and tomato. Many plants can synthesize vitamin C themselves, but humans are powerless and must be taken from food. Vitamin C is involved in sugar metabolism and redox processes in the body. In addition, it can also be used as an antioxidant in food. If you lack vitamin C in your body, you will get scurvy. The symptoms are mainly skin damage, loose teeth, and gum decay.

vitamin C

13.4　Disaccharides and Oligosaccharide

Most carbohydrates in nature contain more than one monosaccharide unit. Those that contain two units are called disaccharides, those that contain three units are called trisaccharides, and so forth. The general term oligosaccharide is often used for carbohydrates that contain from four to ten monosaccharide units. Carbohydrates containing larger numbers of monosaccharide units are called polysaccharides. In a disaccharide, two monosaccharide units are joined together by a glycosidic bond between the anomeric carbon of one unit and an -OH of the other. Three important disaccharides are **sucrose** (蔗糖), **lactose** (乳糖), and **maltose** (麦芽糖).

13.4.1　Nonreducing disaccharide

The glycosidic bond forming a disaccharide is formed by condensation and dehydration of the hemiacetal or hemiketal hydroxyl of two sugar molecules, then the disaccharide will no longer have glycosidic hydroxyl, and there is no reciprocal equilibrium between oxygen ring and open chain in the aqueous solution, so there is no reducibility, osazone formation and no mutarotation phenomenon. Disaccharides with this structure are called nonreducing sugars.

1. Sucrose　Sucrose (table sugar) is the most abundant disaccharide in the biological world. It is obtained principally from the juice of sugarcane and sugar beets. In sucrose, carbon 1 of α-D-glucopyranose is joined to carbon 2 of β-D-fructofuranose by an α-1,2-glycosidic bond.

Both monosaccharide units in sucrose are present as acetals, or glycosides. Neither ring is in equilibrium with its open-chain aldehyde or ketone form, so sucrose does not reduce Tollens reagent and it cannot mutarotate. Because both units are glycosides, the systematic name for sucrose can list either of the two glycosides as being a substituent on the other. Both systematic names end in the -*oside* suffix, indicating a nonmutarotating, nonreducing sugar. Like many other common names, sucrose ends in the -*ose* ending even though it is a nonreducing sugar. Common names are not reliable indicators of the properties of sugars. Sucrose is hydrolyzed by enzymes called invertases, found in honeybees and yeasts, that specifically hydrolyze the β-D-fructofuranoside linkage. The resulting mixture of glucose and fructose is called invert sugar because hydrolysis converts the positive rotation [+66.5°] of sucrose to a negative rotation that is the average of glucose [+52.7°].

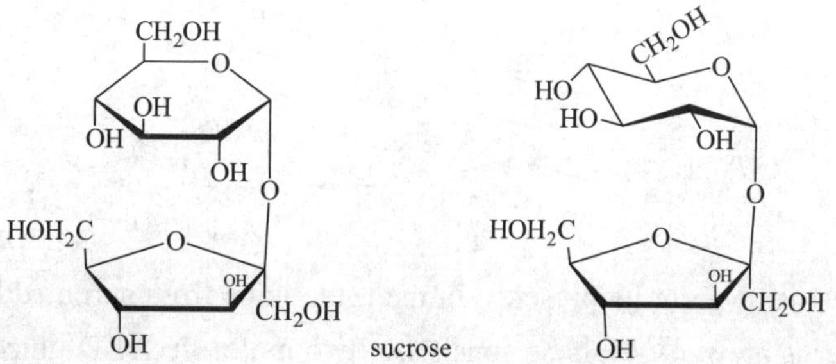

sucrose

2. Trehalose（海藻糖）　Fucose, also called yeast sugar, is rich in yeast. Trehalose is a non-reducing disaccharide formed by removing one molecule of water between two α-D-glucopyranoside hydroxy groups. The full name is α-D-glucopyranosyl-α-D-glucopyranose glycoside, its structure is as follows:

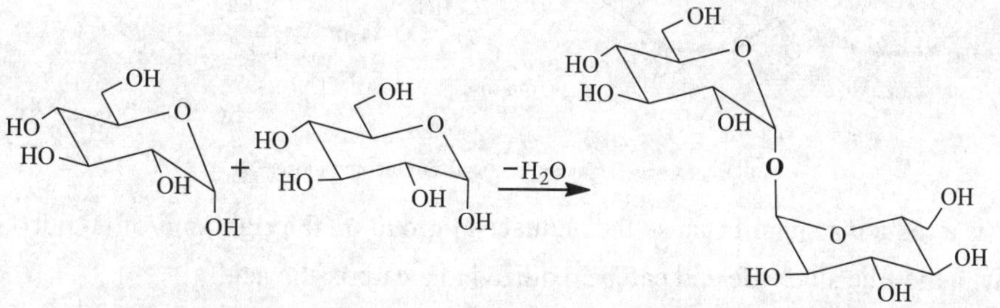

Trehalose is a disaccharide of great development value. The sweetness of trehalose is only 45% of sucrose, and the taste is light, but compared with sucrose, the sweetness is easy to penetrate, leaving no aftertaste after eating, and it is not easy to cause dental caries. Trehalose has the function of protecting biological cells, and can prevent biologically active substances (such as various proteins, enzymes, etc.) from being damaged under stress such as dehydration, drought, high temperature, radiation, and freezing. At present, trehalose is widely used as a new type of food, pharmaceutical and cosmetic additives.

13.4.2　Reducing disaccharide

If a disaccharide is formed by the shrinkage of one molecular glycoside hydroxyl group and another molecular sugar ordinary hydroxyl group, the sugar glycoside hydroxyl group remains in the molecule, and the oxygen ring and open chain interconversions still exist in the water equilibrium, so it has the phenomenon of mutarotation, can form osazone, and still has reducibility. Disaccharides with this structure are called reducing disaccharide. Such as maltose and **cellobiose**（纤维二糖）.

1. Maltose　Maltose is a disaccharide formed when starch is treated with sprouted barley, called malt. This malting process is the first step in brewing beer, converting polysaccharides to disaccharides and monosaccharides that ferment more easily. Like cellobiose, maltose contains a 1,4' glycosidic linkage between two glucose units. The difference in maltose is that the stereochemistry of the glucosidic linkage is α rather than β.

Maltose derives its name from its presence in malt, the juice from sprouted barley and other cereal grains (from which beer is brewed). Maltose consists of two molecules of D-glucopyranose joined by an α-1,4-glycosidic bond between carbon 1 (the anomeric carbon) of one unit and carbon 4 of the other unit. Following are representations for β-maltose, so named because the -OH on the anomeric carbon of the glucose unit on the right is β.

maltose

Maltose,4-O-(α-D-glucopyranosyl)-D-glucopyranose

Maltose is a reducing sugar because the hemiacetal group on the right unit of D-glucopyranose is in equilibrium with the free aldehyde and can be oxidized to a carboxylic acid.

When starch is partially hydrolyzed in dilute acid, (+)-maltose can be obtained, and (+)-maltose can also be obtained in the process of producing ethanol by fermentation. In addition to being hydrolyzed by acid, maltose can also be hydrolyzed by maltase. Maltase is an enzyme that specifically hydrolyzes α-glycosidic bonds and does not work on β-glycosidic bonds. Maltose is present in germinated barley. Since malt contains amylase, it can hydrolyze starch to maltose.

2. Cellobiose Cellobiose, the disaccharide obtained by partial hydrolysis of cellulose, contains a 1,4′ linkage. In cellobiose, the anomeric carbon of one glucose unit is linked through an equatorial (β) carbon–oxygen bond to C4 of another glucose unit. This β-1,4′ linkage Cl from a glucose acetal is called a glucosidic linkage.

The complete name for cellobiose, 4-O-(β-D-glucopyranosyl)-β-D-glucopyranose, gives its structure. This name says that a β-D-glucopyranose ring (the right-hand ring) is substituted in its 4-position by an oxygen attached to a (β-D-glucopyranosyl) ring, drawn on the left. The name in parentheses says the substituent is a β-glucose and the-*syl* ending indicates that this ring is a glycoside. The left ring with the -*syl* ending is an acetal and cannot mutarotate, while the right ring with the -*ose* ending is a hemiacetal and can mutarotate. Because cellobiose has a glucose unit in the hemiacetal form (and therefore is in equilibrium with its open-chain aldehyde form), it is a reducing sugar. Once again, the -*ose* ending indicates a mutarotating, reducing sugar. Mutarotating sugars are often shown with a wavy line to the free anomeric hydroxyl group, signifying that they can exist as an equilibrium mixture of the two anomers. Their names are often given without specifying the stereochemistry of this mutarotating hydroxyl group, as in 4-*O*-(β-D-glucopyranosyl)-D-glucopyranose.

3. Lactose Lactose is the principal sugar present in milk. It makes up about 5%~8% of human milk and 4%~6% of cow's milk. It consists of D-galactopyranose bonded by a *β*-1,4-glycosidic bond to C4 of D-glucopyranose. Lactose is a reducing sugar.

Lactose is similar to cellobiose, except that the glycoside (the left ring) in lactose is galactose rather than glucose. Lactose is composed of one galactose unit and one glucose unit. The two rings are linked by a bond of the galactose acetal to the 4-position on the glucose ring: a *β*-1,4- *galacto*sidic linkage.

lactose,4-O-(β-D-galactopyranosyl)-D-glucopyranose

Hydrolysis of lactose requires a enzyme (sometimes called lactase). Some humans synthesize a but others do not. This enzyme is present in the digestive fluids of normal infants to hydrolyze their mother's milk. Once the child stops drinking milk, production of the enzyme gradually stops. In most parts of the world, people do not use milk products after early childhood, and the adult population can no longer digest lactose. Consumption of milk or milk products can cause digestive discomfort in *lactose-intolerant* people who lack the enzyme. Lactose-intolerant infants must drink soybean milk or another lactose-free formula.

Among the disaccharides, lactose is less water-soluble, non-hygroscopic, and relatively stable, so it is often used as an excipient for tablets, capsules, or granules.

13.4.3 Cyclodextrin

Cyclodextrin (CD) (环糊精) is a type of cyclic oligosaccharide formed by 6, 7, or 8 D-glucopyranose through *α*-1,4-glycoside bonds. According to the number of cyclic glucose, they are called *α*, *β* and *γ*-cyclodextrin, respectively, referred to as *α*, *β* and *γ*-CD. As a new type of drug carrier, cyclodextrin is widely used, especially *β*-CD is most commonly used. *β*-CD is a tubular compound formed by 7 glucose through *α*-1,4-glycoside bonds. The chemical structure is as follows.

The molecular structure of *β*-CD is relatively special. The hydroxyl groups of glucose are distributed at both ends of the tube and outside. The inside is hydrophobic. The cavity of *β*-CD can selectively contain a variety of structures and their matching fat-soluble compounds, and form a host-guest inclusion compound through intermolecular forces. This property is of great significance in drug preparation, complexation catalysis, simulation enzyme and other aspects.

After the formation of clathrates, the physicochemical properties of the clathrates can be changed, such as reducing volatility, improving water solubility and chemical stability, etc. Therefore, the inclusion technology is used in medicine, pesticides, food, chemicals, and organic synthesis and catalysis. There are many applications. The volatile oil of traditional Chinese medicine is easy to volatilize and hardly soluble in water, which brings a lot of inconvenience to the processing and storage of the preparation. When it is made into a CD inclusion compound, the above defects can be significantly improved. In organic synthesis, the addition of CD can often increase the reaction speed and reaction selectivity.

Such as the chlorination reaction of anisole under the action of hypochlorous acid, in the absence of CD, 33% of *o*-chloro products and 67% of *para*-chlorine products are generally produced. However, when *β*-CD is added, only the *para* position of the benzene ring entering the CD cavity is not shielded by CD, so the reaction can selectively occur in the *para* position, producing 96% of parachloroanisole.

The host-guest relationship between CD and the inclusion complex is very similar to the role of enzymes and substrates, so CD and its derivatives have become one of the most widely studied mimic enzymes.

PPT

13.5 Polysaccharides

Polysaccharides are carbohydrates that contain many monosaccharide units joined by glycosidic bonds. They are one class of biopolymers, or naturally occurring polymers. Smaller polysaccharides, containing about four to ten monosaccharide units, are sometimes called oligosaccharides. Most polysaccharides have hundreds or thousands of simple sugar units linked together into long polymer chains. Except for units at the ends of chains, all the anomeric carbon atoms of polysaccharides are involved in acetal glycosidic links. Therefore, polysaccharides give no noticeable reaction with Tollens reagent, and they do not mutarotate. Three important polysaccharides, all made up of glucose units, are **starch** (淀粉), **cellulose** (纤维素), and **glycogen** (糖原).

13.5.1 Starch

Starch is used for energy storage in plants. It is found in all plant seeds and tubers and is the form in which glucose is stored for later use. Starch can be separated into two principal polysaccharides: **amylose** (直链淀粉) and **amylopectin** (支链淀粉).

amylose

amylopectin

The helical structure of amylose also serves as the basis for an interesting and useful reaction. The inside of the helix is just the right size and polarity to accept an iodine molecule. When iodine is lodged within this helix, a deep blue starch-iodine complex results. This is the basis of the starch-iodide test for oxidizers. The material to be tested is added to an aqueous solution of amylose and potassium iodide. If the material is an oxidizer, some of the iodide is oxidized to iodine which forms the blue complex with amylose.

13.5.2　Glycogen

Like amylopectin, **glycogen** (糖原) is a branched polysaccharide of approximately 10^6 glucose units joined by α-1,4- and α-1,6-glycosidic bonds. The total amount of glycogen in the body of a well-nourished adult human is about 350 g, divided almost equally between the liver and muscle.

Glycogen is the carbohydrate that animals use to store glucose for readily available energy. A large amount of glycogen is stored in the muscles themselves, ready for immediate hydrolysis and metabolism. Additional glycogen is stored in the liver, where it can be hydrolyzed to glucose for secretion into the bloodstream, providing an athlete with a "second wind". The structure of glycogen is similar to that of amylopectin, but with more extensive branching. The highly branched structure of glycogen leaves many end groups available for quick hydrolysis to provide glucose needed for metabolism.

13.5.3　Cellulose

Cellulose (纤维素), the most widely distributed plant skeletal polysaccharide, constitutes almost half of the cell wall material of wood. Cotton is almost pure cellulose. Cellulose is a linear polysaccharide of D-glucose units joined by β-1,4-glycosidic bonds. It has an average molecular weight of 400,000 g/mol, corresponding to approximately 2200 glucose units per molecule.

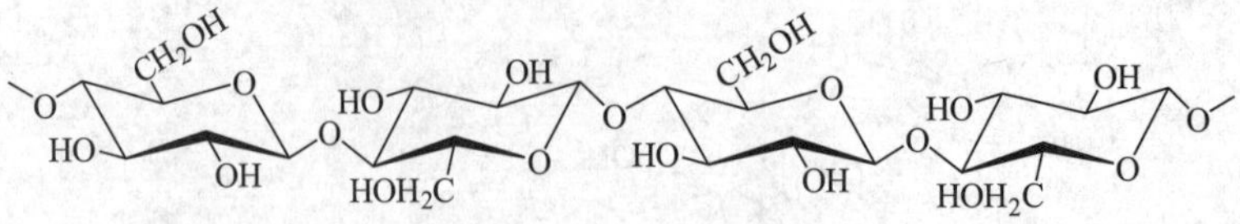

celluose

13.5.4　Chitin and Chitosan

Chitin (甲壳质) is a natural animal fiber widely existing in biological tissues such as crustacean shells, arthropod skins, lower animal cell membranes, and higher plant cell walls. The third largest biological resource that is being developed after cellulose, the amount of biosynthesis in nature reaches 100 billion tons per year. Chitin is a linear polysaccharide made from 2-acetylglucosamine through β-1,4 glycosidic bonds. The chemical structure is as follows.

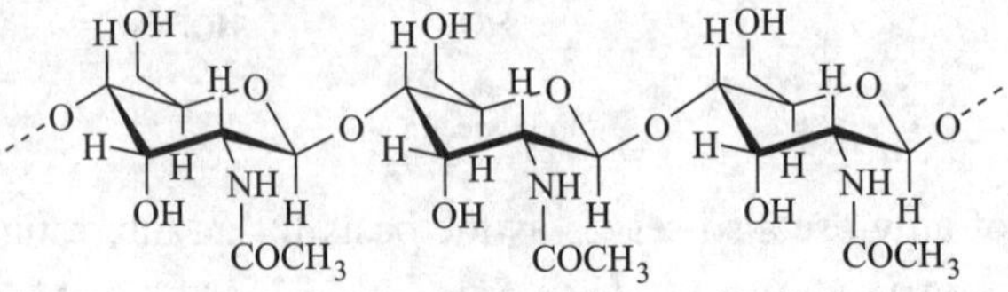

chitin

After deacetylation of chitin, the resulting product is called **chitosan** (壳糖胺).

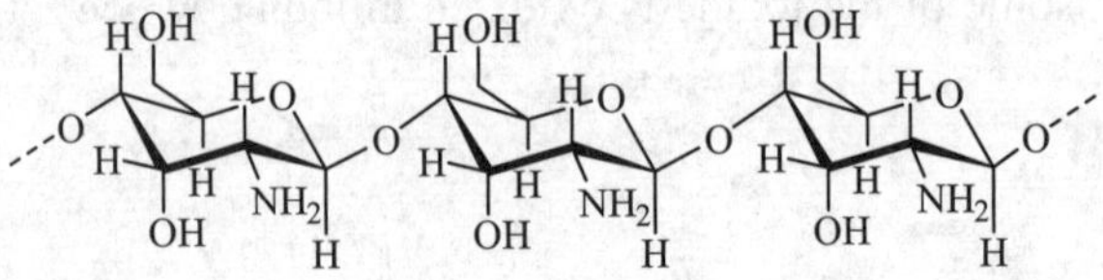

chitosan

Compared with cellulose, the only difference is that chitosan only replaces the C2-OH of glucose in cellulose with -NH_2. In other words, chitosan is an animal cellulose formed from 2-glucosamine through β-1,4 glycosidic bonds.

Modern pharmacological research shows that chitosan and its hydrolysates or partial hydrolysates have various physiological and pharmacological activities. For example, chitosan has the function of regulating the physiological and biochemical functions of the human body, can enhance the human body's immunity, inhibit tumors, reduce blood sugar, blood lipids and cholesterol, can promote wound healing and broken bone regeneration, and has functions such as detoxification. The hydrolysis product of chitosan is a necessary biologically active substance in human cells or tissues, and has good biocompatibility with human tissues. At present, their research, development and utilization have become a hot spot in polysaccharide research.

13.6　Applications of Carbohydrates in the Pharmacy Study

PPT

Pharmaceutically, glucose is probably the most important of all gular monosaccharides. A solution of pure glucose has been recommended for use by subcutaneous injection as a restorative after severe operations, or as a nutritive in wasting diseases. It has also been used to augment the movements of the uterus. Glucose is added to nutritive enemata for rectal alimentation. Its use has also been recommended for rectal injection and by mouth in delayed chloroform poisoning.

Malt consists of the grain of barley, Hordeum distichon (family Gramineae), partially germinated and dried. Maltose is the major carbohydrate of malt and malt extracts. Pharmaceutically, extract of malt is used as a vehicle for the administration of cod-liver oil, and the liquid extract is given with haemoglobin, extract of cascara and various salts.

Lactose has a sweetish taste, and is used extensively in the pharmaceutical industry. It is the second most widely used compound and is employed as a diluent, filler or binder in tablets, capsules and other oral product forms. Alactose is used for the production of lactitol, which is present in diabetic products, low calorie sweeteners and slimming products. As lactose is only 30 percent as sweet as sugar it is used as a sugar supplement, and also in food and confectionery. It is used in infant milk formulas.

In recent years, the results of sugar chemistry research have continuously shown to the world that the relationship between sugar and humans is not just food, clothing, and transportation. The medicinal value of sugar is even more popular. Sugar is closely related to human life activities and has good biocompatibility with human tissues.

Great progress has been made in the study of sugar compounds and biology, and it has also aroused the interest of pharmacists in this class of compounds, which has led to the development of sugar-based drugs. The successful cloning of glycosylase and its understanding of properties, the improvement of oligosaccharide synthesis technology and cell surface glycan analysis technology have made people more clearly understand the important pharmaceutical value of this type of biological molecules. Generally, carbohydrates are used to treat diabetes, epilepsy, thrombosis and arthritis. In the future, carbohydrates are very promising to be used as remedies for reperfusion injury, flu and cancer. Carbohydrates can also be used to modify other drugs (Including recombinant protein drugs), compared with the parent drug, the modified hybrid molecule has improved activity and reduced side effects. In addition, the specific targeting properties of carbohydrates for certain types of cells, and the targeted transport of drugs and genes are also the current research hotspots in biopharmaceutics and gene therapy, and several candidate molecules have been undergoing clinical trials.

重点小结

一、单糖

1. **单糖的分类** 单糖可根据分子中所含碳原子的数目分为戊糖、己糖等。也可根据官能团分为醛糖和酮糖。自然界中存在最广泛的单糖是葡萄糖(多羟基醛)、果糖(多羟基酮)和核糖。

2. **单糖的物理性质** 纯的单糖都是无色结晶,有甜味,易溶于水,不溶于弱极性或非极性溶剂。由于单糖溶于水后存在开链式与环状结构之间的互变,存在变旋光现象。

3. **己碳糖的开链结构和相对构型** 实验证明,葡萄糖具有五羟基醛的开链结构,有四个手性碳原子,糖的构型一般用费歇尔式表示。糖类的构型习惯用D／L名称进行标记。即编号最大的手性碳原子构型与甘油醛比较,与D-甘油醛相似,-OH在右边的为D型;与L-甘油醛相似,-OH在左边的为L型。与葡萄糖构造式相同的己醛糖中有八个D型化合物、八个L型异构体。

二、单糖的环状结构

1. **己碳糖的氧环式结构和α,β-异构体** D-葡萄糖分子内同时存在羟基和醛基,可以发生分子内的加成反应,生成环状半缩醛结构。葡萄糖是以C_5羟基与C_1醛基发生加成反应,生成六元的氧环式结构。当开链的D-葡萄糖分子中C_5羟基与醛基加成后,C_1变成了手性碳原子,有两种构型,一种是C_1的羟基(即半缩醛羟基,也称苷羟基)与决定构型的C_5羟基在同侧,称之为α-异构体;另一种C_1的羟基与C_5羟基分占碳链两侧,称之为β-异构体。葡萄糖的α-和β-构体,其差别仅在于C_1的构型相反,它们互为C_1差向异构体,是一对非对映体,也称作端基差向异构体或异头物。

2. **糖的哈沃斯结构式** Haworth式的构型判断:①**绝对构型**:根据六碳吡喃糖的C_5(五碳呋喃糖的C_4)上取代基的取向,向上的为D型,向下的为L型。②**相对构型**:C_1半缩醛羟基与六碳吡喃糖的C_4(五碳呋喃糖的C_4)上取代基在环的同一侧为β构型,异侧为α构型。

3. **己碳糖的构象以及异头效应** 吡喃糖的构象:$^4C_1$和$^1C_4$两种形式,D-系吡喃糖以N-式存在,为优势构象;但α-D-吡喃艾杜糖除外,其A-式为优势构象。当C_1苷羟基成为$-OCH_3$、$-OCOCH_3$、$-X$时,处于a键往往是优势构象,我们把异头物中C_1位上较大取代基处于a键为优势构象的反常现象,称为异头效应或端基效应。

三、单糖的反应

1. **差向异构化反应** 稀碱处理单糖时,通过烯醇式中间体发生重排,一部分转化成酮糖,另一部分成为一对差向异构体,这一过程叫做差向异构化。

2. **氧化反应**

(1)**碱性溶液中的氧化** 醛糖与酮糖都能被像多伦试剂或费林试剂这样的弱氧化剂氧化,糖分子的醛基被氧化为羧基。 凡是能被上述弱氧化剂氧化的糖,都称为还原糖。果糖也是还原糖。果糖具有还原性的原因:差向异构化作用——酮基不断地变成醛基。

（2）酸性和中性溶液中的氧化　醛糖可被溴水氧化成葡萄糖酸，而酮糖不能被弱氧化剂溴水氧化。在温热的稀硝酸作用下，醛糖的醛基和伯醇羟基可同时被氧化生成糖二酸。酮糖在同样条件下发生$C_2\sim C_3$链的断裂生成小分子二元酸。

（3）高碘酸的氧化　对于多羟基化合物的氧化产物，可以简单地看作是醇羟基碳原子之间的键断裂，断裂部分各与一个羟基结合，然后失水。

（4）氧化到醛酸　葡萄糖在酶的作用下，可以生成葡萄糖醛酸。

3. 糖苷的生成　糖分子中的活泼半缩醛羟基与其他含羟基的化合物（如醇、酚）或含氮杂环化合物作用，失水而生成缩醛的反应称为成苷反应；形成的化学键称苷键。其产物称为糖苷，又称配糖体。苷分子是由糖和非糖部分通过苷键而形成的。

4. 成脎反应　单糖与过量的苯肼反应生成黄色糖脎的结晶。该反应是α-羟基醛或酮的特有反应，糖脎的形成常作为单糖的定性鉴别和制备衍生物用。

5. 还原反应　单糖的羰基可经硼氢化钠等还原得到相应的醇，这类多元醇通称为糖醇。

6. 成酯反应　糖分子富含羟基，和普通醇一样容易被有机或无机酸酯化，醇的磷酸酯具有重要生物学意义。

7. 环状缩醛和缩酮的形成　处于糖环上的顺式邻二羟基可与醛或酮生成环状的缩醛或缩酮，该性质常用于某些合成反应中糖上羟基的保护。反式邻二醇不反应。

8. 脱水和显色反应　单糖在强酸（HCl或H_2SO_4）作用下脱水生成糠醛或糠醛衍生物。生成的化合物可与酚类或芳胺类缩合，生成有色化合物，常利用该性质进行糖的鉴别。

9. 重要的单糖　D-核糖和D-脱氧核糖以及氨基糖。

四、二糖和低聚糖

1. 非还原性双糖　形成双糖的糖苷键由两个糖分子的半缩醛或半缩酮羟基缩合脱水而成，则这种双糖就不再有苷羟基，在水溶液中不存在氧环式与开链式的互变平衡，故无还原性，不能成脎，无变旋光现象。

（1）蔗糖　(+)-蔗糖是由α-D-吡喃葡萄糖的半缩醛羟基与β-D-呋喃果糖的半缩酮羟基缩去一分子水形成的双糖，没有游离的苷羟基，无变旋现象，不能成脎，不能还原多伦试剂和班氏试剂，故蔗糖是非还原糖。蔗糖的化学名称为α-D-吡喃葡萄糖基-β-D-呋喃果糖苷，或β-D-呋喃果糖基-α-D-吡喃葡萄糖苷。

（2）海藻糖　又称为酵母糖，在酵母中含量丰富。(+)-海藻糖具有稳定生物膜（细胞膜）和蛋白质结构及抗干燥的作用。

2. 还原性双糖　形成双糖的糖苷键由一分子糖的苷羟基与另一分子糖的普通羟基缩水而成，在这种糖的分子中仍然留有苷羟基，在水溶液中依然存在氧环式与开链式的互变平衡，因而具有变旋光现象，能够成脎，仍具有还原性。

（1）麦芽糖　4-O-(α-D-吡喃葡萄糖基)-D-吡喃葡萄糖。淀粉在稀酸中部分水解时，可得到(+)-麦芽糖，淀粉发酵成乙醇的过程中也可得到(+)-麦芽糖。

（2）纤维二糖　4-O-(β-D-吡喃葡萄糖基)-D-吡喃葡萄糖。是纤维素的结构单位，是纤维素经一定方法处理后部分水解的产物。(+)-纤维二糖不能被麦芽糖酶（专一性水解α-糖苷键）水解，只能被苦杏仁酶（专一性水解β-糖苷键）水解。

（3）乳糖　(4-O-(β-D-吡喃半乳糖基)-D-吡喃葡萄糖，可用苦杏仁酶水解，这说明(+)-乳糖分子中的糖苷键为β-型。

3. **环糊精**　淀粉经某种特殊酶(浸解杆菌淀粉酶)水解得到的环状低聚糖称为环糊精 (cyclodextrin, 缩写CD)。由6~8个D-葡萄糖基通过α-1,4-糖苷键结合而成，根据所含葡萄糖单位的个数 (6，7或8)，分别称为α-、β-或γ-环糊精 (α-、β-或γ-CD)。组成环糊精的葡萄糖单位不同，其孔腔大小各异。不同的孔腔能选择性地包合多种结构与其匹配的脂溶性化合物，通过分子间特殊的作用力形成主体-客体包合物，这一特性在药物制剂、络合催化、模拟酶等方面颇有意义。

五、多糖

1. **淀粉**　可看作是葡萄糖的聚合物，由α-D-葡萄糖通过α-1,4苷键和α-1,6苷键连接形成的多糖。直链淀粉的螺旋通道适合插入碘分子，并通过范德华力吸引在一起，形成深蓝色淀粉-碘包合物，所以直链淀粉遇碘显蓝色。支链淀粉的葡萄糖单位除了以α-1,4-糖苷键相连外，还有的以α-1,6-糖苷键相连。每隔20~25个葡萄糖单位就会出现一个α-1,6-糖苷键相连的分支。

2. **纤维素**　是自然界中分布最广的有机物，它在植物中所起的作用就像骨骼在人体中所起的作用一样，作为支撑物质。纤维素是D-葡萄糖以β-1,4-苷键连接而成的直链多糖。

3. **糖原**　又称糖原或动物淀粉，是人和动物体内，经一系列酶催化反应，将多个葡萄糖组合而成的分支多糖，是生物体内葡萄糖的一种贮存形式。

4. **甲壳质和壳糖胺**　甲壳质是由2-乙酰氨基葡萄糖通过β-1,4糖苷键连接而成的直链多糖。壳糖胺是由2-氨基葡萄糖通过β-1,4糖苷键连接而形成的动物性纤维素。

六、葡萄糖胺聚糖

葡萄糖胺聚糖是很多重复的二糖组成的线性多糖。这个碳水化合物家族的成员包括透明质酸、肝素、硫酸软骨素和硫酸角蛋白，它们是软骨、肌腱和其他结缔组织的组成部分，而硫酸皮聚糖是皮肤细胞外基质的组成部分。这类多糖的一个普遍特征是一种重复的双糖，由糖醛酸和氨基己糖组成，氨基己糖和醛酸之间有1,4-糖苷键。

七、糖类化合物在药学中的应用

在药物化学领域,糖类药物研究也愈来愈被重视,而糖类化合物也被认为是当前发现药物先导物的重要来源之一。利用糖类化合物对某些类型细胞的特异性寻靶性质,进行药物和基因的定向转运,也是当前生物药剂学和基因治疗学的研究热点。寻找糖类化合物的新药用价值,获益的不仅仅是药物科学,这个过程对人们更好地了解生命过程也是大有益处的。

Problems
目 标 检 测

1. Rewrite the following monosaccharides into Haworth structure or conformation, and write the corresponding names.

将下列几个单糖改写成Haworth结构式或构象式，并写出相应的名称。

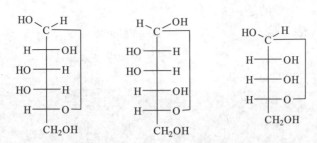

2. Write the Haworth structure and dominant conformation of the following compounds.

写出下列化合物的Haworth结构式及优势构象式。

(1) *α*-D-glucopyranose; (2) *β*-D-mannosylglucoside; (3) *β*-D-fructanose

3. Which of the following compounds can be oxidized by Fehling reagents?

下列化合物哪些能被斐林试剂氧化?

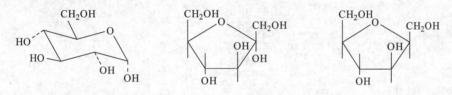

4. How to distinguish the following compounds by chemical method?

如何用化学方法鉴别下列化合物?

(1) Glucose and fructose; (2) Maltose and sucrose; (3) Glucose and galactose; (4) Starch and cellulose; (5) Glucose and methylglucoside

5. Write down several other sugars that can produce the same kind of osazone as D-glucose.

写出能与D-葡萄糖生成同一种糖脎的其他几种糖。

6. Explain the following nouns.

解释下列名词。

(1) Mutarotation phenomenon; (2) Epimers; (3) Anomers; (4) Reducing sugars, non-reducing sugars; (5) osazone; (6) Glycosides, aglycones, glycoside bonds, glycoside hydroxyls

7. Among the following sugars, the group that produces the same osazone is (　　　).

下列糖中，生成相同糖脎的一组是(　　　)。

　　A. mannose, glucose, fructose 　　　　　　B. galactose, glucose, mannose

　　C. galactose, mannose, fructose 　　　　　D. maltose, galactose, fructose

8. Among the sugars that cannot be oxidized by the Fehling reagent are (　　　).

下列糖类化合物中不能被斐林试剂氧化的是(　　　)。

　　A. Glucose 　　　　　　　　　　　　　　B. Fructose

　　C. Mannose 　　　　　　　　　　　　　　D. Sucrose

　　E. Maltose

9. D-glucose and D-fructose are mutually (　　　).

D-葡萄糖和D-果糖互为(　　　)异构体。

　　A. Optical isomers 　　　　　　　　　　　B. Positional isomers

　　C. Functional group isomers 　　　　　　　D. Carbon chain isomers

　　E. Enantiomers

10. Which of the following compounds is not a monosaccharide?

下列化合物中不属于单糖的是（　　　）。

    A. glyceraldehyde   B. galactose

    C. mannose   D. lactose

11. Complete the reaction equation.

完成反应式。

$$\text{(pyranose structure)} + CH_3OH \xrightarrow{HCl}$$

12. Complete the reaction equation.

完成反应式。

$$\begin{array}{c} CHO \\ H \longrightarrow OH \\ HO \longrightarrow H \\ H \longrightarrow OH \\ H \longrightarrow OH \\ CH_2OH \end{array} \xrightarrow{NaBH_4}$$

Discussion Topic

Gastrodia is a saprophytic herb belonging to Orchidaceae and Gastrodia. It is one of the precious Chinese medicinal materials. Gastrodia has been used for medicine for more than a thousand years. Gastrodin is an extract of Gastrodia elata and is commonly used in clinical treatment. Gastrodin tablets can restore the imbalance between the cerebral cortex excitement and inhibition, and have central inhibitory effects such as sedation, sleep and analgesia. It can also increase cerebral blood flow and relieve cerebral vasospasm. At the same time, it can lower blood pressure and improve human immunity.

Gastrodin

Find the glycosides, aglycones and glycoside bonds in the Gastrodin.

Studies in recent years have shown that many traditional Chinese medicines with strong tonic and immune functions are mostly composed of polysaccharides. Please consult the information and literature and summarize.

（赵　群）

第十四章 杂环化合物
Chapter 14 Heterocyclics

 学习目标

1. **掌握** 五元和六元杂环化合物的结构及主要理化性质。
2. **熟悉** 杂环化合物的命名方法及亲电取代定位规律;杂环化合物分类及互变异构现象。
3. **了解** 杂环化合物在医药领域中的作用。

Heterocycles (杂环化合物) are cyclic compounds in which one or more ring atoms are not carbon (that is hetero atoms). Although heterocycles are known to incorporate many kinds of elements into the structures (e.g., N, O, S, Al, Si, P, Sn, As, Cu), only some of common **hetero atoms** (杂原子) N, O, or S are included in most of the structures. In most heterocyclic compounds, the heterocyclic ring is generally referred to as "**aromatic heterocyclic ring** (芳香杂环)" because of the conformity to the Hückel rule to show the **aromaticity** (芳香性).

Heterocycles are conveniently grouped into classes of nonaromatic and aromatic. The nonaromatic compounds have physical and chemical properties that are typical of hetero atom. Cyclic ether, lactone, lactide, lactam, cyclic anhydride, etc. have been introduced previously. They are also heterocyclic structures, but are not usually included in heterocyclic compounds because they're not aromatic.

cyclic ether lactone   lactide   lactam   cyclic anhydride

14.1 Structures and Classifications of Heterocyclics

PPT

There are three main **classification** (分类) methods of heterocyclic compounds.

Based on the number of rings, heterocyclic compounds can be usefully classified into **single ring** (单杂环) and **fused ring** (稠杂环). The single ring can be divided into five-membered ring, six-membered ring and seven-membered ring depending on the size of the ring. The singlering can also be subdivide by the number of heteroatoms in the ring. A single ring fused to benzene ring is benzene-fused heterocyclic rings, and non-benzene heterocyclic ring is called fused heterocycle.

Classified by the heteratom species on the ring: Oxygen-containing heterocycles, nitrogen-containing heterocycles and sulfur-containing heterocycles, etc.

Classified by **π-electronic cloud density** (π电子云密度) on the ring: Multi-π and less-π aromatic heterocycles. Of the three categories, the first is the most common.

14.2　Nomenclature of Heterocyclics

The **nomenclature** (命名) of heterocyclic compounds is complicated. Currently, transliteration is widely used based on the nomenclature principles of IUPAC (2013), while combining the characteristics of Chinese characters.

The **phonetic** (语音的) Chinese characters with "mouth (口)" were selected to form the **transliteration** (音译) name, which was revised according to the principles of nomenclature of *organic chemistry (2017 edition)*. At present, there are still 45 basic heterocyclic compounds with specific common and semi-common names as the basis for naming (Table 14.1).

Table 14.1　The structure, name and number of common basic heterocyclic parent nuclei*

表 14-1　常见基本杂环母核的结构、名称及编号 *

Classification		The structure, name and number of basic heterocyclic parent nuclei
Single rings	Five-member rings — Compounds with one heteroatom	呋喃 Furan　　噻吩 Thiophene　　吡咯 Pyrrole
	Five-member rings — Compounds with two heteroatoms	咪唑 Imidazole　吡唑 Pyrazole　噻唑 Thiazole 异噻唑 Isothiazole　噁唑 Oxazole　异噁唑 Isoxazole
	Six-member rings — Compounds with one heteroatom	吡啶 Pyridine　　吡喃 Pyran(e)
	Six-member rings — Compounds with two heteroatoms	嘧啶 Pyrimidine　吡嗪 Pyrazine　哒嗪 Pyridazine

(continued)

Classification		The structure, name and number of basic heterocyclic parent nuclei
Fused rings	Benzene-fused heterocyclic rings	吲哚 Indole　　喹啉 Quinoline　　异喹啉 Isoquinoline　　酞嗪 Phthalazine　　喹喔啉 Quinoxaline 咔唑 Carbazole　　吖啶 Acridine　　吩嗪 Phenazine　　菲咯啉 Phenanthroline
	Fused heterocycle	Purine　　Pteridine　　Indolizine　　Naphthyridine

*Table 14.1 A part of basic heterocyclic parent nuclei. *Nomenclature of Organic Chemistry (2017 edition)*.

Numbering (编号) of **heterocyclic stem nuclei** (杂环母核) mostly follows the specific rules.

(1) When there is only one heteroatom in the ring, the heteroatom is numbered 1, e.g., furan.

(2) Heterocycles with two or more identical heteroatoms are numbered clockwise (or counterclockwise) according to **the Lowest Sequence Principle** (最低系列原则), keeping the heteroatom number as small as possible. Heteroatom with substituent (or H atoms) is first numbered as 1 and then numbered clockwise (or counterclockwise) by the Lowest Sequence Principle, e.g., pyrazole. If the substituent and hydrogen atoms are present simultaneously, the heteroatom with the substituent has the priority number.

(3) A heterocycle with two or more different heteroatoms, its priority is 1 according to the order of O → S → NH → N. Then it is numbered clockwise (or counterclockwise) according to the Lowest Sequence Principle, e.g., thiazole.

(4) The number can also be identified with α、β. The *ortho* position of the heteroatom is α, the next is β、γ...... e.g., furan.

(5) Fused heterocycle is usually numbered according to the corresponding aromatic ring.

Purine is one of the exceptions to the above rules (Table 14.1).

Heterocyclic compounds, like other compounds, also have **isomerism** (同分异构) due to the migration of the H atom (labelled hydrogen) linked to the heteroatom. In order to distinguish these isomers, it is usually necessary to indicate **the labelled hydrogen** (标记氢), as following.

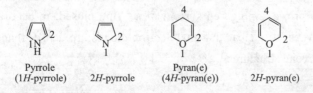

Pyrrole (1*H*-pyrrole)　　　2*H*-pyrrole　　　Pyran(e) (4*H*-pyran(e))　　　2*H*-pyran(e)

PPT

14.3 Properties of Five-membered Heterocycles

Depending on the number of heteroatoms as well as the fused ring, the five-membered heterocycles can be further classified as following.

Five-membered heterocycles
- Compounds with one heteroatom
  - Single heterocycle e.g.:
  - Fused heterocycle e.g.:
- Compounds with two heteroatoms e.g.:

14.3.1　Furan, Pyrrole and Thiophene

Furan, pyrrole and thiophene that contain a heteroatom are very important five-membered heterocyclic compounds. Their **derivatives** (衍生物) are widely found in nature, also, a few have important physiological functions, e.g., heme, chlorophyll and Vitamin B_{12}. Some form the basic ring of a drug, such as **Furazolidone** (痢特灵), **Cephalothin** (头孢噻吩). Some are important raw material for the chemical industry, such as **furfuraldehyde** (呋喃甲醛).

1. Structure　The five-membered ring heterocyclic compounds furan, thiophene, pyrrole, and imidazole are also aromatic.

In these **planar** (平面) compounds, each heteroatom is sp^2 hybridized, and its unhybridized 2p orbital is part of a closed loop of five 2p orbitals. In furan and thiophene, one unshared pair of electrons of the heteroatom lies in the unhybridized 2p orbital and is a part of the π system. The other unshared pair of electrons lies in a sp^2 hybrid orbital perpendicular to the 2p orbitals and is not part of the π system. In pyrrole, the unshared pair of electrons on nitrogen is part of the π system. In imidazole, the unshared pair on one nitrogen is part of the aromatic sextet; the unshared pair on the other nitrogen is not (Figure 14.1).

These compounds form a conjugated system of π_5^6 ring closed, in accordance with Hückel rule to show aromatic properties. It is clear that the electron clouds are denser than benzene and belong to Multi-π aromatic heterocyclic structures.

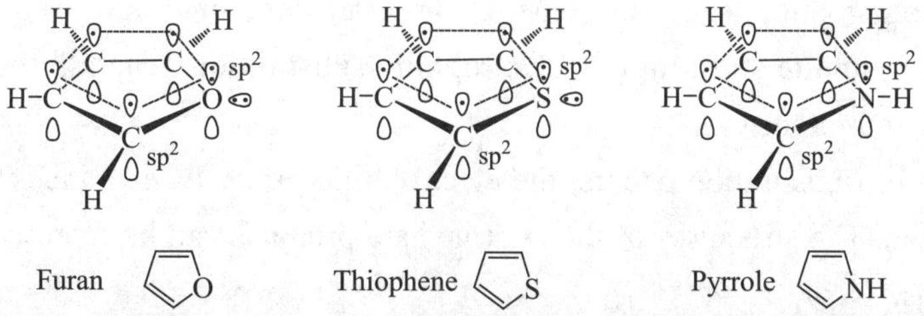

Figure 14.1　Structures of furan, thiophene and pyrrole
图14-1　呋喃，噻吩和吡咯的结构

2. Physical Properties　Furan and thiophene are both clear and colorless liquids at room temperature. While furan is extremely volatile and highly flammable with a boiling point (31.4℃) close to room temperature, relative density 0.9360 (20/4℃). Thiophene coexisted with benzene in coal tar. It possesses a mildly pleasant odor with b.p. 84.2℃ and **relative density** (相对密度) 1.0649 (20/4℃). The b.p. of thiophene is too close to benzene to separate from benzene. Pyrrole was originally isolated from bone oil with a colorless and odorous liquid, b.p.130 ~ 131℃, relative density 0.9691 (20/4℃), and rapidly turning yellow in the air.

All three are poorly soluble in water because the lone pair on the heteroatom is involved in conjugation, greatly weakening the ability to form hydrogen bond with water. Its aqueous solubility (V/V) is approximately: pyrrole (1∶17) > furan (1∶35) > thiophene (1∶700).

3. Chemical Properties

(1) Furan and pyrrole react with the pine slices soaked with hydrochloric acid, displaying green and red，respectively. The blue occurs when thiophene interacts with indigo red and sulfuric acid. These properties can be used to distinguish the three compounds.

(2) Stability of ring　The bond lengths of pyrrole, furan and thiophene molecules are not completely equalized. They still retain the properties of **conjugated diene** (共轭二烯) to a certain extent, and the stability order of the ring is: benzene > thiophene > pyrrole > furan. Furan and pyrrole are not stable. The cyclic structure can be destroyed by acid or **oxidizer** (氧化剂); Thiophene is more stable.

Furan tends to polymerize slowly in the air and is often inhibited by hydroquinone or other phenols. In case of H^+, protonation occurs, cations are generated, and the conjugated system is destroyed to **polymerize** (聚合作用) or open the ring. Pyrrole and furan are roughly the same: under the action of acid, easy to form polymer; the ring can be oxidized by O_2 in the air and slowly opened.

Thiophene, under the action of medium strength proton acid, neither polymerization, nor hydrolysis, is more stable. But it's also unstable when it comes to strong acids.

(3) Electrophilic substitution reaction　Furan, thiophene, and pyrrole belong to "multi-π aromatic heterocyclic ring" structure, therefore, like benzene, can occur electrophilic substitution reaction, and the reactivity is stronger than benzene ring. The sequence of activities is roughly as follows.

Pyrrole > Furan > Thiophene >> Benzene

Because the three structures are not stable enough under the condition of strong acid, the **nitration** (硝化反应) and **sulfonation** (磺化反应) are mainly conducted under mild aprotic conditions, avoiding the use of strong acid. And because the reactivity is very high, the reaction is done at low temperature.

Pyrrole, furan and thiophene undergo electrophilic substitution reactions. However, the reactivity of this reaction varies significantly among these heterocycles. The ease of electrophilic substitution is usually furan > pyrrole > thiophene > benzene. Clearly, all three heterocycles are more reactive than benzene towards electrophilic substitution. Electrophilic substitution generally occurs at C-2, i.e., the position next to the hetero-atom.

When electrophilic substitution occurs, the electrophilic group E^+ can attack the C-2 position and C-3 position of pyrrole. The structures of the intermediate products can be represented by the following resonance expressions:

C-2 position attack:

C-3 position attack:

When attacking the C-2 position, the positive charge can be spread over three atoms. When attacking the C-3 position, the positive charge can only be spread over two atoms. Intermediates are more stable when attacking C-2 position, so C-2 position substituents are mainly generated in electrophilic substitution.

1) **Halogenation reaction (卤代反应)**   Furan, thiophene, pyrrole can occur halogenation reaction, the reaction is usually carried out under mild conditions, such as low temperature and dilute solvent.

2) Nitrification reaction   Furan, thiophene, pyrrole can be nitrated, but cannot be nitrated directly with nitric acid. The strong acidity and oxidation of nitric acid can destabilize the ring. Thus, Nitration is usually carried out with aprotic acetic nitric anhydride at a low temperature.

A freshly prepared **nitric acetic anhydride (硝乙酐)** with acetic anhydride and nitric acid is needed in reaction.

The reaction scheme:

pyrrole + $CH_3COO^- \overset{+}{NO_2}$ $\xrightarrow[-10°C]{(CH_3CO)_2O}$ 2-nitropyrrole (51%) + 3-nitropyrrole (13%)

3) **Sulfonation reaction**　Furan and pyrrole are not stable enough to be sulfonated directly with sulfuric acid. It is obtained by sulfonating with an **aprotic** (非质子) sulfonating reagent (e.g., **pridine trioxide admixture** (吡啶三氧化硫加合物)) and then hydrolyzing under acidic condition.

furan + pyridine·SO_3^- $\xrightarrow[\text{sulfonated}]{CH_2Cl_2\ \text{room temperature}}$ furan-SO_3^-·$\overset{+}{H}N$pyridine $\xrightarrow[\text{hydrolysis}]{HCl}$ furan-SO_3H

Thiophene is relatively stable and can be sulfonated directly with sulfuric acid.

thiophene + H_2SO_4(strong) $\xrightarrow{\text{room temperature}}$ thiophene-SO_3H

This reaction can be used to remove thiophene impurity in benzene. Crude benzene from coal tar usually contains thiophene impurity. The fractional distillation does not remove thiophene easily because thiophene (bp 84.2°C) and benzene (bp 80.1°C) have approximate boiling points. If concentrated H_2SO_4 is added to crude benzene, benzene does not react with concentrated H_2SO_4 at room temperature. However, thiophene reacts with concentrated H_2SO_4 to produce α-thiophene sulfonic acid, which is dissolved in concentrated H_2SO_4. By **oscillation** (震荡), **delamination** (分层) and separation, α-thiophene sulfonic acid can be removed to obtain the refined benzene.

4) **Friedel-Crafts reaction**　Furan can undergo the typical Friedel-Crafts acylation reaction. The Friedel-Crafts acylation of thiophene will proceed under the controlled conditions. The Friedel-Crafts of pyrrole can occur on both carbon and nitrogen atoms.

furan + $CH_3\overset{O}{\overset{\|}{C}}O\overset{O}{\overset{\|}{C}}CH_3$ $\xrightarrow{BF_3}$ furan-$\overset{O}{\overset{\|}{C}}$-$CH_3$ (75%-92%)

furan + $CH_3\overset{O}{\overset{\|}{C}}O\overset{O}{\overset{\|}{C}}CH_3$ $\xrightarrow{BF_3}$ pyrrole(N-H)-$\overset{O}{\overset{\|}{C}}$-$CH_3$
$\xrightarrow{CH_3COONa}$ pyrrole(N-$\overset{O}{\overset{\|}{C}}$-$CH_3$)

It is difficult for the alkylation of furan, thiophene and pyrrole to remain in **the first substitution phase** (一取代阶段), and most of time mixtures with little significance can be obtained.

The above electrophilic substitution reactions of furan, thiophene and pyrrole mostly occur at the α-carbon, which indicated that the α-carbon is more active than the β-carbon.

Their electrophilic substitution reaction mechanisms are similar to that of benzene ring. First, the positively charged electrophilic group E^+ attacks the aromatic heterocycle to form the π complex, which is then transformed into a stable σ complex (see chapter 6). If E^+ attacks α-carbon, σ complex (Ⅰ) is formed, which exists three different resonance structures. And attraction at β carbon leads to the formation of complex (Ⅱ), which only have two different resonance contributors. According to the resonance theory, the more forms of the resonance contributors, the more stable the structure is. Therefore (I) is more stable than (II), and α-carbon is more active than β-carbon.

σ complex

(4) Addition reaction   Furan, thiophene and pyrrole are considered weakly aromatic compared with benzene ring.The aromaticity is in the following order.

The six carbon atoms of the benzene ring are completely delocalized. The resonance theory accounts for the much greater stability of benzene (resonance energy) when compared with other aromatic structures. Furan, thiophene and pyrrole can be regarded as carbon atoms on the benzene ring are replaced by heteroatoms. The less the difference in electronegativity between heteroatom and carbon atom is, the aromaticity is.

The instability of pyrrole and furan leads to the ease of their addition reaction.

1) Catalytic hydrogenation   Furan and pyrrole can react with hydrogen by heating or catalyzing. Thiophene containing sulfur atom will make catalyst poisoning, so its catalytic hydrogenation is more difficult.

2) Diels-Alder reaction   Furan with the least aromatization can also undergo Diels-Alder reaction, showing similar properties to conjugated diolefins.

(5) Acidity and basicity of pyrrole   Because the lone pair of nitrogen are part of the aromatic sextet of electrons, pyrrole is much less basic ($pK_a = 0.4$) than tetrahydropyrrole ($pK_a = 11.11$). Because the nitrogen has certain electronegativity, pyroole can be deprotonated and is aweak acid ($pK_a = 17.5$).

$$\text{pKa } 0.4 \qquad \text{pKa } 11.1$$
$$\text{pKa } 17.5 \qquad \text{Tetrahydropyrrole}$$

4. Derivatives

(1) Furan derivatives　The most important furan derivative is **furfural** (糠醛).

Furfural, i.e., α-furfuraldehyde, is obtained by treating rice bran, corn cob, sorghum stalk, peanut shell and other crops with dilute acid.

Pure furfural is a colorless and toxic liquid, b.p.161.8℃, soluble in water. It is readily polymerized in light, heat and air to turn yellowish brown. Furfural shows a deep purple color when it comes to aniline acetate solution, which is a common method to identify furfural (and other pentoses).

The property of furfural is similar to that of benzaldehyde, which can undergo disproportionation, oxidation and condensation of aromatic aldehyde. The reactions are represented as shown below.

strong NaOH Δ → CH$_2$OH ＋ CHO (Disproportionation reaction)

Furfuryl alcohol(α-furfuryl alcohol)　Furoic acid(α-furanoic acid)

Ag(NH$_3$)$_2^+$ Δ → COOH　(Oxidation reaction)

(CH$_3$CO)$_2$O CH$_3$COONa → CH=CH–CHO (Perkin Condensation reaction)

Furfural is widely used in the syntheses of drugs such as **furazolidone** (痢特灵), **furacillin** (呋喃西林), phenolic resin and pesticide, etc.

O$_2$N—⟨ ⟩—CH＝N—⟨ ⟩=O

Furazolidone
(Furazolidone, antimicrobials)

O$_2$N—⟨ ⟩—CH＝N–NHCONH$_2$

Nitrofurazone
(Antimicrobials, widely used to inhibit and even kill bacteria)

In addition to furfural, furan derivatives also take place in plants and microorganisms.

Tetrahydro-3,6-dimethylbenzofuran
(It's found in peppermint oil)

Rose furan　(Exist in rose oil)

(2) Pyrrole Derivatives　Most pyrrole derivatives exist in **porphyrin (卟吩)** ring. Porphyrin ring refers to the stable and complex conjugated system composed of "α-C" of four pyrrole rings linked by four methine groups (— CH ＝). Its derivatives are called porphyrins. Porphyrins are found abundantly in plants and animals. The N atoms in the ring are readily to form complex compounds with metals (Mg and Fe, etc.) to give a color product. The important porphyrins are heme, vitamin B$_{12}$ and chlorophyll.

(3) Thiophene derivatives　They exist in fungi and compositae (e.g., 2, 2'-dithiophene derivatives). A lot of synthetic drugs also have thiophene rings (e.g., tiprofenoic acid, methorphenilene).

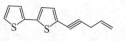

2,2'-dithiophene derivatives
(Can kill nematodes)

Methaphenilene (Antihistamine)

Tiaprofenic Acid
(Anti-inflammatory drugs)

14.3.2 Imidazole, Pyrazole and Thiazole

Heterocyclic aromatic compounds can also have two or more hetero-atoms. If one of the hetero-atoms is a nitrogen atom, and the compound has a five-membered system, their names all end in *-azole*, and the rest of the name indicates other hetero-atoms. According to the relative position of the two heteroatoms, they can be divided into 1, 3-azole and 1, 2-azole.

1, 3-azole   Oxazole   Imidazole   Thiazole

1, 2-azole   Isoxazole   Pyrazole   Isothiazole

1. Structure   Oxazole, imidazole and thiazole systems contain a five-membered ring and two hetero-atoms, one of which is a nitrogen atom. The hetero-atoms are separated by a carbon atom in the ring. The second hetero-atoms are oxygen, nitrogen and sulphur for oxazole, imidazole and thiazole systems, respectively. These compounds are isomeric with the 1,2-azoles, e.g., isoxazole, pyrazole and isothiazole. The aromatic characters of the oxazole, imidazole and thiazole systems arise from delocalization of a lone pair of electrons from the second hetero-atom (sp^2), which meets the Hückel criteria for aromaticity.

1, 3-azole (Z = O, S, N)   1, 2-azole (Z = O, S, N)

2. Physical Properties   Oxazole and isoazole are colorless liquids with b.p. 69 ~ 70℃ and 94.5℃, respectively. Thiazole and isothiazole are also colorless liquids with b.p. 117℃ and 113℃, respectively. Both midazole and pyrazole are colorless crystals with m.p. 90 ~ 91℃ and 69 ~ 70℃, respectively. Both of them are solid by hydrogen bonds at room temperature. All the six azoles have different unpleasant odors and are soluble in water. The lone pair electrons of the N-2 or N-3 position do not participate in conjugation, but exist in an "exposed" formation form hydrogen bond (see 14.4.1). Some of the physical properties of these compounds are presented below (Table 14.2).

Table 14.2　Physical constants of azole compounds

表 14-2　唑类化合物的物理性质

structure					
b.p. (℃)	186～188	257	69～70	95～96	116.8
m.p. (℃)	69～70	90～91	–	–	–
water solubility	1 : 2.5	1 : 0.65	∞	1 : 6	
pK_a	2.5	7.0	0.8	-2.03	2.4

3. Chemical Properties

(1) Acidity and alkalinity　Among these 1, 3-azoles, imidazole is the most basic compound. The increased basicity of imidazole can be accounted for from the greater electron-releasing ability of two nitrogen atoms relative to a nitrogen atom and a hetero-atom of higher electronegativity.

Besides, imidazole and pyrazole also exhibit weak acidity because of NH in their structures.

(2) **Electrophilic substitution** (亲电取代反应)　Oxazole, imidazole and thiazoles are not very reactive towards aromatic electrophilic substitution reactions. They are less reactive than furan, thiophene and pyrrole, even including benzene. The N-2 or N-3 position atom in the azole ring are also involved in the formation of arornatic π system, but the N atom can attract electrons more than the C atom, which leads to a lower electron cloud density on the ring.

It is observed from the above two cases that the electrophilic substitution of azoles can be in either the C-4 position or the C-5 position.

(3) Tautomerism Imidazole and pyrazole undergo 1, 3-migration due to their NH structure, and it is called tautomerism.

Actually, the identical C-4 position or C-5 position of imidazole is not likely to be separated.

5-methyl imidazole　　　4-methyl imidazole

(4) Azole derivatives

1) Imidazole derivatives   Imidazole is both weak acid and weak base, and with $pK_a = 14.5$ close to the physiological pH $= 7.35$.

This unique property enables imidazole ring to play an important role in proton transfer in organisms and is widely used as imidazole drugs.

Metronidazole
(antiamoeba and trichomonas drugs)

In addition, a lot of naturally occurring compounds also contain imidazole ring systems. For example, histidine, pilocarpine, etc.

Histidine
(Exist in proteins)

Pilocarpine (Alkaloid)
(Exist in pilocarpine plants, can treat glaucoma)

2) Other azole derivatives   Other kinds of azoles besides imidazole are involved in some important compounds as shown below.

Pimprinin
(Isolation from Streptomycessp)

2-(4-chlorophenyl)-thiazole-4-acetic acid
(Anti-inflammatory agent)

Isothiazole derivative
(Inhibit virus growth)

Eflunomide
(Antirheumatic and rheumatic drugs)

Dibenzimidazole
(Analgesic, anti-inflammatory and antipyretic effects)

14.4 Properties of Six-membered Heterocycles

PPT

Another major class of heterocyclic compounds is six-membered Heterocycles. Based on the number of heteroatoms and species of fused rings, six-membered heterocycle can be further classified.

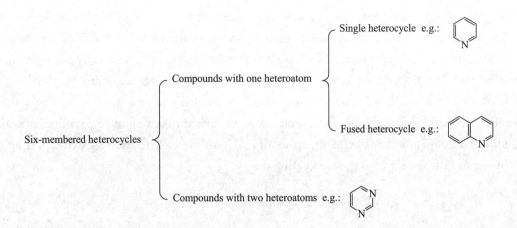

Six-membered heterocycles
- Compounds with one heteroatom
  - Single heterocycle e.g.:
  - Fused heterocycle e.g.:
- Compounds with two heteroatoms e.g.:

14.4.1 Pyridine

1. Structure Pyridine is an aromatic nitrogen analogue of benzene. Each of the five sp^2-hybridized carbons has a p orbital perpendicular to the plane of the ring. Each p orbital has one π electron. The nitrogen atom is also sp^2-hybridized and has one electron in the p orbital. So, there are six p electrons in the ring. The nitrogen lone pair electrons are in an sp^2 orbital and are not a part of the aromatic π system. The C-C bond length is 139 pm and the C-N bond length is 137 pm, which is between the common C-N single bond (147 pm) and the C=N double bond (128pm).

Since pyridine bas six π electrons to comply with (4n+2) π electron rule, it is aromatic. But in this aromatic ring, since N is more electronegative than C, N atom behaves as an electron withdrawing group. In conclusion, pyridine is an aromatic heterocyclic ring, and its electrophilic activity is weaker than that of benzene.

微课

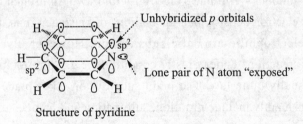

Unhybridized *p* orbitals

Lone pair of N atom "exposed"

Structure of pyridine

Comparing the structure of pyridine and pyrrole, although both pyrrole and pyridine have N atoms, they have different characteristics. Pyrrole N (-NH-) donates electrons and is mostly acidic. Pyridine N is basic and electron withdrawing to the ring. These two types of N atoms also occur in other heterocycles.

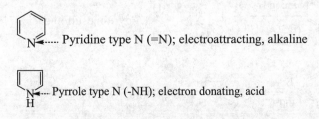

Pyridine type N (=N); electroattracting, alkaline

Pyrrole type N (-NH); electron donating, acid

2. Physical Properties   Pyridine is a colorless liquid (m.p.-42°C, b.p.115°C) with an unpleasant smell. The inhalation of its vapor can easily damage **the nervous system** (神经系统). Pyridine is a polar aprotic solvent and is miscible with both water and organic solvents. The dipole moment of pyridine is 1.57 D. Pyridine is an excellent donor ligand in metal complexes. It is highly aromatic and moderately basic in nature, with a pK_a of 5.23, i.e. a stronger base than pyrrole but weaker than alkylamines. The lone pair of electrons on the nitrogen atom in pyridine is available for bonding without interfering with its aromaticity. Pyridine can be used as extraction agent or a developing solvent for extracting and isolating drugs.

AlkylR: Soluble in organic solvents
Lone pair electrons: form hydrogen bond with water, soluble in water

3. Chemical Properties

(1) Pyridine   Pyridine has an available pair of nonbonding electrons, which can be combined with H^+ to present basicity ($pK_a = 5.2$). The pyridinium ion is still aromatic because the additional proton has no effect on the electrons of the aromatic sextet: It simply bonds to pyridine's nonbonding pair of electrons.

Pyridine hydrochloride

Tertiary amine   Quaternary ammonium salt

(2) Electrophilic substitutions   Pyridine's electron-withdrawing nitrogen causes the ring carbons to have significantly less electron density than the carbons of benzene. Thus, Pyridine is less reactive than benzene towards electrophilic aromatic substitution. However, Pyridine undergoes some electrophilic substitution reactions under drastic conditions, e.g., high temperature, and the yields of these reactions are usually quite low. The main substitution takes place at C-3. The important electrophilic substitutions mainly include nitration, sulfonation and halogenation.

(β-nitropyridine, 15%)

(β-pyridinesulfonicacid, 71%)

(β-brominepyridine)

(3) **Nucleophilic substitutions (亲核取代反应）**　Under the influence of strong electron absorption of N atom, pyridine ring is more prone to nucleophilic substitution reaction, most of which will occur on the *ortho* position to the N atom.

(4) Oxidation　Pyridine ring is harder to oxidize than benzene because it is π electron-deficient aromatic heterocyclic ring. Therefore, it is often used as a complex with oxidizer.

However, when pyridine rings have side chains, the side chains can be oxidized and the rings are unaffected, as with toluene.

(5) Reduction reaction　Pyridine is more readily to undergo reduction reaction than benzene ring under a milder catalytic hydrogenation condition. The saturated hexahydropyridine (piperidine) can be obtained by catalytic hydrogenation.

Hexahydropyridine is actually a cyclic secondary amine to be more basic (pK_a 11.2).

(6) Pyridine derivatives　Niacin and niacinamide: Niacin or β-pyridine formic acid is one of the B vitamins, also known as nicotinic acid, vitamin B_3 or PP, anti-pellagra factor. Niacin acid is colorless, slightly sour acicular crystal with m.p. 236℃; It is soluble in boiling water and hot alcohol, stable in nature. Niacin is present as niacinamide in vivo. Nicotinamide is white crystalline powder, odorless, bitter, soluble in water and ethanol with m.p.128 ~ 131℃ and b.p.150 ~ 160℃. The lack of niacinic acid in the body is prone to pellagra. The main incidence areas are China's xinjiang province and Egypt, where corn is the staple food

β-pyridinecarboxamide (Nicotinamide)　　β-picolinic acid (Nicotinic acid)

Vitamin B_6: One of the B vitamins is widely found in food, including pyridoxine, pyridoxal and pyridoamine, which can be transformed each other in vivo. Their hydrochlorides are colorless crystal, soluble in water, sparingly soluble in ethanol and acetone. Pyridoxine is the representative of Vitamin B_6 because it was the first to be found in the three substances.

Pyridoxine   Pyridoxal   Pyridoamine

Vitamin B_6 is bound up with the metabolism of amino acids. The lack of vitamin B_6 may result in dermatitis, spasm and anemia, etc.

14.4.2 Pyrimidine

1. Structure and Physical Properties   Pyrimidine is a six-membered aromatic heterocyclic compound that contains two nitrogen atoms, separated by a carbon atom. In the planar compound, each heteroatom is sp^2-hybridized and is a part of the π system which meets the Hückel criteria for aromaticity. In pyrimidine, neither unshared pair of electrons of nitrogen is part of the π system. The resonance energy of pyridine is estimated to be 134 kJ·mol^{-1}, slightly less than that of benzene. Pyrimidine is a hygroscopic solid (m.p.22℃, b.p.124℃) and soluble in water.

2. Chemical Properties

(1) Alkalinity   Pyrimidine (pK_b = 12.7) is a weaker base than pyridine (pK_b = 8.8) because of the presence of the second nitrogen as an electron-withdrawing group (the equivalent of -NO$_2$).

(2) Electrophilic substitutions   Pyrimidine is more difficult to undergo electrophilic substitution than pyridine because of the electron withdrawing N atom. Its reactivity is approximately equivalent to 1, 3-dinitrobenzene or 3-nitropyridine. However, if there are other electron-donating groups on the ring, the aromatic ring can be activated for the electrophilic the reactions.

(3) Nucleophilic substitutions   Pyrimidine is readily to undergo nucleophilic substitution because of the electron withdrawing N atom. The reaction is easily to take place in the C-2 position and C-4 position.

In addition, pyrimidine rings, like pyridine rings, can undergo reduction reactions. Side chain of pyrimidine can also be oxidized.

(4) Derivatives   Pyrimidine derivatives, such as cytosine, uracil and thymine, etc. are important components of nucleic acids.

Cytosine (C)

Uracil (U)

14.5   Properties of Fused Heterocyclics

14.5.1   Indole

1. Structure   Indole contains a benzene ring fused with a pyrrole ring at C-2/C-3, and can be described as benzopyrrole. Indole is a ten π electron aromatic system achieved from the delocalization of the lone pair of electrons on the nitrogen atom. Benzofuran and benzothiaphene are very similar to benzopyrrole (indole), with different hetero-atoms, oxygen and sulphur respectively.

| Benzopyrrol Indole | Benzofuran | Benzothiophene |

2. Physical Properties   Indole is colorless or yellowish flake crystal (m.p.52.2℃, b.p.253℃) at room temperature, and possesses an intense faecal smell. However, at low concentrations it has a flowery smell. Indole is slightly soluble in water, but readily soluble in organic solvents, e.g., ethanol, ether and benzene.

3. Chemical Properties   Indole is slightly more aromatic and stable than pyrrole. Indole is not **sensitive (敏感)** to acid, alkali and oxidant. The acidity and basicity of indole are close to pyrrole. Acidity: indole (pK_a 16.2), pyrrole (pK_a 17.5). Basicity: indole (pK_b 17.6), pyrrole (pK_b 13.6).

Electrophilic aromatic substitution of indole occurs on the fifive-membered pyrrole ring, because it is more reactive towards such reaction than a benzene ring. As an electron-rich heterocycle, indole undergoes electrophilic aromatic substitution primarily at C-3 (β position), for example bromination of indole.

$$\xrightarrow[\text{dioxane, 0℃}]{\text{Br}_2}$$ (70%)

Intermediate (Ⅰ) formed by attacking the α position is more stable than the intermediate (Ⅱ) formed by attacking the β position.

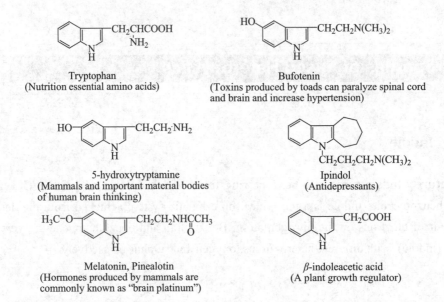

4. Indole Derivatives   Indole derivatives are numerous and wide uses, such as:

Tryptophan
(Nutrition essential amino acids)

Bufotenin
(Toxins produced by toads can paralyze spinal cord and brain and increase hypertension)

5-hydroxytryptamine
(Mammals and important material bodies of human brain thinking)

Ipindol
(Antidepressants)

Melatonin, Pinealotin
(Hormones produced by mammals are commonly known as "brain platinum")

β-indoleacetic acid
(A plant growth regulator)

Indole derivatives are abundant in traditional Chinese medicine, including indigo, azurin B, reserpine and so on.

Indigo

Azurin B

14.5.2   Quinoline and Isoquinoline

1. Structures   Quinoline and isoquinoline, known as benzopyridines, are two isomeric heterocyclic compounds that have two rings, a benzene and a pyridine ring, fused together. In quinoline this fusion is at C2/C3, whereas in isoquinoline this is at C3/C4 of the pyridine ring. Like benzene and pyridine, these benzopyridines are also aromatic in nature.

Quinoline   Isoquinoline

2. Physical Properties　Quinoline is a colorless liquid (m.p. −15.6℃, b.p. 238℃) with **special odor** (特殊气味). Isoquinoline is colorless solid (m.p. 26℃, b.p. 243℃) with benzaldehyde odor. Quinoline and isoquinoline are basic in nature. Like pyridine, the nitrogen atom of quinoline and isoquinoline is protonated under the acidic conditions. The conjugate acids of quinoline and isoquinoline have similar pK_a values (4.85 and 5.14, respectively). Quinoline is only slightly soluble in water but dissolves readily in many organic solvents. Isoquinoline crystallizes to platelets and is sparingly soluble in water but dissolves well in ethanol, acetone, diethyl ether and other common organic solvents. It is also soluble in dilute acids as the protonated derivative.

3. Chemical Properties

(1) Oxidation reaction　Quinoline and isoquinoline are stable. For most oxidants, they do not react, but $KMnO_4$ can oxidize quinoline to quinolinic acid.

(2) Substitution reaction　Quinoline and isoquinoline undergo electrophilic aromatic substitution on the benzene ring, because a benzene ring is more reactive than a pyridine ring towards such reaction. Substitution generally occurs at C-5 and C-8.

Nucleophilic substitutions in quinoline and isoquinoline occur on the pyridine ring because a pyridine ring is more reactive than a benzene ring towards such reaction. While this substitution takes place at C-2 and C-4 in quinoline, isoquinoline undergoes nucleophilic substitution only at C-1.

(3) Reduction reaction　Quinoline and isoquinoline can undergo reduction reaction by **catalytic hydrogenation** (催化氢化), usually pyridine ring is reduced first.

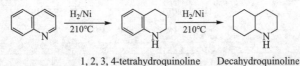

1, 2, 3, 4-tetrahydroquinoline　　Decahydroquinoline

14.5.3　Purine

1. Purine　Purine consists of a pyrimidine ring fused to an imidazole ring. There are two tautomers of 7H-purine and 9H-purine. The former mainly exists in drugs and the latter mainly exists in organisms.

Purine is colorless **needle crystal** (针状结晶) (m.p. 216 ~ 217℃) and soluble in water, ethanol, while insoluble in nonpolar organic solvent. It is aromatic and quite stable. Free purines are rarely found in nature. But purine derivatives are widely found in animals and plants, and many drugs contain the structural skeleton of purines.

2. Purine Derivatives

(1) Uric acid　Uric acid, also named 2, 6, 8-trihydroxy purine, exists in the excreta of some animals and a small amount in human body. It is called uric acid because it was originally found in **urinary stones** (尿结石). There are the following tautomeric isomers.

Uric acid is white crystal. It is sparingly soluble in water. It is in the form of sodium salt in body that can be soluble in water. urinary stones in the joint precipitation will cause **joint pain** (关节痛).

(2) Caffeine　Caffeine is white needle-shaped crystal with bitter taste and exists in tea and coffee. Caffeine can stimulate the heart, brain and induce diuresis.

Caffeine
(Caffeinum, 1, 3, 7-trimethylxanthine)

3. Adenine and Guanine　Both adenine (6-aminopurine) and guanine (2-amino-6-hydroxypurine) have tautomers (equaliam lies to the right).

Adenine and guanine, like uracil, cytosine and thymine, are bases constituting nucleotides. Nucleic acids are formed by the polymerization of nucleotides, which are important living substances.

14.6　Applications of Heterocyclics in the Pharmacy Study

PPT

1. Pyrrole Drugs　Imidazoles and triazoles belong to pyrrole antifungal drugs. People often use ketoconazole, miconazole and clotrimazole. The latter two are mainly used. Fluconazole and itraconazole belong to triazoles, and are mainly used to treat deep **mycosis** (真菌病).

Ketoconazole

Fluconazole

2. Furan Drugs　Furan drugs belong to chemical synthetic drugs. It is used as chemotherapy drugs in the early 1940s. They can act on the **enzyme system** (酶系统) of bacteria, interfere with the sugar metabolism and inhibit bacteria. At present, there are more than 10 kinds of furan drugs, commonly are nitrofurantoin, furazolidone, nitrofurazone, etc. Nitrofurantoin has widely antimicrobial spectrum, inhibits many **Gram-positive and Gram-negative bacteria** (革兰阳、阴性细菌), except Pseudomonas aeruginosa. Because of this, it often used to treat urinary tract infection. Furazolidone can inhibit various bacteria in digestive tract. It can also inhibit trichomonad, and mainly act on digestive tract to treat enteritis, dysentery and typhoid fever.

微课

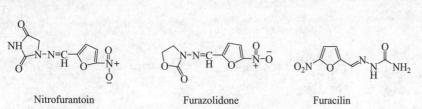

Nitrofurantoin　　　　　　　Furazolidone　　　　　　　Furacilin

3. Thiophene Drugs　α-Thiophene derivatives are widely used in synthetic medicine. Antibiotics with thiophene ring have better curative effect than phenyl homologs. Some new **antiphlogistic and analgesic drugs** (镇痛消炎药), such as methamphetamine, sulfene, tioblonic acid, tiolofenic acid, pizotifen and sufentanil, etc. have significant curative effect.

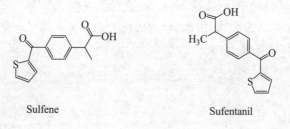

Sulfene　　　　　　　　　　Sufentanil

4. Pyridine Drugs   Pyridine is readily to undergo nucleophilic substitution reaction, so pyridine and pyridine derivatives can prepare many important pyridine compounds with important biological activities. It mainly includes antibacterial, anti-depression, anti-infection, proton pump inhibitors, anti-tuberculosis, vasodilation, and central nervous excitement.

Pasiniazide also is known as Baisheng hydrazine and isoniazid aminosalicylic acid, which is a strong **anti-tuberculosis drug** (抗结核药). It is prepared by oxidizing and acylating pyridine and then adding *p*-aminosalicylic acid.

Pasiniazide

Betahistine hydrochloride, chemically known as *N*-methyl-2-pyridineethylamine dihydrochloride, is a kind of **histamine** (组胺) vasodilators, which can be used to treat coronary artery insufficiency and ischemic cerebrovascular disease, such as dizziness, dizziness and tinnitus which caused by cerebral infarction and high blood pressure.

Betahistine hydrochloride

5. Quinoline Drugs   Nicotinic acid drugs include nicotinamide, cardiotonic, stimulant and taeniasis drugs. The 8-hydroxyquinoline can be mode into the drug for treating amoeba, wound disinfectants, fungicide and textile auxiliaries, etc. Primoquinine, quinine chloride and hydroxylaminine are synthetic effective drugs for **malaria** (疟疾).

Nicotinamide   8-hydroxyquinoline   Quinine

重 点 小 结

　　所谓杂环化合物 (heterocyclic compounds) 是指环上含有杂原子的环状化合物。这里所说的杂原子,是指除碳原子外的其他原子,最常见有 O、S、N 等。在大多数杂环化合物中,因杂原子参与共轭,满足休克尔规则,而表现出芳香性,因此,一般所说的杂环多是指"芳香性杂环"。前面学过的环醚、内酯、交酯、内酰胺、环状酸酐等,虽然也是含有杂原子的环状结构,但没有芳香性,通常也不列入杂环化合物中。

一、杂环化合物的命名

杂环化合物的命名比较复杂。现今广泛使用的是"音译法"，即在IUPAC（2013）命名原则的基础上，结合汉字的特点，选用同音汉字加"口"字旁组成音译名而得，并根据《有机化学命名原则》（2017年版）进行修正。目前还保留有特定的45个基本杂环化合物的俗名和半俗名，并以此作为命名的基础。

基本杂环母核的编号大多有特定的规律：①环上只有一个杂原子时，杂原子编为1号，如呋喃。②环上有两个或多个相同的杂原子的杂环按照最低系列原则顺时针（或逆时针）编号，使杂原子位次尽可能小。有取代基（或H原子）的杂原子优先编为1号，再依"最低系列"法顺时针（或逆时针）编号，如吡唑。如果取代基、氢原子同时存在，则有取代基的杂原子优先编号。③环上有两个或多个不同的杂原子时，依 $O \rightarrow S \rightarrow NH \rightarrow N$ 的顺序，优先编为1号，再依"最低系列"法顺时针（或逆时针）编号，如噻唑。④ 编号亦可采用 α、β 标识：杂原子的邻位为 α，次位为 β 位、γ 位……如呋喃。⑤稠杂环一般根据相应芳环的编号方式编号。当然，也有不遵循上述规则的特例，如嘌呤。

杂环化合物同其他化合物一样，有时也存在同分异构现象。这是由环状共轭的核体系中的H原子发生了共振迁移所致，这样的H原子称为标记氢。为了区别这些异构体，通常要指出标记氢的位置。

二、五元杂环化合物的性质

呋喃环中的C原子和O原子均为 sp^2 杂化。且C—C、C—O、C—H之间均以 σ 单键相连。由于 sp^2 杂化的平面构型，使得C、O、H各原子共平面。每个C原子有一个未杂化的p轨道，垂直于该平面，其上有一个电子；O原子上也有一个未杂化的p轨道，垂直于该平面，其上有两个电子。这样，四个C原子和一个O原子上的五个p轨道相互平行，侧面重叠，形成 π_5^6 环状闭合的共轭体系，符合休克尔规则，有芳香性。显然，与苯相比，π电子云密度更大，属于"多π芳杂环"结构。

三、六元杂环化合物的性质

吡啶可以看成是苯分子中的一个CH被N原子取代后形成的化合物。这个N原子与C原子一样，采取 sp^2 杂化，所以吡啶同苯一样，是共平面分子。在N原子的三个 sp^2 杂化轨道中，有两个分别与相邻的碳原子形成 σ 键，另一个则被一对孤对电子占住。这一对孤对电子没有参与形成大π键，"外露"于共轭环系。另外，N原子上还有一个未参与杂化的p轨道，带有一个电子。这个p轨道垂直于分子平面，与五个碳原子的p轨道平行。因此，它们彼此侧面重叠，形成环状共轭大π键，在这个共轭大π键中，一共有6个π电子，符合4n+2规则，所以具芳香性。但是，在这个芳香环中，由于N电负性比C强，因此π电子云流向N原子。所以，N原子表现为吸电子，使得芳环电子云密度降低。总之，吡啶是"缺π芳杂环"的结构，亲电反应活性比苯弱。

Problems
目 标 检 测

1. Rank the alkalinity of the following compounds from strong to weak, which one is correct?
   按照从强到弱的顺序排列下列化合物的碱性，以下选项正确的是（　　）。

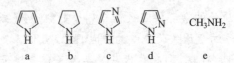

   A. a>b>c>d>e
   B. d>c>a>b>e
   C. b>e>c>d>a
   D. d>e>a>c>b

2. Which is the main position that 2-aminopyridine can process nitration reaction?
   2-氨基吡啶能在温和的条件下进行硝化反应，硝化的位置主要发生在（　　）。
   A. 2- position（2位）
   B. 3-position（3位）
   C. 3-position（4位）
   D. 5- position（5位）

3. Rank the alkalinity of the following compounds from strong to week.
   按照碱性从强到弱的顺序排列下列化合物（　　）。

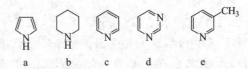

   A. d>e>c>b>a
   B. b>e>c>d>a
   C. a>b>c>d>e
   D. e>d>c>b>a

4. Chose the correct answer belongs to multi-π aromatic heterocycles?
   下列芳香族化合物中属于多π芳杂环的是（　　）。
   A. thiophene (噻吩)
   B. pyridine (吡啶)
   C. N-methylpyridine（N-甲基吡啶）
   D. Cyclooctatetraene anion (环辛四烯二负离子)
   E. benzene (苯)

5. Which of the following structures without aromaticity?
   下列化合物没有芳香性的是（　　）。
   A. Cyclooctatetraene anion (环辛四烯二负离子)
   B. cyclopentadiene anion (环戊二烯负离子)
   C. pyridine (吡啶)
   D. pyrane (吡喃)
   E. Cyclooctatetraene (环辛四烯)

6. Which of the following structures with aromaticity?
   下列化合物具有芳香性的是（　　）。
   A. cyclopentadiene (环戊二烯)
   B. furan (呋喃)
   C. Cyclooctatetraene (环辛四烯)
   D. tropilidene anion (环庚三烯负离子)

E. cyclopentadiene anion (环戊二烯正离子）

7. Which is not prone to electrophilic substitution reactions in the following compounds?
下列芳香族化合物不容易发生亲电取代反应的是（　　　　）。

A.  ⬡　　　B.  ⬠S　　　C.  ⬠O　　　D. 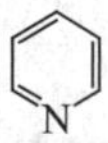 ⬡N　　　E.  ⬠N

8. The most stability of those compounds is （　　　）.
下列化合物的芳香性（环的稳定性）最大的是（　　　　）。

A. thinphone (噻吩）　　　B. benzene (苯）　　　C. furan (呋喃）
D. pyrrole (吡咯）　　　E. pyridine (吡啶）

9. The best mothed to remove thiophene from benzene is （　　　）.
从苯中除去噻酚的最好办法是（　　　　）。

A. 层析法　　　B. 蒸馏法　　　C. 硫酸洗涤法
D. 溶剂提取法　　　E. 减压蒸馏

10. The order of stability of furan (a), pyrrole (b), thiophene (c) and benzene (d) is （　　　）.
呋喃（a），吡咯（b），噻吩（c）和苯（d）的稳定性大小次序是（　　　　）。

A. d>c>b>a　　　B. a>c>d>b　　　C. b>c>a>d
D. a>b>c>d　　　E. c>b>a>d

11. Which can react with KOH crystal in the following compounds?
能与固体KOH反应的是（　　　　）。

A. Pyridine (吡啶）　　　B. Pyrrole (吡咯）
C. tetrahydropyrrole (四氢吡咯）　　　D. aniline (苯胺）
E. acetone (丙酮）

12. Which of the following structures without aromaticity?
下列化合物中，不具有芳香性的是（　　　　）。

A. 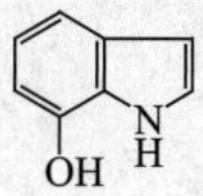 ⬡-fused-ring OH,N-H　　　B. 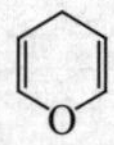 ⬡O　　　C. 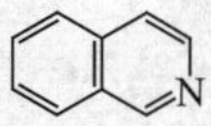 isoquinoline N　　　D. 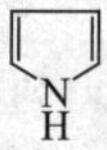 ⬠N-H　　　E.  ⬠⊖

13. Which position is able to process electrophilic substitution easily on ⬠N-H ?
⬠N-H 亲电取代最容易发生的位置是（　　　　）。

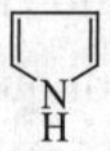

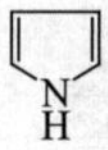

A. N　　　B. the H atom on N atom（N原子上的氢）
C. α-carbon（α-碳）　　　D. β-carbon（β-碳）
E. γ-carbon（γ-碳）

14. Compare the alkalinity of different N in the following structure.
下列化合物中不同N原子的碱性强弱比较为（　　　　）。

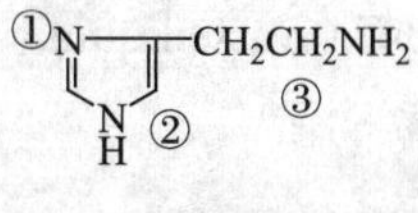

①N⸺CH₂CH₂NH₂③
N-H②

A. ① > ② > ③　　　B. ③ > ② > ①　　　C. ③ > ① > ②
D. ② > ① > ③　　　E. ② > ③ > ①

15. What is the structure of pyridine?

下列吡啶的结构是（　　　　）。

a. 　　b. 　　c. 　　d.

A. a　　　　　　B. b　　　　　　C. c　　　　　　D. d

16. Which compound has the most alkalinity in the following options?

下列化合物碱性最强的是（　　　　）。

A. pyrrole　　　　　　　　　B. methylamine　　　　　　　C. pyridine

D. ammonia　　　　　　　　E. thiophene

17. Rank those compounds from strong to week, according to aromaticity.

下列化合物芳香性顺序为（　　　　）。

A. a>c>d>b　　　　　　　　B. a>b>c>d　　　　　　　　C. b>d>c>a

D. d>b>c>a　　　　　　　　E. c>d>b>a

18. What is the reaction of aniline, glycerin, strong H_2SO_4, nitrobenzene to form quinoline?

苯胺、甘油、浓H_2SO_4、硝基苯共热生成喹啉的反应是（　　　　）。

A. Cannizzaro reaction　　　　　　　　B. Fisher reaction

C. Skraup reaction　　　　　　　　　　D. Hantzsch reaction

19. What method can be used to make the reaction of furan and bromine stays the first phase?

为了使呋喃或噻吩与溴反应得到一溴代产物，采用（　　　　）。

A. high temperature

B. high pressure

C. high temperature and Pressure

D. Solvent dilution and low temperature

20. In the reaction of quinolone，Br_2，Ag_2SO_4，H_2SO_4，where the Br atom mainly atack?

喹啉与Br_2，Ag_2SO_4，H_2SO_4反应，Br原子主要进入（　　　　）。

A. benzene ring

B. pyridine ring

C. benzene and pyridine rings

D. N atom

Discussion Topics

1. Application of heterocyclic compounds in pharmacy and medicine.

2. Treatment of heterocyclic compounds in waste liquid.

（杨　静）

第十五章 基础有机化学反应机理
Chapter 15 Basic Mechanisms of Organic Reactions

 学习目标

1. **掌握** 有机化学反应机理的分类及表达方法;碳正离子、碳负离子和碳自由基的结构特征以及影响稳定性的因素;有机化学反应的类型以及分类依据。
2. **熟悉** 取代反应、加成反应和消除反应的机理。
3. **了解** 周环反应和重排反应的机理。

Mastering of organic chemistry can be a challenging job for most of the students. The main problem comes from the requirement of memorizing a large number of reactions that seem following no rational pattern. However if you understand how reactions work, you may be able to group reactions into only several types according to the common characters of the reactions. In fact, organic reactions feature highly logic structural relationship of compounds, which is called **mechanism** (机理). A mechanism can tell you in a transformation which bonds are broken, which bonds are formed and in what order. In this chapter, we will study the methods for establishing and understanding organic reaction mechanisms. When you master this material, you will even be able to predict what will happen in completely unfamiliar reactions and organic chemistry may become easy and fun.

15.1 Types of Mechanism and How to Present it

PPT

15.1.1 Types of Mechanism of Organic Reactions

An organic reaction mechanism is a step-by-step description of bond-breaking and bond-forming processes, as well as an understanding of the nature of any **intermediates** (中间体) that may be a part of the process.

When organic reactions happen, one or more covalent bonds break and one or more new covalent bonds form. Organic mechanisms can thus be divided into three types based on how the bonds break and form.

1. **Heterolysis (异裂)** the bond breaks in such a way that both electrons remain with one fragment.

Such reactions usually will involve the formation of ionic intermediates.

$$\text{Heterolytic bond breaking:} \quad A:B \longrightarrow A^+ + :B^- \text{ (polar)}$$

2. **Homolysis (均裂)**   the bond breaks in such a way that each fragment gets one electron. Free radicals are formed as intermediates.

$$\text{Homolytic bond breaking:} \quad A:B \longrightarrow A\cdot + B\cdot \text{ (radical)}$$

3. **Pericyclic (周环)**   the bond breaking and forming at the same time and electrons move in a closed ring. No intermediates, ions, or free radicals are formed.

Pericyclic bond breaking and making at the same time: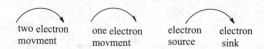

15.1.2   How to Present a Mechanism

All of organic reactions involve the bond breaking and bond formation, and thus the flow of electrons. Organic chemists use a technique called arrow pushing to describe the movement of electrons during chemical reactions. Arrow pushing helps chemists keep track of the way in which electrons and their associated atoms redistribute as bonds are made and broken. So with the arrow pushing technique, a mechanism can be presented. When arrow pushing is used, there are several rules should be kept in mind.

1. **First Rule   Curved arrows (弯箭头)** show movement of electrons and it must start at an **electron source (电子源头)** and end at an **electron sink (电子井)**.

The curved arrows can have a double-sided arrow head or single-sided arrow head. A double-sided arrow head is used to indicate the movement of two electrons, while a single-sided arrow head is used for single electron movement.

two electron movment   one electron movment   electron source   electron sink

Learning to identify the electron sources and sinks is the key to learning organic chemistry reaction mechanisms. The electron source is a site of higher electron density (e.g., lone pairs, bond) and the electron sink is a site of lower electron density [e.g., positively or partially positively charged (带有部分正电) atoms]. Each arrow should point toward where that pair of electrons will be in the products, either forming a bond to another atom, or else residing as an unshared pair on an atom. For example (see Figure 15.1), for arrows that depict the formation of a new σ bond between two molecules, the electron source is often readily identified as being a lone pair on the most electron-rich atom of one molecule or ion and the electron sink is readily identified as the most electron-poor atom of the other molecule or ion. Thus, the prediction of many of the most important electron sources and sinks comes down to telling the differences in electron negativity between atoms.

Meanwhile, we will also introduce here two terminologies that are related to electron source and electron sink, which are **nucleophile (亲核试剂)** and **electrophile (亲电试剂).** Nucleophile originates from the Greek meaning nucleus loving, seeking a region of low electron density. Nucleophiles are electron sources that can **donate** (提供) a pair of electrons to form a new covalent bond. Most of time, they are molecules that have lone pairs or relatively electron-rich π bonds. Electrophile originates from the Greek meaning electron loving, seeking a region of high electron density. It is an electron sink that can **accept** (接受) a pair of electrons to form a new covalent bond. Electrophiles are molecules with relatively electron-poor atoms that serve as sinks.

Analogously, a molecule (or region of a molecule) that is a source for a curved arrow is called nucleophilic, while a molecule or region of a molecule that is a sink for these arrows is referred to as being electrophilic. It is also applied to reactions in which new σ bonds form between electron-rich and electron-poor regions of molecules. The term nucleophile is also analogous to **Lewis bases** (路易斯碱), and the term electrophile is analogous to **Lewis acids** (路易斯酸). The choice of terminology depends upon context.

2. **Second Rule**　making a new bond to an electron sink often requires the simultaneous breaking of one of the bonds connected to the sink atom to avoid overfilling its valence orbitals, a situation referred to as **hypervalence** (超共价).

In this case, the electron source for the arrow is the bond being broken and the sink is an atom able to accommodate the electrons as a lone pair, generally an electronegative atom such as an oxygen or a halogen atom. If an ion is created, that ion is often stabilized by resonance delocalization or other stabilizing interactions. For example, the proton transfer reaction between acetic acid and hydroxide, we can summarize as simple one-step mechanism as shown in Figure 15.1.

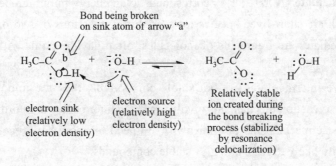

**Figure 15.1　Mechanism of the reaction between acetic acid and hydroxide to
show the application of curved arrow**
图15-1　以乙酸和氢氧化钠的反应为例展示弯箭头的正确画法

3. **Third Rule**　Arrows are never used to indicate the movement of atoms directly.

The arrows only show atom movement indirectly as a consequence of electron movement when covalent bonds are made and broken. One of the right and wrong contrast examples of reaction between acetic acid and hydroxide is shown in Figure 15.2.

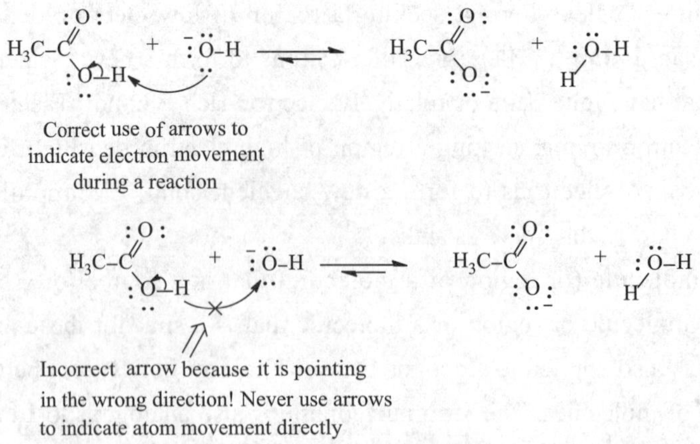

Correct use of arrows to
indicate electron movement
during a reaction

Incorrect arrow because it is pointing
in the wrong direction! Never use arrows
to indicate atom movement directly

**Figure 15.2　Right and wrong contrast examples of the reaction
mechanism between acetic acid and hydroxide**
图15-2　正确与错误对比展示弯箭头的画法

15.2　Types of Intermediates of Organic Reactions

A **reactive intermediate** (反应中间体) corresponds to an energy minimum between two **transition states** (过渡态), which are short-lived species forming during reactions of two or more steps. In most cases, reactive intermediates are fragments of molecules, often having atoms with unusual numbers of bonds.

Although reactive intermediates are not stable compounds, they are important to our study of organic chemistry. Most reaction mechanisms involve reactive intermediates. In order to understand these mechanisms and propose mechanisms of your own, you need to know how reactive intermediates are formed and how they are likely to react and give stable compounds.

Since organic chemistry is mainly about the chemistry of carbon, the character of some of the common reactive intermediates containing carbon atoms are summarized here. These species have only two or three bonds, compared with carbon's four bonds in its stable compounds.

1. Carbocations (碳正离子)　Carbocation has a trigonal planar shape. The carbon atom is sp^2 hybridized with three bonds and an empty p orbital.

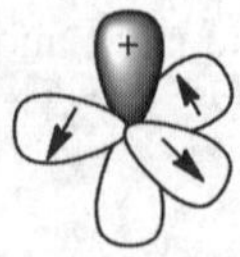

carbocation

Based on the structure of the carbocation, the **stability** (稳定性) of it can be influenced by the following factors.

Stabilized by inductive electron donors, so 3°> 2°>1°.

3°	2°	1°	
Tertiary	**Secondary**	**Primary**	**Methyl**
（叔碳）	（仲碳）	（伯碳）	

Stabilized by **resonance** (共振); **allyl** (稀丙基) and **benzyl** (苄基) cations are stabilized.

allylic (烯丙基的) carbocations　　**benzylic** (苄基的) carbocations

Stabilized by sharing electrons with atoms bearing lone pairs (O, N, S).

As we have seen in chapter six, aromatic carbocations, such as **tropylium ion** (䓬离子), is especially stable.

tropylium ion

Destabilized by electron-withdrawing atoms or groups.

Cations that are constrained not to be planar generally are not stable. For example the cation in the following picture can not be prepared because the bridge head carbon is constrained to a pyramidal structure with an empty sp^3 orbital.

In organic reactions, a carbocation behaves as an electrophile. It seeks a negatively charged center to neutralize its positive charge and to obtain a stable octet of electrons. The stability of carbocation has great effect on the reactions with carbocation as intermediate. The more stable the carbocation, the faster it is formed and thus the faster the reaction. For example, carbocations are formed in the heterolytic cleavage of the carbon-oxygen bond in alcohol. The ease of this reaction depends on the stability of the carbocation produced. So the reaction to give RCH_2^+ is difficult, because the primary carbocation is rather unstable, but that to give R_3C^+ is easy, because the tertiary carbocation is much more stable.

$$R\text{--}\ddot{O}H \xrightarrow{H^+} R\text{--}\overset{H}{\underset{H}{O^+}} \longrightarrow R^+ + H_2O$$

$$RCH_2\text{---}O^+H_2 \longrightarrow RCH_2^+ \quad \text{Difficult}$$

$$R_3C\text{---}O^+H_2 \longrightarrow R_3C^+ \quad \text{Easy}$$

Carbocations also often undergo rearrangements, producing more stable ions. For example, when we try to prepare propylbenzene by the **Friedel-Craft alkylation** (付氏烷基化) the main product is iso-propylbenzene instead. The primary cation is formed initially, but it rapidly rearranges by hydride migration to give the more stable secondary cation.

primary carbocation   secondary carbocation

2. Carbon radicals (碳自由基)   Carbon radicals have seven electrons around the central carbon atom—thus three electron pairs and one half-filled orbital. Generally, like the carbocation, it is planar with sp^2 hybridization and a single electron in the unhybridized porbital.

carbon radicals

The stability of carbon radicals mostly parallels that of cations (3° > 2° > 1°), but overall stability is usually lower. They can also be stabilized by resonance (e.g., allyl and benzyl radicals) as shown in the followings. Also notice that for radicals, we use single-headed arrows to move one electron at a time in generating the resonance forms.

allylic radicals

benzylic radicals

The stability of carbon radical has great effect on the reaction mechanism. For example, as we have learned in Chapter 2, the regioselectivity in halogenations of propane can be accounted for in terms of the relative stabilities of radicals (3°> 2°> 1°. methyl). A **regioselective** (区域选择性的) reaction is a reaction in which one direction of bond forming or bond breaking occurs in preference to all other directions.

3. Carbanions (碳负离子)　Carbanions have eight electrons with three bonds and a **lone pair** (孤对电子). Therefore, most of the time , unlike carbocations and carbon radicals, repulsion of the four pairs of electrons will make carbanions to form sp^3 hybridization and **pyramidal** (三角锥型) in geometry, which is associated with much of the reactivity of carbanions.

Carbanions

Carbanions are stabilized by **inductive electron withdrawers** (诱导拉电子基团), so $[CCl_3]^-$ is relatively stable. The effect of **inductive donors** (诱导给电子基团) such as alkyl groups are predictably, exactly the opposite of their effect on carbocations and carbon radicals, so tertiary carbanions are the least stable, and primary carbanions are the most stable.

Carbanions can also be stabilized by resonance; allyl, benzyl, and particularly **enolate ions** (烯醇负离子) are stabilized. Enolate ions are especially stable because in one of the resonance structure, the negative charge is on the more electronegative oxygen atom. Note that when carbanions are in conjugation systems, they are sp^2 hybridized. Carbanions can also be stable when they are aromatic (e.g., cyclopentadienyl anion).

allylic carbanions　　　enolate ions　　　Cyclopentadienyl anion

In reactions, a carbanion reacts as a nucleophile. It seeks a positively charged center to neutralize its negative charge. The stability of carbanions also relates closely to the reaction mechanism. For example, as we have learned in Chapter 11, **β-dicarbonyl** (β–二羰基化合物) compounds are relatively easier to be dehydrogenated than ketones and thus can work as good nucleophiles and be alkylated with alkyl halides. The relatively higher stability of β-dicarbonyl compound comes from the electron withdrawing effect of the two carbonyl groups.

15.3　Types of Organic Reactions

PPT

Organic reactions can be classified in many different ways based on different criteria. In this chapter, we will discuss two major classification methods when reaction mechanism is studied.

15.3.1　Classification by Structural Difference

Based on the difference between reactants and products, there are four basic types of organic reactions: **additions** (加成反应), **eliminations** (消除反应), **substitutions** (取代反应) and **rearrangements**

(重排反应). There are of course some other kinds of changes happen between reactants and products, like ring opening/closure and polymerization; however they are beyond the range of discussion of basic organic chemistry.

Addition reactions are defined as two reactants add together to form the product and no other small molecule is formed. When an addition reaction happens, a π bond is broken and two new σ bonds are formed. So if carbon–carbon multiple bond or carbon–hetero multiple bond is present in the starting material, you may predict a possible addition reaction to happen under certain conditions.

$$A=B + Y-W \longrightarrow A-B\overset{W}{\underset{Y}{}} \quad \text{(Addition)}$$

Elimination reactions are defined as a small molecule being removed from the adjacent two atoms of the starting material and a new multiple bond being formed in the product. Elimination is the converse process of the addition. When an elimination reaction happens, two σ bonds are broken and a new π bond is formed. So if it is possible to remove a stable small molecule from the adjacent two atoms of a starting material, an elimination reaction may be predicted. And when a new multiple bond is formed in the product, you may guess the mechanism of this reaction is elimination.

$$A-B\overset{W}{\underset{Y}{}} \longrightarrow A=B + Y-W \quad \text{(Elimination)}$$

Substitution reactions are defined as an atom or group of atoms in the starting material being replaced by different atoms or group of atoms. The group that has been replaced is called leaving group. So if you find some part in the starting material has become the product, then you may guess substitution might be the mechanism.

$$A-X + Y \longrightarrow A-Y + X \quad \text{(Substitution)}$$

When a single reactant undergoes a reorganization of bonds and atoms to yield a single isomeric product, rearrangement reactions occur. So if you find the product have the same set of atoms as the starting material, but only differ in the arrangement of atoms, you may guess it might be a rearrangement mechanism.

$$\overset{W}{}A-B \longrightarrow A-B\overset{W}{} \quad \text{(Migration)}$$

15.3.2 Classification by Bond-broken Type

Based on how the bond is broken, there are four basic types of organic reactions: **nucleophilic reaction** (亲核反应), **electrophilic reaction** (亲电反应), **free radical reaction** (自由基反应) and **pericyclic reaction** (周环反应).

If in a reaction, a bond is broken heterolytically with carbon taking the positive charge and an electron pair donor (the nucleophile, Nu) is involved, then the reaction is nucleophilic. If in a reaction, a bond is broken heterolytically with carbon taking the negative charge and an electron pair accepter (the electrophile, E) is involved, then the reaction is electrophilic and in most situation carbocations are formed as the intermediate. If in a reaction, carbon-carbon bond is broken homolytically and carbon radical is formed as the intermediate, the reaction is a free radical reaction. If in a reaction, the bonds break and form at the same time without any intermediate and if the reaction goes through a cyclic transition state, then it is pericyclic.

In order to clarify a reaction mechanism, both the structural difference and the bond-broken type should be considered. It means that the addition reaction can be classified as nucleophilic, electrophilic, free-radical or pericyclic addition; substitution reaction can be classified as nucleophilic, electrophilic or free-radical substitution; in most of elimination reactions nucleophiles are included in the mechanism and it can be seen as the opposite direction process of electrophilic addition, but exceptions do exist; rearrangement reactions can also be classified as nucleophilic, electrophilic, free-radical or pericyclic rearrangement. In the next four sections, we will discuss some basic organic reactions mechanism by combining these two standards. And for the ease of cross reference, the location of reactions examples in the former chapters relating to the mechanism discussed will be shown.

15.4 Addition Reaction

PPT

15.4.1 Electrophilic Addition

Carbon-carbon double bond (or triple bond) is the typical functional group for you to predict an electrophilic addition (see Chapter 4.5.2), because the π electrons are loosely held together and they are also stuck out the plane of the molecule to be easily caught by the electrophile. As shown in the Figure 15.3, the general mechanism of electrophilic addition involves two steps, with a carbocation as the intermediate. In the first step, an electrophile, such as a hydrogen halide, halogen or water is added to the carbon–carbon π bond of the alkene to form a carbocation. This is the rate-determining step in the mechanism. In the second step, the negative part of the electrophile, usually X^-, HO^- and so on, attacks the carbocation to form the product.

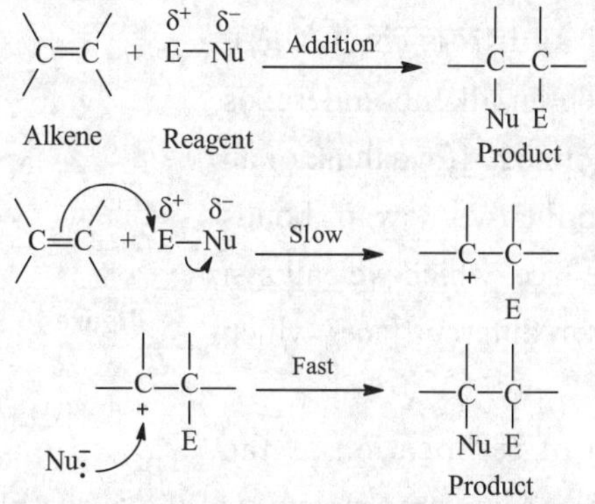

Figure 15.3 General mechanism of the electrophilic addition reaction
图15-3 亲电加成反应机理

When you predict a mechanism for a specific reaction, **regioselectivity** (区域选择性) and **stereoselectivity** (立体选择性) are two things you should notice. Regioselectivity is the favoring of a reagent to one atom over all other possible atoms in a substrate. Stereoselectivity is the property of a chemical reaction in which a single reactant forms an unequal mixture of stereoisomers during a non-stereospecific creation of a new stereocenter or during a non-stereospecific transformation of a pre-existing one.

For electrophilic addition, regioselectivity matters when the alkene has two nonsymmetrical sp^2 carbons. One carbon initially bonds to the electrophile, and the other carbon will become carbocation. You should always choose to make the more stable carbocation of the two possible carbocations. Figure 15.4 illustrates this. Depending on which of the two carbons the proton adds to, we obtain either a tertiary or a secondary carbocation. The reaction that actually occurs is invariably that forms the more stable, in this case the tertiary, carbocation. This is often described as Markovnikov's rule, and this name is used to define any electrophilic addition that proceeds via the most stable carbocation.

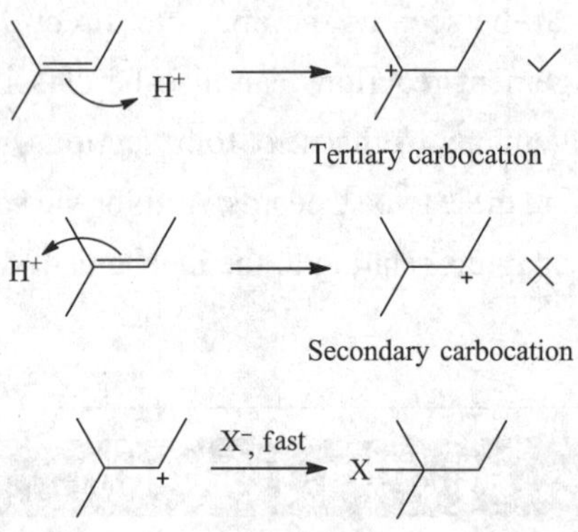

Figure 15.4   Addition of HX to a nonsymmetrical alkene
图15-4   卤化氢的与非对称烯烃的亲电加成

Few electrophilic addition reactions of HX are stereoselective; this is consistent with the mechanism of formation of a planar carbocation intermediate that can be attacked from either face (Figure 15.5a). Here we make the assumption that the proton initially approaches the top face of the molecule (though both faces are equally likely). When the intermediate carbocation is attacked by chloride anion either face can be attacked, so we get two **diastereoisomeric** (非对映异构体) products (and both **enantiomers** (对映异构体) of each diastereomer will be obtained). There are two stereochemistry related terms about addition— **syn addition** (顺式加成) and **anti addition** (反式加成). As shown in Figure 15.5b, when an alkene undergoes addition, two new σ bonds are formed. If we think of an alkene as having two faces, then the two new σ bonds can either both form on the same face, which we call syn addition, or they can be formed on different faces which we call anti addition.

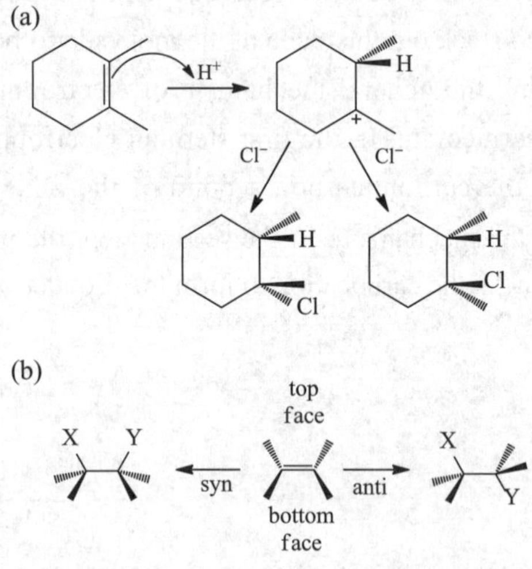

Figure 15.5   Stereochemistry of electrophilic addition reactions
图15-5   亲电加成反应的立体化学

Because of the formation of carbocation as the intermediate, there are another two common characters of the electrophilic addition reactions. First, the product with rearrangement of carbocation can often be observed. One example of rearrangement reaction and the mechanism of the reaction are shown in Figure 15.6.

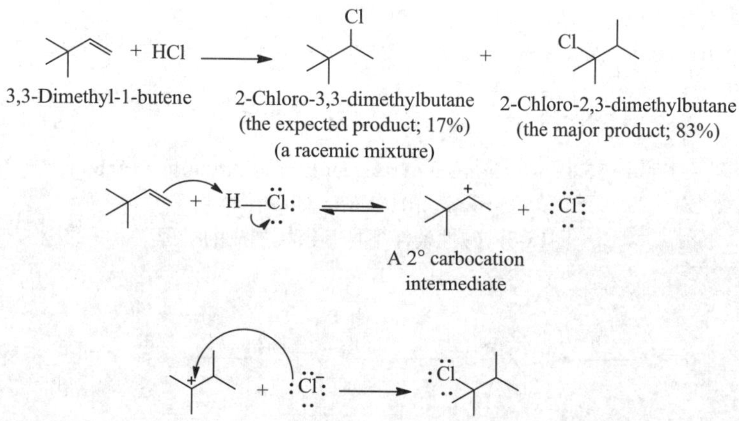

Figure 15.6   Carbocation rearrangement in electrophilic addition reaction
图15-6   亲电加成反应中碳正离子的重排

Second, you may also observe that alkenes substituted by **electron-donating** (供电子基团) groups react faster with electrophiles, which implies that the first step, the addition of the electrophile, is the rate determining step. For example, due to the stabilization by resonance donation of oxygen lone pair, the relative rate of acid-catalyzed addition of water to alkene with structure of $CH_3OCH=CH$ is 5×10^4 faster than simple ethylene.

15.4.2   Nucleophilic Addition

Compared to electrophilic addition, multiple bonds bear atom capable of accepting electron density, such as **carbonyl** (羰基), **nitro** (硝基), or **nitrile** (氰基), are typical functional groups for nucleophilic addition reactions (see Chapter 9.4.1). In basic organic chemistry, nucleophilic addition reactions dominate the chemistry of **aldehydes** (醛) and **ketones** (酮) and are largely specific to these types of carbonyl compound and their close relations.

The reactivity of the carbonyl group arises from the electronegativity of the oxygen atom and the resulting polarization of the carbon–oxygen double bond. The charge-separated resonance forms of aldehydes and ketones (Figure 15.7) clearly show the reason why the π bond in carbonyl group is easily attacked by nucleophile. The electrophilic carbonyl carbon atom is hybridized and flat, leaving it relatively unhindered and open to attack from either face of the double bond. And at the end of the addition, nucleophiles add to the carbon and electrophiles add to the oxygen.

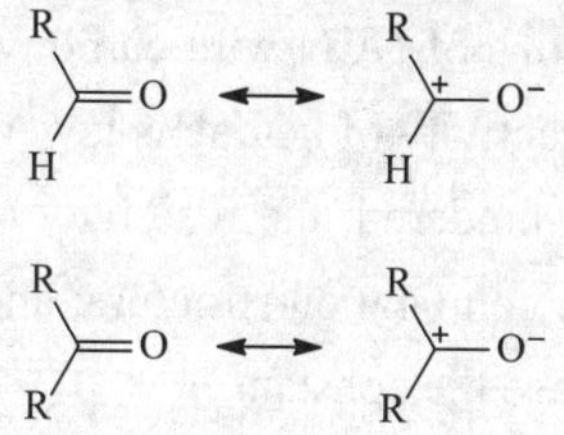

Figure 15.7   Resonance forms of aldehydes and ketones
图15-7   醛酮分子的共振式

Reactions of nucleophiles with aldehydes and ketones can occur under either acidic or basic conditions. The reactions in base generally involve a negatively charged nucleophile, which attacks at the carbon of the carbonyl, with the electrons of the π-bond being displaced to oxygen. The process is completed when the negatively charged oxygen picks up a proton, either from protonated nucleophile or solvent (Figure 15.8). In the acid-catalyzed process (Figure 15.9), the reaction is initiated by protonation of the oxygen of the carbonyl group. The nucleophile (usually neutral) then attacks the carbon atom, and the π-electrons are displaced to oxygen neutralizing the positive charge. The final step is loss of a proton from the nucleophile. You should have noticed that in both processes, base

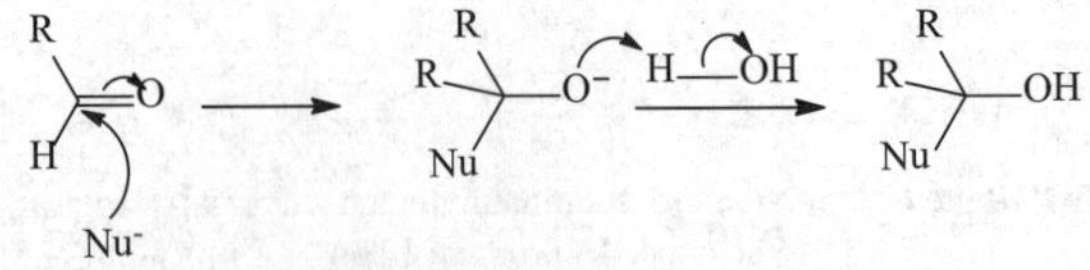

Figure 15.8 Mechanism of nucleophilic addition to a carbonyl group under basic condition

图15-8 碱性条件下羰基亲核加成的机理

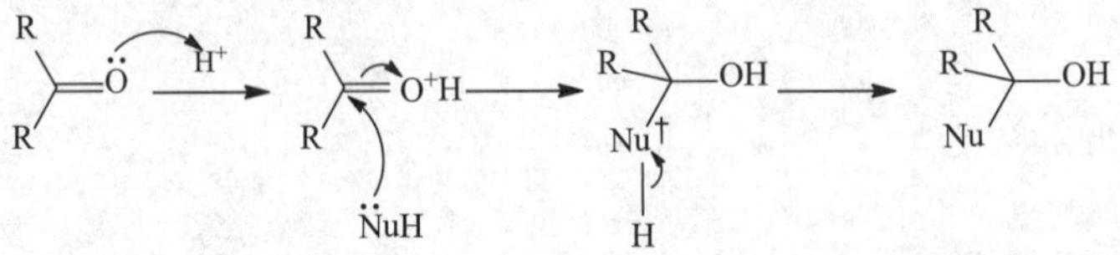

Figure 15.9 Mechanism of nucleophilic addition to a carbonyl group under acidic condition

图15-9 酸性条件下羰基亲核加成的机理

or acid is just a catalyst—the proton or hydroxyl group is regenerated at the end of the reaction. As we will see, this means that the rate of this type of reaction is strongly dependent on solution pH. Under acidic conditions, the carbonyl is rendered more electrophilic by protonation (or coordination to a Lewis acid). Under basic conditions, an attacking species is more likely to be deprotonated and hence more nucleophilic.

The nucleophilic addition reactivity of carbonyl groups depends on both the electronic and steric factors. As to the electronic factors, the electron withdrawing groups will make the starting carbonyl compound less stable and more **reactive (反应活性)**; however the conjugation will make the starting carbonyl compound more stable and less reactive. When the steric factors are taken into consideration, the nucleophile must approach the carbon, so if the carbon site is more steric hindered, the reactivity will be lower. We will take the reactions with HCN as examples. The reactions with HCN are reversible and a bigger equilibrium constant indicates a higher reactivity. The **equilibrium constants (平衡常数)** for the reactions of HCN with carbonyl groups in different compounds are shown in Table 15.1. As we can see the reactivity of ketone is lower than that of aldehyde, because the alkyl groups are electron-donating and steric-hindered; the reactivity of **benzaldehyde (苯甲醛)** is much lower than **acetaldehyde (乙醛)** due to the conjugation structure; and the CF_3 group, a strongly electron-withdrawing group, can substantially increase the constant.

Table 15.1 Equilibrium constants for cyanohydrin formation in aqueous solution

表15-1 氰醇形成反应的平衡常数

Carbonyl compound	CH_2O	CH_3CHO	Me_2CO	PhCHO	PhCOMe	$PhCOCF_3$
$K(M^{-1})$	1.6×10^7	7100	280	200	0.8	3.4×10^5

The stereoselectivity of the nucleophilic addition reaction of carbonyl group is similar as the electrophilic reaction of alkene group with respect to that most of the time the nucleophile can attack the carbonyl group from either side of the π bond plane. Therefore, if the carbon atom adjacent to the carbonyl group is not chiral, nucleophile attacks planar carbonyl form either side at the same rate to yield a racemic pair of enantiomers. However, if the carbonyl compound has a chiral center adjacent to carbonyl group, the nucleophile will experience more steric hindrance from one side, leading to unequal synthesis of the two diastereomers. And the prediction of the major product can follow the Cram's Rule (克拉姆规则). Cram's rule includes following items and is illustrated with Figure 15.10.

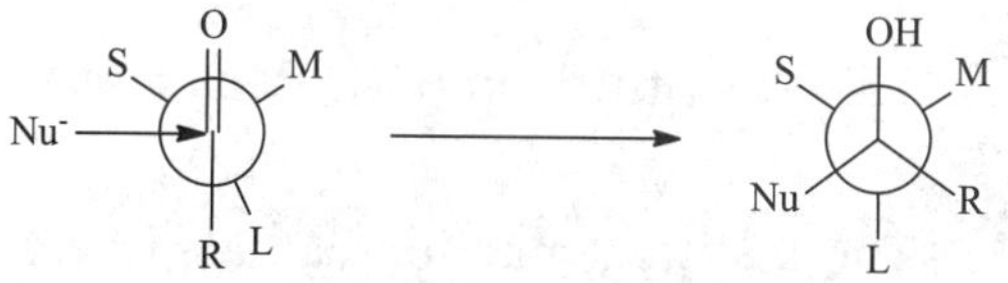

Figure 15.10 Illustration of Cram's rule
图15-10 克拉姆规则

(1) The existing asymmetric center would have a small, medium and large group, denoted S, M and L respectively.

(2) In the reactive conformation, the carbonyl group would orient itself in such a way that it will rest between the small group and the medium group.

(3) The attacking nucleophile would prefer to attack from the side of the small group, resulting in the predominant formation of one diastereomer in the product.

15.4.3 Free Radical Addition

Under the radical forming conditions, such as heating and with dibenzoyl peroxide as initiator, the addition of HBr to alkenes may give the anti-Markovnikov product. The detailed mechanism is shown in Figure 15.11. From the mechanism, we can explain the formation of anti-Markovnikov product. In fact, in free radical addition, it is the bromine radical instead of proton (as in the electrophilic addition reactions) that adds first to the sp^2 carbon and a carbon radical is formed as the intermediate (see Chapter 4.5.2). So now the path which a more stable radical will be formed is favored and the hydrogen atom will at last add to the carbon radical to form the product.

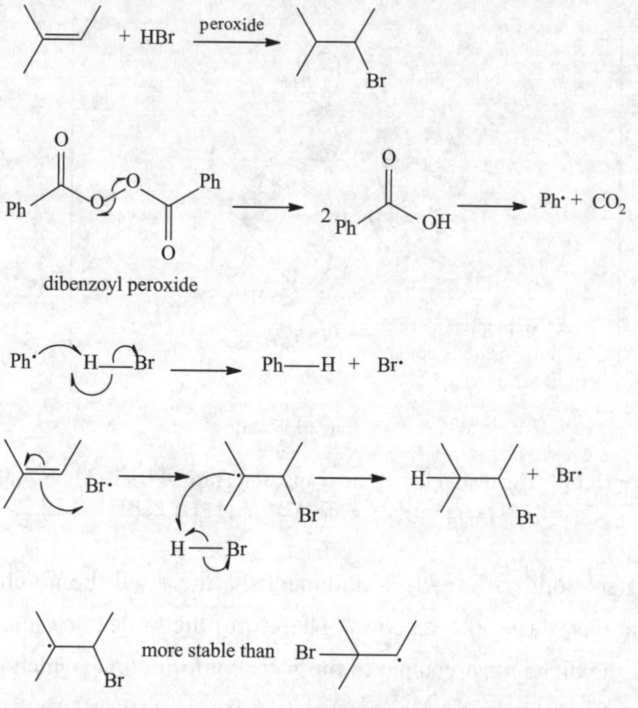

Figure15.11 Free radical addition of HBr to alkene
图15-11 烯烃与HBr的自由基加成反应

385

15.4.4 Addition with Pericyclic Process

The addition reaction can also be a **pericyclic** (周环的) process which means two or more bonds are formed **simultaneously** (同时地) without the formation of intermediate. The general process is shown in Figure 15.12.

Figure 15.12 Synchronous addition of A–B to an alkene
图15-12 烯烃的周环加成反应

The most famous and typical pericyclic addition is the Diels-Alder reaction. Diels-Alder reaction belongs to cycloaddition reaction. Cycloaddition can be classified based on the number of atoms in each component that we are adding together to make a new cyclic compound. And Diels-Alder reaction can be called as a 4 + 2-**cycloaddition** (环加成) because it involves the addition of a **conjugated diene** (共轭双烯烃) with an alkene called **dienophile** (亲双烯体) as shown in Figure 15.13. The new bonds are formed simultaneously, and the transition state is cyclic.

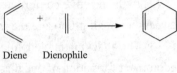

Diene   Dienophile

Transition state

Figure 15.13 Diels–Alder reaction
图15-13 Diels–Alder反应

As a pericyclic process, when Diels-Alder reaction happens, the three pairs of electrons involved in the reaction must move simultaneously, and the transition state must have a geometry that allows **overlap** (重叠) of the two end p orbitals of the diene with those of the dienophile. Figure 15.14 shows the required geometry of the transition state. Due to the special requirement of the transition state, there are two stereochemistry characteristics of Diels-Alder reactions.

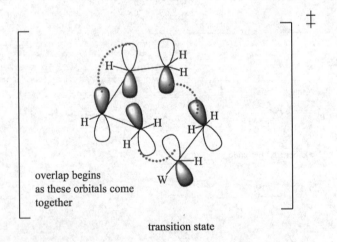

transition state

Figure 15.14 The geometry of the transition state of Diels-Alder reaction
图15-14 Diels-Alder反应的过渡态结构

First, because the new bonds are made simultaneously, there will be no chance for the molecules to change the conformations during the reaction. Therefore, the molecules must come together in the correct orientation. And the diene must change to the *s-cis* conformation (which often is the unfavorable conformation) in order to form the product in a pericyclic process. The *s-cis* and *s-trans* conformations of butadiene are shown in Figure 15.15.

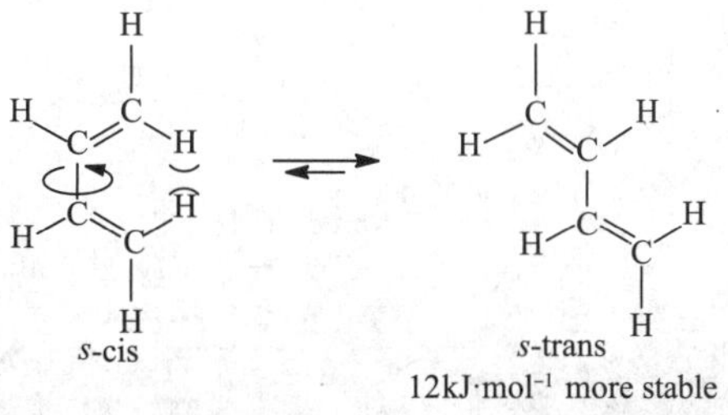

Figure 15.15　The conformation of butadiene
图15-15　丁二烯的构象

Second, the geometry of the Diels–Alder transition state also explains the stereochemistry of the products. The reaction is generally **stereospecific** (立体专一的), which means the original geometry of the alkenes is **preserved** (保留) in the product. Or we can say that Diels–Alder reaction is a **syn addition** (顺式加成) with respect to both the diene and the dienophile. The dienophile adds to one face of the diene, and the diene adds to one face of the dienophile. The following examples in **Figure 15.16** show how the configurations of **dimethyl fumarate** (富马酸二甲酯) and **dimethyl maleate** (马来酸二甲酯) are kept in the syn addition.

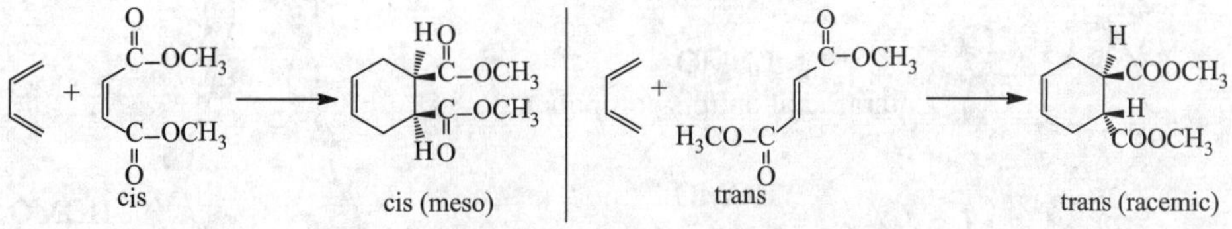

Figure 15.16　Stereospecific syn addition of Diels-Alder reactions
图15-16　Diels-Alder反应顺式加成的立体专一性

Diels-Alder reactions have favorable **enthalpy change** (焓变) because you are losing two π-bonds and gaining two stronger σ-bonds. However, because you are going from two molecules to one, degrees of freedom are reduced, the **entropy change** (熵变) is unfavorable. Thus synthetically, more practical processes are achieved with the dienophile (the alkene) bears at least one electron-withdrawing group such as carbonyl, nitro, or cyanide and the diene normally bears electron-donating group. When the starting diene and alkenes are substituted, the Diels-Alder reactions are also highly region specific. We can predict the major products of unsymmetrical Diels–Alder reactions simply by remembering that the electron-donating groups of the diene and the electron-withdrawing groups of the dienophile usually bear either a 1,2-relationship or a 1,4-relationship in the products, but not a 1,3-relationship as shown in Figure 15.17. These results can be explained with molecular orbital theory, which is beyond the scope of this book and will not be discussed in detail.

You also need to know that not all kinds of cycloaddition will happen with heating. For example the 2+2-cycloaddition is **heat forbidden** (热禁阻) reaction. You could write apparently sensible curly arrows for 2 + 2-cycloadditions, but they don't readily occur on heating. The answer lies in considering the molecular orbitals that are involved in making the new bonds. The outcome of all pericyclic reactions depends on the symmetry of the orbitals involved. In order to form the product, the symmetry of the orbitals of the diene and dienophile must be correct for two bonds to be made simultaneously. The

D
+
△
D
W
W
but not
D
W
1,4-product
1,3-product
D
+
△
W
W
D
D
but not
W
1,2-product
1,3-product

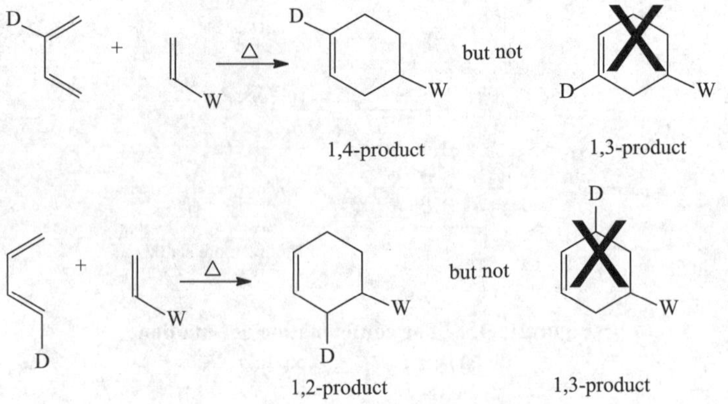

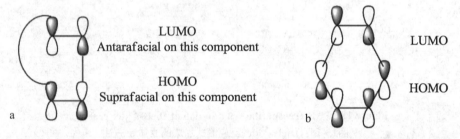

Figure 15.17 Regioselectivity of Diels-Alder reactions
图15-17 Diels-Alder反应区域选择性

HOMO–LUMO approach considers these orbitals as the ones we will use to make the new bonds in a thermal reaction. HOMO is the **highest occupied molecular orbital** (最高已占据轨道) and LUMO is the **lowest unoccupied molecular orbital** (最低非占据轨道). The symmetry of the HOMO and LUMO in the 2+2 and 4+2 cycloaddition reactions are shown in Figure 15.18.

LUMO
Antarafacial on this component
HOMO
Suprafacial on this component
a
LUMO
HOMO
b

Figure 15.18 a. HOMO + LUMO for a 2 + 2-cycloaddition b. HOMO + LUMO for a 2 + 4-cycloaddition
图15-18 a. 2 + 2-环加成的HOMO + LUMO b. 2 + 4-环加成的HOMO + LUMO

15.5 Substitution

PPT

15.5.1 Electrophilic Substitution

Electrophilic substitution reaction is typical for aromatic compounds and it often be called as electrophilic aromatic substitution (see Chapter 6.4.1). As you may remember from Section 15.4.1, when you see a carbon-carbon double bond, you may predict an electrophilic addition mechanism for the reaction. So when you see the double bond in the aromatic compound, you may wonder is it possible that an electrophilic addition reaction may happen. However we also have learned in chapter six that the aromatic compounds have special stability and under normal conditions the aromatic system will not be broken. In fact, when an electrophile meets with an aromatic compound, it still wants to get the π electrons, but the double bond will be quickly restored by losing the proton. The mechanism of the reaction is a two-step process—addition of an electrophile, followed by loss of a proton (Figure 15.19).

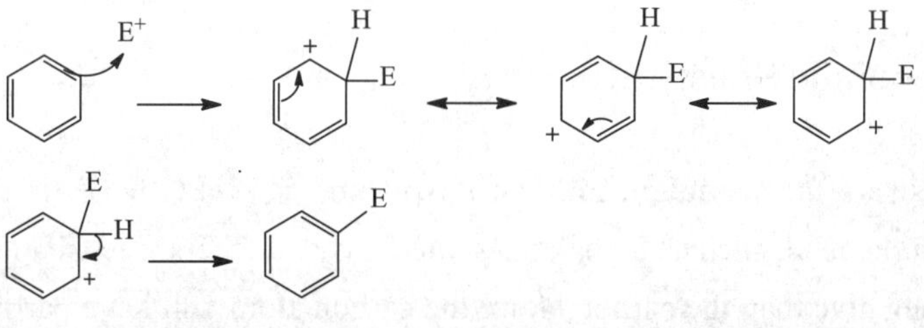

Figure 15.19　Mechanism of the electrophilic substitution reaction
图15-19　亲电取代反应的机理

Almost all the reactions we will meet in this section go by identical mechanisms, so that once we have mastered one, we know them all. The first step, when the intermediate cation is formed, is the rate determining step. And the carbocation produced is called an σ -complex.

This kind of reactions may happen with any kind of aromatic compound, such as benzene or benzene derivatives and even aromatic heterocyclic compound, and many kinds of electrophile can attack the aromatic compound, such as, carbocation, halogens, **nitronium ion** (硝基正离子) and **sulfonic acid** (磺酸). The detailed reaction conditions are introduced in Chapter 6 and Chapter 14 and will not be taught here.

When the aromatic compounds are substituted, the substituent will have a **directing effect** (定位效应) (someone call it orienting effect), which means the substituents already on the aromatic ring will direct the position and also affect the activity of the next substitution. This effect has also been described in detail in chapter six. Here we will give one example of the electrophilic substitution mechanism of toluene to refresh your memory. As shown in Figure 15.20, the substitution of toluene was mainly on the *ortho-* and *para-* position. This directing effect can be explained by Figure 15.21. We can see clearly that when the substitution is on the *ortho-* or *para-* position, a stable tertiary carbocation can be formed. However when substitution is on *meta-* position, no special stable carbocation can be formed.

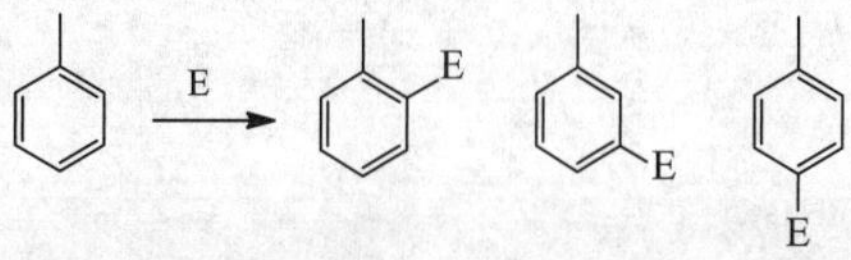

	ortho	meta	para
HNO₃, H₂SO₄	59	4	37
Cl₂	60	0	40

Figure 15.20　Regiochemistry of nitration of toluene
图15-20　甲苯亲电取代反应中的区域选择性

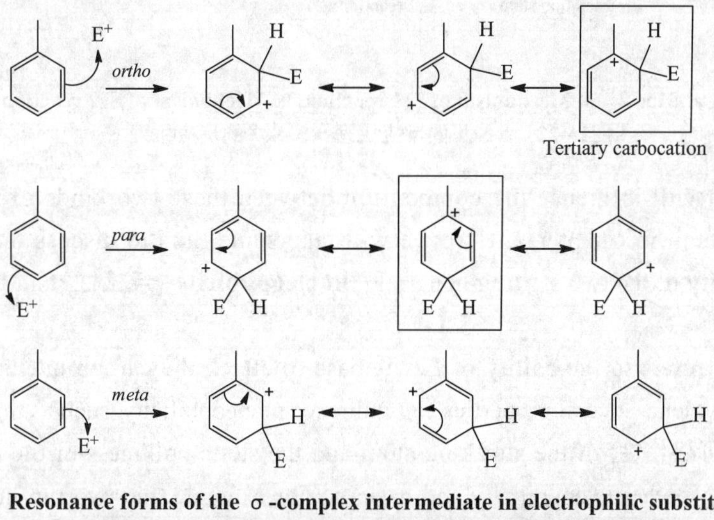

Tertiary carbocation

Figure 15.21　Resonance forms of the σ -complex intermediate in electrophilic substitution of toluene.
1) the *ortho*-position; 2) the *para-* position; 3) the *meta-* position
图15-21　甲苯亲电取代反应的中间体
1）邻位；2）对位；3）间位

15.5.2 Nucleophilic Substitution

The typical substrate for a **nucleophilic substitution** (亲核取代反应) is the compound with a carbon-heteroatom single bond, such as haloalkanes, alcohols and ethers (see Chapter 7.4.1). If the hetero atom is more electronegative than the carbon atom, the carbon atom will have partial positive charge and can be attacked by a nucleophile. And if at the same time, the heteroatom itself or the functional group it belongs to is able to leave with a pair of electron, and then a nucleophilic substitution may happen. The general substitution reaction is shown as following.

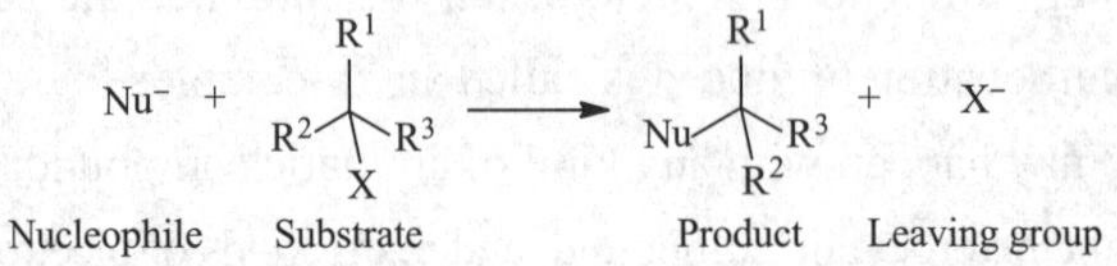

$$Nu^- + \underset{\substack{\text{Nucleophile} \qquad \text{Substrate}}}{\overset{R^1}{\underset{X}{R^2-C-R^3}}} \longrightarrow \underset{\substack{\text{Product} \qquad \text{Leaving group}}}{\overset{R^1}{\underset{R^2}{Nu-C-R^3}} + X^-}$$

Generally, there are two kinds of possible mechanisms for nucleophilic substitution, which are S_N1 and S_N2. S_N1 stands for **s**ubstitution, **n**ucleophilic, **uni**molecular. S_N2 stands for **s**ubstitution, **n**ucleophilic, **bi**molecular. The examples of these two kinds of mechanisms are shown in Figure 15.22.

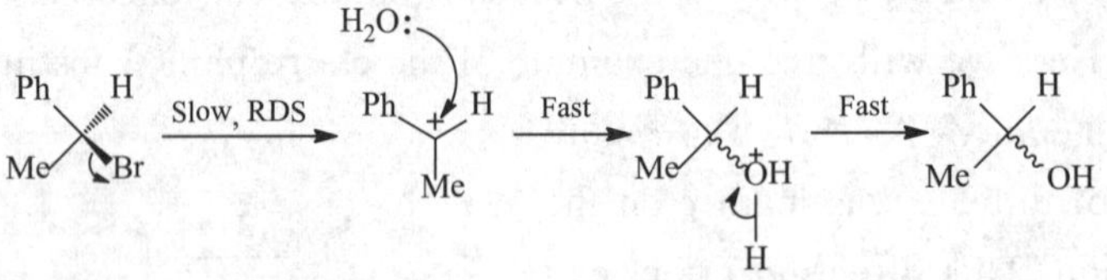

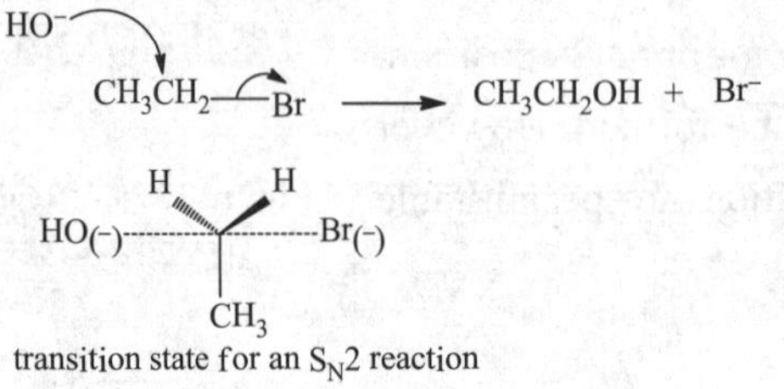

Figure 15.22　a. Mechanism of S_N1 reaction. b. Mechanism of S_N2 reaction
图15-22　a. S_N1 的反应机理　b. S_N2 的反应机理

The factors that will influence the competition between these two kinds of mechanisms have been discussed in detail in Chapter 7. Here we will take some time to discuss the factors that will influence the reactivity of the two starting material—**nucleophilicity** (亲核性) and **leaving ability** (离去能力).

Nucleophilicity refers to the ability of Lewis base to attack the carbon nucleus. It is similar as basicity, but not equivalent. Several properties can influence nucleophilicity, such as the electronegativity, the **polarizability** (可极化性) of the attacking atom and the steric bulkiness of the nucleophile. Some trends are common to both nucleophilicity and basicity. For example, the two run in parallel when the attacking atom is the same. Thus, for nucleophilicity, $EtO^- > HO^- > PhO^- > AcO^- > H_2O$. Similarly, if the attacking atom is in the same period of the periodic table, the nucleophilicity and basicity are with the same trend, eg. $H_3C^- > H_2N^- > HO^- > F^-$. However, when goes down a group, trends of nucleophilicity do not parallel basicity. If we consider the halide ions, fluoride is the most basic and iodide the least so. But in terms of nucleophilicity the reverse is true; iodide is the best nucleophile, and fluoride the least good; Similarly, RO^- is more basic, but much less nucleophilic, than RS^-. This is related to the polarizability of the atom. With second and third row elements, the outer electrons are further away from the nucleus. Thus, when they approach a charge or dipole, the electron cloud is more easily squashed out of shape, in preparation for the formation of the new bond. Steric effects are also important in determining nucleophilicity. The reaction between a base and a proton is sterically undemanding. However, reaching of the carbon nucleus reaction is sterically much more challenging. For example, bulky species, such as Ph_3C^- and *tert*-BuO^-, are good bases but very poor nucleophiles.

Another factor that will affect the substitution reaction is the ability of the leaving group on the substrate. And there is one key message here—good leaving groups are the anions/conjugate bases of strong acids. They are "happy" to leave with a pair of electrons and are "satisfied" (stable) with the extra electrons. So iodide (pK_a HI = −9), bromide (pK_a HBr = −7), and H_2O (pK_a H_3O^+ = −1.7) are good leaving groups, but cyanide (pK_a HCN = 9.1) and hydroxyl (pK_a H_2O = 15.7) are poor ones.

The stereochemistry of nucleophilic substitution reactions depends on the mechanism. There is no intermediate in the S_N2 reaction, and the stereochemistry involves inversion of configuration, rather like an umbrella blowing inside out. The intermediate in the S_N1 reaction is a carbocation. As you will remember carbocations are planar and trigonal, sp^2 hybridized, with an empty p_z orbital. Because they are planar, they can be attacked by nucleophiles from either face, so that the product should be racemic.

In some very special situations, nucleophilic substitution may happen with an aromatic compound. And we call this kind of reaction nucleophilic aromatic substitution. For example, in the presence of strong electron-withdrawing group *ortho*- and/or *para*- to the chlorine atom, aryl chloride can be capable of reacting with nucleophilic substitution mechanism as shown in Figure 15.23.

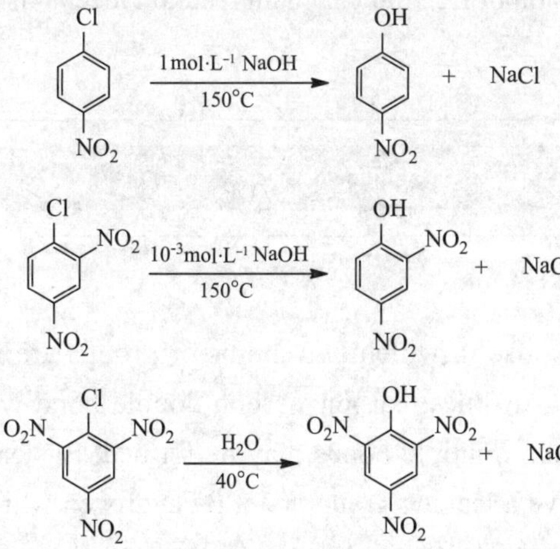

Figure 15.23   Examples of nucleophilic aromatic substitution reactions
图15-23   芳香亲核取代反应的例子

This is surprising, since aryl halides are generally incapable of reacting with substitution because of the strong carbon halide bond. In fact the nucleophilic aromatic substitution does not go by either a S_N1 or S_N2 pathway. It goes by a two-step process shown in Figure 15.24. The two step mechanism is characterized by initial addition of nucleophile to the aromatic ring, followed by loss of a halide anion form the negatively charged intermediate. And this mechanism is called S_NAr.

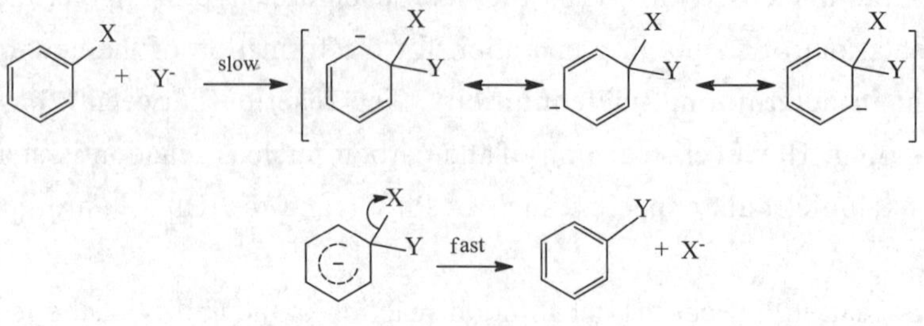

Figure 15.24   Mechanism of S_NAr
图15-24 S_NAr反应机理

15.5.3   Free Radical Substitution

The typical substrate for free radical substitution is an alkane (see Chapter 2.1.5). The hydrogen atoms in alkanes can be substituted with halogen atoms. Since carbon-hydrogen bonds are basically nonpolar and most of the time, can only be broken homolytically, it will be attacked by a radical.

The mechanism of free radical substitution is a chain reaction which includes three steps in general: ①The initiation steps, which generate a reactive intermediate of free radical. ② Propagation steps, in which the reactive intermediate reacts with a stable molecule to form a product and another reactive intermediate, allowing the chain to continue until the supply of reactants is exhausted or the reactive intermediate is destroyed. ③ Termination steps, side reactions that destroy reactive intermediates and tend to slow or stop the reaction.

And for the detailed discussion of free radical chain reaction mechanism please refer to Chapter 2.

PPT

15.6   Elimination

Elimination reactions are those in which two atoms or groups are removed from a molecule to generate a multiple bond. Most of times, carbon–carbon double bond will be formed, and sometime triple bonds or carbon–heteroatom multiple bonds may be formed. Basically, a typical substrate for an elimination reaction needs to have a leaving group and a β- hydrogen which can be removed by a base. So, elimination and substitution processes are sister reactions. They can happen at the same time with the same starting materials, and they are side reactions to each other. Although we have good methodologies for inducing reactions to go in one direction or the other, we should recognize that there may always be some competition between these two reactions.

15.6.1   Elimination Reactions

The two common mechanisms of elimination reactions (see Chapter 7.4.2), E1 and E2, have strong similarities to the S_N1 and S_N2 processes described in the previous section and are generally favored by similar conditions to these analogues. Examples of E1and E2 are shown in Figure 15.25. The mechanism of E1 has strong similarities to the S_N1 process—indeed, the rate-determining step（RDS）is identical. In RDS a leaving group is lost, taking the electrons from the bond being broken with it to give a carbocation. However, the next step is not capture of the carbocation by a nucleophile but a rapid loss of a proton to give an alkene. While E2 is analogous to the S_N2 reaction; it is a single-step, bimolecular process, involving both the base, B^-, and the substrate. The factors that will influence the competition between these two kinds of mechanisms have also been discussed in detail in Chapter 7.

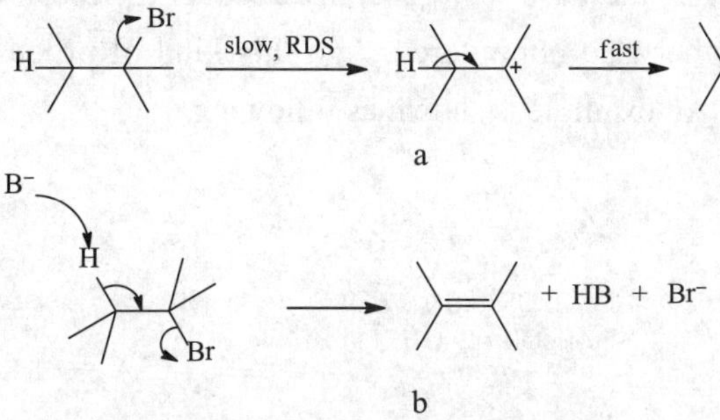

Figure 15.25   a. Mechanism of E1 reaction   b. Mechanism of E2 reaction
图15-25   a. E1 反应机理   b. E2 反应机理

Stereochemistry for elimination is generally a less critical issue than for substitution, Only E2 mechanism has an absolute requirement that the two bonds being broken are coplanar, and there is a preference for the two bonds being **antiperiplanar** (反式共平面) which is just a fancy way of saying that the reaction works best when the two bonds are on opposite sides of the developing double bond. And this requirement may lead to a specific stereochemistry outcome of the product. Examples of antiperiplanar elimination of HBr are shown in Figure 15.26.

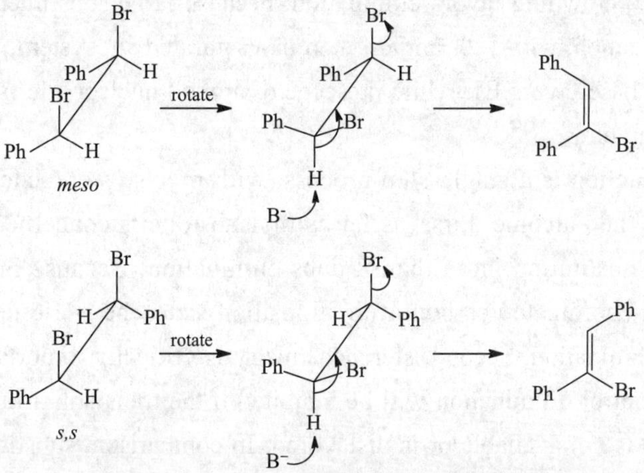

Figure 15.26   Antiperiplanar elimination of HBr
图15-26   HBr消除反应中的反式共平面

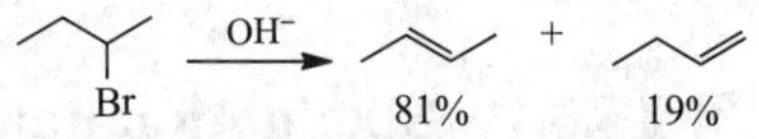

The regiochemistry of elimination reaction is related to the mechanism. However, broadly speaking, most of the time, no matter what mechanism it goes through, the formation of a more stable alkene, generally the most substituted one, will predominate. This is referred to as **Zaitsev elimination** (扎伊采夫消除反应)

Figure 15.27   Zaitsev elimination
图15-27   扎伊采夫消除反应

and an example is shown in Figure 15.27. Let us analyze the outcome with respect to the mechanism. In the E1 process, the formation of the product from the intermediate carbocation is fast. This means that the stability of the developing alkene is the only factor that we need to consider. And in the E2 process, the transition states are characterized as being "alkene-like", thus the formation of the most stable alkene is also favored.

So far we have looked mainly at single-atom leaving groups, where steric constraints are limited for the regioselectivity. However, if either the base or the leaving group is very large, steric hindrance may be an issue. Where this is the case, the base will generally abstract the most accessible proton, usually one that is either terminal or otherwise unhindered. And this kind of process is called as **Hofmann elimination** (霍夫曼消除反应). An example is shown as following.

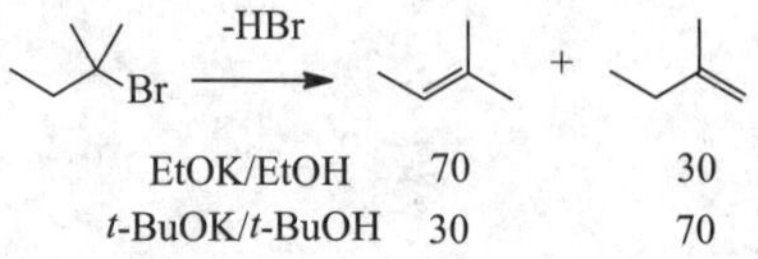

EtOK/EtOH	70	30
t-BuOK/t-BuOH	30	70

15.6.2   Elimination versus Substitution

Both S_N1 and E1 reactions are favored by polar solvents, good leaving groups, the formation of a stable carbocation, and relief of steric strain. Both S_N2 and E2 reactions are favored by nonpolar solvents, good leaving groups, and a relatively uncrowded molecule. So how can we predict what will happen?

The choice between S_N1 and E1 reactions is the more difficult to control, because the rate-limiting step is the same for both of them, and the product is formed in a fast step, with relatively low-energy requirements. In general, crowding favors elimination, because large substituents are further apart in a product with sp^2 hybridization, with 120° angles, than in a saturated sp^3 system, with 109.5 ° separations. The presence of a good base favors E1, while presecce of a good nucleophile favors S_N1. Elimination is favored by higher temperatures.

Because the E2 reaction is a single-step process, with precise stereochemical requirements, we can do more to control the outcome. First, as far as the substrate is concerned, branching at α - and/ or β -positions slows substitution more than it does elimination. Because in substitution, there is a requirement for a "five-coordinate" carbon at the transition state, the steric demand of this process is much higher than for elimination. Second, stereochemistry is critically important to E2 reactions, so it can be used to do the control. Elimination will be inhibited if the trans-anti arrangement of the atoms or groups to be eliminated is not available or is disfavored. In comparison, substitution will be unaffected. For example, as shown in Figure 15.28a, the conformation that would be required for elimination would place both of the very **bulky croups** (大基团) in **axial** (竖直) positions on the six-membered ring, which would be very much disfavored. By contrast, Figure 15.28b undergoes mainly elimination (92 %) under the same conditions, with only 8% of the S_N2 process occurring. In the six-membered ring, one of the two bulky substituents is forced to occupy an axial position; they are of comparable size, so there is no strong preference. Having the leaving group axial is ideal for E2 elimination, and removal of the very bulky group reduces steric **congestion (拥挤)** in the product.

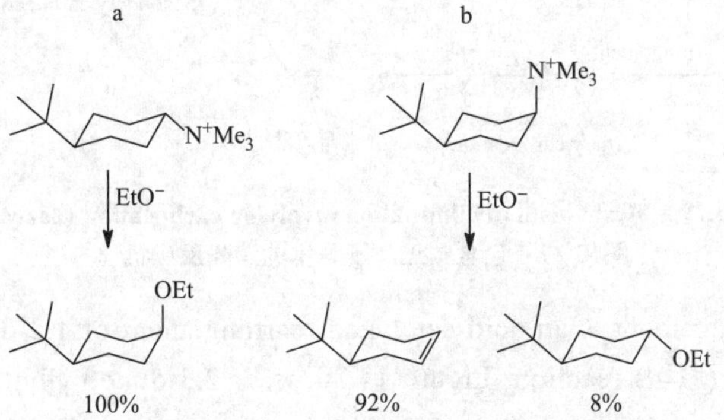

Figure 15.28   Elimination versus substitution controlled by the stereochemical requirement of E2 process

图15-28　通过E2消除反应的立体化学要求控制消除与取代反应的竞争

15.7   Rearrangement Reactions

As we have studied a range of organic reactions, we have seen a number of rearrangements, most often as unwanted side reactions of the process we intended. Many of these could be characterized as carbocations rearrangement. In this section, we will systematize these processes and get a clear understanding of their mechanisms.

15.7.1   Nucleophilic Rearrangement

Nucleophilic rearrangements normally consist of the following steps: creation of an electron deficient center, such as carbocation, is followed by migration of a group with one electron pair and formation of a new electron deficient center and at last further reaction happen on the new electron deficient center. We have previously met carbocation rearrangements when we studied substitution, elimination, and addition reactions. When we write the mechanism for these reactions, in each case, we produce a carbocation, and it rearranges to a more stable carbocation. And this is almost invariably the driving force for this type of rearrangement. For example, Figure 15.29 shows the mechanism of a carbocation rearrangement in the process of elimination reaction.

We have also met two kinds of nucleophilic rearrangement in the previous chapters which have some kind of synthetic applications: **pinacol rearrangement** (频哪醇重排) and **Beckmann rearrangement** (贝克曼重排).

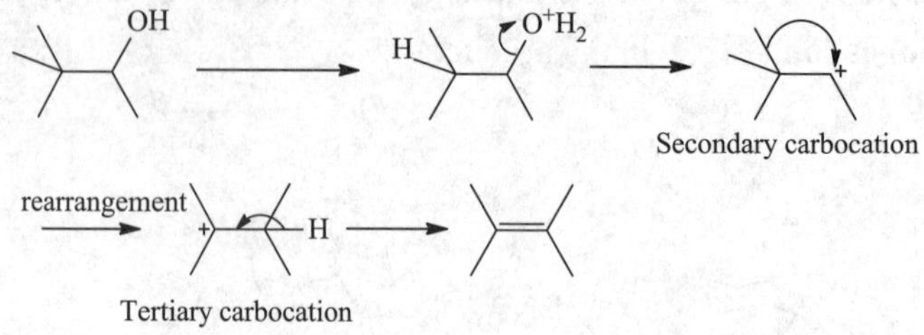

Figure 15.29 Mechanism of elimination involving carbocation rearrangement
图15-29　包含碳正离子重排的消除反应机理

The pinacol rearrangement is an acid-catalyzed rearrangement of 1,2-diols which may make a ketone. The **prototype** (原型) reaction, Figure 15.30, is of 2,3-dimethylbutane-2,3-diol, which has the common name of pinacol. The product, 3,3-dimethyl-2-butanone, goes by the common name of **pinacolone** (频哪酮). Although the reaction mechanism is drawn initially as concerted, and the reaction is usually concerted with the migrating group being trans and anti to the departing water, we could dissect it into three steps. Water departs, leaving a carbocation. The methyl group migrates to leave a carbocation stabilized by its interaction with the lone pair of electrons at oxygen. Finally, a proton is lost to give the ketone.

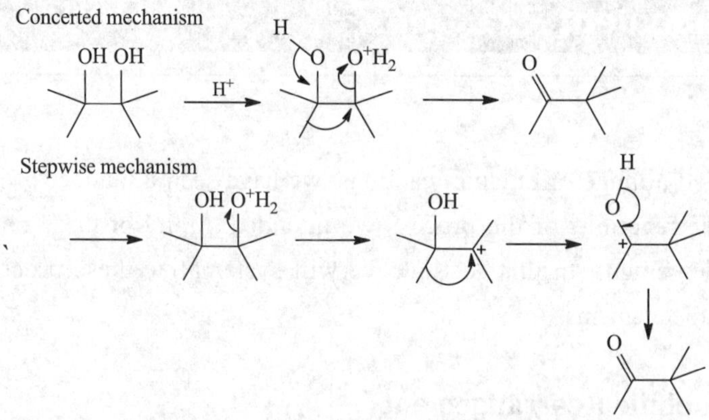

Figure 15.30   Mechanism of Pinacol rearrangement
图15-30　频呐醇重排反应机理

Beckmann rearrangement (Figure 15.31) involves a nitrogen electron deficient center and can be used to synthesize amides. It is a rearrangement of **oximes** (肟) (produced from aldehydes and ketones using hydroxylamine). The OH of the oxime is protonated to make it into a leaving group, and migration of the group that is trans to the departing water gives an intermediate nitrilium ion. This is attacked by water, and the intermediate tautomerizes to give an amide.

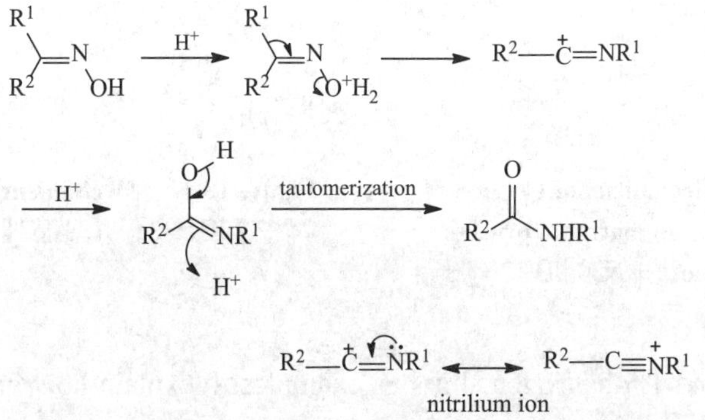

Figure 15.31　Mechanism of the Beckmann rearrangement
图15-31　贝克曼重排反应机理

15.7.2　Electrophilic Rearrangement

Anionic rearrangements are somewhat less common than those involving cations, and we have not met any example in the chapters before. We will only give one example here-the **benzylic acid** (苯甲酸) rearrangement. When **benzil** (偶苯酰) reacts with strong base, migration of the phenyl group is induced by attack of the base on the carbonyl group, and the process is rendered irreversible by the proton transfer in the final step. The mechanism is shown in Figure 15.32.

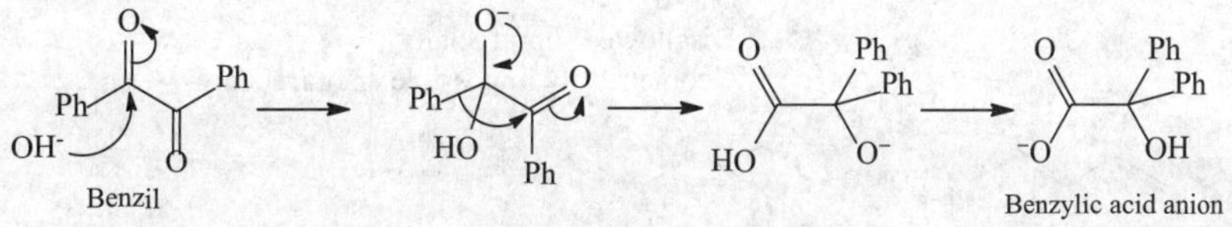

Figure 15.32　Mechanism of the benzylic acid rearrangement
图15-32　苯甲酸重排反应机理

15.7.3　Neutral Rearrangement

Although most of this section will deal with rearrangements of neutral species, we may see a few anions and cations—the key common feature of this group of reactions is that, like Diels-Alder and other cycloaddition processes, they are controlled by orbital symmetry.

1. Sigmatropic Rearrangements　**Sigmatropic rearrangements** (σ-迁移) are a class of reaction in which an σ-bond is moved effectively "across" a π-system to a new position. Figure 15.33 shows a **Cope rearrangement** (科普重排反应) and is a 3,3- Sigmatropic process (note this is not an equally balanced equilibrium; the material with the disubstituted double bond is more stable). Figure 15.34 shows a 1,5-sigmatropic shift, of a hydrogen atom, from one end of the molecule to the other. The numbering of these rearrangements often seems confusing. The simplest way to approach the problem is to number the atoms from the σ-bond that is being broken toward its new position (ignore substituents). The numbers at the site of the new σ-bond give the so-called order of the reaction. So for the Cope rearrangement, we see where the 3,3-designation arises. This also works for the 1,5-proton shift—there is only one atom, the hydrogen, to be numbered in the migrating unit, and it moves to atom 5 in the chain.

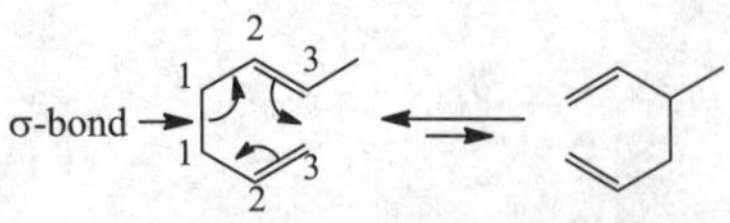

Figure 15.33   Mechanism of Cope rearrangement, a 3,3-sigmatropic process
图15-33   Cope重排反应机理

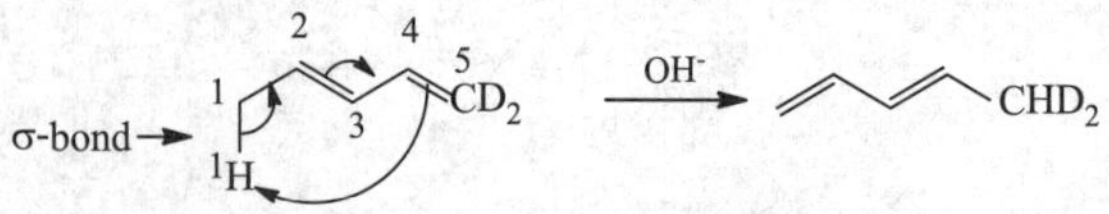

Figure 15.34   Mechanism of 1,5-hydride shift
图15-34   1, 5-氢迁移反应机理

We will take 1,3- and 1,5-hydrogen shifts as examples to explain how orbital symmetry affects the reactions. The experimental result shows that the 1,5-shift works well, however the 1,3-hydogen shift is disallowed thermally. In analyzing these reactions, the convention is that we consider one LUMO and one HOMO, and the HOMO is always the HOMO of the hydride ion, a 1s orbital. The LUMO is the lowest unoccupied molecular orbital of the allyl cation. Drawing these out as in Figure 15.35 can explain the result. The hydrogen cannot move across the face of the allyl cation—a **suprafacial** (同面的) process, because the orbitals are out of phase, and the "formally allowed" **antarafacial** (异面的) process, where it would move from one face to the other, is geometrically impossible. If we do the same analysis for the 1,5-hydride shift, however, all is well.

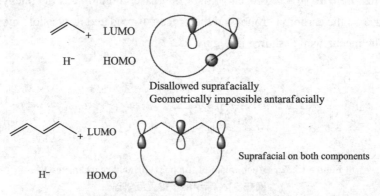

Disallowed suprafacially
Geometrically impossible antarafacially

Suprafacial on both components

Figure 15.35   Orbital symmetry analysis of 1,3- and 1,5-hydrogen shifts
图15-35   1,3-和1, 5-迁移反应中轨道对称性分析

2. Electrocyclic Reactions   An electrocyclic reaction is the formation of a σ-bond between the terminal atoms of a linear conjugated system of π electrons. Two examples are shown in Figure 15.36. And when we compare these two reactions, we can find that butadiene and hexatriene follow different stereochemistry rules. Let us analyze the results with orbital symmetry together as followings.

a   b

Figure 15.36   Electrocyclic reactions
(a) a substituted butadiene; (b) a substituted hexatriene
图15-36   电环化反应
（a）取代丁二烯；（b）取代己三烯

For thermal reactions, we need to consider the HOMO. If we look at the HOMO of butadiene and consider how we need to rotate the p orbitals to make the new bond, we see that to bring them together in phase, we must rotate them in the same direction, as drawn here, both clockwise (Figure

15.37). This process is described as **conrotatory** (顺旋的), and conrotatory closure is the norm for all thermal, 4n, electron processes. The effect of this conrotatory ring closure becomes clear when we look at a substituted butadiene. The reactions are completely stereospecific—no mistakes, ever. We now consider the HOMO of hexatriene, where the closure reaction is a six-electron process, to bring the orbitals together to make the new σ-bond, the two orbitals must now rotate in opposite directions,

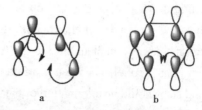

Figure 15.37　Orbital analysis of electrocyclic reactions
(a) butadiene; (b) hexatriene
图15-37　电环化反应中轨道对称性分析
（a）丁二烯；（b）己三烯

one clockwise and one counterclockwise. This is a **disrotatory** (对旋的) process, and this rotation applies to all 4n + 2 electrocyclic reactions. As always, it is easier to see what is happening, when there are substituents. In photochemical reactions, we consider the LUMO (the light photon promotes electrons into the LUMO); the symmetry of this is invariably opposite to that of the HOMO, so 4e processes are disrotatory, and 6e processes are conrotatory, exactly the opposite of the thermal reaction.

15.8　Reaction Types Include Two Kinds of Mechanisms

PPT

Some of the mechanism you have seen in the previous chapters may seem to be complicated and cannot be characterized as any of the above simple mechanism. At the first sight, it may make you frown. However, as we have said earlier, once you understand the basic mechanistic principles, you CAN tackle and understand new types of reactions. We will take two examples—substitution of carboxylic acid derivatives and aldo condensation to show you that once you are familiar with typical changes of atoms/groups, the arrow pushing is relatively easy.

First, let us see the substitution of carboxylic acid derivatives and its general mechanism is shown in Figure 15.38. We may get this mechanism by the following analysis: first we know that it cannot be a simple or direct substitution—there are no S_N2 reactions at sp^2 carbon atoms. And then after checking the starting material it has a C=O bond, so it may go through a nucleophilic addition and it also have a group that is possible to leave. So the reaction includes two steps—nucleophilic addition followed by elimination.

Tetrahedral intermediate

Figure 15.38　Mechanism of the substitution of carboxylic acid derivatives
图15-38　羧酸亲核取代反应机理

Figure 15.39 shows the **aldol condensation** (羟醛缩合) of reaction of **acetophenone** (苯乙酮) and *p*-**chlorobenzaldehyde** (对氯苯甲醛). If you compare the structures of the product and starting materials, you can find that a new carbon-carbon bond is formed between the β carbon of acetophenone and the carbon on the C=O bond of *p*-chlorobenzaldehyde. We should also know that acetophenone can change to enol form and act as a nucleophile, so the first reaction is like the nucleophilic addition of ketone which will result in an alcohol. Then you can find the hydroxy group is lost and a new double bond formed. So the second step should be elimination of the alcohol.

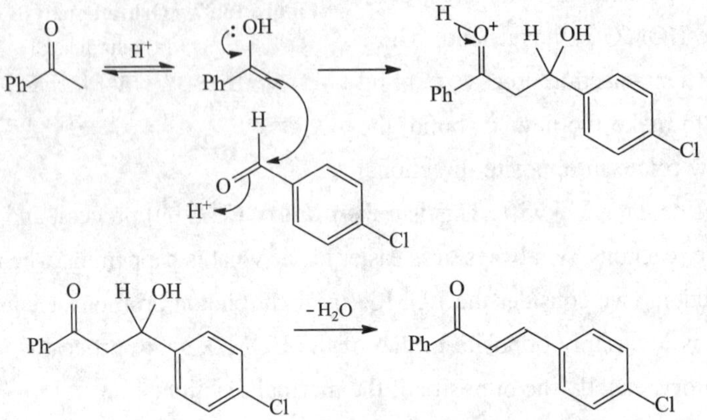

Figure 15.39   Mechanism of the aldol condensation
图15-39   羟醛缩合反应机理

重 点 小 结

一、有机反应机理类型及表达方式

  反应物转变为产物的具体途径叫做反应机理。反应物转变成产物的全过程,包括试剂的进攻、反应物中旧键的断裂、反应中间体的形成以及产物中新键的形成。 根据化学键断裂及形成的方式可将反应机理分成三大类:异裂,均裂和周环反应。共价键发生异裂,形成正负离子,为离子型反应。共价键均裂时产生两个自由基,为自由基反应。共价键的断裂与形成同时(协同)进行,则没有中间体的生成,如果反应经由一个环状的过渡态,此类反应称为周环反应。

  反应机理可以通过表示电子移动的弯箭头表达出来。 弯箭头的末端是电子的来源地,而箭头的头部表示接受电子的地方。在画箭头表示机理时要注意三个问题。①电子来源于电子云密度大的地方,指向电子云密度小的地方。②在箭头落到两个原子之间时,表示新键的形成,此时为了避免超过八隅体的状况出现,很有可能需要同时断开一个旧键。③箭头只表示电子的移动,而不能表示原子的移动。

二、有机反应中间体的类型

  反应中间体是化学反应的中间产物,一般不稳定,但对反应机理的判断,有着重要的影响。多数反应具有中间体,在这类反应机理的判断过程中要涉及对中间体的类型和结构,中间体的稳定性和中间体如何转化为产物的思考过程。 因为在基础有机化学中主要涉及含碳的化合物,此章节中主要讲述了碳正离子,碳负离子和碳自由基三类中间体。碳正离子带正电,采取sp^2杂化,形成三个共价键和一个空的p轨道。给电子基团使碳正离子稳定。碳负离子带负电,与饱和键相连时采取sp^3杂化,孤对电子在一个杂化轨道上;而与具有拉电子能力的不饱和键相连时则采取sp^2杂化。拉电子基团使碳负离子稳定。碳自由基具有一个单电子,采取sp^2杂化,单电子在p轨道上。由于还未满足八隅体结构,自由基也处于缺电子状态,给电子基团使自由基稳定。

三、有机反应类型及机理简介

　　根据分类依据的不同,有机反应可以分成很多种类型。在本章中主要讲解了两种分类依据。一是根据反应物与产物之间的结构差别,有机反应可分为加成反应、消除反应、取代反应和重排反应。加成反应是两种反应物加和形成产物,没有小分子的形成。 消除反应是从反应物中相邻的两个原子上去除两个基团,形成一个小分子,同时形成一个新的π键。取代反应是一个反应物(或反应物中的一部分)取代了另一个反应物中的一部分,从而生成了一个新的化合物,同时被取代的部分成为另一个产物,被称为离去基团。 重排反应是指一个单一反应物通过化学键的重新组合形成另一个异构体产物。二是根据断键类型将有机反应分为亲核反应、亲电反应、自由基反应和周环反应。如果反应中涉及化学键的异裂且带正电的碳离子受到电子供体(即亲核试剂)的进攻,则反应为亲核性。如果反应中涉及化学键的异裂且带负电的碳离子进攻电子受体(即亲电试剂),则反应为亲电性,且一般会有碳正离子中间体的生成。如果反应中涉及化学键的均裂且有自由基中间体的生成,则为自由基反应。如果化学键的断裂和生成同时发生,没有中间体生成,且经过一个环状过渡态,则为周环反应。

　　当我们去分析一个反应的机理时,既要去分析它的反应物和产物之间的变化,也要考虑到断键的形式和中间体的结构。 例如我们在谈论芳香环上的卤代反应的时候,我们不能仅仅知道是用卤素取代了氢原子,还要搞清楚电子是如何移动的。我们从苯环的结构分析出 π电子伸出平面外且流动性较大可以给出去,因此这是一个亲电反应,有碳正离子中间体生成,我们就可以正确的画出弯箭头是从双键出发落到了亲电试剂上。 同时我们也可以分析出苯环上有给电子基团时这个反应的活性会加强,而且具体取代的位置也应该是在电子云密度相对较大的碳原子上。 而在第二步的反应中由于苯环的芳香结构具有特殊的稳定性,所以亲核试剂不是把电子给了碳正离子,而是给到了β–H,恢复了双键和芳香结构,完成了取代反应。通过这个例子,我们可以看出机理的分析是掌握结构与性质之间关系的手段。而且反应机理是由一些简单的基本元素构成的,当我们掌握这些基本元素后,即使是复杂的从没有见过的反应,我们也能够分析出反应机理,并了解其构性关系,甚至能够通过变换反应条件控制反应产物,"痛苦"的有机化学的学习过程应该也有了回报,一切枯燥和烦琐可以变得有趣起来。在本章中,以反应物产物之间的变化和断键类型这两条线索为依据,讲述了这些基本元素。希望可以为你的有机化学学习之旅提供有力的帮助。

Problems
目 标 检 测

1. Classify each of the following reactions as one of substitution, addition, elimination, condensation or rearrangement.

　　请把以下反应按照取代反应、加成反应、消除反应、缩合反应、重排反应分类。

(a) 2 ⌀̈ $\xrightarrow{H^+}$ ⌀̈ + H_2O

(b)

$$CH_3\overset{O}{\underset{}{C}}Cl + H_2O \longrightarrow CH_3\overset{O}{\underset{}{C}}OH + HCl$$

(c)

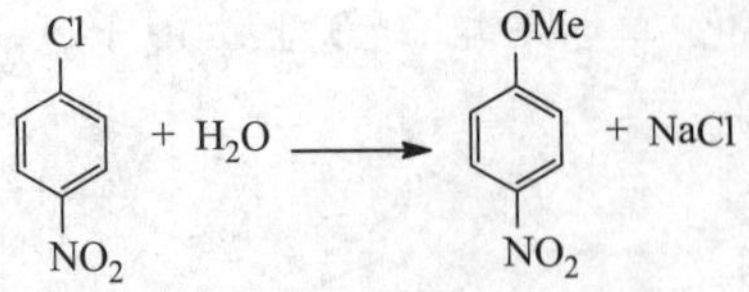

(d)

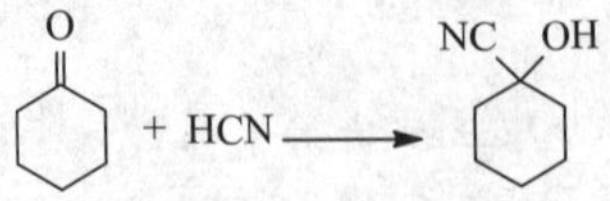

(e)

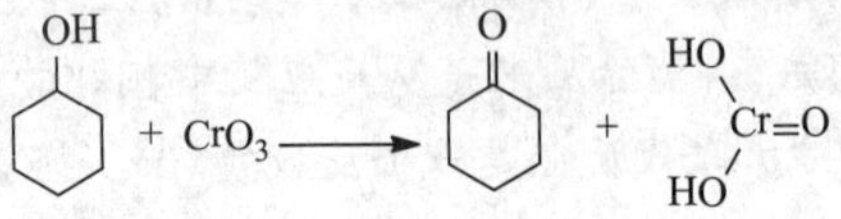

(f)

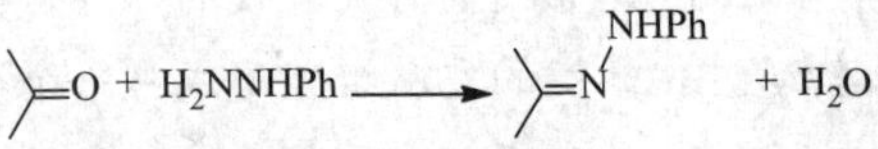

(g)

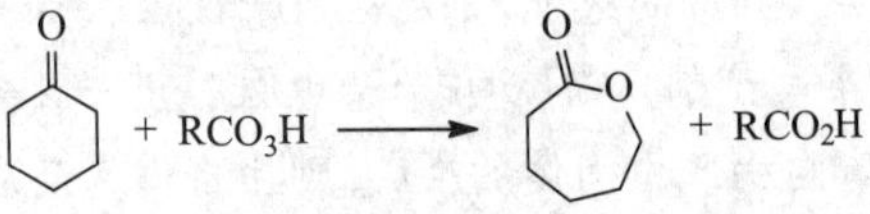

2. Provide curly arrows to show the mechanisms of the following reactions.

请为以下反应画出箭头解释机理。

(a)

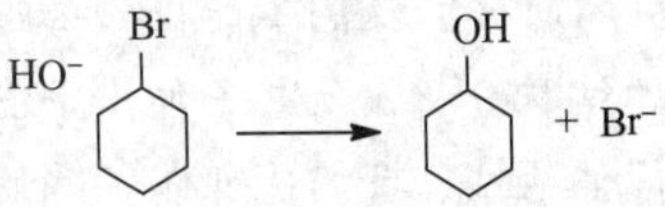

(b)

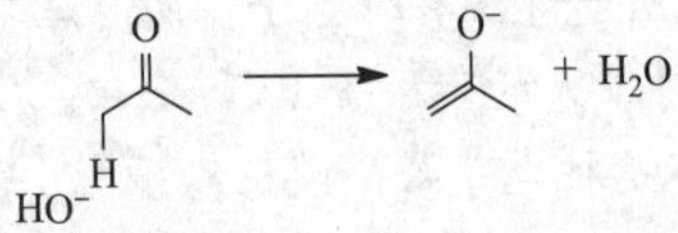

(c)

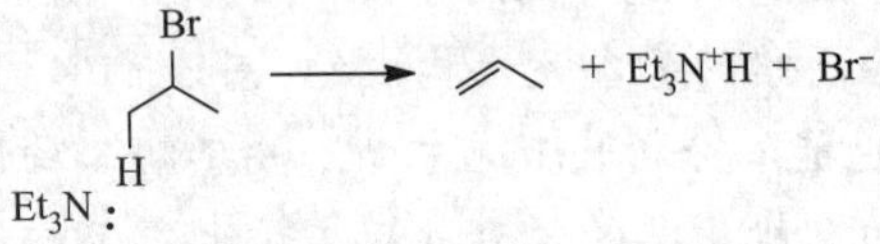

(d)

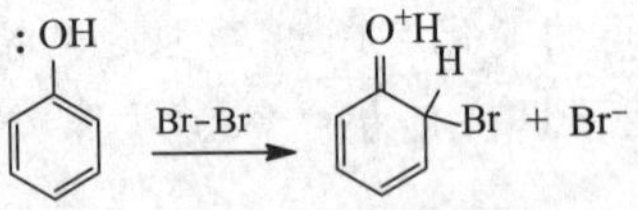

(e)

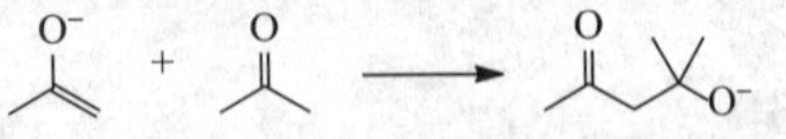

(f)

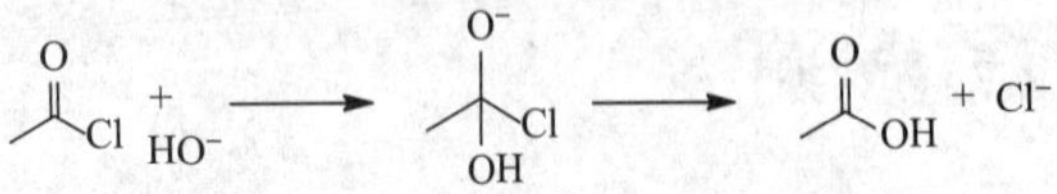

Discussion Topic

What is the significance of the analysis of intermediate; discuss it with some specific examples.

（寇晓娣　尹　飞）

词 汇 表
（按英文字母顺序排列）

1,3-diaxial interaction　1,3-二直立键的相互作用

2,4-pentanedione　乙酰丙酮

absolute configuration　绝对构型

acceptor　受体

acetal　缩醛

acetaldehyde　乙醛

acetic acid　醋酸

acetone　丙酮

acetophenone　苯乙酮

acetylenic hydrogen　炔氢

achiral　非手性的

acid decomposition　酸式分解

acidity　酸性

activation energy　活化能

active iodine　活性碘

acyclic　非环状的

acylase　酰化酶

acylating agent　酰化剂

acyl chloride　酰氯

acyl halide　酰卤

acyl-oxygen bond cleavage　酰氧键断裂

addition reaction　加成反应

adenine　腺嘌呤

adrenalin　肾上腺素

alcohol　醇

alcohol acid　醇酸

alcoholysis　醇解

aldehyde acid　醛酸

alditol　糖醇

aldohexose　己醛糖

aldol condensation　羟醛缩合

aldonic acid　醛糖酸

aldopentose　五碳醛糖

alicyclic compound　脂环化合物

aliphatic　脂肪族

aliphatic compound　脂肪族化合物

alkali　碱

alkaloid　生物碱

alkane　烷烃

alkene　烯烃

alkoxy group　烷氧基

alkyl　烷基

alkyl halide　卤代烃

alkyl-oxygen bond cleavage　烷氧键断裂

alkyne　炔烃

allyl halide　烯丙型卤烃

alphabetical order　字母顺序

aluminum trichloride　三氯化铝

amide　酰胺

amidol　酰胺醇

amine　胺

amino acid　氨基酸

amino group　氨基

aminolysis　氨解反应

ammonia　氨

ammonium cyanate　氰酸铵

ammonium ion　铵离子

ammonium salt　铵盐

ammonium sodium tartrate　酒石酸铵钠

amylopectin　支链淀粉

amylose　直链淀粉

analgesic　止痛药

anesthesia　麻醉

anesthetic　麻醉剂

angle strain　角张力

anhydride　酸酐

anhydrous ethyl ether　无水乙醚

aniline　苯胺

anomeric carbon　异头碳

anomer　异头物

antarafacial　异面的

anthracene　蒽

anti-addition　反式加成

anti　反式

anti-asthma　平喘

antibacterial action　抗菌活性

anti-bonding orbital　反键轨道

anticholinergic drug　抗胆碱能药物

antiestrogen effects　抗雌激素作用

anti-staggered conformation　对位交叉构象

arachidonic acid　花生四烯酸

arene　芳烃

aromatic acid　芳香酸

aromatic character　芳香性

aromatic compound　芳香化合物

aromatic nitrile　芳香腈

aromatic　芳香族

aryl halide　芳基卤烃

aryl　芳基

aspirin　阿司匹林

astragalus　黄芪

asymmetric alkane　不对称烷烃

asymmetric carbon　不对称碳原子

asymmetric center　不对称中心

atomic radius　原子半径

atropine sulfate　硫酸阿托品

aturated fatty acid　饱和脂肪酸

axial bond　直立键

azo　偶氮的

bactericidal effect　杀菌效果

banana bond　香蕉键

barium sulfate　硫酸钡

Bayer strain　拜尔张力

Beckmann rearrangement　贝克曼重排

benzaldehyde　苯甲醛

benzene halides　卤苯

benzene　苯

benzenoid　苯类

benzil　偶苯酰

Benzodiazepine　苯二氮䓬类

benzyl alcohol　苯甲醇, 苄醇

benzyl group　苄基

benzyl halide　苄基型卤烃

benzylic acid　苯甲酸

bimolecular elimination　双分子消除

bimolecular nucleophilic substitution reaction mechanism　双分子亲核取代反应机理

biochemistry　生物化学

boat conformation　船式构象

boiling point　沸点

bond angle　键角

bond energy　键能

bond length　键长

bond line formula　键线式

bond order　键级

bonding orbital　成键轨道

bovine insulin　牛胰岛素

branched- chain alkane　支链烷烃

bridged alkane　桥环烃

bridgehead carbon　桥碳原子

broad-spectrum antibiotic　广谱抗菌素

broken wedge　虚楔线

bromoalkane　溴代烷

bromomethane　溴甲烷

butanol　丁醇

butanone diacid　丁酮二酸

caffeine　咖啡碱

Cahn-Ingold-Prelog sequence　CIP顺序规则

camphor　樟脑

camptothecin　喜树碱

carbanion　碳负离子

carbocation　碳正离子

carbocyclic compound　碳环化合物

carbohydrate　碳水化合物

carbohydrate　糖类

carbon radical　碳自由基

carbon tetrachloride　四氯化碳

carbon–carbon triple bond　碳碳叁键

carbonyl acid　羰基酸

carbonyl　羰基

carboxylate　羧酸盐

carboxylic acid derivative　羧酸衍生物

carboxylic acid　羧酸

carboxylic group　羧基

carcinogenic　致癌

catalyst　催化剂

catalytic hydrogenation　催化氢化

catharsis　导泻

cellobiose　纤维二糖

cellulose　纤维素

central carbon atom　中心碳原子

chain initiation　链引发

chain propagation　链增长

chain termination　链终止

chair conformation　椅式构象

chiral carbon　手性碳原子

chiral center　手性中心

chiral molecular　手性分子

chiral　手性的

chitosan　甲壳素

chitosan　壳糖胺

chloramphenicol　氯胺苯醇

chlorane　氯烷

chloric acid　盐酸

chloroalkane　氯代烷

chloroamphenicol　氯霉素

chloroform　氯仿

chlorohydrocarbons　氯代烃

chlorophyll　叶绿素

chlorpromazine　氯丙嗪

chromatography　色谱

chromic acid　铬酸

cinnamic acid　肉桂酸

cis-butenedioic acid　顺丁烯二酸

cis-trans isomer　顺反异构

citric acid　柠檬酸

Claisen ester condensation reaction　克莱森酯缩合

Claisen rearrangement　克莱森重排

Clemmensen reduction　克莱门森还原反应

clofibrate　安妥明

combustibility　可燃性

combustion　燃烧

common names　普通命名法

complex product　复杂产物

concerted process　协同过程

condensed structural formula　结构简式

configurational isomer　构型异构体

configuration　构型

conformational analysis　构象分析

conformational isomer　构象异构体

conformation　构象异构

conjugate acid　共轭酸

conjugate base　共轭碱

conjugate effect　共轭效应

conjugate system　共轭体系

conjugated diene　共轭二烯烃

conjugation energy　共轭能

conrotatory　顺旋的

constitutional formula　构造式

constitutional isomer　构造异构体

coplanar　共平面的

copper acetylide　炔化亚酮

Corey-House synthesis　科瑞-郝思合成法

coronary artery　冠状动脉

coupling reaction　偶联反应

covalent bond　共价键

crossamide　交酰胺

crown ethers　冠醚

cumulated diene　累积二烯烃

cuprous chloride　氯化亚铜

cuprous iodide　碘化亚铜

cyanohydrin　氰醇

cyano group　氰基

cyclic alcohol　环醇

cyclic ether　环醚

cyclic transition state　环状过渡态

cycloaddition　环加成反应

cycloalkane　环烷烃

cycloalkanone　环烷酮

cyclodextrin (CD)　环糊精

cyclohexadienone　环己二烯酮

cyclohexanone　环己酮

cytosine　胞嘧啶

Darzens reaction　达森反应

dashed lines　虚线

D-deoxyribose　脱氧核糖

dean-stark trap　分水器

decarboxylation　脱羧反应

deformation　变形

degrees of Unsaturation　不饱和度

dehydration　脱水

dehydrohalogenation of dihalide　二卤代烷脱卤化氢

dehydrohalogenationreactions　脱卤化氢反应

delocalization energy　离域能

delocalization　离域

density　密度

deoxyadrenalin　去氧肾上腺素

desulfonation　脱磺化

Deuterium　氘

dextrolactic acid　右旋乳酸

dextrorotatory　右旋

dextrose　右旋糖

dialkyl copper-lithium　二烷基铜铝

diastereomer　非对映异构体

diazonium ion　重氮离子

diene　二烯烃

dienophile　亲双烯体

diethyl ether　乙醚

diethyl malonate　丙二酸二乙酯

dimethyl ether　甲醚

dimethyl fumarate　富马酸二甲酯

dimethyl maleate　马来酸二甲酯

dimethyl sulfoxide　二甲亚砜,DMSO

dipole moment　偶极矩

dipole solvent　偶极溶剂

dipole　偶极

dipole-dipole force　取向力

directing effect　定位效应

disinfectant　消毒剂

dispersion force　色散力

disrotatory　对旋的

distillation　蒸馏

DNA deoxyribonucleic acid　脱氧核糖核酸

donor　供体

D-ribose　核糖

dynamic inductive effect　动态诱导效应

eclipsed　重叠式

eclipsed conformation　重叠构象

electrocyclic　电环化

electron delocalization　电子离域

electron pairing　电子配对

electron sink　电子井

electron source　电子源头

electron-donating conjugated effect　给电子共轭效应

electron-donating inductive effect　给电子诱导效应

electron-donating　供电子基团

electronegativity　电负性

electronic formula　电子式

electrophile　亲电试剂

electrophilic addition　亲电加成

electrophilic aromatic substitution　亲电芳香取代

electrophilic reaction　亲电反应

electropositivity　正电性

electron-releasing　供电子的

electron-withdrawing conjugation effect　吸电子共轭效应

electron-withdrawing group　吸电子基

electron-withdrawing inductive effect　吸电子诱导效应

elimination reaction　消除反应

enantiomer　对映异构体

endothermic　吸热

energy similarity　能量近似

enolate ion　烯醇负离子

enol　烯醇

enthalpy change　焓变

entropy change　熵变

enzyme　酶

epoxide　环氧化合物

equatorial bond　平伏键

equilibrium　平衡

equilibrium constant　平衡常数

equilibrium control　平衡控制

equilibrium reaction　平衡反应

erythritol　赤藓糖醇

ester　酯

esterification　酯化反应

estradiol　雌二醇

estrogen stimulant　雌激素兴奋剂

estrogen　雌激素

estrone　雌激素酮

ethanal　乙醛

ether　醚

ethyl acetoacetate　乙酰乙酸乙酯

ethyl ether　乙醚

ethyl p-aminobenzoate　对氨基苯甲酸乙酯

ethyl p-nitrobenzoate　对硝基苯甲酸乙酯

ethyne　乙炔

ethynyl estradiol　乙炔雌二醇

etrahedral　四面体的

even number　偶数

exothermic　放热的

expectorant　祛痰剂

extinguishing agent　灭火剂

Fehling's solution　斐林试剂

field effect　场效应

first-order reaction in kinetics　动力学一级反应

first-order unimolecular elimination　单分子消除反应

Fischer projection　费歇尔投影式

fluorohydrocarbons　氟代烃

fluoroquinolone　氟喹诺酮类

formalin　福尔马林

formate ion　甲酸根

formic acid　甲酸

free radical　自由基

free radical chain reaction　自由基连锁反应

free radical reaction　自由基反应

free-radical addition　自由基加成

Friedel-Craft alkylation　付克烷基化

fructose　果糖

functional group　官能团

furan　呋喃

galactaric acid　半乳糖二酸

galactose　半乳糖

ganoderma　灵芝

gastric acid secretion　胃酸分泌

gastric ulcer　胃溃疡

gauche conformation　邻位交叉构象

geminal dihalide　偕二卤代烷

geminal diol　偕二醇

general anesthetic　全身麻醉剂

genetic information　遗传信息

geometric isomers　几何异构体

ginseng　人参

glimepiride　格列美脲

glucaric acid　葡萄糖二酸

glucitol　葡萄糖醇

glucosaminoglycans　葡萄糖胺聚糖

glucose　葡萄糖

glyceraldehyde　甘油醛

glyceryl trinitrate　甘油三硝酸酯

glycogen　糖原

glycoside　糖苷

glycosidic bond　糖苷键

gram-positive and negative bacteria　革兰阳性、阴性细菌

greatest orbital overlap principle　轨道最大重叠原则

Grignard reagent　格氏试剂

guanine　鸟嘌呤

half chair conformation　半椅式构象

halogen exchange reaction　卤素交换反应

halogenated acid　卤代酸

halogen　卤素

halohydrins　卤代醇

halonium ion　卤鎓离子

heat forbidden　热禁阻

heat of combustion　燃烧热

heat of hydrogenation　氢化热

heavy metals antidote　重金属解毒剂

Hell-Volhard-Zelinski reaction　赫尔-乌尔哈-泽林斯基反应

hematopoietic　造血

heme　血红素

hemiacetal　半缩醛

hemiketal　半缩酮

hemostatic　止血的

heparin　肝素
heteroatom　杂原子
heterocyclic compound　杂环化合物
heterogeneous reaction　非均相反应
heterolysis　异裂
hexose　己糖
histidine　组氨酸
Hofmann elimination　霍夫曼消除反应
highest occupied molecular orbital　最高已占据轨道
homolog　同系物
homologous series　同系列
homolysis　均裂
Hund rule　洪特规则
hyaluronic acid　透明质酸
hybrid orbital　杂化轨道
hybrid orbital theory　杂化轨道理论
hybridization　杂化
hydrazine　肼
hydrazones　腙
hydroboration　硼氢化反应
Hydroboration-Oxidation　硼氢化-氧化
hydrocarbon　烃
hydrochloric acid　盐酸
hydrogen bond acceptor　氢键受体
hydrogen bond　氢键
hydrogen bond donor　氢键供体
hydrogen halide　卤化氢
hydrogen sulfide　硫化氢
hydrolysis　水解
hydroperoxidation　氢过氧化反应
hydroperoxide　氢过氧化物
hydrophobic　疏水的
hydroxy acid　羟基酸
hydroxyl　羟基
hyperconjugated　超共轭的
hypervalence　超共价
hypoglycemic agent　降血糖药
hypolipidemic effect　降血脂作用
imidazole derivative　咪唑衍生物
imidazole　咪唑
index of Hydrogen Deficiency　缺氢指数

indole　吲哚
induced dipole moment　诱导偶极矩
induced effect　诱导效应
induction force　诱导力
inductive donor　诱导给电子基团
inductive electron withdrawer　诱导拉电子基团
inert gas　惰性气体
infrared spectroscopy　红外光谱
inhaled anesthetics　吸入麻醉剂
inner ester　内酯
inorganic esters　无机酸酯
insect repellent　杀虫剂
instantaneous dipole　瞬时偶极
interconversion　互变现象
intermediate　中间体
intermolecular　分子间的
intermolecular hydrogenbonding　分子间氢键
internal salt　内盐
intestinal tract　肠道
inversion of configuration　构型翻转
iodic acid　碘酸
iodine tincture　碘酊
iodoalkane　碘代烷
ionic bond　离子键
isobutane　异丁烷
isoelectric point　等电点
isomer　同分异构体
isomeric　同分异构的
isopropyl　异丙基
isoproterenol　异丙肾上腺素
isoquinoline　异喹啉
isothiazole　异噻唑
K. Fries rearrangement　傅瑞斯重排
Kekulé Structure　凯库勒结构式
ketoacid　酮酸
keto–enol tautomerism　酮式-烯醇式互变异构
ketohexose　己酮糖
ketone　酮
ketone decomposition　酮式分解
kinetic control　动力学控制
kinetic energy　动能
kinetic product　动力学产物

Kolbe-Schmidt reaction　科尔贝-许密特反应

lactam　内酰胺

lactic acid　乳酸

lactide　交酯

lactones　内脂

lactose　乳糖

lauric acid　月桂酸

leaving ability　离去能力

leaving group　离去基团

levofloxacin　左氧氟沙星

levorotatory　左旋

Lewis acid　路易斯酸

Lewis base　路易斯碱

like dissolves like　相似相溶

Lindlar's catalyst　林德拉催化剂

linear combination　线性组合

linoleic acid　亚油酸

litmus paper　石蕊试纸

local anesthetic procaine　局部麻醉剂普鲁卡因

lone pair electron　孤对电子

lowest unoccupied molecular orbital　最低非占据轨道

magnesium ion　镁离子

malic acid　苹果酸

malonic acid　丙二酸

maltose　麦芽糖

mannose　甘露糖

Markovnikov's rule　马氏规则

maximum overlapping principle　最大重叠原则

mechanism　机制

melting and boiling point　熔点与沸点

menthol　薄荷醇

mercaptan　硫醇

mercuric ion　汞离子

metal alkoxide　金属醇盐

metallic bond　金属键

metallic compound　金属化合物

meta-　间位

methyl ether　甲醚

migration　迁移

minimum energy principle　能量最低原理

misoprostol　米索前列醇

mixed ether　混合醚

m-methoxybenzoic acid　间甲氧基苯甲酸

m-nitrobenzoic acid　间硝基苯甲酸

molar mass　摩尔质量

molecular orbital theory　分子轨道理论

molecular orbital　分子轨道

monochromatic　单色光

monosaccharide　单糖

morphine　吗啡

mucinic acid　黏酸

muscle relaxant　肌松剂

mutarotation　变旋现象

N,N-dimethyl formamide（DMF）　N,N-二甲基甲酰胺

naphthalene　萘

naphthyric acid　萘啶酸

n-butane　正丁烷

neighboring participation　邻基参与

neopentane　新戊烷

nervous system　神经系统

Newman projection　纽曼投影式

nicotinamide　烟酰胺

nitrate Esters　硝酸酯

nitrile　氰基

nitrite　亚硝酸盐

nitro　硝基

nitrogen　氮气

nitronium ion　硝基正离子

nitrosylation reaction　亚硝化反应

nomenclature　命名法

non-bonding orbital　非键轨道

non-coincidence　不重合

non-irritating　无刺激

nonpolar covalent bond　非极性共价键

nonpolar molecule　非极性分子

nonpolar solvent　非极性溶剂

nonprescription laxative　非处方泻药

non-protonic solvent　非质子溶剂

non-steroidal antiestrogen　非甾体抗雌激素药

normal alkane　正烷烃

n-pentane　正戊烷

nuclear magnetic resonance　核磁共振

nuclei 原子核

nucleic acid 核酸

nucleophile 亲核试剂

nucleophilic addition-elimination reaction 亲核加成-消除反应

nucleophilic addition 亲核加成

nucleophilic reaction 亲核反应

nucleophilic substitution 亲核取代

nucleophilicity 亲核性

nucleophilic 亲核的

nutmyl acid 肉豆蔻酸

octet 八隅体

odd number 奇数

ofloxacin 氧氟沙星

ointment 油膏

olefin 烯烃

oligosaccharides 低聚糖

open chain compound 开链化合物

optical activity 光学活性

optical isomerism 旋光异构

optically active 光学活性的

organic chemistry 有机化学

organic compound 有机化合物

organic synthesis 有机合成

organolithium 有机锂化物

organometallic 有机金属

organophosphorus pesticide poisoning 有机磷农药中毒

ortho- 邻位

osazone formation 成脲反应

osazone 糖脲

overlap 重叠

ovulation 排卵

oxalic acid 草酸

oxalosuccinic acid 草酰琥珀酸

oxidation-reduction 氧化还原反应

oxidative cleavage 氧化断键

oximes 肟

oxirane 环氧乙烷

ozonolysis 臭氧化分解

painkiller 止痛

palladium 钯

palmitic acid 软脂酸

panaxynol 人参炔醇

para position 对位

paraffins 石蜡

paramagnetism 顺磁性

paratyphoid 副伤寒

parsalmide 帕莎米特

Pauli exclusion principle 泡利不相容原理

p-chlorobenzaldehyde 对氯苯甲醛

pellagra 糙皮病

penehyclidine hydrochloride 盐酸戊乙奎醚

penicillin 青霉素

pericyclic reaction 周环反应

periodate 高碘酸盐

periodic acid 高碘酸

peroxide effect 过氧化物效应

peroxy acid 过氧酸

perspective drawing 透视图

perspective formula 透视式

pesticide 农药

petrochemical 石油化学

petrolatum 石油冻

pharmacodynamic 药效的

pharmacokinetic 药物动力学

pharmacological activity 药理活性

pharmacy 药学

phenanthrene 菲

phenol 苯酚

phenol 酚

phenolic acid 酚酸

phenolphthalein 酚酞

phenyl group 苯基

phenylacetic acid 苯乙酸

phenylsulfonic acid 苯磺酸

phosphate ester 磷酸酯

phosphorus halide 卤化磷

phosphorus pentachloride 五氯化磷

phosphorus tribromide 三溴化磷

pilocarpine 毛果芸香碱

pinacol rearrangement 频哪醇重排

pinacolone 频哪酮

pipemidic acid 吡哌酸

piperidine　哌啶

piro atom　螺原子

plane of polarized light　平面偏振光

plane triangle configuration　平面三角形构型

plastic　塑料

polar covalent bond　极性共价键

polar molecule　极性分子

polarimeter　旋光仪

polarity　极性

polarizability　极化性

polarization　偏振方向

polarized light　偏振光

polarizing filter　偏振光滤光器

polyene　多烯烃

polysaccharides　多聚糖

polyvidoneiodine　聚维酮碘

porphin　卟吩

porphyrin　卟啉

Potassium　钾

potassium dichromate　重铬酸钾

potassium iodide　碘化钾

potassium permanganate　高锰酸钾

prefix　前缀

preservative　防腐剂

primary　伯

primary alcohol　伯醇

primary alkyl halide　一级卤代烷

primary amine　伯胺

priority　优先级

procaine hydrochloride　盐酸普鲁卡因

product　产物

projected bond　投影键

propan-2-ol　异丙醇

propane　丙烷

propanone　丙酮

protic solvent　质子溶剂

protocatechuic acid　原儿茶酸

protonation　质子化

protonic solvent　质子型溶剂

prototype　原型

purine　嘌呤

pyramidal　三角锥型

pyramidoamine　吡多胺

pyrazole　吡唑

pyridine　吡啶

pyrimidine　嘧啶

pyridoxal　吡哆醛

pyridoxine　吡哆醇

pyrrole　吡咯

pyruvate　丙酮酸

p-π conjugate　p-π共轭

qualitative analysis　定性分析

qualitative and quantitative analysis　定性和定量分析

quantitative analysis　定量分析

quaternary　季

quaternary ammonium hydroxide　季铵碱

quercetin　槲皮素

quinine　奎宁

quinine　金鸡纳碱

quinolone　喹诺酮类

quinone　醌

racemate　外消旋体

racemic　外消旋的

racemization　外消旋化

radical　自由基

rate-determining step　决速步骤

rayon　人造纤维

reaction rate　反应速率

reaction with Halogens: Halogenation　卤化反应

reactive intermediate　反应中间体

rearrangement　重排

recrystallization　重结晶

reflux　回流

Reformatsky reaction　雷福尔马斯基反应

refrigerant　制冷剂

regioselective　区域选择性的

regular polygon　正多边形

regular tetrahedral geometry　正四面体结构

relative configuration　相对构型

relative density　相对密度

relative displacement　相对位移

resolution　拆分

resolvable　可拆分的

resolving agent　拆分剂

resonance energy　共振能

resonance hybrid　共振杂化

resonance stabilization　共振稳定

resonance theory　共振理论

retention of configuration　构型保持

reversible reaction　可逆反应

rhubarb　大黄

ring strain　环张力

RNA ribonucleic acid　核糖核酸

saccharic acid　糖二酸

salicylic acid　水杨酸

salmonella　沙门菌

saturability　饱和性

saturated　饱和的

saw frame projection　锯架式

sawtooth-shaped　锯齿状

Schiff base　希夫碱

scopolamine　东莨菪碱

sec-butyl　仲丁基

secondary alcohol　仲醇

secondary amine　仲胺

secondary halide　二级卤代烷

secondary ionization constant　二级电离常数

secondary　仲

second-order reaction in kinetics　动力学二级反应

sedative　镇静剂

sequence of amino acids　氨基酸序列

shiitake mushrooms　香菇

side reaction　副反应

sigmatropic rearrangements　σ-迁移

silver acetylide　炔化银

silver bromide　溴化银

silver halide precipitation　卤化银沉淀

silver nitrate　硝酸银

simple ether　简单醚

single molecule nucleophilic substitution reaction mechanism　单分子亲核取代反应机理

single nucleophilic substitution reaction　单分子亲核取代反应

skin disinfectant　皮肤消毒剂

sodium acetylide　乙酰化钠

sodium alkynide　炔基钠

sodium bicarbonate　碳酸氢钠

sodium D line　钠D线

sodium ethoxide　乙醇钠

sodium hydroxide　氢氧化钠

sodium lamp　钠光灯

sodium methoxide　甲醇钠

sodium　钠

solid wedge　实楔线

solvation　溶剂化

sorbitol　山梨醇

specific rotation　比旋度

specificity　专一性

spiro alkane　螺环烃

stabilization energy　稳定能

staggered conformation　交叉构象

starch　淀粉

static induction effect　静态诱导效应

steam distillation　水蒸气蒸馏

stearic acid　硬脂酸

stereocenter　立体中心

stereochemistry　立体化学

stereoeffect　立体效应

stereoisomerism　立体异构

stereoselective　立体选择性

stereospecific　立体择向的

steric configuration　空间构型

steric hindrance　空间位阻

steric strain　空间张力

straight-chain alkane　直链烷烃

structural formula　结构式

structural isomers　结构异构体

structure theory　结构理论

sublimation　升华

substituted aromatic acid　取代芳香酸

substituted carboxylic acid　取代羧酸

substitution　取代反应

substrate　底物

succinic acid　琥珀酸

sucrose　蔗糖

suffix　后缀

sulfate ester　硫酸酯

sulfonic acid　磺酸

sulphuric acid　硫酸

suprafacial　同面的

symmetrical matching principle　对称性匹配原则

symmetrical　对称的

syn addition　顺式加成

syn　顺式

systematic nomenclature　系统命名法

system energy　体系能量

tamoxifen　他莫西芬

tartaric acid　酒石酸

tautomerism　互变异构

tcarbonate　碳酸盐

tereochemical　立体化学的

terminal alkyne　末端炔烃

tertachlormethane　四氯甲烷

tert-butyl　叔丁基

tertiary　叔

tertiary alkylhalide　三级卤代烷

tertiary amine　叔胺

tetrahedral configuration　四面体构型

tetrahedron　四面体

tetrahydrofuran　四氢呋喃

the potential energy　势能

thermodynamic　热力学的

thermodynamic control　热力学控制

thermodynamic product　热力学产物

thiazole　噻唑

thionyl chloride　亚硫酰氯

thiophene　噻吩

thymol　百里香酚或麝香草酚

Tollens reagent　多伦试剂

toluene　甲苯

torsional energy　扭转能

torsional strain　扭转张力

totally eclipsed conformation　全重叠构象

traditional Chinese medicine　中药

trans-butenedioic acid　反丁烯二酸

transesterification　酯基转移作用

transition state　过渡态

trehalose　海藻糖

tricarboxylic acid cycle　三羧酸循环

triol　三醇

trivial name　俗名

tropane　莨菪烷

tropyliumion　䓬离子

twist boat conformation　扭船式构象

typhoid　伤寒

ultraviolet spectroscopy　紫外光谱

unconjugated diene　非共轭二烯烃

unsaturated　不饱和的

unsaturated alkyl halide　不饱和卤代烃

unshared electron pair　未共用电子对

uracil　尿嘧啶

urea　尿素

uric acid　尿酸

urinary tract　泌尿道

uronic acid　醛酸

valence　化合价

valence bond theory　价键理论

valence shell　价电子层

van der Waals force　范德华力

van der Waals strain　范德华张力

vapor density　蒸气密度

vaseline　凡士林

veterinary drugs　兽药

vicinal　连位的

vicinal diol　邻二醇

vinyl halide　乙烯基型卤烃

vinylogy rule　插烯规则

vitamin B_6　维生素B_6

vitamin B_{12}　维生素B_{12}

volatile　挥发性的

volatility　挥发性

Walden inversion　瓦尔登转化

Wangner-Meerweinrearrangements　瓦格涅尔-麦尔外因重排

water solubility　水溶性

wedge　楔形线

wedge perspective formula　楔形式

Wolff-Kishner reduction　沃尔夫-凯惜纳尔还原反应

work-up　后处理

Wurtz synthetic method　武兹合成法

xanthine　黄嘌呤

X-ray diffraction measurement　X射线衍射

yield　产率

Zaitsev elimination　扎伊采夫消除反应

Zaitsev's rule　扎伊采夫规则

zwitterion　两性离子

α-D-glucopyranose　吡喃葡萄糖

α-thiophenesulfonic acid　α-噻吩磺酸

β-carotene　β-胡萝卜素

β-dicarbonyl　β-二羰基化合物

β-diketones　β-二酮

β-ketoacid esters　β-酮酸酯

参 考 答 案

第一章

1. 有机化合物中同分异构现象非常普遍，这是造成有机化合物众多的主要原因。

2. sp^3杂化轨道

3. sp^3杂化轨道的电子云密度强于p轨道电子云密度，因此形成的σ键更稳定。

4.

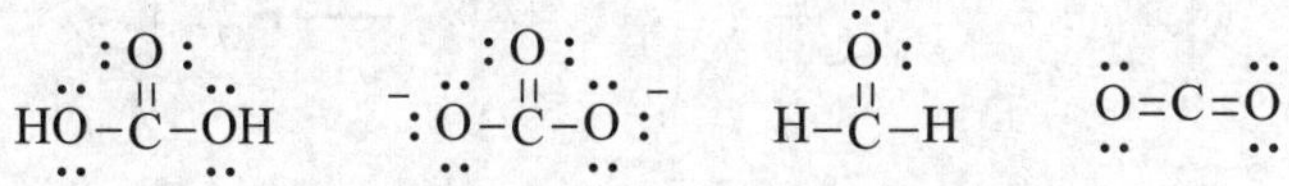

$$HO-\overset{\overset{\displaystyle :O:}{\|}}{\underset{\cdot\cdot}{C}}-\overset{\cdot\cdot}{\underset{\cdot\cdot}{O}}H \qquad \overset{\cdot\cdot}{\underset{\cdot\cdot}{:O}}-\overset{\overset{\displaystyle :O:}{\|}}{C}-\overset{\cdot\cdot}{\underset{\cdot\cdot}{O}}: \qquad H-\overset{\overset{\displaystyle \overset{\cdot\cdot}{O}:}{\|}}{C}-H \qquad \overset{\cdot\cdot}{\underset{\cdot\cdot}{O}}=C=\overset{\cdot\cdot}{\underset{\cdot\cdot}{O}}$$

5. 不一定，CCl_4是含有极性共价键的非极性分子。

6. Br_2, CH_4, CO_2是非极性分子，HCl, CH_3OH, CH_3NH_2, H_2O, CH_3OCH_3是极性分子。

7. 氯仿的偶极矩更大。

8. σ键有更好的轨道重叠。

9. 共轭酸的酸性与其对应碱的碱性相反。碱性越弱，说明其阴离子越不容易得到质子，而其分子更容易电离出质子，因此对应的共轭酸的酸性越强，因此H_2O的酸性强于NH_3。

10. a与d是亲电试剂，b是亲核试剂，c两者都是。

第二章

1. D 2. A 3. D 4. B 5. B 6. A 7. D 8. A 9. B 10. B 11. A 12. D 13. C 14. C 15. C 16. D 17. B 18. C 19. B

讨论题：优势构象：

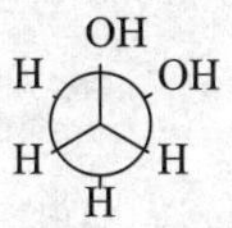

因为两个羟基处于邻位时可以形成分子内氢键，有利于构象的稳定。

第三章

1. C 2. B 3. C 4. B 5. D 6. C 7. C 8. A 9. D 10. D

第四章

1.

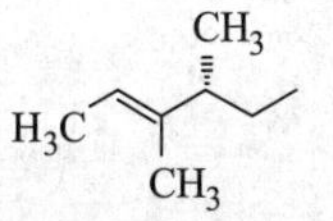

(R,E)-3,4-dimethylhex-2-ene

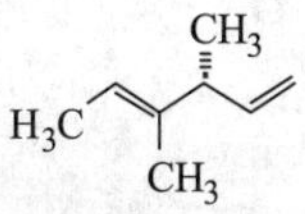

(R,E)-3,4-dimethylhexa-1,4-diene

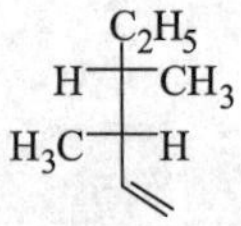

(3*S*,4*S*)-3,4-dimethylhex-1-ene 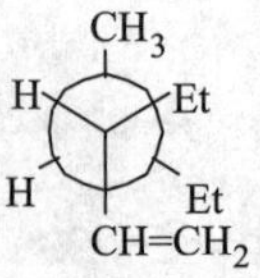 (3*S*,4*S*)-3-ethyl-4-methylhex-1-ene

2.

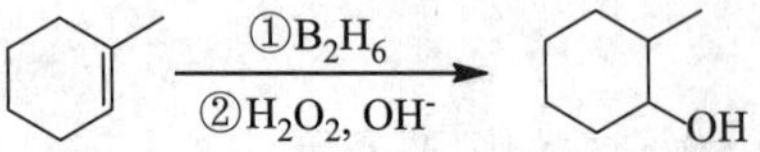

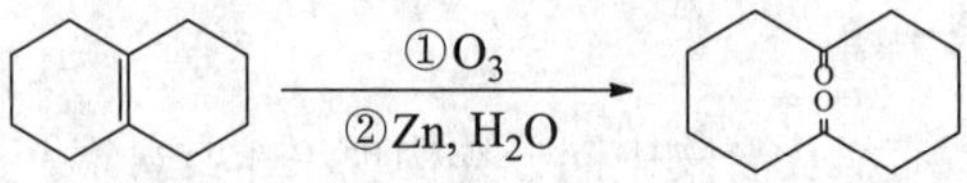

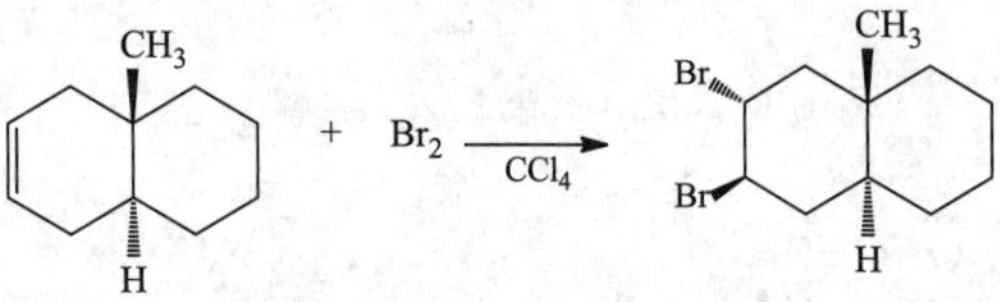

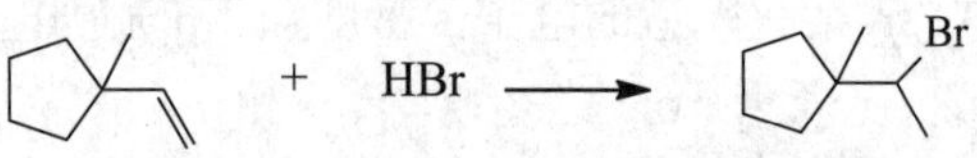

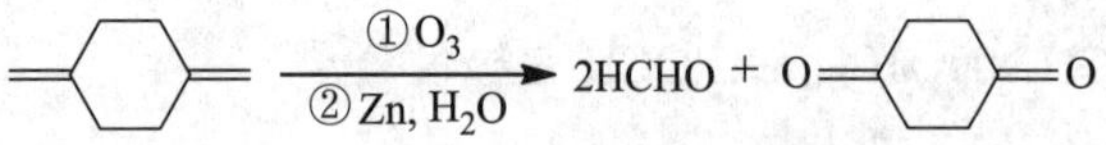

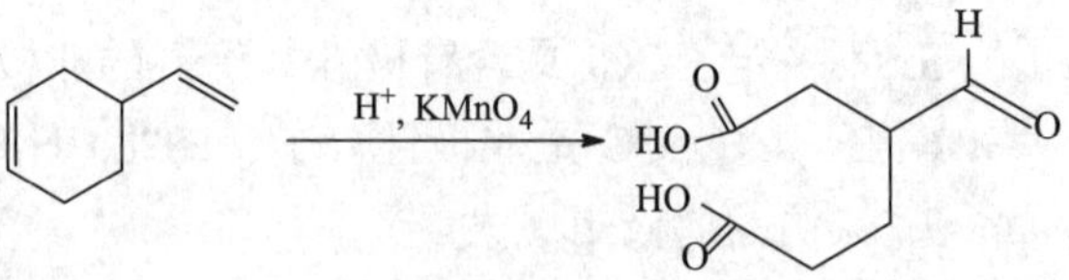

3.

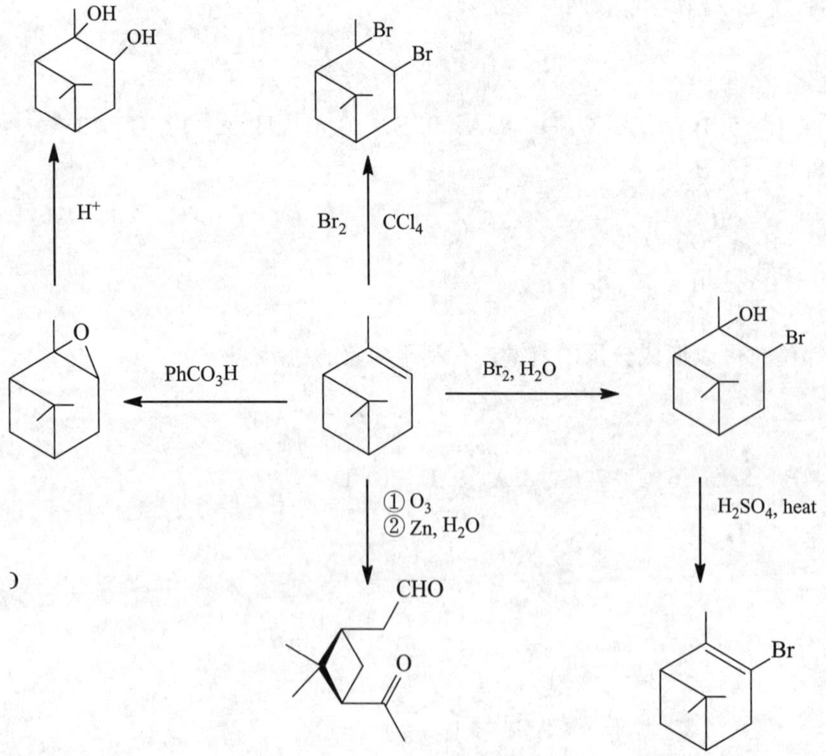

第五章

1. B   2. C   3. A   4. C   5. D

第六章

1. C   2. D   3. B   4. C   5. C   6. C   7. B   8. C

9.

Ortho attack

Meta attack

Para attack

第七章

1. (1)（E）-4-bromo-2,3-dimethylhex-3-ene   (2) 5-chloro-2-methylspiro[3.4]octane

(3) (trans)-1-bromo-3-t-butylcyclohexane   (4) (R)-2-chlorohex-3-ene

(5)（2R, 3S）-3-bromo-2-chloropentane   (6) 5-bromobicycle[2.2.1]hept-2-ene

2. (1)   (2) $(CH_3)_2CHCH_2Br$

(3)   (4)

3. (1) D>A>B>C   (2) B>C>D>A   (3) A>C>D>B   (4) D>C>A>B

4.

(1)

(2)

5. (1) $CH_3CH_2CH_2OCH_2CH_3$ (2) $(CH_3CH_2)_2C = CHCH_3$ (3) $CH_3CH_2CH_2CHCH_3$ / OH

(4) ; (5) $CH_3CH_2CH_2C≡CCH_3$

6. D   7. B   8. A   9. C

第八章

1. (1) 2-methyl-l-phenylbutan-1-ol   (2) 1-cyclopropyl-3,3-dimethylbutan-1-ol
   (3) 5-chlorobenzene-1,3-diol   (4) ethynyl vinyl ether

2.

(1)   (2)   (3)

(4)   (5)   (6)

第九章

1. (E)-hex-2-enal
2. $CH_3CH_2CH_2CH(CH_3)MgBr+CH_3CH_2CHO$ 或
   CH_3CH_2MgBr $CH_3CH_2CH_2CH(CH_3)CHO$

3.

4. 醛酮的羰基自身极性较大，碳原子缺乏电子云的屏蔽而易引发亲核进攻，而烯烃电子云分布均匀，烯键碳受到屏蔽，但π电子云可极化性强，因此易于引发亲电反应。

5.

6.

7.

8.

9.

10.

11.

12.

13.

14.

15.

16.

17.

18.

19.

第十章

1. C　2. A　3. A　4. B　5. D　6. A　7. C　8. D　9. B　10. C

讨论题

顺式，可拆分　　反式，内消旋体，不可拆分

第十一章

1. C　2. B　3. A　4. D　5. B　6. B　7. C　8. B　9. B　10. C　11. B　12. B　13. C
14. C　15. D　16. D　17. C　18. D　19. A　20. C

第十二章

1. D 2. B 3. D 4. B 5. D 6. C 7. C 8. D 9. C 10. C

11. A. B. C.

D. E.

第十三章

7. A 8. D 9. C 10. D

第十四章

1. C 2. D 3. B 4. A 5. DE 6. B 7. D 8. B 9. C 10. A 11. B 12. B 13. C
14. C 15. C 16. B 17. A 18. C 19. D 20. A

第十五章

1. (a) Condensation   (b) Substitution   (c) Substitution   (d) Addition   (e) Elimination
   (f) Condensation   (g) Rearrangement

参 考 文 献

［1］赵骏，杨武德.有机化学.北京：中国医药科技出版社，2017.

［2］Robert J. Ouellette, J. David Rawn. Principles of organic chemistry. USA: Elsevier, 2015.

［3］Richard B. Silverman, Mark W. Holladay. The organic chemistry of drug design and drug action, Third edition. USA: Academic Press of Elsevier, 2014.

［4］Satyajit D. Sarker, Lutfun Nahar. Chemistry for pharmacy students: general, organic, and natural product chemistry. England: John Wiley & Sons Ltd, 2007.